Lecture Notes in Computer Science 4799

Commenced Publication in 1973
Founding and Former Series Editors:
Gerhard Goos, Juris Hartmanis, and Jan van Leeuwen

Andreas Holzinger (Ed.)

HCI and Usability for Medicine and Health Care

Third Symposium of the Workgroup
Human-Computer Interaction and Usability Engineering
of the Austrian Computer Society, USAB 2007
Graz, Austria, November 22, 2007
Proceedings

 Springer

Volume Editor

Andreas Holzinger
Medical University Graz (MUG)
Institute of Medical Informatics, Statistics and Documentation (IMI)
Research Unit HCI4MED
Auenbruggerplatz 2/V, 8036 Graz, Austria
E-mail: andreas.holzinger@meduni-graz.at

Library of Congress Control Number: 2007938906

CR Subject Classification (1998): H.5, D.2, J.3, J.4, K.4

LNCS Sublibrary: SL 2 – Programming and Software Engineering

ISSN   0302-9743
ISBN-10   3-540-76804-1 Springer Berlin Heidelberg New York
ISBN-13   978-3-540-76804-3 Springer Berlin Heidelberg New York

Springer is a part of Springer Science+Business Media

springer.com

© Springer-Verlag Berlin Heidelberg 2007
Printed in Germany

Typesetting: Camera-ready by author, data conversion by Scientific Publishing Services, Chennai, India
Printed on acid-free paper   SPIN: 12191642   06/3180   5 4 3 2 1 0

Preface

The work group Human–Computer Interaction & Usability Engineering (HCI&UE) of the Austrian Computer Society (OCG) serves as a platform for interdisciplinary exchange, research and development. While human–computer interaction brings together psychologists and computer scientists, usability engineering is a discipline within software engineering.

It is essential that psychology research must be incorporated into software engineering at a systemic level. The aspect of integration of human factors into informatics is especially important, since it is here that innovations take place, systems are built and applications are implemented.

Our 2007 topic was "Human–Computer Interaction for Medicine and Health Care" (HCI4MED), culminating in the third annual Usability Symposium USAB 2007 on November 22, 2007 in Graz, Austria (http://www.meduni-graz.at/imi/usab-symposium).

Medical information systems are already extremely sophisticated and technological performance increases exponentially. However, *human cognitive evolution does not advance at the same speed.* Consequently, the focus on interaction and communication between humans and computers is of increasing importance in medicine and health care. The daily actions of medical professionals must be the central concern, surrounding and supporting them with new and emerging technologies.

Information systems are a central component of modern knowledge-based medicine and health services, therefore knowledge management needs to continually be adapted to the needs and demands of medical professionals within this environment of steadily increasing high-tech medicine. information processing, in particular its potential effectiveness in modern health services and the optimization of processes and operational sequences, is of increasing interest.

It is particularly important for medical information systems (e.g., hospital information systems and decision support systems) to be designed with the end users in mind. Within the context of this symposium our end users are medical professionals and justifiably expect the software technology to provide a clear benefit; they expect to be supported efficiently and effectively in their daily activities.

This is a highly interdisciplinary field, producing specific problems; especially for younger researchers, who, being new to their field and not yet firmly anchored in one single discipline, are in danger of "falling between two seats." It is much easier to gain depth and acknowledgement within a scientific community in one single field. While innovation and new developments often take place at the junction of two or more disciplines, this requires a broader basis of knowledge and much more effort.

Working in an interdisciplinary area, one needs the ability to communicate with professionals in other disciplines and a willingness to accept and incorporate their points of view. USAB 2007 was organized in order to promote this closer collaboration between software engineers, psychology researchers and medical professionals.

USAB 2007 received a total of 97 submissions. We followed a careful and rigorous two-level, double-blind review, assigning each paper to a minimum of three and maximum of six reviewers. On the basis of the reviewer's results, 21 full papers (≥ 14 pages), 18 short papers, 1 poster and 1 tutorial were accepted.

USAB 2007 can be seen as a bridge, within the scientific community, between computer science and psychology. The people who gathered together to work for this symposium have displayed great enthusiasm and dedication.

I cordially thank each and every person who contributed toward making USAB 2007 a success, for their participation and commitment: the authors, reviewers, sponsors, organizations, supporters, the team of the Research Unit HCI4MED of the Institute of Medical Informatics, Statistics and Documentation (IMI), and all the volunteers. Without their help, this bridge would never have been built.

Finally, we are grateful to the Springer LNCS Team for their profound work on this volume.

November 2007   Andreas Holzinger

Organization

Programme Committee

Patricia A. Abbot, Johns Hopkins University, USA
Ray Adams, Middlesex University London, UK
Sheikh Iqbal Ahamed, Marquette University, USA
Henning Andersen, Risoe National Laboratory, Denmark
Keith Andrews, TU Graz, Austria
Sue Bogner, LLC Bethesda, USA
Noelle Carbonell, Université Henri Poincare Nancy, France
Tiziana Catarci, Università di Roma La Sapienza, Italy
Wendy Chapman, University of Pittsburgh, USA
Luca Chittaro, University of Udine, Italy
Matjaz Debevc, University of Maribor, Slovenia
Alan Dix, Lancaster University, UK
Judy Edworthy, University of Plymouth, UK
Peter L. Elkin, Mayo Clinic, Rochester, USA
Pier Luigi Emiliani, National Research Council, Florence, Italy
Daryle Gardner-Bonneau, Western Michigan University, USA
Andrina Granic, University of Split, Croatia
Eduard Groeller, TU Wien, Austria
Sissel Guttormsen, University of Bern, Switzerland
Martin Hitz, University of Klagenfurt, Austria
Andreas Holzinger, Med. University of Graz, Austria (Chair)
Timo Honkela, Helsinki University of Technology, Finland
Ebba P. Hvannberg, University of Iceland, Reykjavik, Iceland
Julie Jacko, Georgia Institute of Technology, USA
Chris Johnson, University of Glasgow, UK
Anirudha N. Joshi, Indian Institute of Technology, Bombay, India
Erik Liljegren, Chalmers Technical University, Sweden
Zhengjie liu, Dalian Maritime University, China
Klaus Miesenberger, University of Linz, Austria
Silvia Miksch, Donau University Krems, Austria
Lisa Neal, Tufts University School of Medicine Boston, USA
Alexander Nischelwitzer, FH Joanneum Graz, Austria
Shogo Nishida, Osaka University, Japan
Hiromu Nishitani, University of Tokushima, Japan
Nuno J Nunes, University of Madeira, Portugal
Anne-Sophie Nyssen, Université de Liege, Belgium
Erika Orrick, GE Healthcare, Carrollton, USA
Philipe Palanque, Université Toulouse, France

Helen Petrie, University of York, UK
Margit Pohl, TU Wien, Austria
Robert W. Proctor, Purdue University, USA
Harald Reiterer, University of Konstanz, Germany
Wendy Rogers, Georgia Insitute of Technology, USA
Anxo C. Roibas, University of Brighton, UK
Anthony Savidis, ICS FORTH, Heraklion, Greece
Albrecht Schmidt, Fraunhofer IAIS/B-IT, Uni Bonn, Germany
Andrew Sears, UMBC, Baltimore, USA
Ahmed Seffah, Concordia University, Montreal, Canada
Ben Shneiderman, University of Maryland, USA
Katie A. Siek, University of Colorado at Boulder, USA
Daniel Simons, University of Illinois at Urbana Champaign, USA
Christian Stary, University of Linz, Austria
Constantine Stephanidis, ICS FORTH, Heraklion, Greece
Zoran Stjepanovic, University of Maribor, Slovenia
A Min Tjoa, TU Wien, Austria
Manfred Tscheligi, University of Salzburg, Austria
Berndt Urlesberger, Med. University of Graz, Austria
Karl-Heinz Weidmann, FHV Dornbirn, Austria
William Wong, Middlesex University, London, UK
Panayiotis Zaphiris, City University London, UK
Jürgen Ziegler, Universität Duisburg Essen, Germany
Ping Zhang, Syracuse University, USA
Jiajie Zhang, University of Texas Health Science Center, USA

Organizing Committee

Marcus Bloice, Med. University of Graz
Maximilian Errath, Med. University of Graz
Regina Geierhofer, Med. University of Graz
Christine Haas, Austrian Computer Society
Martin Hoeller, Student
Andreas Holzinger, Med. University of Graz (Chair)
Birgit Jauk, Med. University of Graz
Sandra Leitner, Austrian Computer Society
Thomas Moretti, Med. University of Graz
Elisabeth Richter (Student Volunteers Chair)
Gig Searle, Med. University of Graz
Elisabeth Waldbauer, Austrian Computer Society

Members of the WG HCI&UE of the Austrian Computer Society

Ahlstroem, David
Aigner, Wolfgang
Albert, Dietrich
Andrews, Keith
Baillie, Lynne
Baumann, Konrad
Bechinie, Michael
Benedikt, Eckhard
Bernert, Christa
Biffl, Stefan
Binder, Georg
Bloice, Marcus
Breiteneder, Christian
Burgsteiner, Harald
Christian, Johannes
Debevc, Matjaz
Dirnbauer, Kurt
Ebner, Martin
Edelmann, Noelle
Ehrenstrasser, Lisa
Erharter, Dorothea
Errath, Maximilian
Ferro Bernhard
Flieder, Karl
Freund, Rudi
Frühwirth, Christian
Füricht, Reinhard
Geierhofer, Regina
Gorz, Karl
Grill, Thomas
Gross, Tom
Haas, Christine
Haas, Rainer
Haberfellner, Tom
Hable, Franz
Hailing, Mario
Hauser, Helwig
Heimgärtner, Rüdiger
Hitz, Martin
Hoeller, Martin
Holzinger, Andreas
Hruska, Andreas
Huber, Leonhard
Hyna, Irene
Jaquemar, Stefan

Jarz, Thorsten
Kainz, Regina
Kempter, Guido
Kickmeier-Rust, Michael
Kingsbury, Paul
Kittl, Christian
Kment, Thomas
Koller, Andreas
Költringer, Thomas
Krieger, Horst
Kriegshaber, Ursula
Kriglstein, Simone
Kroop, Sylvana
Krümmling, Sabine
Lanyi, Cecilia
Leeb, Christian
Leitner, Daniel
Leitner, Gerhard
Leitner, Hubert
Lenhart, Stephan
Linder, Jörg
Loidl, Susanne
Maier, Edith
Makolm, Josef
Mangler, Jürgen
Manhartsberger, Martina
Meisenberger, Matthias
Messner, Peter
Miesenberger, Klaus
Miksch, Silvia
Motschnig, Renate
Musil, Sabine
Mutz, Uwe
Nemecek, Sascha
Nischelwitzer, Alexander
Nowak, Greta
Oppitz, Marcus
Osterbauer, Christian
Pesendorfer, Florian
Pfaffenlehner, Bernhard
Pohl, Margit
Purgathofer, Peter
Putz, Daniel
Rauhala, Marjo
Reichl, Peter

Richter, Helene
Riener, Andreas
Sahanek, Christian
Scheugl, Max
Schreier, Günther
Schwaberger, Klaus
Searle, Gig W.
Sefelin, Reinhard
Seibert-Giller, Verena
Seyff, Norbert
Sik-Lanyi, Cecilia
Spangl, Jürgen
Sproger, Bernd
Stary, Christian
Stenitzer, Michael
Stickel, Christian
Stiebellehner, Johann
Stjepanovic, Zoran
Thümer, Herbert

Thurnher, Bettina
Tjoa, A Min
Tscheligi, Manfred
Urlesberger, Bernd
Vecsei, Thomas
Wagner, Christian
Wahlmüller, Christine
Wassertheurer, Siegfried
Weidmann, Karl-Heinz
Weippl, Edgar
Werthner, Hannes
Wimmer, Erhard
Windlinger, Lukas
Wöber, Willi
Wohlkinger, Bernd
Wolkerstorfer, Peter
Zechner Jürgen
Zeimpekos, Paris
Zellhofer, Norbert

Sponsors

We are grateful to the following companies and institutions for their support in our aims to bridge science and industry. Their logos are displayed below.

Table of Contents

Formal Methods in Usability Engineering

User-Centered Methods Are Insufficient for Safety Critical Systems   1
  Harold Thimbleby

Improving Interactive Systems Usability Using Formal Description
Techniques: Application to HealthCare 21
  Philippe Palanque, Sandra Basnyat, and David Navarre

Using Formal Specification Techniques for Advanced Counseling
Systems in Health Care 41
  Dominikus Herzberg, Nicola Marsden, Corinna Leonhardt,
  Peter Kübler, Hartmut Jung, Sabine Thomanek, and Annette Becker

System Analysis and Methodologies for Design and Development

Nurses' Working Practices: What Can We Learn for Designing
Computerised Patient Record Systems? 55
  Elke Reuss, Rochus Keller, Rahel Naef, Stefan Hunziker, and
  Lukas Furler

Organizational, Contextual and User-Centered Design in e-Health:
Application in the Area of Telecardiology 69
  Eva Patrícia Gil-Rodríguez, Ignacio Martínez Ruiz,
  Álvaro Alesanco Iglesias, José García Moros, and
  Francesc Saigí Rubió

The Effect of New Standards on the Global Movement Toward Usable
Medical Devices 83
  Torsten Gruchmann and Anfried Borgert

Usability of Radio-Frequency Devices in Surgery 97
  Dirk Büchel, Thomas Baumann, and Ulrich Matern

BadIdeas for Usability and Design of Medicine and Healthcare
Sensors 105
  Paula Alexandra Silva and Kristof Van Laerhoven

Physicians' and Nurses' Documenting Practices and Implications for
Electronic Patient Record Design 113
  Elke Reuss, Rahel Naef, Rochus Keller, and Moira Norrie

Ambient Assisted Living and Life Long Learning

Design and Development of a Mobile Medical Application for the
Management of Chronic Diseases: Methods of improved Data Input for
Older People   119
    *Alexander Nischelwitzer, Klaus Pintoffl, Christina Loss, and
Andreas Holzinger*

Technology in Old Age from a Psychological Point of View   133
    *Claudia Oppenauer, Barbara Preschl, Karin Kalteis, and
Ilse Kryspin-Exner*

Movement Coordination in Applied Human-Human and Human-Robot
Interaction   143
    Anna Schubö, Cordula Vesper, Mathey Wiesbeck, and Sonja Stork

An Orientation Service for Dependent People Based on an Open
Service Architecture   155
    *A. Fernández-Montes, J.A. Álvarez, J.A. Ortega,
Natividad Martínez Madrid, and Ralf Seepold*

Competence Assessment for Spinal Anaesthesia   165
    *Dietrich Albert, Cord Hockemeyer, Zsuzsanna Kulcsar, and
George Shorten*

Visualization and Simulation in Medicine and Health Care

Usability and Transferability of a Visualization Methodology for
Medical Data   171
    Margit Pohl, Markus Rester, and Sylvia Wiltner

Refining the Usability Engineering Toolbox: Lessons Learned from a
User Study on a Visualization Tool   185
    Homa Javahery and Ahmed Seffah

Interactive Analysis and Visualization of Macromolecular Interfaces
Between Proteins   199
    Marco Wiltgen, Andreas Holzinger, and Gernot P. Tilz

Modeling Elastic Vessels with the LBGK Method in Three
Dimensions   213
    *Daniel Leitner, Siegfried Wassertheurer, Michael Hessinger,
Andreas Holzinger, and Felix Breitenecker*

Usability of Mobile Computing and Augmented Reality

Usability of Mobile Computing Technologies to Assist Cancer
Patients   227
    *Rezwan Islam, Sheikh I. Ahamed, Nilothpal Talukder, and
Ian Obermiller*

Usability of Mobile Computing in Emergency Response
Systems – Lessons Learned and Future Directions   241
    Gerhard Leitner, David Ahlström, and Martin Hitz

Some Usability Issues of Augmented and Mixed Reality for e-Health
Applications in the Medical Domain   255
    Reinhold Behringer, Johannes Christian, Andreas Holzinger, and
    Steve Wilkinson

Designing Pervasive Brain-Computer Interfaces   267
    Nithya Sambasivan and Melody Moore Jackson

Medical Expert Systems and Decision Support

The Impact of Structuring the Interface as a Decision Tree in a
Treatment Decision Support Tool   273
    Neil Carrigan, Peter H. Gardner, Mark Conner, and John Maule

Dynamic Simulation of Medical Diagnosis: Learning in the Medical
Decision Making and Learning Environment MEDIC   289
    Cleotilde Gonzalez and Colleen Vrbin

SmartTransplantation - Allogeneic Stem Cell Transplantation as a
Model for a Medical Expert System   303
    Gerrit Meixner, Nancy Thiels, and Ulrike Klein

Framing, Patient Characteristics, and Treatment Selection in Medical
Decision-Making   315
    Todd Eric Roswarski, Michael D. Murray, and Robert W. Proctor

The How and Why of Incident Investigation: Implications for Health
Information Technology   323
    Marilyn Sue Bogner

Research Methodologies, Cognitive Analysis and Clinical Applications

Combining Virtual Reality and Functional Magnetic Resonance
Imaging (fMRI): Problems and Solutions   335
    Lydia Beck, Marc Wolter, Nan Mungard, Torsten Kuhlen, and
    Walter Sturm

Cognitive Task Analysis for Prospective Usability Evaluation in
Computer-Assisted Surgery   349
    Armin Janß, Wolfgang Lauer, and Klaus Radermacher

Serious Games Can Support Psychotherapy of Children and
Adolescents   357
    Veronika Brezinka and Ludger Hovestadt

Development and Application of Facial Expression Training System 365
    Kyoko Ito, Hiroyuki Kurose, Ai Takami, and Shogo Nishida

Usability of an Evidence-Based Practice Website on a Pediatric
Neuroscience Unit... 373
    Susan McGee, Nancy Daraiseh, and Myra M. Huth

Ontologies, Semantics, Usability and Cognitive Load

Cognitive Load Research and Semantic Apprehension of Graphical
Linguistics 375
    Michael Workman

An Ontology Approach for Classification of Abnormal White Matter in
Patients with Multiple Sclerosis..................................... 389
    *Bruno Alfano, Arturo Brunetti, Giuseppe De Pietro, and
    Amalia Esposito*

The Evaluation of Semantic Tools to Support Physicians in the
Extraction of Diagnosis Codes 403
    Regina Geierhofer and Andreas Holzinger

Ontology Usability Via a Visualization Tool for the Semantic Indexing
of Medical Reports (DICOM SR) 409
    Sonia Mhiri and Sylvie Despres

Agile Methodologies, Analytical Methods and Remote Usability Testing

Fostering Creativity Thinking in Agile Software Development.......... 415
    Claudio León de la Barra and Broderick Crawford

An Analytical Approach for Predicting and Identifying Use Error and
Usability Problem... 427
    Lars-Ola Bligård and Anna-Lisa Osvalder

User's Expertise Differences When Interacting with Simple Medical
User Interfaces 441
    Yuanhua Liu, Anna-Lisa Osvalder, and MariAnne Karlsson

Usability-Testing Healthcare Software with Nursing Informatics
Students in Distance Education: A Case Study 447
    Beth Meyer and Diane Skiba

Tutorial: Introduction to Visual Analytics........................... 453
    Wolfgang Aigner, Alessio Bertone, and Silvia Miksch

Author Index.. 457

User-Centered Methods Are Insufficient for Safety Critical Systems

Harold Thimbleby

Director, Future Interaction Technology Laboratory
Swansea University
Wales, SA2 8PP
h.thimbleby@Swansea.ac.uk

Abstract. The traditional approaches of HCI are essential, but they are unable to cope with the complexity of typical modern interactive devices in the safety critical context of medical devices. We outline some technical approaches, based on simple and "easy to use" formal methods, to improve usability and safety, and show how they scale to typical devices. Specifically: (*i*) it is easy to visualize behavioral properties; (*ii*) it is easy to formalize and check properties rigorously; (*iii*) the scale of typical devices means that conventional user-centered approaches, while still necessary, are insufficient to contribute reliably to safety related interaction issues.

Keywords: Human–Computer Interaction, Interaction Programming, Usability Engineering, Safety Critical Interactive Devices.

1 Introduction

It is a commonplace observation that interactive devices are not easy to use, nor even always safe — cars and the use of entertainment electronics in cars being a familiar example. While we all like mobile phones, it is probably true that nobody fully understands their phone. Users may not know everything about a phone, but it is sufficient that phones do enough for their users. Although there is no reason why every user needs to be able to do everything, even with the subset of features each user knows and feels comfortable with there are still usability issues.

The regular experience of usability problems stimulates the user into coveting the latest model, which promises to solve various problems, and anyway provides many new tempting features. Evidently, the successful business model and user experience of consumer devices is different to what is appropriate for the design of safety critical and medical devices. To contrast it explicitly: it is no use a nurse having problems with a defibrillator and therefore wishing to buy a new one! In this case, the nurse has a very limited time to work out how to use the device; its incorrect use may be fatal. Of course, the nurse should be well-trained, another difference between medical and consumer device design.

Devices have to work and hence, necessarily, they have to be specified and programmed. Some evidence, unfortunately, suggests that medical device designers

A. Holzinger (Ed.): USAB 2007, LNCS 4799, pp. 1–20, 2007.

apply consumer device practices rather than safety critical practices. Medical devices are complex. They are not just more complex than their users can handle but they are also more complex than the manufacturers can handle. User manuals for medical devices often contain errors, suggesting weaknesses in the design process — and also bringing into question the reliability of user training, since it is based on the manufacturer's accurate models of the systems. For some concrete examples see [17].

The traditional HCI response to these well-recognized usability problems is to concentrate on the problems as they face the user. Working with users, we get insight into training, user misconceptions, user error, recovery from error, and so on. This is a sufficient challenge for research in HCI, but in industrial development the insights from evaluation need feeding back into the design process to improve designs: this is *iterative design*. Iterative design properly informed by the user experience is called *user-centered design* (UCD).

The arguments for UCD and iterative design have been widely made; outstanding references being Landauer [10] and Gould & Lewis [5]. Gould & Lewis is now a classic paper, proposing the importance of three key ideas: (*i*) early and continual focus on users; (*ii*) empirical measurement of usage; (*iii*) iterative design whereby the system (simulated, prototype, and real) is modified, tested, modified again, tested again, and the cycle is repeated again and again (i.e., iterative design). More recent work [3] emphasizes the important role of *design* in medical device design, though this is essentially ergonomics and industrial design.

The continued problems with usability has led to an explosion in alternative UCD methods: task analysis, contextual design, activity theory, cognitive walkthrough, heuristic evaluation, questions-options-criteria, ecological methods; more theoretically-motivated methods such as information foraging; and numerous concepts, from scenarios and diary techniques to grounded theory. Methods may be theoretical (e.g., any of the numerous extensions and variations of GOMS-like approaches), done by experts without users (*inspection methods*), or involve users (*test methods*). All HCI textbooks cover a representative range of such techniques; [8] is a convenient review.

The hurdles in the way of the adoption of UCD methods in industry has led to much work in the politics of usability: how does an industry that is driven by technology come to terms with UCD methods? How can usability experts have the political power in order to ensure their empirically-informed insights are adopted? According to some commentators, just providing technical designers with actual observations of use (such as videos) is enough to make them "squirm" *sic* [20] and hence will motivate them to adopt or support UCD methods within their company. Indeed, words like "squirm" are indicative of the problems usability professionals perceive: technologists seem to be the problem; they need to understand the value of usability work; and they need to be told what to do by the people who understand users [10].

1.1 Overview of This Paper

In this paper we review the role of UCD applied to a simple safety critical device. **The device is simple but it is beyond conventional UCD techniques to manage.** Our conclusion is that UCD is necessary but is far from sufficient; indeed the

Fig. 1. The Fluke 114, showing the LCD, five buttons and the knob. Two test leads are plugged into the bottom of the device. The device is 75×165mm — easily held in one hand.

conventional emphasis on UCD diverts attention from technical problems that must also be solved. Technologists' lack of concern for UCD is not the scapegoat. The bottleneck in design is the failure to use professional programming methodologies. Ironically, while this skills bottleneck is ignored, emphasizing more UCD — the standard remedy — will only worsen usability and safety problems because UCD is not reliable, and because it creates a mistaken stand-off between human factors and computing people. Programming methodologies that support usability (that is, interaction programming [18]) have received scant attention; indeed, one view is that they are suppressed within the broad field of HCI because the majority of people working in HCI are human-oriented and unwilling and often unable to acknowledge the effectiveness of technical approaches [19]. This view reinforces the perception [3] that incidents are caused not by poor design but by users and hence should be blamed on users. Many incident analyses (such as [9,12]) fail to explore problems in program design at all.

Some work has been done on developers' understanding of usability [6], showing developers had knowledge of 38% of usability issues prior to empirical studies. This paper was concerned with user interfaces to complex GUI systems, where many usability problems might be expected to be unique to the systems. In contrast, in this paper we are concerned with relatively simple, mostly pushbutton style devices, typical of interactive medical devices. Here, the industry has a long record of incremental development: the paper [6] probably under-estimates developer knowledge in this context. On the other hand, merely knowing about usability issues is quite different from being able and willing to fix them, or even to be able identify them specifically enough to be able to reprogram them out. Bad programmers resist UCD, not because UCD issues are unexpected, but because bad programmers have difficulty accommodating *any* revision.

The present paper provides a case study of a relatively simple interactive device. The analysis of the case study supports the paper's argument: better interaction programming contributes to improving the usability and the safety of interactive systems. The final section, 6, puts the approach into the broader perspective of the full design cycle.

2 Choice of a Case Study

We need a case study that is simple enough to explain and explore in detail, yet complex enough to be representative of the problems we are trying to exhibit. It would be convenient to have evidence of actual design problems, yet not such serious problems that our discussion might raise legal issues. Since we want to demonstrate that the methods proposed are plausible and can scale up to real systems, we need a case study that is fully-defined, and clearly a real device rather than an abstraction (or an inappropriate abstraction) or simplification of a real device, which while making the paper easier would compromise the grounds of the argument that appropriate methods are being discussed. Furthermore, we want to choose a representative device. If we chose, say, a ventilator, would the insights generalize to other medical devices?

For scientific reasons, we wish to do work that is rigorous and replicable. This is particularly important in work that claims to be safety related: readers of this paper should be able to try and test the ideas proposed against the actual case study device. The device chosen must therefore be a current product and readily available. The Appendix to this paper provides an overview of the device definition; the complete source code of everything demonstrated in this paper is available from a web site. The point is that the results presented are rigorously obtained from the actual model.

The Fluke 114 digital handheld multimeter meets these requirements nicely. It is representative because it is *not* a specific medical device, but it has comparable complexity to basic medical devices, and it has a range of features that are broadly similar.

The Fluke 114 is a mid-range device in a collection of similar multimeters, the 111 to the 117. It is a current 2007 model and costs around $100 (*much* cheaper than medical devices!). We can assume the typical user is trained and conversant with relevant electrical theory and good practice, analogous to the medical situation, where the clinicians are assumed to be trained in theory and good practice.

Its user interface has a beeper, a knob, five buttons, and a LCD screen (see figure 1). The purpose of the meter is to measure voltage and resistance (the higher-end models also measure current and frequency, etc), which it does with two probes. The multimeter is safety critical in that erroneous or misleading measurements can contribute to adverse incidents involving the user as well as other people. For example, if we set the multimeter to VAC and measure the UK domestic mains voltage, we get a reading around 240V. This is hazardous, and the meter shows a small (5mm high) $\nmid$-like symbol as a warning. If, however, we set the meter to measure millivolts, the reading should in principle be 240000mV. Set to mVAC the meter flashes OL (i.e., overload) and shows $\nmid$. In mVDC it shows around 40mV, a safe external voltage, and it does not show $\nmid$. The ranges mVDC and mVAC are just one button press apart; perhaps a user could mistake them? Interestingly, the multimeter has a fully automatic range (called AUTO V) where the meter chooses the measurement range itself — in this case, it would choose VAC — and one wonders if the technology can do that, why it does not also at least warn that the user's chosen range is potentially inappropriate? One also wonders why the $\nmid$ symbol is so small, and why the beeper is not also used to make the user more aware of the hazard.

For the purposes of this paper, we shall assume the device probes are shorted: that is, all readings the device makes are zero. This is analogous to discussing, say, a

```
s = S[[1]];
b[r_]:=Button[#,
Print[s=action[#,s]]]
&/@actions[[r]];
b/@{{1,2},{3,4,5,6},
{8,9,10},{7}}
```

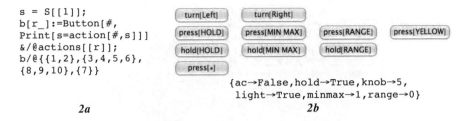

{ac→False,hold→True,knob→5,
light→True,minmax→1,range→0}

2a   *2b*

Fig. 2. A simulation (fig 2a) representing all user actions as button presses and the display as a textual representation of the state (fig 2b). The very little code required (6 lines, including 2 to define button layout) shows the simplicity of creating working simulations from specifications.

syringe pump without starting an infusion. This decision means all further insights of this paper have nothing *per se* to do with electrical measurement, but are general insights equally applicable to the medical field. We shall also ignore "start up" options in our discussion (for example, it is possible to disable the beeper on start up). This is analogous to ignoring technician settings on medical devices. Finally, we ignore the battery and energy-saving features (the device normally *silently* auto-powers off after a period of user inactivity) and assume it is always working or able to work.

There is evidence that the manufacturers do not fully understand the device. The user manual has separately printed addenda, suggesting it went to print before the technical authors had fully understood it. The Fluke web site has a demonstration of the device (actually the top-end 117) [4] which has errors. The web site behaves as if the LCD backlight will stay on when the device is set to Off. The real device does not work like this (the battery would go flat). Of course, the web site is merely to market the device and in itself it is not safety critical, but this error implies the designers of the web site simulation did not understand the device. Put another way, the Fluke 114 seems to be of sufficient complexity to be a challenge to understand even for the manufacturers — who in principle have full information about the design. Technical authors, trainers and users have opportunities to be misled too.

2.1 Defining the Device

We need a formal model of the Fluke 114. As it happens, we obtain this by reverse engineering whereas the manufacturer obviously has the program code in order to manufacture the device.

We assume the device can be represented by a finite state machine (FSM) and that the LCD panel tells us everything we need to know to identify the state of the device. For example, when the LCD is blank, the device is in the state Off.

For concreteness, we use *Mathematica*; Java or C# would be fine, as would other systems like MathCad. However, an interactive system with a library of resources (e.g., to draw graphs) is helpful, and *Mathematica* also has the advantage of being a stable standard: everything this paper shows works as shown.

The full code needed only runs to 12 statements (see Appendix). The definitions define a small FSM with 425 states and 4250 transitions (425 states, 10 user actions).

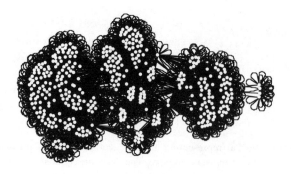

Fig. 3. Transition diagram of the Fluke 114 FSM. The picture is essentially useless and making it larger won't help: contrast this with figures 4, 5 and 7, which are examples of projecting the same device into simpler components, and are easy to interpret. All diagrams, in this and other figures, are exactly as *Mathematica* generates them. For reasons of space, they are too small to read some details that, while important for analysts, are not relevant for the purposes of this paper, such as specific action and state names.

In reality there is no reason to suppose Fluke themselves programmed the 114 as an explicit FSM. In fact, an FSM can be built automatically by systematic state-space exploration from a program implementing the device, provided the program is able to "see" the state. In other words, however Fluke programmed the device, it should be trivial to proceed with the sorts of analysis discussed in this paper. Ideally, the program should also be able to set the state, as this makes the state space search much more efficient (here, by a factor of about 4), but it is not necessary to be able to do so. **If a manufacturer develops a device in such a way that this is not possible, then it would be fair to ask how they can achieve adequate quality control.**

The first property we test to sanity-check the device is that `action` defines a strongly connected FSM. If a device is strongly connected, then every state can be reached by the user from any state. (In a device that is not strongly connected, some states may be impossible to reach at all, or some states may not be reachable if others are visited first — for example, once an infusion is started, it is not possible to reach any state where the patient has not started the infusion.) The Fluke 114, however, is expected to be strongly connected, and a single line of *Mathematica* confirms it (or rather its model) is. Other very basic properties to check are that every user action (e.g., every possible button press) is well-defined in every state (see [18] for more properties). It is worth pointing out that a Cardinal Health Alaris GP infusion pump is sold with 3 out of 14 of its front panel buttons not working — even though the manufacturers claim to use best human factors design principles [2] in design, and that the infusion pump has the *same* user interface as another device [1] where all buttons work. Cardinal Health apparently failed to notice the buttons not working as they claimed. Had they used formal methods it would have been easy to ensure the product worked as claimed (or that the claims were modified).

In a full design process (which this brief paper cannot hope to emulate), one would also consider models of the user, error recovery and other properties of the design [18]. We shall see what can be done *without* user models...

2.2 The Bigger Picture

Our case study defines a finite state machine. Most real devices are not finite state, but have time dependent features, numbers, and (often in medical devices) continuous calculation. Actually, our approach generalizes easily to handle such issues, but it would make this paper more complex to consider them in depth.

Another sort of bigger issue to consider is the relation of this work to HCI more generally. We return to these issues particularly in section 3 (next) and section 6. (Due to space limitations, the purpose of this paper is not to review the significant literature of formal methods in HCI; see, for example, the *Design, Specification and Verification of Interactive Systems* (DSVIS) series of conferences.) DSVIS covers many formal methods, such as theorem proving, model checking, process algebras; other alternatives include the lesser-known scenario-based programming [7], and many proprietary methods. FSMs, however, are free, simple and easy to use: the examples given in this paper are readily achieved with no more programming skill than required for building a device—and they make the point of this paper more forcefully.

3 The Insufficiency of UCD

Suppose we take a working device and analyze it with a user or a group of users. Can they tell us anything useful? Of course, users will notice obvious features of the design such as the YELLOW button that only works in four states: why is it needed when mostly it does nothing? Indeed, YELLOW changes the mV range to AC and DC, but the knob is otherwise used for VAC and VDC. This is a trivial incompatibility a user evaluation might easily spot. This analysis then implies considering design tradeoffs: does this design decision have safety implications or does it have offsetting advantages? Possibly the YELLOW button has advantages for the manufacturer — it is not labeled and can be used to extend any feature without additional manufacturing costs. The higher-end meters use the YELLOW button for adding numerous other features that are irrelevant to the 114. Arguably the 114 considered alone would be better without the YELLOW button.

The requirement that a device be strongly connected (see previous section) makes a focused illustration of this section's central claim: UCD (while necessary) is insufficient. It is essential that the strongly connected is confirmed; for the Fluke 114 it is essential that the property is true — and in general, a designer would need to know the strongly connected components match the design's requirements. Checking strong connectivity by hand, e.g., in a usability lab, requires getting users to visit every state and to ensure that from each state it is possible to reach all others. Since there are 425 states for the Fluke, that requires exploring up to $425^2 = 180625$ pairs of states — and every pair needs the user to work out a sequence of actions to get from one to the other; that is an implausible workload for a human. Contrast that amount of work with the effort of typing a single command to *Mathematica* and getting a reliable answer in just seconds!

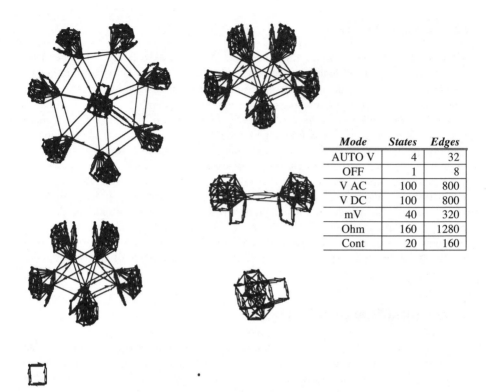

Mode	States	Edges
AUTO V	4	32
OFF	1	8
V AC	100	800
V DC	100	800
mV	40	320
Ohm	160	1280
Cont	20	160

Fig. 4. Transition diagram of the Fluke 114 FSM, projected to ignore self-loops and all left/right knob turn transitions. Note that there is a single state (the dot at the bottom middle of the diagram) with no transitions: this is Off.

For safety critical evaluation it is arguable that a user study should explore the entire device. There may be hazards in its design anywhere in the state space, or some states may be problematic in some way. The question arises: how long would a user study take to adequately explore a device?

This question can easily be answered. If the user is systematic and makes no errors,[1] a full exploration may take 10390 user actions, though the exact number depends on the user's strategy — for example, turning the knob tends to change to another "mode" and therefore makes systematic exploration of the state space harder. Requiring a user to follow 10^4 steps systematically is unreasonable. If the user is somewhat random in their exploration, then the exploration effort increases though the cognitive load decreases. If we estimate that 1 state is defective, then the probability of a sample test of one state finding no problem is 1–1/425=0.998. If the user tests 100 random states in a laboratory study, the probability of finding no problem is 0.8; in other words usability testing is unlikely to find problems. Worryingly, such simplistic estimates are unrealistically optimistic: there is no way

---

[1] The user being "systematic" isn't as easy as it sounds. The program that simulates a systematic user took me a morning to code correctly.

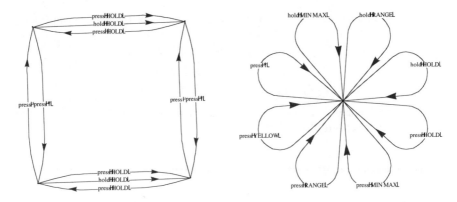

Fig. 5. Exploring some subspaces in more detail, here the AUTO V mode (figure 4 diagram, bottom left), and the Off mode (figure 4 diagram, bottom right) but now with self-loop transitions shown (it is so pretty!) — which visually confirms that in Off nothing (other than knob actions, which are not shown) does anything

(unless we build a special test rig that assists the user) that a user can sample "random" states. Since we have the actual FSM, we can calculate exactly from it, and doing so we find that if the user does random actions with equal probability then 10^7 actions are expected to explore only 90% of the state space — spending this astronomical effort on user testing has a 1 in 10 chance of missing a defective state! Figure 6 draws a graph of the proportion of states expected to be visited by a random user after so-many actions. "Hard to access" states do not get much easier to reach as time goes by, at least on this device. We should conclude: (certainly) some tests should be done formally rather than by UCD testing, and that (possibly) a redesign of the device would make states more accessible to user testing than is the case here.

If the user is told what the intended device model is, and so has an accurate model, and checks it by exactly following an optimal "recipe," then 10004 actions are required — if there is an error in the device (or the recipe) then on average the user will require half that effort; but to confirm there are no errors requires all the actions to be undertaken by the user without error. Determining an optimal recipe to do the exploration is a complex problem, even beyond many programmers [15]: in other words, empirical evaluation of a device is doomed unless it is supported by some other techniques that guarantee coverage of the design.

4 Visualizing Device Behavior

There are many alternative ways to explore the state space of a device, rather than the impossibly laborious manual exploration. First, we consider visualization.

Figure 3 shows a transition diagram of the entire FSM; it is so dense it is clearly not very informative. Figure 3 makes obvious a reason why FSMs are not popular: anything more complex than a few states makes a diagram too dense to comprehend. Instead it is more informative to draw projections of the FSM; thus, if we ignore all actions that do nothing (i.e., self-loops) and all knob turning, we obtain figure 4.

It is worth emphasizing how easy it is to get the visualization shown in figure 4. The single instruction:

```
GraphPlot[DeleteCases[fsm,{_,turn[_]}],
          DirectedEdges→True,EdgeLabeling→False,
          SelfLoopStyle→None]
```

is *all* that is required. *Mathematica* then automatically identifies the six strongly connected components of the FSM and draws them separately.

5 From Visualization to Formal Evaluation

Pictures evidently clarify many features of the design, particularly when appropriate projections of the device are chosen. *Mathematica* makes it very easy to select specified parts of a device to explore any design criterion and visualize them directly, but for a safety critical device we need to be *sure*.

The message of this section is that simple programming methods can ensure reliable device design for a wide and insightful range of criteria. This section gives many examples, using first year undergraduate level computer science techniques. In an industrial design environment, the design insights could be generated automatically, and the designer need have no understanding or access to how those insights are generated. Readers may prefer to skim to section 6, below.

A FSM can be represented by a transition matrix, and as [16] shows, each user action can be represented as an individual matrix, as if we are considering the subset of the FSM that has transitions that are only that action. Hence if A is the transition matrix for user action A, with elements 0 and 1, and s a vector representing the current state (i.e., all 0 except 1 at s_k representing the device in state k), then sA is the state following action A. The paper [16] calls such matrices *button matrices*, but we note that the matrices correspond to any defined user action on the FSM, which may be more general than a button press. In particular for the Fluke 114, the 10 actions we have defined over the FSM are pressing one of its 5 buttons, holding one of 3 buttons down for a few seconds (this has no effect on 2 buttons), and turning the knob left or right (clockwise or anticlockwise).

Because of the associativity of matrix multiplication, if A, B, C are user action matrices, then $M=ABC$ is a matrix product that represents the sequence of actions A then B then C, and sM is the state after doing that sequence of actions. Evidently, we can explore the behavior of a user interface through matrix algebra, and in a system like *Mathematica*, the algebra is as simple to do as it looks.

We can either explore behavior of states individually (as in sM) or we can talk of user actions in any state. The matrix M defines what the user action M does in *every* state, and hence matrix algebra allows us to explore very easily properties of an interactive device in all states — something empirical work with users would find problematic with any but the simplest of devices. To give a simple example, is $sAB=sBA$ then we know that in state s, it does not matter which order the user does actions A and B. If however we show $AB=BA$, we then know that it does not matter which order A and B are done in any state. Notice that two efficient matrix

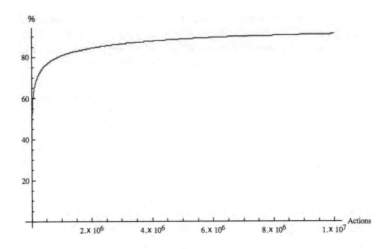

Fig. 6. A cost of knowledge graph: the percentage of states a user is expected to visit after a given number of actions. Here, 10^7 actions provides about 90% coverage of the state space. The graph is based on the method of [13].

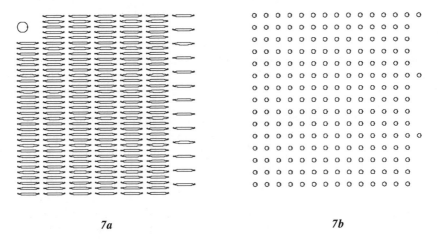

7a   7b

Fig 7. Visualizations of all states but projected to show only the LIGHT button transitions. In figure 7a, there is one state (top left, which we can assume is Off) where the LIGHT button does nothing, otherwise the visualization clearly shows all states grouped into pairs. This visualization is informative and checks more information than a usability study could realistically cover. In figure 7b (where what looks in this paper like circles are in fact states with a self-looping arrow) the visualization additionally merges states that *only* differ in the status of the light component; although the state Off is no longer distinguished, we are now certain that the LIGHT button *only* changes the state of the light (if at all). Indeed, the visualizations are exactly what we expect, but (even together) they don't rigorously convince us the properties of the LIGHT button are correct. However, together with figure 5 (which shows LIGHT really does nothing in Off), we can see that pressing LIGHT *always* flips the state of the device light, except when the device is Off (when it does nothing), and it has *no* other side-effect in any state.

Table 1. Exact and partial laws of length up to 2. Matrices 'crossed out' represent the matrix of the corresponding button press but held down continuously for several seconds. You can see, for example, that holding MINMAX twice is the same as holding it once. Notice that holding MINMAX behaves differently when followed by a left or right knob turn.

Exact laws	*Partial laws*
$HOLD^2 = I$	HOLD LEFT $\approx$ LEFT
MINMAX LEFT = LEFT	HOLD RIGHT $\approx$ RIGHT
$LIGHT^2 = I$	MINMAX RIGHT $\approx$ RIGHT
~~MINMAX~~ LEFT = LEFT	RANGE $\approx I$
~~MINMAX^2~~ = ~~MINMAX~~	YELLOW9$\approx I$
~~HOLD^2~~ = ~~HOLD~~	~~MINMAX~~ RIGHT $\approx$ RIGHT
$YELLOW^2 = I$	~~RANGE~~ $\approx I$
YELLOW LEFT = LEFT	~~HOLD~~ LEFT $\approx$ LEFT
YELLOW RIGHT = RIGHT	~~HOLD~~ RIGHT $\approx$ RIGHT
~~RANGE~~ RIGHT = RIGHT	~~HOLD~~ ~~MINMAX~~ $\approx$ ~~MINMAX~~

multiplications and a test for equality are sufficient to verify this fact for all states (we do *not* need to check s*AB*=s*BA* for every **s**).

In our discussion, we will show that exploring device properties is straightforward, and moreover, we also show that finding interesting device properties (that are readily found using matrix algebra) is infeasible for empirical techniques.

For example, if *M* is singular (a simple matrix property), then the device cannot provide an undo for the sequence of actions (or, more precisely, it cannot provide an undo that always works). For the F114 only the matrices for HOLD, YELLOW and the STAR button are non-singular, and these all share the property that $M^2=I$, for *M* any of the three matrices.

We may think that the STAR button switches the light on and light off. We can check visually, as in figure 7, that the STAR button pairs all states apart from Off, as we expect. Figure 7 *looks* right. Even though we could easily generate many more visualizations each refining our knowledge of the device design, we should be more rigorous: if *L* is the matrix corresponding to the user action of pressing STAR, then *Mathematica* confirms that $L^2=I$. In other words, in every state, pressing STAR twice does nothing: if the light was on initially, it's still on, and if it was off, it's still off.

We can manually explore such laws, for instance that if *H* is the hold action matrix that *HL=LH*. (In *Mathematica's* notation, we evaluate H.L==L.H, and *Mathematica* responds True.) This simple equation confirms that in *all* states, it does not matter which order the user sets the hold mode (or unsets it) and sets (or unsets) the light. Of course, it turns out that some laws we expect to be true fail to hold in all states. For example, we may imagine the hold action "freezes" the display, so we expect HOLD followed by pressing the YELLOW button should have no effect, so we expect *HY=H*, that is, the *Y* after *H* should change nothing. In fact *HY≠H*.

Mathematica can summarize the exceptions to this putative law; here, the exception occurs only when the knob is in position 5 and hold is already on. We overlooked that the HOLD action does not ensure the device is in the hold mode; actually it flips hold/no hold, and in knob position 5 only does pressing the YELLOW button change the meter from AC to DC when the hold mode is off.

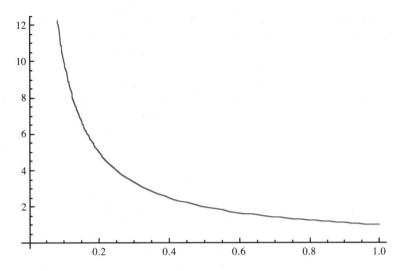

Fig. 8. Graph showing how user testing becomes more efficient assuming we know what issue is being looked for. Thus, the cost at the left (0 knowledge) represents an initial test, when users knows nothing to guide their search; and the cost at the right (100% optimal knowledge) represents a retest performed with the user given optimal instructions to test the issue.

A partial law such as this is potentially of great interest for safety critical user interface design. Here, we found a law that holds in 99.7% of device states, but from their experience a user is likely to believe this law is exactly true. After all, as we can see from figure 6, even with a history of tens of millions of user actions — a considerable practical experience of device use — a typical user will have only explored around 90% of the state space (real users will do particular tasks often, and their state space coverage would be less). Would they have visited a problematic state and noticed its problem? Clearly it is important for a designer to identify partial laws and to decide whether they have safety or usability implications, and, if so, how to deal with the implications — for instance, by redesign or by user training. It is too unreliable to leave it to empirical work alone.

(Although we will not explore it here, it is easy to define projections of the state space as matrices to explore partial laws that apply within specified subspaces. For instance we might wish to explore partial laws ignoring Off.)

To spot that the "law" $HY=H$ fails, a user would have to visit one of the states where the law fails and then confirm that indeed the law breaks: they'd have to press YELLOW in one of those states. There are four such states: the backlight can be on or off, and the "yellow" mode can be on or off (the "yellow" mode is whether the meter is set to AC or DC volts — but the electrical details do not matter for our discussion). Now we happen to know that the "law" fails and we know where it fails. Suppose as would be more realistic for a usability study that we are looking for a potential design problem like this, but of course not knowing in advance what we were looking for. How long would a user study take to find the same issue?

Imagine a user in a laboratory exploring a device looking for similar issues, but of course not knowing what or where they are *a priori*. Such a user would be behaving essentially randomly. We can therefore set up a Markov model of user behavior and work out the expected time to get from (say) Off to the any state where a sufficiently perceptive and observant user would have noticed the issue. Given the matrices for the device's user actions — for the Fluke 114, we have ten such matrices M_i each 425×425 — we define a stochastic matrix representing the user doing action i with probability p_i by $T = \sum p_i M_i$. (Equivalently, we can view action matrices M_i obtained from a given stochastic matrix of user action by setting the probability of the action i to 1.) Given T then sT^n is the expected probability distribution of state occupancy after n user actions. This is easy to evaluate (one line of *Mathematica* code) and gives another indication of the ease and efficiency of analyzing devices in this way.

Given a stochastic matrix standard Markov techniques routinely obtain the expected number of actions [14]: on average it would be 316 actions. Markov models also obtain many other properties that we do not have space to explore here.

The more the user knows about how to reach a state, the faster they should be. We can draw a graph showing how the user gets faster from a state of "uniform ignorance" (where all actions have equal probabilities) to an expert user (the optimal action has probability 1). Figure 8 visualizes this.

Ironically, if we know the class of state we want the user to find, we could analyze the design issue formally and not involve users at all. We need not waste user time if we can characterize a class of design fault, for instance from previous experience with similar devices. There should be a list of laws we wish to check hold (or fail, as appropriate) for all designs of a certain sort.

Rather than thinking of possible laws and testing them, a short program can systematically search for all laws, say breadth-first in increasing length until some condition, because laws of sufficient complexity will have little significance for users.

A program was written to find all laws of the form $A=I$, $AB=I$, $AB=BA$, $A^2=A$, $A^2=I$. Some laws (e.g., $AB=I$ imply others, such as $AB=BA$) so a list of laws can be reduced (however, $AB \approx B$ and $A \approx B$ does not imply $AB \approx A$, etc). Also, we are interested in approximate laws; for example, we are unlikely to find any user action A that has $A=I$, but $A \approx I$ suggests the provision of the feature A is probably not necessary, or that the few things it does must be justified. The program finds 12 exact laws and 18 partial laws; a sample are shown in Table 1.

The partial law criterion was arbitrarily set at 10% of states. In a thorough analysis, states need not have equal weight in the definition of partiality: we migth choose to ignore Off altogether (since a user is presumably aware a device will behave differently when it is off) and we might weigh states according to how often, or how long, they are used for a representative suite of tasks.

If we look for laws when the knob is in particular positions, it is apparent that the structure of the device changes dramatically with knob position — for example pressing YELLOW only works in one knob position. Perhaps the YELLOW button is provided on the Fluke 114 not because it is useful or effective, but because the 114 is one of a range of devices and YELLOW has better use on the other models? An unlabelled button like YELLOW is obviously useful for providing different features in different devices, because it works across the entire range without incurring additional manufacturing costs. Considering the 114 alone, the button appears to be more confusing than useful: it means that VAC, VDC and mVAC and mVDC work in uniquely different ways. Either YELLOW should have worked consistently with

VDC/VAC — hence reducing the number of knob positions by one — or the RANGE feature could have been used so mV was an extended range of V (which is how some other Fluke meters, such as the model 185, work).

6 Putting the Approach into a Wider Perspective

The received wisdom in HCI can be summarized as UCD, user-centered design. It is informative to put UCD in perspective by drawing a typical iterative design cycle, as in figure 9.

For a device to work well, all parts of the cycle must function adequately. Using Reason's swiss cheese model [11], we might say that each arrow done properly is a defense against one or another sort of design or use error, and that all must be done well and by appropriate methods. The appropriate methods in each case are very different, and come from different disciplinary backgrounds with different styles of working.

UCD concentrates on the lefthand side (and often the lower lefthandside), the "external" or human side of the cycle. The diagram makes it clear that UCD is necessary, but also that it is not sufficient. This paper has shown that the righthand, the "internal" or systems side of the cycle has problems being done well in industry, but that there are techniques (such as those covered in this paper) that can address some of the righthand side issues.

Industry needs to get *both* sides right, both external and internal. To date, HCI as a community has more-or-less ignored the internal, and hoped to fix problems by even better UCD. Unfortunately the professional skills required for UCD are very different than the skills required for systems, which exacerbates the isolation of UCD from technical approaches.

This paper argued not only can the internal be improved, but that it is essential to address HCI issues rigorously internally because safety critical devices cannot be handled thoroughly by conventional UCD, and certainly not by UCD uninformed by formal methods.

6.1 Research Versus Industry

Industry and research have different goals, and in HCI the different emphases are easy to confuse, especially as "usability" is a proper part of HCI but which is specifically concerned with effective products rather than with effective research concepts. In industry, the iterative cycle of Figure 8 is probably only run around once per product: once anything is good enough to demonstrate or evaluate, it is probably good enough to market. *Every* step of the iterative cycle must be good enough, though it is probably distributed over a series of products, each cycle around generating a new product, or an enhancement to an existing product.

From the different perspective of research, improving *any* part of the iterative cycle is worthwhile, or finding out how to better conceptualize parts of the cycle — making contributions to how HCI is done, rather than to any specific product. Yet despite the clear distinction between research and industrial practice, the HCI community often wants to have research cover all aspects of the cycle. In this paper, we only addressed

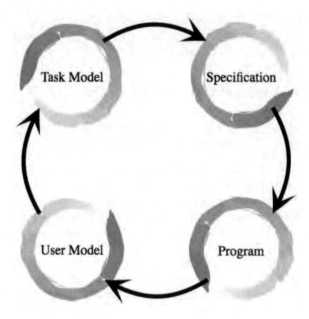

Fig. 9. A schematic of the full iterative design cycle. For the device to work, it must have a program. For the user to understand how to use the device, they must have a model. To perform the task, the user must have a task model, which depends on the user having the user model to use the device to achieve the task goals. The task model defines the requirements for the specification. The specification defines the program. The left side of the diagram, the external side, is the main concern of UCD, the right side, internal side, is the main concern of software development.

"internal" issues, but we showed how they can contribute to better and safer design. As the cycle makes clear, improving internal issues will make UCD, which follows internalist development, easier and more reliable. And, of course, if the cycle is pursued, better UCD in turn improves internal approaches. Each follows the other.

6.2 Interactive Exploration

A separate aspect of our chosen case study is that we analyzed it using *Mathematica*, a popular computer algebra system [21]. Everything described in this paper works in *Mathematica* exactly as described; *Mathematica* makes exploring a device extremely easy, and because *Mathematica* is interactive, a team of people can explore a device testing how it works as they wish. As they discuss the device and what *Mathematica* shows about it, they will have further insights into the design.

In product development, a specification of the device will be used and refined to a program, and designers and users will explore prototypes at various stages of development. For this paper, we had access to no such specification; instead we took the device and its user manual to be an expression of the specification. We then built a model of it, and explored the model. As problems were identified, we either fixed the model, or revised our mental models of what we thought it was doing. Here, I wish to acknowledge the help of Michael Harrison (Newcastle University, UK) and José Campos (University of Minho, Portugal), who sat through demonstrations and helped with bug fixes. It was they who noted how useful exploration of a device's design could be using an interactive tool such as *Mathematica*.

In a sense, then, our development of the model in this paper is closely analogous to how a model might be developed in industry using our techniques, differing only in the starting point: we started with a finished product; industry would start with the idea (somehow expressed) of the product to be finished.

For research, *Mathematica* is ideal, as it provides a vast collection of sophisticated and powerful features. For industrial development, of course *Mathematica* could still be used, though this presumes a certain level of familiarity with it. In the future the appropriate features of *Mathematica* that are useful to device design and analysis will be packaged and made as easy to use as typical web site development tools. Indeed, one would then consider features for managing design, not just exploring it. For example, in safety critical design it is important to guarantee certain features or properties are fixed, or once approved are not subsequently changed without due process.

7 Conclusions

Both industry and the HCI research community struggle with the complexity of modern interactive devices. This unmanaged complexity is no more apparent and worrying than in the area of interactive medical devices. This paper provided evidence that conventional UCD methods are faced with state spaces that are too large to evaluate empirically, and it also showed that basic technical methods can contribute to the usability analysis. In particular, the paper showed visualizations and formal methods based on finite state machines and matrix algebra. In the future, the techniques could be embedded in programming tools, and designers need have no special expertise to use the techniques effectively as is currently necessary.

The HCI community has to date emphasized user-centered approaches, including empirical evaluation, to compensate for the poor state of interactive system usability. This is important but is not sufficient for ensuring safety critical systems are usable, safe or effective. For this, analytic techniques such as those presented in this paper are required. This paper showed how visualization and formal methods work together, and with a suitable interactive analysis tool they support exploratory dialogue in the design team. This paper has shown that essentially elementary formal methods can have a rigorous, considerable and insightful impact on the design process and hence on the quality of interactive safety critical devices.

Note: The definition of the Fluke 114 used in this paper along with all calculations and graphs referred to is at www.cs.swansea.ac.uk/~csharold/fluke114.

Acknowledgements. The author thanks the referees for their insightful and helpful comments.

References

1. Cardinal Health, MX-4501N_20060929_104509.pdf (2007)
2. Cardinal Health (May 20, 2007), http://www.cardinal.com/uk/alaris/solutions/medicationsafety/IVsystems/
3. Department of Health and The Design Council. Design for Patient Safety (2003)
4. Fluke 117 Virtual Demo accessed August (2007), http://us.fluke.com/VirtualDemos/117_demo.asp,
5. Gould, J.D., Lewis, C.: Designing for Usability: Key Principles and What Designers Think. Communications of the ACM 28(3), 300–311 (1985)
6. Høegh, R.T.: Usability Problems: Do Software Developers Already Know? In: Proceedings ACM OZCHI, pp. 425–428 (2006)
7. Harel, D., Marelly, R.: Come, Let's Play. Springer, Heidelberg (2003)
8. Holzinger, A.: Usability Engineering for Software Developers. Communications of the ACM 48(1), 71–74 (2005)
9. Institute for Safe Medication Practice Canada: Fluorouracil Incident Root Cause Analysis, (May 8, 2007), http://www.cancerboard.ab.ca/NR/rdonlyres/D92D86F9-9880-4D8A-819C-281231CA2A38/0/Incident_Report_UE.pdf
10. Landauer, T.: The Trouble with Computers. MIT Press, Cambridge (1995)
11. Reason, J.: Human Error: Models and Management. British Medical Journal 320, 768–770 (2000)
12. Scottish Executive: Unintended Overexposure of Patient Lisa Norris During Radiotherapy Treatment at the Beaston Oncology Centre, Glasgow (January 2006), http://www.scotland.gov.uk/Publications/2006/10/27084909/22
13. Thimbleby, H.: Analysis and Simulation of User Interfaces. In: Proceedings BCS Conference on Human Computer Interaction 2000, vol. XIV, pp. 221–237 (2000)
14. Thimbleby, H., Cairns, P., Jones, M.: Usability Analysis with Markov Models. ACM Transactions on Computer-Human Interaction 8(2), 99–132 (2001)
15. Thimbleby, H.: The Directed Chinese Postman Problem. Software — Practice & Experience 33(11), 1081–1096 (2003)
16. Thimbleby, H.: User Interface Design with Matrix Algebra. ACM Transactions on Computer-Human Interaction 11(2), 181–236 (2004)
17. Thimbleby, H.: Interaction Walkthrough: Evaluation of Safety Critical Interactive Systems. In: Doherty, G., Blandford, A. (eds.) DSVIS 2006. LNCS, vol. 4323, pp. 52–66. Springer, Heidelberg (2007)
18. Thimbleby, H.: Press On. MIT Press, Cambridge (2007)
19. Thimbleby, H., Thimbleby, W.: Internalist and Externalist HCI. Proceedings BCS Conference on Human Computer Interaction 2, 111–114 (2007)
20. Udell, J.: Lights, Camera, Interaction. InfoWorld (June 23, 2004)
21. Wolfram, S.: The Mathematica Book, 3rd edn. Cambridge University Press, Cambridge (1996)

Appendix: Definition of the Fluke 114 FSM

This appendix demonstrates how concise and flexible a FSM definition is; it provides details of the definition of the Fluke 114 used in this paper. Programming a FSM in any other high level language would differ notationally but would require similar effort to using *Mathematica*: that is, not much. Documentation and complete code is provided on this paper's web site.

A state is represented as a tuple of components, such as {ac→False, hold→False, knob→2, light→False, minmax→0, range→0}, which in this case means the meter is set to DC (AC is false), the hold feature is off, the knob is in position 2, the light is off, the minmax feature is disabled (it can take 4 other values) and the range is automatic (it can take 7 other values). This notation means the state can be easily read by the programmer. Within a program, knob/.s extracts from s the value of the knob term. Our notion of state can be extended with more components, such as doserate→2.3, units→"ml/hr", totaldose→54.7 and so on; then to get *exactly the same* results as discussed in this paper, we would then project down to a finite space ignoring such components.

We define the device by writing a function action that maps the user's action and the current state to the next state.

In *Mathematica*, the order of rules matters. First we say that if the device is Off, no button pressing or continuous holding has any effect:

```
action[(press|hold)[_],state_]:=state/;(2==knob/.state)
```

This is to be read as "if the knob is in position 2 (i.e., off) then any press or hold action leaves the state unchanged."

If the device is On, then pressing the STAR button changes the state of the LCD backlight. Here is the rule expressing this:

```
action[press["*"],state_]:=
      override[light→!(light/.state),state]
```

This is to be read as "pressing the STAR button when in state inverts the value of the light component of the state." Specifically, "!" means logical *not*, and override means replace components of the state (here, the light) with new values. The remaining rules are as follows:

```
KnobTurned[state_]:= (* any turn resets hold to False, etc *)
      override[{hold→False,minmax→0,range→0},
               {ac→(3==knob/.state)},
               {light→(If[2==knob,False,light]/.state)},state]
action[turn["Right"],state_]:=
      KnobTurned[override[knob→(knob+1/.state),state]]
               /;(knob!=7/.state)
action[turn["Left"],state_]:=
      KnobTurned[override[knob→(knob-1/.state),state]]
               /;(knob!=1/.state)
action[(press|hold)[_],state_]:=state/;(2==knob/.state)
action[press["HOLD"],state_]:=
      override[hold→!(hold/.state),state]
action[hold["HOLD"],state_]:=
      override[hold→False,state]
action[press["MIN MAX"],state_]:=
      override[minmax→
```

```
    Mod[minmax/.state,4]+1,state]
    /;((!hold||minmax!=0)&&1!=knob/.state)
action[hold["MIN MAX"],state_]:=
    override[{hold→False,minmax→0},state]
    /;((!hold||minmax!=0)&&1!=knob/.state)
action[press["YELLOW"],state_]:=
    override[ac→!(ac/.state),state]
    /;(minmax==0&&!hold&&5==knob/.state)
action[press["RANGE"],state_]:=
    override[range→
    (Mod[range,If[6==knob,7,4]]/.state)+1,state]
    /;(minmax==0&&!hold&&(3==knob||4==knob||6==knob)/.state)
action[hold["RANGE"],state_]:=
    override[range→0,state]/;(!hold&&minmax==0/.state)
action[_,state_]:=state
```

Improving Interactive Systems Usability Using Formal Description Techniques: Application to HealthCare

Philippe Palanque, Sandra Basnyat, and David Navarre

LIIHS-IRIT, Université Paul Sabatier – Toulouse III
118 route de Narbonne, 31062, Toulouse Cedex 4
{palanque,basnyat,navarre}@irit.fr

Abstract. In this paper we argue that the formal analysis of an interactive medical system can improve their usability evaluation such that potential erroneous interactions are identified and improvements can be recommended. Typically usability evaluations are carried out on the interface part of a system by human-computer interaction/ergonomic experts with or without end users. Here we suggest that formal specification of the behavior of the system supported by mathematical analysis and reasoning techniques can improve usability evaluations by proving usability properties. We present our approach highlighting that formal description techniques can support in a consistent way usability evaluation, contextual help and incident and accident analysis. This approach is presented on a wireless patient monitoring system for which adverse event (including fatalities) reports are publicly available from the US Food and Drug Administration (FDA) Manufacturer and User Facility Device Experience (MAUDE) database.

Keywords: Human–Computer Interaction, Incident and Accident Investigation, Formal Description Techniques, Medical Informatics, Patient Monitoring.

1 Introduction

The advances of healthcare technology have brought the field of medicine to a new level of scientific and social sophistication. They have laid the path to exponential growth of the number of successful diagnoses, treatments and the saving of lives. On the hand, technology has transformed the dynamics of the healthcare process in ways which increase the distribution & cooperation of tasks among individuals, locations and automated systems. Thus, technology has become the backbone of the healthcare process. However, it is usually the healthcare professionals who are held responsible for the failures of technology when it comes to adverse events [8]. It has been estimated that approximately 850,000 adverse events occur within the UK National Health Service (NHS) each year [26]. A similar study in the United States arrived at an annual estimate of 45,000-100,000 fatalities [17]. While functionality of medical technology goes beyond imaginable, the safety and reliability aspects have lagged in comparison with the attention they receive in other safety-critical industries. The discussion of human factors in medicine has centered on either compliance with government standards or on analyzing accidents, a posteriori.

A. Holzinger (Ed.): USAB 2007, LNCS 4799, pp. 21–40, 2007.

In this paper we are interested in medical systems that offer a user interface and require operators interaction while functioning. Due to their safety-critical nature there is a need to assess the reliability of the entire system including the operators. The computer-based part of such systems is quite basic (with respect to other more challenging safety-critical systems such as command and control systems (cockpits, Air Traffic Management, ...)) and thus current approaches in the field of systems engineering provide validated and applicable methods to ensure their correct functioning[1]. Things are more complicated as far as the user interface and operators are concerned. Human-computer interaction problems can occur because of numerous poor design decisions relating to the user interface (UA). For example, poor choice of color, a mismatch between the designer's conceptual model and the user's mental model or insufficient system feedback to allow the user to understand the current state of the system which is known as mode confusion. Mode confusion refers to a situation in which a technical system can behave differently from the user's expectation [9]. Operator assessment is even harder to perform due to the autonomous nature, independence and vulnerability to environmental factors like stress, workload.

Lack of usability has been proved to be an important source of errors and mistakes performed by users. Faced with poor user interfaces (designed from a non-user centered point of view) users are prone to create alternative ways of using the applications thus causing hazardous situations. In the worst cases, lack of usability may lead users to refuse to use the applications, potentially resulting in financial loss for companies with respect to the need for redesign and possibly additional training. For detecting and preventing usability problems it is important to focus the design process from the point of view of people who will use the final applications. The technique called User-Centered Design (UCD) [30] is indeed the most efficient for covering user requirements and for detecting usability problems of user interfaces.

Usability evaluation, both formative and summative, can improve these kinds of issues by identifying interaction problems. However, after the design iteration, another round of usability evaluation is required in order to verify whether there is any improvement. This can be costly in terms of time and resources. A further way to improve design is to apply interface design criteria and guidelines during the design process. Ergonomic criteria have been proved to increase the evaluation performance of experts [6][1]. Safety critical systems have been for a long time the application domain of choice of formal description techniques (FDT). In such systems, where human life may be at stake, FDTs are a means for achieving the required level of reliability, avoiding redundancy or inconsistency in models and to support testing activities. In this paper, we will illustrate our approach using a FDT based on Petri nets. In this paper, we show that FDTs can be used to support a selection of usability related issues including:

- Formative evaluation
- Summative evaluation
- Contextual help for end users
- Investigation of incidents and accidents by providing formal descriptions of the behavior of the system

---

[1] As pointed out in 14 and 15 there is an increasing integration of new technologies in medical application that will raise new issues in their reliability assessment.

The next section presents an overview of usability evaluation, including formative and summative evaluation and presents several ergonomic/usability criteria and guidelines for interface design including some that are targeted at touch screen interfaces. Section 3 provides an informal presentation of formal description techniques and the Interactive Cooperative Objects (ICOs) formalism we will use in this paper, before presenting the case study in section 4. Section 5 illustrates the approach by discussing ways in which formal specification of the behavior of the system using ICOs and its environment (Petshop) can improve formative and summative evaluation, provide contextual help for end users and support the investigation of incidents and accidents. We then conclude and provide suggestions for future work.

2 Usability Aspects for Interactive Systems

Usability addresses the relationship between systems, system interfaces, and their end users. Interactive systems, or systems requiring humans in the loop of control must be designed from a UCD approach in order for the intended users to accomplish their tasks efficiently, effectively and as expected by the user. A system with poor usability can cost time and effort, and can greatly determine the success or failure of a system (See [13] for guide on usability engineering and [14] for interface design for medical applications). In this section, we firstly present a brief overview of usability testing, distinguishing between the two main forms of testing, formative and summative (2.1). We then discuss usability criteria and guidelines as a means of improving usability (2.2). This is extended in the third section (2.3) by describing guidelines for a specific kind of interface, touch screens. The fourth section (2.4) describes the importance of providing contextual helps as a means of improving reliability while the final section (2.5) introduces a slightly different dimension in which we discuss the ways in which formal description techniques (FDTs) can assist in improving the design of safety-critical interactive systems, such as medical applications, by helping to avoid incidents and accidents.

2.1 Usability Testing

The usability of a system can be tested and evaluated. Typically, the testing serves as either formative or summative evaluation.

Formative tests are carried out during the development of a product in order to mould or improve the product. Such testing does not necessarily have to be performed within a lab environment. Heuristic Evaluation (HE) [27] is an established formative Usability Inspection Method (UIM) in the field of HCI. The analyst, who must have knowledge in HCI, follows a set of guidelines to analyze the user interface. The technique is inexpensive, easy to learn and can help predict usability problems early in the design phase if prototypes are available. Examples of HEs include [35], [29] and [6]. While this paper focuses on presenting how formal description techniques (FDTs) can support and improve the usability of interactive safety-critical systems, in [8] we have shown how they can support usability testing.

Summative tests in contrast are performed at the end of the development in order to validate the usability of a system. This is typically performed in a lab environment and involves statistical analysis such as time to perform a task.

2.2 Usability Criteria and Guidelines

Heuristic evaluation is a common usability inspection method achieved by performing a systematic inspection of a user interface (UI) design for usability. The goal of heuristic evaluation is to find the usability problems in the design so that they can be attended to as part of an iterative design process. Heuristic evaluation involves having a small set of evaluators examine the interface and judge its compliance with recognized usability principles (the "heuristics") [29]. Bastien and Scapin [6] evaluate the usefulness of a set of ergonomic criteria for the evaluation of a human-computer interface with results indicating that usability specialists benefit from their use. Since their early publications concerning ergonomic guidelines for interface design to improve usability [36], Neilsen and Molich coined the term "heuristic evaluation" [27]. Since the original publication of heuristics, Nielsen refined them to derive a set of heuristics with maximum explanatory power. The following lists present Bastien and Scapin's [6] and Nielsen's guidelines [28].

Table 1. Criteria for interface design to improve usability

Ergonomic criteria – from Table 1 in Bastien and Scapin [6]		Nielsen's Heuristics guidelines [28].	
1	Guidance	1	Visibility of system status
2	User workload	2	Match between system and the real world
3	User explicit control	3	User control and freedom
4	Adaptability	4	Consistency and standards
5	Error management	5	Error prevention
6	Consistency	6	Recognition rather than recall
7	Significance of codes	7	Flexibility and efficiency of use
8	Compatibility	8	Aesthetic and minimalist design
		9	Help users recognize, diagnose, and recover from errors
		10	Help and documentation

Based on the ergonomic criteria set out by Bastien and Scapin, they further define guidelines relating to each criterion. The following is an example of the "user explicit control" criteria. The guideline is called "user control" (see Table 2 in [6])

1   Allow users to pace their data entry, rather than having the pace being controlled by the computer processing or by external events
2   The cursor should not be automatically moved without users' control (Except for stable and well known procedures such as form-filling)
3   Users should have control over screen pages

4   Allow users to interrupt or cancel a current transaction or process
5   Provide a 'cancel' option that will have the effect of erasing any changes just made by the user and restoring the current display to its previous version.

We have already shown how formal description techniques can support systematic assessment of user interface guidelines in [33] by relating using explicit states representations as a link between guidelines and behavioral description of interactive applications.

The usability guidelines and criteria mentioned so far in this section relate to UIs in general. However, as we will see in our case study, the interaction between human and system with medical applications is migrating towards touch-screen interfaces. The following subsection presents guidelines relating specifically to these kinds of devices.

2.3   Guidelines for Specific Devices: The Case of Touch Screen

Tactile interfaces have, in addition to the above mentioned usability criteria and guidelines, their own interface design challenges. For example, the use of a touch screen of an outdoor cash machine in winter by a user wearing gloves may not be effective.   Though standards for principles and recommendations as well as specifications such as ISO13406, ISO 14915, ISO 18789 exist for interface and interaction, they do not specifically target tactile interfaces for neither typical walk-up-and-use interfaces, nor domain specific interfaces such as those for the healthcare domain.

In a recent conference called Guidelines on Tactile and Haptic Interactions (GOTHI), Fourney and Carter [11] review existing international standards on tactile/haptic interactions and provides a preliminary collection of draft tactile/haptic interactions guidelines based on available guidance. The guidelines are categorized under the following headings:

- Tactile/haptic inputs, outputs, and/or combinations
- Tactile/haptic encoding of information
- Content specific Encoding
- User Individualization of Tactile / Haptic Interfaces

We provide below a selection of the guidelines proposed Fourney and Carter [11] that particularly relate to the research and case study presented in this paper, that is the improvement of usability of a safety-critical interactive touch screen system. See [11] for the complete list of guidelines.

- Provide navigation information
- Provide undo or confirm functionality
- Guidance on combinations with other modalities
- Make tactile messages self descriptive
- Mimic the real world
- Use of apparent location

- Keep apparent location stable
- Provide exploring strategies
- Use size and spacing of controls to avoid accidental activation
- Avoid simultaneous activation of two or more controls

[1] has also studied usability of touch screen interfaces and suggests that menus and buttons that are often selected should be located at the bottom of the interface rather than the top to avoid covering the screen (and in the case of a safety-critical application, vital information) with one's arm while accessing the menus. Additional references to research on guidelines for touch screens can be found [15] and [16].

Furthermore, critical interfaces, such as those used in to monitor patients in the healthcare domain provide vital information. This means, if selecting an option within the application forces a new window to cover the initial screen, there is a risk that it will be unintentionally missed during that short period of time (see Fig. 1). A lot of work has been done on the visibility of a workspace, with techniques designed to provide contextual awareness at all times even while the user focuses on one aspect of the interface (see [21]).

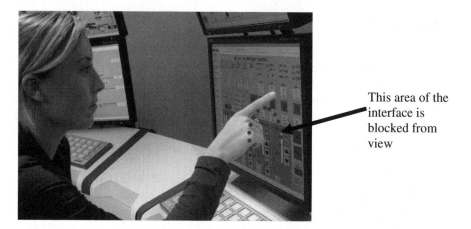

This area of the interface is blocked from view

Fig. 1. Interaction with a touchscreen

Coutaz et al, [10] describe "CoMedi", a media space prototype that addresses the problem of discontinuity and privacy in an original way. It brings together techniques that improve usability suggested over recent years. The graphical user interface of CoMedi is structured into three functional parts: at the top, a menu bar for non frequent tasks (see Fig. 2 and Fig. 3). In the center, a porthole that supports group awareness. At the bottom, a control panel for frequent tasks. The "The fisheye porthole" supports group awareness. Though the example in the paper is of a collaborative tool in a research laboratory, the way in which the fisheye porthole is designed could be useful in a medical environment. The porthole may have the shape of an amphitheatre where every slot is of equal size. When the fisheye feature is on, selecting a slot, using either the mouse or a spoken command, provokes an animated distortion of the porthole that brings the selected slot into the centre with progressive enlargement [10].

The idea of context loss while zooming or focusing on a particular aspect of the interface has been argued and researched by Bob Spence with his database navigation approach [37], Furnas, with his generalized fisheye views [12]and Mackinlay et al.,

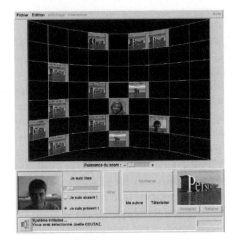 

Fig. 2. The graphical user interface of CoMedi

Fig. 3. The UI of CoMedi with Porthole when the fisheye view is activated

[20] with their perspective wall his perspective wall approach and by John Lamping with his hyperbolic browser [18].

2.4 Importance of Contextual Help

Contextual help is an important feature that interactive systems should include. It refers to the available of help at any point in time relative to the given state of the system. For example, if a user expects an option or a button to be available and it is not (i.e. a grayed out button in a menu), the contextual help would indicate why this option is not available at that moment. Formal description techniques (FDTs) provide a way to have a complete and unambiguous representation of all states of the system and state changes as well as a means for reasoning about these states and state changes. However, without formal methods fully describing the behavior of the system, providing this kind of advice is very difficult. By simply analyzing a user interface, it is almost impossible (except with a very basic system) to know exactly what state the system is in. In order for the formal description of the system to be directly usable for the end user (i.e. the operator), it would certainly need additional annotations. It is unlikely that medical personnel would be able to read such a model without training and it is not useful to train them in such a way. The main point is that the formal model supports designers and developers' activities by for instance identifying why users are experiencing a problem and how to overcome it. Furthermore, analyzing lines of code to determine the system state is extremely cumbersome and providing developers with an abstract view of the system can support them. We show in section 5.2 that a formal graphical description of the behavior of the system can provide information on the current state, why the option/button is not available and what must be performed by the user for that option to be available. More information on how such contextual help can be supported by FDT can be found in [34].

2.5 Avoidance of Incidents and Accidents

The heading incident and accident investigation can be misleading in a usability-related article, however we believe and described in the introduction, that poor usability of a system can potentially lead to an adverse event.

In terms of supporting their investigation, we have devised an approach exploiting two complementary techniques for the design of safety-critical interactive systems. More precisely we address the issue of system redesign after the occurrence of an incident or accident. The techniques used are Events and Causal Factors analysis, used in accident investigation to identify the path of events and contributing factors leading to an accident, and secondly, Marking Graphs, an analysis technique available after formally modeling the interactive system, using Petri nets for example. Marking graphs are used to systematically explore all possible scenarios leading to the accident based on a formal system model. The formal description techniques (FDTs) allow us to exploit and represent all system states. The techniques serve two purposes. Firstly, it is intended to ensure that the current system model accurately models the sequence of events that led to the accident according to information provided by Events and Causal Factors Analysis. Secondly, our approach can be used to reveal further scenarios that could eventually lead to similar adverse outcomes. The results can in turn be used to modify the system model such that the same accident is unlikely to recur. This is achieved by identifying system states that are undesirable and may want to be avoided. By identifying the state changes leading to such undesirable states, it is possible to adapt the system model to make these states impossible by making the paths (combination of state changes leading to that state) impossible. This is supported by simulating the identified scenarios on the adapted system model. This will be exemplified in section 5.3, on a simple example using the PatientNet case study. The interested reader can see a complete example in [4].

3 Formal Specification of Interactive Systems

The previous sections have described usability evaluation techniques, usability guidelines targeted specifically at touch screen interfaces as well as relating usability to incident and accidents. In this section, we present a formal description technique (FDT) for describing the behavior of interactive systems. Most FDT could provide the kind of support for usability we are describing but we present rapidly here the FDT ICOs (Interactive Cooperative Objects) dedicated to the specification of interactive systems based on Petri nets that we will use for the case study in section 4.

Formal verification of models can only be achieved if the formalism is based on mathematical concepts. This allows reasoning on the system models in addition to empirical testing once the system has been implemented. Verification of models has economical advantages for resolving problems in terms of consumption of resources. Further advantages include the possibility to locate an error (though perhaps not the exact causes for the error).

One of the aims of the formal analysis is to prove that there is no flaw in the models. Using the ICO formalism for describing the models, the analysis is done by using the mathematical tools provided by the Petri net theory. Using those tools, one can prove general properties about the model (such as absence of deadlock) or semantic domain related properties (i.e. the model cannot describe impossible behavior such as the light is on and off at the same time).

Modeling systems in a formal way helps to deal with issues such as complexity, helps to avoid the need for a human observer to check the models and to write code. It allows us to reason about the models via verification and validation and also to meet three basic requirements notably: reliability (generic and specific properties), efficiency (performance of the system, the user and the two systems together (user and system) and finally to address usability issues.

3.1 ICOs as an Concrete Example of Formal Description Techniques

The aim of this section is to present the main features of the Interactive Cooperative Objects (ICO) formalism that we have defined and is dedicated to the formal description of interactive systems. We encourage the interested reader to look at [23] for a complete presentation of this formal description technique as we only present here the part of the notation related to the behavioral description of systems. This behavioral description can be completed with other aspects directly related with the user interface that are not presented here for space reasons.

Interactive Cooperative Objects Formalism (ICO)
ICOs are dedicated to the modeling and the implementation of event-driven interfaces, using several communicating objects to model the system, where both behavior of objects and communication protocol between objects are described by the Petri net dialect called Cooperative Objects (CO). In the ICO formalism, an object is an entity featuring four components: a cooperative object which describes the behavior of the object, a presentation part (i.e. the graphical interface), and two functions (the activation function and the rendering function) which make the link between the cooperative object and the presentation part.

Cooperative Object: Using the Cooperative Object formalism, ICO provides links between user events from the presentation part and event handlers from the Cooperative Objects, links between user event availability and event handler availability and links between state in the Cooperative Objects changes and rendering.

Presentation part: The presentation of an object states its external appearance. This presentation is a structured set of widgets organized in a set of windows. Each widget may be a way to interact with the interactive system (user → system interaction) and/or a way to display information from this interactive system (system → user interaction).

Activation function: The user → system interaction (inputs) only takes place through widgets. Each user action on a widget may trigger one of the Cooperative Objects event handlers. The relation between user services and widgets is fully stated by the activation function that associates each event from the presentation part with the event handler to be triggered and the associated rendering method for representing the activation or the deactivation.

Rendering function: the system → user interaction (outputs) aims at presenting the state changes that occurs in the system to the user. The rendering function maintains the consistency between the internal state of the system and its external appearance by reflecting system states changes.

ICOs are used to provide a formal description of the dynamic behavior of an interactive application. An ICO specification fully describes the potential interactions that users may have with the application. The specification encompasses both the "input" aspects of the interaction (i.e. how user actions impact on the inner state of the application, and which actions are enabled at any given time) and its "output" aspects (i.e. when and how the application displays information relevant to the user).

An ICO specification is fully executable, which gives the possibility to prototype and test an application before it is fully implemented [24]. The specification can also be validated using analysis and proof tools developed within the Petri net community and extended in order to take into account the specificities of the Petri net dialect used in the ICO formal description technique. This formal specification technique has already been applied in the field of Air Traffic Control interactive applications [25], space command and control ground systems [32], or interactive military [5] or civil cockpits [2]. The example of civil aircraft is used in the next section to illustrate the specification of embedded systems.

To summarize, we provide here the symbols used for the ICO formalism and a screenshot of the tool.

- States are represented by the distribution of tokens into places 

- Actions triggered in an autonomous way by the system are represented as 

and called transitions
- Actions triggered by users are represented by half bordered transition 

ICOs are supported by the Petshop environment that makes it possible to edit the ICO models, execute them and thus present the user interface to the user and support analysis techniques such as invariants calculation.

3.2 Related Work on Formal Description Techniques and Usability

The use of formal specification techniques for improving usability evaluation has been explored by several authors.

Loer and Harrison [19] provide their view on how Formal Description Techniques (FDTs) can be used to support Usability Inspection Methods (UIM) such as the formative techniques discussed in the introduction of this paper. In accordance with our views, the authors argue that the costs of using FDTs are justified by the benefits gained. In their paper, the authors exploit the "OFAN" modeling technique [10], based on statecharts with the statemate toolkit for representing the system behavior. Each of the usability heuristics is formalized and tested however, the authors are limited to functional aspects of the system only.

In closer relation to the approach we present in this paper, Bernonville et al [7] combine the use of Petri nets with ergonomic criteria. In contrast to Loer & Harrison's approach of assisting non-formal methods experts to benefit from FDTs, Beronville et al argue for the necessity of better communicating to computer scientists of ergonomic data.

Bernonville et al [7] use Petri nets to describe tasks that operators wish/should perform on a system as well as the procedure provided/supported by the software application. By using ergonomic criteria proposed by Bastien & Scapin [6], the method supports the analysis of detected problems. While Petri nets can be used to describe human tasks, dedicated notations and tool support such as ConcurTaskTrees (CTT) and ConcurTaskTree Environment (CTTe) [22] exist for the description of operator tasks, this would probably be more practical considering that "ErgoPNets" proposed in their paper currently has no tool support and is modeled using MS Visio.

Palanque & Bastide [31] have studied the impact of formal specification on the ergonomics of software interfaces. It is shown that the use of an object oriented approach with Petri nets ensures software and ergonomic quality. For example, by ensuring predictability of commands, absence of deadlocks and by offering contextual help and context-based guidance.

Harold Thimbley provides a worked example of a new evaluation method, called Interaction Walkthrough (IW), designed for evaluating safety critical and high quality user interfaces, on an interactive Graseby 3400 syringe pump and its user manual [38]. IW can be applied to a working system, whether a prototype or target system. A parallel system is developed from the interaction behavior of the system being evaluated.

4 Case Study

Before providing details of our approach, an outline of the case study on which we present our approach is provided here. The medical domain differs from the aviation domain (with which we are more familiar) in that it appears that there is more time to analyze a situation, for example, via contextual help. In a cockpit environment, it is

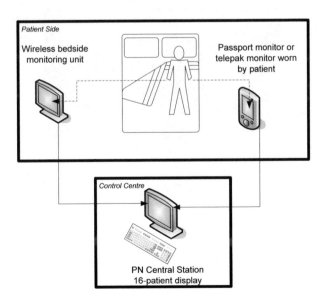

Fig. 4. Simplified layout of the PatientNet system

unlikely that the pilot or co-pilot will have time to access a contextual help service. The case study we have chosen is a telemetry patient monitoring system.

The PatientNet system, operating in a WMTS band, provides wireless communications of patient data from Passport monitors and/or telepak monitor sworn by patients, to central monitoring stations over the same network. Fig. 4 provides a simplified diagram of the layout of the PatientNet system.

Table 2. Summary of MAUDE adverse report search relating to PatientNet system

Patient Outcome Description	Number
Death	10
Unknown (was not specified in report)	8
Other	2
Required intervention	1
Life threatening	1
Total	**22**

Our interest in this system resulted from research on the US Food and Drug Administration (FDA) Manufacturer and User Facility Device Experience (MAUDE) database which has numerous adverse events reported (including fatalities) relating to this device/system. In our search, between the 20/03/2002 and the 08/02/2007, 22 reports were received relating to the PatientNet system including the central station and the passport. Of these 22, 10 resulted in patient death. Table 2 summarizes our findings in terms of patient outcome.

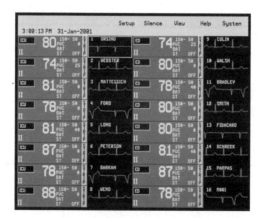

Fig. 5. PatientNet Central Station 16-patient display

The paper does not directly address incident and accident investigation, thus we will not continue our discussion on the adverse events that occurred during human-computer interaction with this system. We simply highlight the importance of such interactions and potential for abnormal human and technical behavior and the impact they may have on human life.

Furthermore, we will not describe the full system in detail; rather focus on the PatientNet central station for which we have screenshots of the application. The central station is a server and workstation where patient information from a variety of different vital sign monitors and ambulatory EGG (Electroglottograph) transceivers is integrated, analyzed and distributed to care providers. Real-time patient data is presented in a common user interface so caregivers have the information they need to respond to critical patient events immediately [39]. In addition to the patient's current status, the Central Station allows for retrospective viewing of patient information, which equips caregivers to make critical decisions and provide consistent, high-quality care to their patients across the enterprise (see Fig. 5 and Fig. 6 for Central Station screenshot).

Features of the central station include: Demographics, alarm limits, bedside view, event storage, ST analysis, full disclosure, trend storage, reports, care group assignments, physiological alarm volumes, print setup, choice of single or duel display.

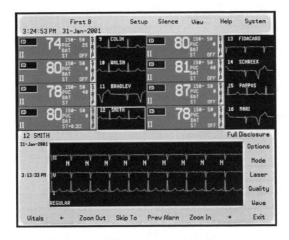

Fig. 6. PatientNet Central Station Full Disclosure

Typical selling points from the product brochure include statements such as "a simple touch of the screen brings up the information you need", "easy-to-read screen focuses attention on vital details" and "find answers to common questions with online Help menus".

5 The Approach

We demonstrate the applicability of this approach and how it could potentially predict and prevent some of the adverse events associated with a telemetry patient monitoring system made by a major medical technology manufacturer and currently used in a number of hospitals in North America.

Using the case study, we provide a simple example showing how formal description techniques and particularly the ICO notation and its Petshop tool, support the three points discussed in section 2 : usability evaluation, contextual help and incident and accident investigation. Each issue is a research domain within its own right and we are therefore limited to what we want to and will show on each point. The aim here is to give an overview of the types of support formal description techniques methods can provide to usability related methodologies/issues.

5.1 Supporting Usability Evaluation

As previously discussed, usability evaluation can be considered as formative or summative depending on the analysis technique used. Here we have decided to describe how the ICO FDT can be used to support a formative evaluation. The formative approach we use to illustrate this part of our approach is the application of ergonomic criteria and guidelines defined in [6], a type of usability inspection method. The selected criterion is "user explicit control" which has 5 guidelines. The selected guideline (number 4) is defined as "Allow users to interrupt or cancel a current transaction or process".

The PatientNet Central Station interface (see Fig. 5a, Fig. 5b) does not have a cancel or undo/redo option[2]. According to Bastien and Scapin's guideline, the user should be able to interrupt or cancel their current transaction. When the Central Station is in "full disclosure mode" (see Fig. 5b), an "exit" button is available to close the window and return to the 16-patient or 8-patient view mode. However, this "exit" button is not a cancel or an undo/redo function, it simply closes the window to change viewing mode.

Typically, if this problem was identified during a usability evaluation, the system would be redesigned, and a second round of user testing would be performed, to verify if all actions now have an interrupt option. Due to the highly interactive nature of such an application, it is impossible to fully verify if all available actions on the system can be interrupted, cancelled, or undone and redone.

The ICO formalism and its dedicated tool Petshop, can support the usability evaluation of such a guideline by providing a means to exhaustively analyze the system. Once the application's user interface has been modified (to provide a cancel or undo/redo button applicable to all transactions), the Petri net, using the ICO formalism is also modified to represent the same behavior. A link can then be made between the two components using the Activation (and/or rendering) function(s). Following this design modification stage, the benefits of FDTs can be exploited.

In addition to a second round of user testing, FDTs, such as ICOs provide a number of benefits for designers who wish to reason about the effects of proposed changes to a system. For example, a marking tree (an analysis technique based on mathematical foundations) can be used to identify the set of reachable states provided that some information is given about the initial status or 'marking' of the system. This technique helps to produce what can be thought of as a form of state transition diagram. Several tools are currently available to support this analysis (i.e. JARP Petri Net Analyzer version 1.1 developed by Sangoi Padilha (http://jarp.sourceforge.net/us/index.html) which has been developed as an auxiliary tool for the Petri net

---

[2] Access to the PatientNet system was not available for this study. We base our approach and examples on documented data. The point here is to illustrate the approach on a medical application.

analyzing tool ARP developed by Alberto Maziero (http://www.ppgia.pucpr.br/~maziero/diversos/petri/arp.html).

Thus in our example, marking graph can be used to trace all possible interactions with the system and **prove** that each action has a cancel or undo/redo option available.

The model provided in Fig. 7 represents the behavior of a small subsection of the PN central station system. The Petri net models the selection of modes (16-view, 8-view or full disclosure) and the availability of the cancel button. Assuming the initial state and default state is the 16-view mode, the cancel buttons are therefore available in 8-view mode and full disclosure mode. Furthermore, the user can switch between 8-view and full disclosure modes. We have modeled the system such that a "cancel" operation will take the system back into 16-view mode.

The places have been labeled p1 – p3 while the transitions have been labeled t1 – t6. Transitions that are fireable are shaded in dark grey. This from the initial state, 1 token in place p1, transitions t1 and t2 are the only two transitions that are fireable.

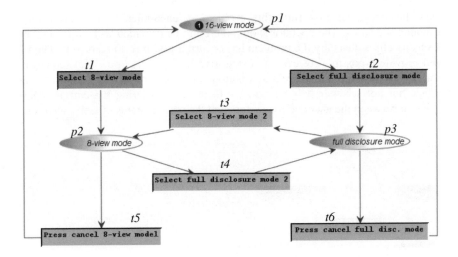

Fig. 7. ICO model proving a cancel option is always available

Since the model is simplistic, it is possible to analyze the model graphically. However, when modeling a complete system the use of systems' generated marking graphs is necessary to prove system properties. Here we provide the marking tree from the current marking of the model provided in Fig. 7. There are 3 possible states:

```
State M0: {p1}, State M1: {p2}, State M2: {p3}
The marking tree (represented in text format) is as follows:

M0: (t1:M1), (t2:M2) - From state M0, transition t1 changes the
current state to state M1, while transition t2 changes the current
state to M2

M1: (t4:M2), (t5:M0) - From state M1, transition t4 changes the
current state to state M2, while transition t5 changes the current
state to M0
```

```
M2: (t3:M1), (t6:M0) - From state M2, transition t3 changes the
current state to state M1, while transition t6 changes the current
state to M0
```

The analysis techniques described here provide exhaustive proofs of the availability of the cancel option in different system modes.

5.2 Supporting Contextual Help

The use of FDTs can provide accurate contextual help at any moment of system interaction. Since the model describes the concise behavior of the system, it is possible to analyze the exact state the system is in. As an example, we refer back to section 2.4, and discuss the case of an operator wishing to press a button to perform an action, though this button has no effect because it is currently "grayed out", inactive. This kind of interaction problem is encountered often with applications such as MS Word. Though after encountering the same problem numerous times, we begin to learn why the button is inactive and correct our actions accordingly. In the medical domain however, the first time this problem is encountered may result in life threatening scenarios. The operator should therefore be provided with contextual help that specifically states why the problem has occurred **and** how to resolve it. The FDT would provide a possible (or even an exhaustive list of all the possible) sequence of actions that would activate the required button, and a way of how to do it on the user interface but not whether it is wise to try to do it considering the current state of interaction between the user and the system and the current status of task execution.

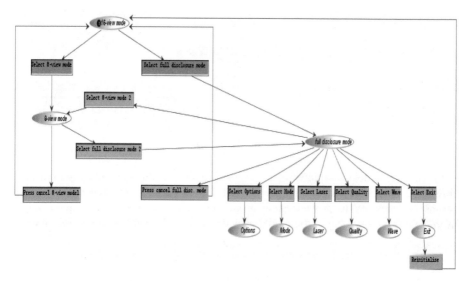

Fig. 8. ICO model with laser button inactive

Using ICOs, it is possible to analyze the Petri net and identify the current state (i.e. the distribution of tokens throughout the set of places) and understand why the button is inactive (i.e. which transition(s) must be fired in order for the place representing the button activation to contain its necessary token).

Fig. 8 and Fig. 9 illustrate a simple Petri net, using the ICO notation, in order to demonstrate this part of the approach. We take the same model illustrated in the previous section, this time adding the options available when in full disclosure mode. These include options, mode, laser, quality, wave and exit. The exit transition takes the system back into the 16-mode view. Fig. 9 indicates that the "laser" button is currently not available, because its associated transition is not fireable.

The only way that the "select laser" transition can become fireable is for place "full disclosure mode" to contain a token. If the user was searching for the laser button while operating the PN Central station and referred to a help file, the Petri net model could provide detailed help on why the button is not available and how to make it available. Fig. 9 therefore shows the same model, this time in the state allowing the "select laser" transition to be fireable.

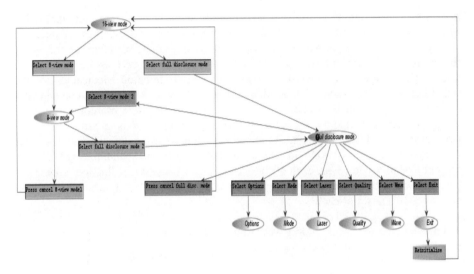

Fig. 9. ICO model with active laser button

5.3 Supporting Incident and Accident Investigation

In order to support the investigation of an incident or accident, such as the reports on the US Food and Drug Administration (FDA) Manufacturer and User Facility Device Experience (MAUDE) database , we advocate the use of system modelling to prove that a given incident or accident cannot be triggered in the new system design. We are not claiming that the mishap will not recur in the real system, but in the model describing the behavior of the system. This model is an abstraction of the real system assuming that system engineers will use that model for designing the improved system.

To be more concrete, we have devised an approach that allows a system to be formally modeled and then formally analyzed to prove that the sequence of events described in the accident report is not able to be triggered again in that model.

The aim is to use formal description techniques to model the complete behavior of the system and identify hazardous states that we would wish to avoid. These states can be avoided by eliminating the sequence of state changes leading to that state.

One of our aims for including data from incident and accident investigations in safety-critical interactive systems design is to ensure that the same incident or accident analyzed will not occur again in the re-modeled system. That is, that an accident place within the network will be blocked from containing a token. Due to space constraints we do not illustrate this part of the approach in this paper, though the interested reader can refer to [3] and [4].

6   Conclusion

This paper has addressed issues of usability for modern medical applications. We have argued that empirical usability evaluations, such as formative and summative evaluations are not sufficient when dealing with complex safety-critical interactive systems that have an extremely large number of system states. We advocate the use of formal methods, particularly those based on Petri nets, such as the Interactive Cooperative Objects (ICOs) formalism, a formal notation dedicated to the specification of interactive systems that can be exploited to provide mathematically grounded analyses, such as marking graphs, to argue and prove usability properties. By producing high-fidelity formally specified prototypes, it is possible to verify whether the new design satisfactorily prevents such ergonomic defects. The paper has given three examples of ways in which Formal Description Techniques (FDTs) can support usability. However, each of these examples (formative evaluation, contextual help and supporting incident and accident investigation) is its own research area, and here we give a taster of each.

Acknowledgments. This work was supported by the EU funded Network of Excellence ResIST http://www.resist-noe.eu.

References

1. Adolf, J.A., Holden, K.L.: Touchscreen usability in microgravity. In: Tauber, M.J. (ed.) CHI 1996. Conference Companion on Human Factors in Computing Systems: Common Ground, ACM Press, New York (1996)
2. Barboni, E., Navarre, D., Palanque, P., Basnyat, S.: Exploitation of Formal Specification Techniques for ARINC 661 Interactive Cockpit Applications. In: HCI Aero 2006. Proceedings of HCI aero conference, Seatle, USA (September 2006)
3. Basnyat, S., Chozos, N., Palanque, P.: Multidisciplinary perspective on accident investigation. Reliability Engineering & System Safety 91(12), 1502–1520 (2006)
4. Basnyat, S., Chozos, N., Johnson, C., Palanque, P.: Redesigning an Interactive Safety-Critical System to Prevent an Accident from Reoccurring. In: 24th European Annual Conference on Human Decision Making and Manual Control (EAM) Organized by the Institute of Communication and Computer Systems, Athens, Greece (October 17-19, 2005)
5. Bastide, R., Palanque, P., Sy, O., Le, D.-H., Navarre, D.: Petri Net Based Behavioural Specification of CORBA Systems. In: ATPN 1999. International Conference on Application and Theory of Petri nets. LNCS, Springer, Heidelberg (1999)

6. Bastien, C., Scapin, D.: International Journal of Human-Computer Interaction. 7(2), 105–121 (1995)
7. Bernonville, S., Leroy, N., Kolski, C., Beuscart-Zéphir, M.: Explicit combination between Petri Nets and ergonomic criteria: basic principles of the ErgoPNets method. In: EAM 2006. European Annual Conf. on Human Decision-Making and Manual Control, Universitaires de Valenciennes (2006) ISBN 2-905725-87-7
8. Bernhaupt, R., Navarre, D., Palanque, P., Winckler, M.: Model-Based Evaluation: A New Way to Support Usability Evaluation of Multimodal Interactive Applications. In: Law, E., Hvannberg, E., Cockton, G. (eds.) Maturing Usability: Quality in Software, Interaction and Quality, Springer, London (2007)(accepted for publication)(to appear)
9. Bredereke, J., Lankenau, A.: A Rigorous View of Mode Confusion. In: Anderson, S., Bologna, S., Felici, M. (eds.) SAFECOMP 2002. LNCS, vol. 2434, pp. 19–31. Springer, Heidelberg (2002)
10. Coutaz, J., Bérard, F., Carraux, E., Crowley, J.L.: Early Experience with the Mediaspace CoMedi. In: Proceedings of the IFIP Tc2/Tc13 Wg2.7/Wg13.4 Seventh Working Conf. on Engineering For Human-Computer interaction, pp. 57–72. Kluwer Academic Publishers, Dordrecht (1999)
11. Fourney, D., Carter, J.: Proceedings of GOTHI-05 Guidelines On Tactile and Haptic Interactions. In: Fourney, D., Carter, J. (eds.) USERLab, Univ. of Saskatchewan (2005), http://userlab.usask.ca/GOTHI/GOTHI-05 Proceedings.html
12. Furnas, G.W.: Generalized Fisheye Views. In: Proc of ACM CHI 1986, pp. 16–23. ACM, New York (1986)
13. Holzinger, A.: Usability Engineering for Software Developers. Communications of the ACM 48(1), 71–74 (2005)
14. Holzinger, A., Errath, M.: Designing Web-Applications for Mobile Computers: Experiences with Applications to Medicine. In: Stary, C., Stephanidis, C. (eds.) User-Centered Interaction Paradigms for Universal Access in the Information Society. LNCS, vol. 3196, pp. 262–267. Springer, Heidelberg (2004)
15. Holzinger, A.: Finger Instead of Mouse: Touch Screens as a means of enhancing Universal Access. In: Carbonell, N., Stephanidis, C. (eds.) Universal Access. Theoretical Perspectives, Practice, and Experience. LNCS, vol. 2615, pp. 387–397. Springer, Heidelberg (2003)
16. Holzinger, A., Sammer, P., Hofmann-Wellenhof, R.: Mobile Computing in Medicine: Designing Mobile Questionnaires for Elderly and Partially Sighted People. In: Miesenberger, K., Klaus, J., Zagler, W., Karshmer, A.I. (eds.) ICCHP 2006. LNCS, vol. 4061, pp. 732–739. Springer, Heidelberg (2006)
17. Kohn, L., Corrigan, J., Donaldson, M.: To Err Is Human: Building a Safer Health System. In: Institute of Medicine.Committee on Quality of Health Care in America, National Academy Press, Washington DC (1999)
18. Lamping, J., Rao, R.: Laying out and visualizing large trees using a hyperbolic space. In: ACM Symp User Interface Software and Technology, pp. 13–14. ACM Press, New York (1994)
19. Loer, K., Harrison, M.: Formal interactive systems analysis and usability inspection methods: Two incompatible worlds? In: Palanque, P., Paternó, F. (eds.) DSV-IS 2000. LNCS, vol. 1946, pp. 169–190. Springer, Heidelberg (2001)
20. Mackinlay, J.D., Robertson, G.G., Card, S.K.: Perspective Wall: Detail and Context Smoothly Integrated. In: Proceedings of SIGCHI 1991, pp. 173–179 (1991)

21. Memmel, T., Reiterer, H., Holzinger, A.: Agile Methods and Visual Specification in Software Development: a chance to ensure Universal Access. In: Coping with Diversity in Universal Access, Research and Development Methods in Universal Access. LNCS, vol. 4554, Springer, Heidelberg (2007)
22. Mori, G., Paternò, F., Santoro, C.: CTTE: Support for Developing and Analyzing Task Models for Interactive System Design. IEEE Transactions on Software Engineering , 797–813 (August 2002)
23. Navarre, D.: Contribution à l'ingénierie en Interaction Homme Machine - Une technique de description formelle et un environnement pour une modélisation et une exploitation synergiques des tâches et du système. PhD Thesis. Univ. Toulouse I (July 2001)
24. Navarre, D., Palanque, P., Bastide, R., Sy, O.: Structuring Interactive Systems Specifications for Executability and Prototypability. In: Palanque, P., Paternó, F. (eds.) DSV-IS 2000. LNCS, vol. 1946, Springer, Heidelberg (2001)
25. Navarre, D., Palanque, P., Bastide, R.: Reconciling Safety and Usability Concerns through Formal Specification-based Development Process. In: HCI-Aero 2002, MIT Press, USA (2002)
26. NHS Expert Group on Learning from Adverse Events in the NHS. An organisation with a memory. Technical report, National Health Service, London, United Kingdom (2000), http://www.doh.gov.uk/orgmemreport/index.htm
27. Nielsen, J., Molich, R.: Heuristic evaluation of user interfaces. In: Proceedings of CHI 1990, pp. 249–256. ACM, New York (1990)
28. Nielsen, J.: Heuristic evaluation. In: Nielsen, J., Mack, R.L. (eds.) Usability Inspection Methods, John Wiley & Sons, New York (1994)
29. Nielsen, J. (2005) http://www.useit.com/papers/heuristic/ last accessed 21/06/2007
30. Norman, D.A., Draper, S.W. (eds.): User-Centred System Design: New Perspectives on Human-Computer Interaction. Lawrence Earlbaum Associates, Hillsdale (1986)
31. Palanque, P., Bastide, R.: Formal specification of HCI for increasing software's ergonomics. In: ERGONOMICS 1994, Warwick, England, 19-22 April 1994 (1994)
32. Palanque, P., Bernhaupt, R., Navarre, D., Ould, M., Winckler, M.: Supporting Usability Evaluation of Multimodal Man-Machine Interfaces for Space Ground Segment Applications Using Petri net Based Formal Specification. In: Ninth International Conference on Space Operations, Rome, Italy (June 18-22, 2006)
33. Palanque, P., Farenc, C., Bastide, R.: Embedding Ergonomic Rules as Generic Requirements in a Formal Development Process of Interactive Software. In: proceedings of IFIP TC 13 Interact 99 conference, Edinburg, Scotland, 1-4 September 1999 (1999)
34. Palanque, P., Bastide, R., Dourte, L.: Contextual Help for Free with Formal Dialogue Design. In: Proc. of HCI International 93. 5th Int. Conf. on Human-Computer Interaction joint with 9th Symp. on Human Interface (Japan), North Holland (1993)
35. Pierotti, D.: Heuristic Evaluation - A System Checklist, Xerox Corporation (1995), Available onlineat http://www.stcsig.org/usability/topics/articles/he-checklist.html
36. Scapin, D. L.: Guide ergonomique de conception des interfaces homme-machine (Rapport de Recherche No. 77). INRIA - Rocquencourt – France (1986)
37. Spence, R., Apperley, M.: Data Base Navigation: An Office Environment for the Professional Behaviour and Information Technology 1(1), 43–54 (1982)
38. Thimbleby, H.: Interaction walkthrough - a method for evaluating interactive systems. In: The XIII International Workshop on Design, Specification and Verification of Interactive Systems, July 26-8, 2006 Trinity College, Dublin(2006)
39. http://www.gehealthcare.com/usen/patient_mon_sys/wireless_and_telemetry/products/tele metry_sys/products/patientnet_centralstation.html

Using Formal Specification Techniques for Advanced Counseling Systems in Health Care

Dominikus Herzberg[1], Nicola Marsden[1], Corinna Leonhardt[2], Peter Kübler[1], Hartmut Jung[2], Sabine Thomanek[3], and Annette Becker[3]

[1] Heilbronn University, 74081 Heilbronn, Germany
Faculty of Informatics, Department of Software Engineering
{herzberg,marsden}@hs-heilbronn.de
[2] Philipps-University Marburg, 35032 Marburg, Germany
Department of Medical Psychology
cleonhar@med.uni-marburg.de
[3] Philipps-University Marburg, 35032 Marburg, Germany
Department of General Practice/Family Medicine
annette.becker@med.uni-marburg.de

Abstract. Computer-based counseling systems in health care play an important role in the toolset available for doctors to inform, motivate and challenge their patients according to a well-defined therapeutic goal. In order to study value, use, usability and effectiveness of counseling systems for specific use cases and purposes, highly adaptable and extensible systems are required, which are – despite their flexibility and complexity – reliable, robust and provide exhaustive logging capabilities. We developed a computer-based counseling system, which has some unique features in that respect: The actual counseling system is generated out of a formal specification. Interaction behavior, logical conception of interaction dialogs and the concrete look & feel of the application are separately specified. In addition, we have begun to base the formalism on a mathematical process calculus enabling formal reasoning. As a consequence e.g. consistency and termination of a counseling session with a patient can be verified. We can precisely record and log all system and patient generated events; they are available for advanced analysis and evaluation.

Keywords: Human-Computer Interaction in Health Care, Counseling Systems, Formal Methods, Usability Engineering.

1 Introduction

In an interdisciplinary team of general practitioners, psychologists and software engineers, we developed a Computer-Based Counseling System (CBCS) for use in health care. Background is that physical activity is a recognized therapeutic principle for patients who suffer from a chronic disease [12] – in our case diabetes mellitus and/or cardiac insufficiency [7, 25]. A lot of research has been done on the question how these patients can be properly motivated and finally activated to work out as part of their therapy [20]. Since computer-based counseling systems have been found to be

A. Holzinger (Ed.): USAB 2007, LNCS 4799, pp. 41–54, 2007.

an effective tool in working with these patients [1, 3], we study its use in practices of general practitioners. In a dialog with the patient, the counseling system concludes the motivational level of the patient based on the trans-theoretical model of behavior change [21] and moves over into an adapted, well-fitted consultation. The system explains the effect of physical activity for the health and well-being of the patient and interactively discusses and proposes strategies to integrate physical activities in daily life.

While computer-based counseling systems are not new in health care [16], our system has some distinguishing features. What sets our system apart from others is that the interaction between the patient and the computer system is based on a formal specification; for practical reasons, we have chosen XML (eXtended Markup Language) as the basis for our specification language. Given an XML-based interaction specification, the actual counseling system is generated out of the specification. On the presentation/interaction layer, the generated system uses well-proven web-based technologies such as HTML (HyperText Markup Language), CSS (Cascading Style Sheets) and JavaScript. This results in some unique features: our system is easily distributable, it supports multi-media (video, audio, images, text), it is extensible and highly adaptable, the generated output can be customized for different devices and carriers (e.g. web browsers, PDAs (Personal Digital Assistants), mobile phones, plain old paper) and all interaction events are logged in real-time, which supports advanced analysis of the actual human-computer interaction.

Furthermore, the formalism is based on a sound mathematical formalism, a process calculus. Especially the fact that we use a process calculus in space *and* time is novel. It enables us to specify an interaction dialog spatially *and* timewise. Thus, we can apply formal reasoning on a complete interaction specification. For example, we can

- verify whether all resources (video, audio etc.) are properly used
- detect inconsistencies between actions in time and slots in space (called styles)
- check whether all paths of interaction are reachable (there are alternatives) and whether an interaction session with a patient will terminate, i.e. will come to an end

The formal basis of our approach is key for producing highly adaptable but still reliable counseling systems.

The system has been implemented and is currently used in a field study. The mathematical formalization of our specification approach has not been finalized. Nonetheless, the above mentioned and some more checks have been implemented to a large degree.

In section 2, we discuss related work. In section 3, we provide an overview of the specification process we developed and explain in detail the XML-based specification style of our counseling system. In section 4, a brief overview of the related mathematical formalism is given. Section 5 comprises some lessons learned from working in an interdisciplinary team. Finally, section 6 closes with some conclusions and an outlook on future research.

2   Related Work

Computer-based counseling systems have mostly focused on software which offers individualized feedback regarding a certain topic – but offer little flexibility as to adapting or changing the system or even transforming it for another device. Systems to persuade people to change their behaviors have been implemented using mobile phones or PDAs [4, 5, 13, 22, 24, 26]. Computer-based systems on a desktop or touchscreen kiosk [17] have typically been web-based, implemented in Flash Action script and HTML: For example the counseling systems based on the transtheoretical model dealing with physical activity [23], obesity [15] or contraceptive use [19]. In addition to the features of the systems described in theses studies, our system will not only allow for individualized feedback and navigation based on the transtheoretical model, but will also log all interaction in real-time, allowing for diagnosis of the patient and usability analysis.

Based on our research, formal specifications, which can be used to facilitate development of information systems [18], have not been used to generate computer-based counseling systems. Yet creating the counseling system directly out of the specification could help to solve a bundle of problems for which solutions had to be hand-tailored in the past: Generate output for different media, e.g. Instant Messaging [24] or for a specific target group, e.g. elderly people remaining at home [11].

Design, development and implementation of patient counseling systems require close collaboration between users and developers. While this is true in any software development process, it can be particularly challenging in the health counseling field, where there are multiple specialties and extremely heterogeneous user groups [8, 10]. Benson [2] has proposed UML (Unified Modeling Language) as stringent specifications in order to not alienate users and being able to communicate within the multidisciplinary development team. We are taking this one step further by formally specifying the human-computer interaction and having the whole system generated on the basis of this formal reasoning.

3   An XML-Based Specification for Counseling Systems

At the heart of our counseling system are two documents, both being specified in XML (eXtensible Markup Language): one document specifies the *intended* user interface in form of logical styles, the other specifies the HCI (Human-Computer Interaction) in form of a script and refers to the logical styles. In further steps, the specifications are enhanced and adapted towards a concrete platform using a specific technology. An overview of this process is given in the following subsection. Subsequent subsections detail information about the above mentioned core documents and the produced counseling system.

3.1   Overview of the Specification Process

In our approach we distinguish three phases in the process of specifying a Computer-Based Counseling System (CBCS), shown in figure 1: a conception phase, a realization phase and a deployment phase.

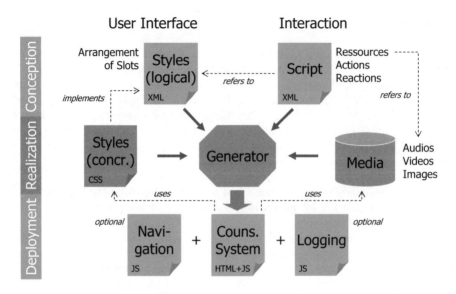

Fig. 1. Overview of the specification process, involved artefacts and tools

In the first phase, the conception phase, an author or – in our case – a team of authors creates two specifications using XML: one document specifies the intended user interface of the counseling system, the other the interaction behavior. Innovative here is that the user interface is specified in form of *logical styles*. A logical style defines an arrangement of placeholders (which we call *slots*) on a dialog page. A slot can be loaded and unloaded with visual content. The arrangement of slots per page reflects just an intention of the layout of the visual elements; the arrangement is not mandatory. In this sense, logical styles are a formalization of the idea of paper prototypes for the design of user interfaces. The specification of the interaction behavior, which we call *script*, decomposes the interaction with a user into so called *dialog pages*. Each page refers to a logical style, it specifies the resources required on the page, it defines the actions the counseling system performs on this page and how the system reacts on user generated events such as pressing a button. This might result in a jump to another dialog page. Taken together, the specification of logical styles and the specification of a script are executable. This can be used for prototyping purposes and early testing of dialogs and user interactions. We did not have the chance to develop any tools for this, so we did not make use of this option – yet such an early execution of specifications would be extremely helpful.

In the second phase, the realization phase, we "put meat to the bones". The specification of logical styles is enhanced by a specification of concrete styles. A concrete style implements, so to speak, the logical style. A concrete style gives a precise definition of the placement of slots on a dialog page and their appearance. How this is done and which technology is to be used for that purpose, is platform dependent and a matter of choice. The specifications of the conception phase are abstract in the sense that they are platform independent (though executable). In the realization phase, decisions have to be made. In our case, we decided to use web-based technology and prepare for distributed use over the Internet. That's why use Cascading Style Sheets (CSS) to substantiate logical styles. In a future release, we plan to use CSS in combination with HTML (HyperText Markup Language) tables for improved flexibility in layout design. In an experimental spin-off of our counseling system we targeted for Flash-based technology, which is an example of another realization platform. In addition to the concretization of logical styles, all the resources, which are listed in the interaction specification, need to be related to "the real thing", namely audios, videos and images. Therefore, the media have to be produced, assembled and stored in a central place. The core element in the realization phase is the generator. The generator takes the specifications of the conception phase, concrete style definitions and the media resources as an input and outputs the actual counseling system for the target platform. Our generated counseling system consists of a number of HTML pages including JavaScript (JS), actually one HTML page per dialog page in the interaction specification. The HTML pages use the concrete style definitions and the media resources. If we think of a key role associated with the realization phase, it is a digital media designer. Due to budget limitations, we had no access to a professional media designer but could afford a professional speaker for the audios. The concrete styles, the videos and audios are produced by our team.

In the deployment phase, the output of the realization phase (the actual counseling system) gets installed on physical hardware and is ready for use in experiments, field studies and the like. To provide extra functionality, modules for navigation and logging (both implemented in JavaScript) are delivered with the counseling system. We installed the counseling system on TabletPCs with a screen resolution of 1024 x 768 pixels and built-in speakers. The screen is touch-sensitive and can be used with a pen only; it does not react on touch with a finger. During a counseling session with a patient, the keyboard is not available; it is covered by the screen. The pen is the only input medium for the patient.

In the following subsections we provide further information about the specification of styles and scripts and the generated counseling system.

3.2 Specification of Styles

In the first place, we describe the layout of an HCI dialog with so-called logical styles. A logical style captures the intention of what a page in an interaction dialog should look like. The focus is on a logical organization of placeholders (called *slots*), which can be loaded or unloaded with content, referring to resources such as text, images, videos and buttons. The concrete realization of a logical style is separate from this.

The logical organization of slots for a style is given by two commands: stacked and juxtaposed. The command "stacked" puts one slot above another (vertical arrangement), "juxtaposed" sets them next to each other (horizontal arrangement). Here is a simple example:

```
<style name="YesNoQuestion">
  <stacked>
    <slot name="text" type="text" class="Text"/>
    <juxtaposed>
      <slot name="yes"  type="button" class="Button Yes"/>
      <slot name="no"   type="button" class="Button No"/>
      <slot name="next" type="button" class="Button Next"/>
```

```
    </juxtaposed>
  </stacked>
</style>
```

In the example, we have a style named "YesNoQuestion". It describes an arrangement of four slots, with a slot called "text" placed above three other slots ("yes", "no" and "next"); these three slots are positioned next to each other. Per slot a type is given, indicating which kind of resource the slot is reserved for. The class-attribute provides abstract information about the appearance of a slot. For instance, slot "yes" is associated with an abstract design called "Button" and refined by a design called "Yes". The class-attribute cascades abstract design information.

For engineering purposes, a "copy"-element can be used in a style definition. This enables you to compose a logical style out of other styles and benefit from reuse (see the following example). Here, a style is composed of a header, some images and a question. The question refers to the style exemplified before.

```
<style name="QuestionWith3Images">
  <stacked>
    <copy ref="Header"/>
    <copy ref="3Images"/>
    <copy ref="YesNoQuestion"/>
  </stacked>
</style>
```

The key point for the specification of styles is to distinguish two phases of user interface design: In the *conception phase*, we first agree on the placeholders and their intended positioning in an interaction dialog. This is primarily a contract about slots available on a dialog page, their intended layout is secondary – this is close to the idea of paper prototypes and wireframes. In a second phase, in the *realization phase* (see also figure 1), we separately specify the concrete positioning and appearance of slots. Concrete styles refine, detail and may even overwrite the intended layout of slots in the logical styles. Concrete styles cannot add or remove (just hide) slots.

The way of specifying concrete styles is dependent on the target platform and the technology used for the counseling system. For web-based platforms CSS is a natural choice.

3.3 Specification of Interactions

The overall structure of an interaction specification resembles the organization of a book or a screenplay. Especially the metaphor of a screenplay helped the involved authors, medical doctors and psychologists, to get used to the idea of specifying interactive counseling dialogs in XML. The interaction specification is organized in chapters, and each chapter is composed of pages, each page representing a self-contained interaction scenario with the patient. The interaction starts with the first page of the first chapter of a script. Besides, some meta-information (authors, revision information etc.) can be supplied. Note the reference to logical styles.

```
<script version="0.9" date="2007-05-13" styles="StilKempkes.xml">
  <title>Counseling System</title>
  <author name="Thomanek">Philipps-University Marburg</author>
  <chapter name="Registration">…</chapter>
  <chapter name="Welcome">…</chapter>
```

```
<chapter name="Questions">…</chapter>
<chapter name="Counseling">…</chapter>
</script>
```

For each page in a chapter a number of resources are declared, which are at disposal for this specific dialog. Resources can be audios, buttons, images, texts, timers and videos. Furthermore, a page is composed of actions. The following actions are available: show, empty, play, halt, jump, eval (evaluate) and log. Actions can run in sequence or concurrently in parallel and specify the systems behavior. In addition to actions, reactions can be specified, which define how the counseling system reacts on user initiated events, such as moving the input device (e.g. a mouse or a pen) or pressing a button on the screen. The body of a reaction is specified with JavaScript; JavaScript is used as an action language. This provides quite some flexibility on how to react on user initiated events. As an example, see the following specification of the dialog page "Question", which refers to the logical style "YesNoQuestion":

```
<page name="Question" style="YesNoQuestion">
  <intention>Reflect emotion</intention>
  <resources>
    <audio id="Question" src="audios/enjoy.mp3">…</audio>
    <text id="Question">Did you enjoy this session?</text>
    <timer id="pause" duration="1"/>
    <button id="Yes">Yes</button>
    <button id="No">No</button>
  </resources>
  <actions run="sequence">
    <actions run="parallel">
      <play res="audio" ref="Question"/>
      <show slot="text" res="text" ref="Question"/>
    </actions>
    <play res="timer" ref="pause"/>
    <actions run="parallel">
      <show slot="yes" res="button" ref="Yes"/>
      <show slot="no" res="button" ref="No"/>
    </actions>
  </actions>
  <reaction res="button" ref="Yes" event="onclick">
    jump("Counseling","Great")
  </reaction>
  <reaction res="button" ref="No" event="onclick">
    jump("Counseling","Bad")
  </reaction>
</page>
```

First comes a mission statement, which describes the intention of this page. Then, five resources are introduced: an audio, a text, a timer for one second and two buttons. On call, the page first plays the audio and at the same time shows the text of the question in slot "text". After that the timer plays thereby delaying further actions by one second. Then the buttons "Yes" and "No" show up in slots "yes" and "no" at the very same time. Note that slot "next" (see logical style "YesNoQuestion") simply remains unused. At any time, the user might do something. If the user clicks on the "Yes" button (it must have shown up in a slot before being accessible), the next page the counseling system jumps to is in chapter "Counseling" called "Great". Likewise, the next page is "Bad" in chapter "Counseling", if the user clicks on button "No".

There is also a copy-attribute available for the specification of pages; the attribute is not used in the example above. It serves the same purpose as the copy-attribute for slots: to facilitate reuse and foster composition of pages out of other pages. A practical use case is the definition of a master page other pages refer to via copy. The master page introduces some standard resources and some standard reactions, which can be overwritten if required.

3.4   Generating the Counseling System

From the specification of logical styles, the specification of a script, concrete specifications for the dialogs' look & feel (concrete styles) and a pool of media resources, the counseling system is automatically generated for a specific target platform. In our case, the system generates HTML-based web pages, which run in full screen mode on a TabletPC. The counseling system is prepared for remote use over the Internet. Figure 2 shows a screenshot of the CBCS in action running within Microsofts Internet Explorer 7. Our system relies on a web browser supporting SMIL (Synchronized Multimedia Integration Language).

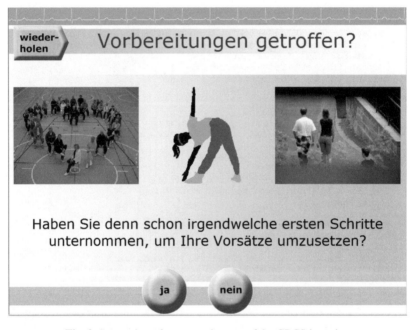

Fig. 2. Screenshot of an example page of the CBCS in action

The generator embeds two modules in the counseling system to provide navigation and logging capabilities. The navigation module enables quick and easy jumps to any dialog page of a counseling session via a navigation menu. It is accessible via a hot key and available for testing purposes only. The logging module records all events in real-time, system generated events as well as user initiated events, and sends them to a logging server. An excerpt is shown below. The first column is a session id out of six digits. Next comes the date and the time with a precision of milliseconds. Then there is the logged event in XML format including a list of parameters.

```
000056 | 2007-07-24 10:26:16,718: <onmouseup clientX="570" clientY="702" button="1" />
000056 | 2007-07-24 10:26:16,734: <onclick res="button" ref="nein" clientX="570"
                                     clientY="702" />
000056 | 2007-07-24 10:26:16,750: <jump chapter="TTM" page="Zukuenftig" />
000056 | 2007-07-24 10:26:16,796: <onend res="audio" ref="einleitung" />
000056 | 2007-07-24 10:26:17,109: <onmousemove clientX="570" clientY="702" button="0" />
000056 | 2007-07-24 10:26:17,187: <show slot="titel" res="text" ref="titel" />
000056 | 2007-07-24 10:26:17,203: <show slot="reload" res="button" ref="reload" />
000056 | 2007-07-24 10:26:17,203: <play res="audio" ref="frage" />
000056 | 2007-07-24 10:26:17,218: <show slot="text" res="text" ref="frage" />
000056 | 2007-07-24 10:26:17,234: <show slot="ja" res="button" ref="ja" />
000056 | 2007-07-24 10:26:17,234: <show slot="nein" res="button" ref="nein" />
000056 | 2007-07-24 10:26:18,625: <onmousemove clientX="569" clientY="702" button="0" />
000056 | 2007-07-24 10:26:18,625: <onmouseover res="button" ref="nein" clientX="569"
                                     clientY="702" />
```

If used over the Internet, real-time logging allows an observer to monitor a counseling session between the system and a patient on a remote computer. This feature might be useful for conducting online experiments with the counseling system.

With the event log on the logging server, counseling sessions can be precisely reconstructed and replayed. The log data can be analyzed and evaluated from various points of view, e.g. how much time a patient spends on a dialog page etc.

3.5 User-Centered Design

The specification process shown in figure 1 can be related to Garrett's "Elements of User Experience" [6], a framework for user-centered design for the Web, which slices user-centered design into five planes. There is a strategy plane, a scope plane, a structure plane, a skeleton plane and a surface plane. For a counseling system, not all aspects apply as for a full-blown web application. Nonetheless, our specification process favors a similar top-down user-centered design process: the user needs and the counseling objectives need to be clarified (strategy); the script has to be broken up into chapters and pages including mission statements (scope); the pages refine the interaction design how the system behaves and responses to the user (structure); the logical styles define the interface and navigation design by means of an arrangement of interface elements (skeleton); concrete styles and the media storage define the visual design (surface). The logging facilities of our tool let us evaluate aspects of user experience. Combined with usability engineering methods [9], such as Thinking Aloud (THA), the specification process can be iteratively repeated for improvements.

Our toolset could also be extended in order to effectively involve users (patients) in the specification process. What is missing in our toolset is a suitable front-end for the conception phase. Even though XML is simple, an authoring environment for the creation and editing of script and style specifications with some visual/graphical support would be beneficial. Such an environment could (a) help enforce the specification process and (b) – more importantly – let users immediately participate in the specification process. This would open up the development of counseling systems to a new level of user involvement in aspiration of user-centered design.

4 A Calculus for Specifications in Space and Time

As was mentioned in the introduction, our approach is based on a mathematical formalism, a process calculus. While this is research work in progress, we can already outline the idea behind combined specifications in time *and* space. For the sake of brevity and due to space constraints, we provide informative definitions and keep the mathematical formalism very short. Since our formalism has its roots in Finite State Processes (FSP), see [14], the interested reader can retrieve complete formal FSP semantics from that source.

FSP as such is not explicitly concerned with the passage of time, even though the aspect of time remains the dominating mindset. FSP is concerned with actions in a certain order without stating *when* an action is performed. We take this to our advantage and use FSP not only to describe actions in a certain order, but also to describe slots in space in an orderly arrangement. Subsequently, we will first define the terminology and concepts for action-based specifications and then we will – by analogy – derive terminology and concepts for slot-based specifications.

4.1 Specifying Processes

For a formal treatment, Magee and Kramer [14] define FSP semantics in terms of Labeled Transition Systems (LTSs). A finite LTS *lts(P)* is represented by a tuple $<S,A,\Delta,q>$, where S is a finite set of states; $A = \alpha P \cup \{\tau\}$ is a set of actions (αP denotes the alphabet of P) including τ denoting an unobservable internal action; $\Delta = S - \{\pi\} \times A \times S$ is a transition relation mapping an action from a state towards another state (π denotes the error state); $q \in S$ is the initial state of P. The transition of an LTS $P = <S,A,\Delta,q>$ with action $a \in A$ into an LTS P', denoted as $P \xrightarrow{a} P'$, is given by $P' = <S,A,\Delta,q'>$, where $q' \neq \pi$ and $(q,a,q') \in \Delta$.

The following three definitions should suffice to see the analogy to our XML-based specification of sequential and parallel actions. All following quotes refer to [14].

Definition 1 (Process Alphabet). "The alphabet of a process is the set of actions in which it can engage."

Definition 2 (Action Prefix). "If x is an action and P a process then the action prefix (x->P) describes a process that initially engages in the action x and then behaves exactly as described by P."

Definition 3 (Parallel Composition). "If P and Q are processes then (P||Q) represents the concurrent execution of P and Q. The operator || is the parallel composition operator."

Formally speaking, given an $lts(E) = <S,A,\Delta,q>$, the action prefix $lts(a \rightarrow E)$ is defined by $<S \cup \{p\}, A \cup \{a\}, \Delta \cup \{(p, a, q)\}, p>$, where $p \notin S$. Given two processes $P = <S_1,A_1,\Delta_1,q_1>$ and $Q = <S_2,A_2,\Delta_2,q_2>$ parallel composition $P\|Q$ is defined by $<S_1 \times S_2, A_1 \cup A_2, \Delta, (q1, q2)>$, where Δ is the smallest relation satisfying rules of interweaving the transitions of both processes. Let a be an element of the universal set of labels including τ the rules are:

$$\frac{P \xrightarrow{a} P'}{P \| Q \xrightarrow{a} P' \| Q} a \notin \alpha Q \qquad \frac{Q \xrightarrow{a} Q'}{P \| Q \xrightarrow{a} P \| Q'} a \notin \alpha P$$

$$\frac{P \overset{a}{\longrightarrow} P', Q \overset{a}{\longrightarrow} Q'}{P \parallel Q \overset{a}{\longrightarrow} P' \parallel Q'} a \neq \tau$$

Two more definitions might help anticipate how reactions (see XML-based specification) may be specified with FSP replacing JavaScript as an action language.

Definition 4 (Choice). "If x and y are actions then (x->P|y->Q) describes a process which initially engages in either of the actions x or y. After the first action has occurred, the subsequent behavior is described by P if the first action was x and Q if the first action was y."

Definition 5 (Guarded Actions). "The choice (**when** B x->P|y->Q) means that when the guard B is true then the actions x and y are both eligible to be chosen, otherwise if B is false then the action x cannot be chosen."

Of course, there is more to FSP such as indexed processes and actions, process parameters, process labeling and re-labeling, hiding and interfacing. Magee and Kramer [14] provide a model-checking tool which understands the above syntax.

4.2 Specifying Slots

In informatics, there is an interesting space-time dualism. We can interpret a pattern in space as a series of events over time and vice versa. A notation to describe scenarios in space or in time can be based on the same formalism. We make use of this insight and describe arrangements of slots with the very same calculus used for processes – just the semantic interpretation is different. By analogy, we conclude that actions in time compare to slots in space, and processes compare to styles. To give you an idea of the changed semantics, we reformulate the first three definitions:

Definition 6 (Style Alphabet). The alphabet of a style is the set of slots which it can position.

Definition 7 (Slot Prefix). If s is a slot and T a style then the slot prefix (s->T) describes a style that initially positions slot s and then chains exactly as described by T.

Definition 8 (Parallel Style Composition). If T and U are styles then (T||U) represents coexistence of T and U in space. The operator || is the parallel composition operator. Be aware that slot prefix and parallel style composition do not match with "stacked" and "juxtaposed" arrangements of slots (see the XML-based specification). There is a subtle but crucial point here. The *slot prefix* operator just specifies one-dimensional, linear orderings of slots. If one is concerned with time, one dimension is all one needs. However, for spatial two-dimensional orderings, we need to introduce two kinds of prefix operators in order to come up to the same level of expressiveness of the XML-based specification style. The *parallel style composition* operator relates styles in the same way the style specification does for a set of styles in the XML-format. Coexisting styles are commutative and associative.

We do not elaborate further on the topic of shared slots and the interrelation of actions and slots, since this is research work in progress. We have pragmatically solved these issues in our current implementation of the tool; a formal underpinning has not been completed yet.

5 Lessons Learned

As mentioned in the introduction, the counseling system has been developed by an interdisciplinary team of general practitioners, psychologists and software engineers. We all worked together in this constellation for the first time. Since each discipline works according to different standards and routines, uses a different lingo, has implicit expectations etc. we had to overcome some obstacles and learned some interesting things. Let us mention two lessons learned.

- *A shared and instructive metaphor may put an interdisciplinary project on auto-pilot.* When we started the project, we had no software for a counseling system and no experience in how to conceive and develop counseling scripts for a computer-based system – we were lacking a structured approach and a method. The software engineers handed out copies of "The Elements of User Experience" [6]. The general practitioners and medical psychologists were overwhelmed with the "strange terminology" and could not see how this book should help them. They made no progress and there was some desperation in the project. Finally, we put the book aside and identified a metaphor all parties understood and which helped us find our way through: Compare the conception and making of a counseling system with a movie. An author writes a script or screenplay and decomposes the movie into scenes (dialog pages). For each scene, the required actors and requisites (resources) are to be identified and the setting is to be defined (logical style). It has to be described what should happen in a scene (actions) and so on. From that moment on, after the metaphor was invented, our project was put on auto-pilot. Everyone had an idea on how to proceed.
- *It is important to speak a common language – so create one and learn to be precise. Formalization may help.* The very first counseling scripts we wrote were inspired by our shared metaphor but we used plain German to describe dialogs. We started to have a common vocabulary and terminology, but suffered from the imprecision of informal prose. Since XML is easy to learn, we organized a workshop and developed a structured description for scripts and styles as presented in this paper. We created an accurate language for our universe of discourse – a very important step as it turned out. The precision gained boosted our productivity and raised a lot of formerly unnoticed problems and questions. The relation to a mathematical formalism helped the software engineers notice design flaws in the language design.

6 Conclusion and Outlook

In this paper, we presented a way to specify and generate a counseling system with a set of unique features. These features, including a systematic specification process, are currently not available in other related approaches. Key is the formal nature of our system. In a next step we plant to work out the details of our calculus, embed the process calculus in our tool and "implant" an inference engine, so that our pragmatic checks are replaced by formal proofs. Another issue is the development of an authoring environment, which will enable a new kind of user-centered design process for counseling systems.

Acknowledgements. The very first software implementations of our CBCS were developed by competing teams of students during summer term 2007 at Heilbronn University, who participated in the LabSWP course. Thanks to all for their hard and fantastic work.

References

1. Andrews, G.: ClimateGP – web based patient education. Australian Family Physician 36(5), 371–372 (2007)
2. Benson, T.: Prevention of errors and user alienation in healthcare IT integration programmes. Informatics in Primary Care 15(1), 1–7 (2007)
3. Bu, D., Pan, E., Walker, J., Adler-Milstein, J., Kendrick, D., Hook, J.M., Cusack, C., Bates, D.W., Middleton, B.: Benefits of Information Technology-Enabled Diabetes Management. Diabetes Care 30(5), 1127–1142 (2007)
4. Consolvo, S., Everitt, K., Smith, I., Landay, J.L.: Design Requirements for Technologies that Encourage Physical Activity. In: CHI 2006, pp. 457–466. ACM Press, New York (2006)
5. Fogg, B.J.: Persuasive Technology: Using Computers to Change What We Think and Do. Morgan Kaufmann Publishers, San Francisco (2003)
6. Garrett, J.J.: The Elements of User Experience: User-Centered Design for the Web. New Riders (2003)
7. Gregg, E.W., Gerzoff, R.B., Caspersen, C.J., Williamson, D.F., Narayan, K.M.: Relationship of walking to mortality among US adults with diabetes. Arch. Intern. Med. 163(12), 1440–1447 (2003)
8. Holzinger, A.: User-Centered Interface Design for Disabled and Elderly People: First Experiences with Designing a Patient Communication System (PACOSY). In: Miesenberger, K., Klaus, J., Zagler, W. (eds.) ICCHP 2002. LNCS, vol. 2398, pp. 33–40. Springer, Heidelberg (2002)
9. Holzinger, A.: Usability Engineering Methods for Software Developers. Communication of the ACM 48(1), 71–74 (2005)
10. Holzinger, A., Sammer, P., Hofmann-Wellenhof, R.: Mobile Computing in Medicine: Designing Mobile Questionnaires for Elderly and Partially Sighted People. In: Miesenberger, K., Klaus, J., Zagler, W., Karshmer, A.I. (eds.) ICCHP 2006. LNCS, vol. 4061, pp. 732–739. Springer, Heidelberg (2006)
11. Hubert, R.: Accessibility and usability guidelines for mobile devices in home health monitoring. SIGACCESS Access. Comput. 84(1), 26–29 (2006)
12. Karmisholt, K., Gøtzsche, P.C.: Physical activity for secondary prevention of disease. Dan. Med. Bull. 52(2), 90–94 (2005)
13. Lee, G., Tsai, C., Griswold, W.G., Raab, F., Patrick, K.: PmEB: A Mobile Phone Application for Monitoring Caloric Balance. In: CHI 2006, pp. 1013–1018. ACM Press, New York (2006)
14. Magee, J., Kramer, J.: Concurrency – State Models and Java Programs, 2nd edn. Wiley, Chichester (2006)

15. Mauriello, L.M., Driskell, M.M., Sherman, K.J., Johnson, S.S., Prochaska, J.M., Prochaska, J.O.: Acceptability of a school-based intervention for the prevention of adolescent obesity. Journal of School Nursing 22(5), 269–277 (2006)
16. Murray, E., Burns, J., See Tai, S., Lai, R., Nazareth, I.: Interactive Health Communication Applications for people with chronic disease. Cochrane Database Syst. Rev. CD004274 (4) (2005)
17. Nicholas, D., Huntington, P., Williams, P.: Delivering Consumer Health Information Digitally: A Comparison Between the Web and Touchscreen Kiosk. Journal of Medical Systems 27(1), 13–34 (2003)
18. Ortiz-Cornejo, A.I., Cuayahuitl, H., Perez-Corona, C.: WISBuilder: A Framework for Facilitating Development of Web-Based Information Systems. In: Conielecomp 2006. Proceedings of the 16th International Conference on Electronics, Communications and Computers vol.00 (February 27 - March 01, 2006)
19. Peipert, J., Redding, C.A., Blume, J., Allsworth, J.E., Iannuccillo, K., Lozowski, F., Mayer, K., Morokoff, P.J., Rossi, J.S.: Design of a stage-matched intervention trial to increase dual method contraceptive use (Project PROTECT). Contemporary Clinical Trials, Epub ahead of print (2007) doi:10.1016/j.cct.2007.01.012
20. Pinto, B.M., Friedman, R., Marcus, B.H., Kelley, H., Tennstedt, S., Gillman, M.W.: Effects of a computer-based, telephone-counseling system on physical activity. American Journal of Preventive Medicine 23(2), 113–120 (2002)
21. Prochaska, J.O., Velicer, W.F.: Behavior Change. The Transtheoretical Model of Health Behavior Change. American Journal of Health Promotion 12(1), 38–48 (1997)
22. Silva, J.M., Zamarripa, S., Moran, E.B., Tentori, M., Galicia, L.: Promoting a Healthy Lifestyle Through a Virtual Specialist Solution. In: CHI 2006, pp. 1867–1872. ACM Press, New York (2006)
23. Singh, V., Mathew, A.P.: WalkMSU: an intervention to motivate physical activity in university students. In: CHI 2007, pp. 2657–2662. ACM Press, New York (2007)
24. Sohn, M., Lee, J.: UP health: ubiquitously persuasive health promotion with an instant messaging system. In: CHI 2007, pp. 2663–2668. ACM Press, New York (2007)
25. Taylor, R.S., Brown, A., Ebrahim, S., Jolliffe, J., Noorani, H., Rees, K., et al.: Exercise-based rehabilitation for patients with coronary heart disease: systematic review and meta-analysis of randomized controlled trials. Am. J. Med. 116(10), 682–692 (2004)
26. Toscos, T., Faber, A., An, S., Gandhi, M.P.: Click Clique: Persuasive Technology to Motivate Teenage Girls to Exercise. In: CHI 2006, pp. 1873–1878. ACM Press, New York (2006)

Nurses' Working Practices: What Can We Learn for Designing Computerised Patient Record Systems?

Elke Reuss[1], Rochus Keller[2], Rahel Naef[3],
Stefan Hunziker[4], and Lukas Furler[5]

[1] Institute for Information Systems, Department of Computer Science, Swiss Federal Institute
of Technology ETH, Haldeneggsteig 4, 8092 Zurich, Switzerland
[2] Datonal AG, Medical Information Systems Research,
Lettenstrasse 7, 6343 Rotkreuz, Switzerland
[3] Departement of Nursing and Social Services, Kantonsspital Luzern,
Spitalstrasse, 6016 Lucerne, Switzerland
[4] Departement of Nursing and Medical Informatics, Kantonsspital Luzern,
Spitalstrasse, 6016 Lucerne, Switzerland
[5] Departement of Nursing, Stadtspital Waid, Tièchestrasse 99, 8037 Zurich, Switzerland
reuss@inf.ethz.ch, rkeller@datonal.com, rahel.naef@ksl.ch,
stefan.hunziker@ksl.ch, lukas.furler@waid.stzh.ch

Abstract. As demonstrated by several studies, nurses are reluctant to use poorly designed computerised patient records (CPR). So far, little is known about the nurses' interaction with paper-based patient records. However, these practices should guide the design of a CPR system. Hence, we investigated the nurses' work with the patient records by means of observations and structured interviews on wards in internal medicine, geriatrics and surgery. Depending on the working context and the nursing tasks and activities to be performed, characteristic access preferences and patterns were identified when nurses interacted with patient records. In particular, we found typical interaction patterns when nurses performed tasks that included all assigned patients. Another important finding concerns worksheets. Nurses use them during their whole shift to manage all relevant information in a concise way. Based on our findings, we suggest a CPR design which reflects the identified practices and should improve the acceptance of CPR systems in the demanding hospital environment.

Keywords: User-centred system design, System analysis, Patient record.

1 Introduction

Healthcare professionals use patient records as their principal information repository. The records are an important management and control tool by which all involved parties coordinate their activities. This implies that all relevant information should be recorded immediately and data should be available ubiquitously. To accomplish these goals, hospitals started to replace their paper records with computerised patient records (CPR) [1].

A. Holzinger (Ed.): USAB 2007, LNCS 4799, pp. 55–68, 2007.
© Springer-Verlag Berlin Heidelberg 2007

Nurses are responsible for a substantial part of the patient record and hence are particularly affected by the computerisation. However, they tend to feel uncertain in their overall computer literacy [2], and several studies have concluded that usability is one of the most critical factors in the nurses' acceptance of a CPR system [3-6].

Only a few studies report on nurses' daily work. Some address aspects such as cognitive workload or stress [7-10], but provide no relevant information on nurses' interaction with patient records. Other focused on the evaluation of existing CPR systems and found a low user acceptance due to a poor fit between the nurses' tasks and the CPR design [11,12]. In mobile contexts, continuous attention to a system is limited to bursts of just four to eight seconds, less than the 16 seconds observed in the laboratory situation [13]. This shows that interaction time with the system should be minimized in hectic and demanding environments.

As previous studies provide no information about the nurses' work with the patient record, we investigated the nurses' daily working routines on the ward and their interaction with the patient record, with the objective of obtaining criteria for CPR design. The study assumes that the paper-based documentation is largely adapted to the nurses' needs.

We first describe the study design and then present the results in detail, which should allow for a comparison of our findings with the nurses' routines in other countries.

2 Study Design

Our study was carried out in two Swiss acute care hospitals with 270 and 680 beds, respectively. Twelve registered nurses from five different wards in internal medicine, geriatrics and surgery volunteered to participate in the study. At both hospitals, nurses worked with paper-based patient records. We observed three nurses during their morning shift, and five during their evening shift for 55.25 hours in total. Each interview took about 50 minutes. Nurses' working experience varied from 10 months to 25 years with an average of 9.5 years. Nine nurses were female and three nurses were male.

The investigation focused on the nurses' interaction with the patient record during a usual shift that lasted about 8.5 hours. The number of patients assigned to a nurse on the observed morning shifts was two to eight, and on the observed evening shifts four to eleven. Initially, we conducted an individual one hour interview with two head nurses to obtain an overview of the nurses' daily work on the ward. Observations and interviews were carried out during normal weekday shifts.

The observation method was designed to be minimally intrusive. Data were recorded chronologically by writing down in a brief note the time and the working context when nurses accessed the patient record. In addition, each interaction with the patient record was captured in a structured way by using a shorthand notation. In this manner, the following data were collected: (1) type of information (such as chart, medications, other prescriptions et cetera) and (2) purpose of access (request or entry of information, respectively).

After the observation part, the researcher interviewed each participant by means of a questionnaire. The questions addressed the nurses' access behaviour in different working contexts, such as the beginning of the shift or the ward round. Participants were also asked how they search for information in patient records, and what information should be accessible during ward rounds or in patient rooms. The verbal answers were tape-recorded and later transcribed into a written report.

3 Results

We first outline the principal structure of patient records and the nurses' work organisation in Switzerland. We then present our findings organised according to the principle shift structure with three identified characteristic phases.

3.1 General Results

In Switzerland, nurses use a separate record for each patient. A diverse set of differently structured forms are organised by source and filed under different tabs in the patient record. Tabs containing frequently used information – such as the chart with vital signs, the medications list and other prescriptions, notes to the next shift or questions to the physician - are placed in the front of the folder. In the back of the folder are information on the progress notes, on the individual psychosocial patient situation and specific nursing assessments, wound documentation as well as care and discharge plans.

Fig. 1. Personal worksheet of an evening shift nurse (on the left) and memo (on the right): The worksheet contains a row per patient, where some information are printed (as the patient name, date of birth or medical diagnosis) and some are handwritten. Information of memos are transcribed to the patient record.

Because of this consistent and set order, information can be accessed quickly and efficiently. All interviewed nurses (n=12) stated that in principle they knew exactly where to find required information within the patient record. If this strategy of information retrieval failed, most of the interviewed (n=7) consulted the daily progress notes or the notes to the next shift as these two information types are used somewhat like a blackboard. Parallel to the patient record, all observed nurses (n=8) used worksheets and memos (see figure 1). Worksheets and memos are temporary documentation tools and are not part of the patient records.

During the morning, evening and night shifts, nurses care for a varying number of patients. At the time of our observations, one nurse in the morning shift usually worked with one or two nursing students and cared for up to 8 patients. In the evening shift, one nurse cared for up to eleven patients, and in the night shift up to 26 patients.

Each shift basically consisted of three phases. At the beginning of the shift, nurses had to become familiar with the current patient status. Subsequently, during the main phase, they started to process all the required nursing tasks. Prior to ending the shift, nurses checked whether all open issues had been processed, completed their entries in the patient record and finished their work. In the following, these three shift phases are described in depth.

3.2 Beginning of the Shift

At the beginning of their shift, nurses had to become familiar with the status of their assigned patients. The main source of information was the patient record, but all observed nurses (n=8) additionally used a worksheet. This sheet contained a list of all patients, organised by number of patient rooms. Each ward used a different column layout and column labelling respectively, i.e. the sheet was tailored to meet specific needs. For example, on one of the surgical wards, the worksheet contained a column for drainage information and on a geriatric ward, a column was dedicated to blood sugar levels. During this shift phase, all observed participants made notes on the worksheet. In one hospital, all nurses on one ward shared a single worksheet that was used during one to up to three shifts. Each nurse from the other hospital maintained her or his own worksheet. Nurses usually kept the worksheet in their pocket so that it was always available. At both hospitals, worksheets were used to present and manage to do's and relevant information in a concise way.

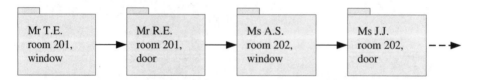

Fig. 2. Instance of an interaction pattern of type 1, where patient records are accessed according to the patients' room order on the ward

When a patient was new to the nurses, they studied the record in detail. With the record of patients with whom the nurses were already familiar, they predominantly focused on the most current and added information, since their last access to the record. The majority of the observed nurses (n=6) stayed in the nurses' office to read the patient record. They sat at the desk and accessed the records one after the other in the same order as the patient rooms (interaction pattern of type 1, see figure 2). On one of the surgical wards, the evening shift nurse walked from one room to another with the respective off-going nurses to meet the assigned patients. For this purpose, the nurses used a mobile cart to take the patient records and the worksheet with them.

The evening shift nurse partly retrieved information from the record and partly communicated verbally with the colleagues. We observed two handovers from evening-to-night shift on a surgical ward. Here, the off-going nurse recited relevant information to the night nurse using his or her worksheet and the patient records. The night shift nurses also used a worksheet for their shift.

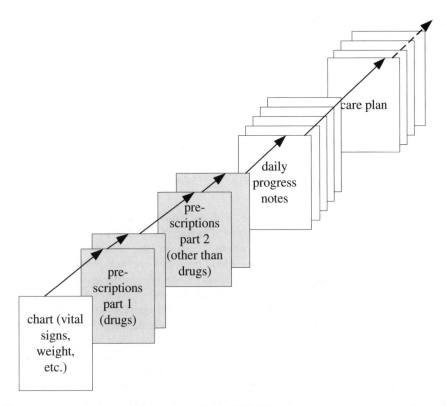

Fig. 3. Instance of an interaction pattern of type 2. Within a patient record, nurses usually went through the record sequentially from the front to the back, studying forms with more frequently needed information (i.e. the chart or prescriptions) in more detail and mostly skimming through or even skipping other forms. If there was something that captured their interest, they stopped to read the corresponding parts.

During the observation at the beginning of the shift, a total of 311 accesses to the patient records were recorded. Normally, nurses went through the record from the front to the back. Infrequently (about 7% out of 311 accesses) they skimmed back to consult a particular form. Most frequently, nurses studied information about prescriptions, i.e. medications (21.2% out of 311 accesses) and other prescriptions (16%), and they consulted the chart (21.5%), daily progress note (10%), notes to the next shift or questions to the physician (9%) and care plans (5%). Other information, for instance on the psychosocial situation of the patient or wound documentation, were studied depending on the patient's problem or the nurse's familiarity with the record. To access the required information from the file, nurses sequentially went from one tab to another and then skimmed through the forms or skipped them (interaction pattern of type 2, see figure 3).

In this shift phase it was rather unusual for a nurse to enter information to the record. Out of the 311 patient record accesses, only eleven concerned data entry, whereas four of them were questions to the physician. However, nurses frequently made brief notes on the worksheet to record relevant information and to do's. For example, the worksheet contained columns to manage information about resuscitation status, mobilisation, vascular accesses or questions to the physician (see figure 1). Worksheets were filled in individually and some contained handwritten notes in almost every column.

These observations were confirmed by the interviewed nurses. All of them stated (n=12) that they went through each record from the front to the back. Hence, there was virtually no information that was not accessed during this phase of the shift. Most frequently, nurses required information about progress notes (n=9), medications (n=9), other prescriptions (n=7) and the chart (n=6). Only three nurses stated that they entered information to the patient record at this time.

3.3 Main Phase of the Shift

After nurses had familiarized themselves with the patient records, they started to carry out required tasks and nursing activities. During this phase of the shift, all the observed nurses (n=8) used both the patient record and the worksheet. Additionally, memos were used to record the patients' blood pressures or notes. Nurses moved a lot between the nurses' office and the patient rooms to complete their tasks. For example, when a patient was in pain, nurses walked to the nurses' office to consult the patient record for analgesics prescribed as required. Subsequently, they prepared the analgesic and administred it to the patient. Information gathered at the patient's room or elsewhere outside the nurses' office, such as blood pressure, was captured on a memo or the nurse tried to memorise the issue to write it to the record later on. When a task was completed, it was checked off on the worksheet.

All patient records were only carried along during the ward round in the morning shift and the last round at the end of the evening shift. At all other times, the records were kept at the nurses' office where they were accessible to all healthcare professionals. During the shift's main phase, all observed nurses had to complete tasks that we categorized as follows:

(A) Prepare and check medications before administration.
(B) Participate in ward rounds or in a consultation (Kardexvisite) with the physician in the afternoon: The nurse and physician exchange information and mainly discuss medications, other prescriptions and questions.
(C) Coordinate with colleagues, i.e. attend brief handovers in connection with a break (lunchtime, etc.) or for other reasons.
(D) Perform assessments, carry out medical prescriptions and collect relevant information in the patient room: Measuring or observing vital signs, weight, elimination patterns, fluid balance; checking vascular accesses and drainages;

performing wound care; assisting patients in positioning themselves or in mobilisation; taking blood samples; checking blood sugar levels; gathering information concerning patient status as nutrition status, skin integrity etc.

(E) Assist patients with activities of daily living, such as getting dressed, bathing, toileting, eating etc. and

(F) Perform organisational and administrative tasks and manage the patient's psychosocial issues: Managing patient admission or discharge; ordering, coordinating or preparing examinations, tests and therapies such as consultations, diagnostic tests (such as laboratory tests, including preparation of taking blood samples), physio therapy etc.; writing or updating care plans; ordering meals; communicating with the patient's relatives etc.

Characteristic interaction patterns were identified when nurses accessed the patient records to perform one of the following tasks: When they prepared medications, participated in ward rounds, consulted with the physician or when they carried out brief handovers. These tasks had to be done for all patients at the same time. Nursing tasks that were completed spontaneously seemed to show no patterns (categories D-F), as they were performed individually depending on the patient status or the organisation of work.

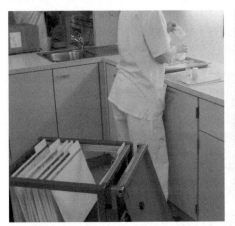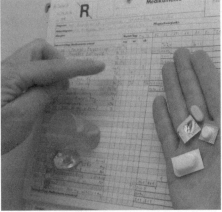

Fig. 4. Nurses prepared the medications in the nurses' office for all patients on the ward using the medication list

To administer medications, one nurse prepared all medications (for day or night) for all patients on the ward. To carry out this task, nurses stood in front of the medicine cupboard which was located in the nurses' office, and used the cart with the patient records (see figure 4). They took each record in the same order as the patient rooms (see interaction pattern of type 1), put it on the desk and consulted the medication list. They then looked up the column for the actual day and prepared the corresponding drugs. Fluids, controlled substances and injection drugs were prepared just before they were administered. This means that the preparation was repeated for these substances. To reduce errors, all medications were controlled by one or two nurses later on. In doing so, the nurses accessed the patient records in the same manner as during the initial preparation. To prepare medications, nurses predominantly retrieved information. During the observation time, only three nurses entered information to the record, such as writing down a question to the physician or to mark something on the list of medications.

On normal weekdays, two regular meetings between nurse and physician take place (category B) to discuss the patient status. The ward round starts in the morning, where the treating physician and the assigned nurse meet the patient at the bedside. During the evening shift, the physician and nurse meet in the nurses' office (called Kardexvisite) to consult on the patient situation and to clarify open questions.

We observed 33 accesses to the patient record during the rounds. Healthcare professionals went from one patient room to the next and therefore accessed the patient records according to the room order. For the most part, nurses and physicians consulted the same form and discussed the patient situation. The interviewed nurses stated that they most frequently access medications (n=12), other prescriptions (n=11) and the questions to the physician (n=7) during morning rounds. Four nurses expressed the need for all the information from the patient record. Information on the psychosocial situation was never used during rounds by four (n=4), and care plans by three (n=3) interviewed nurses. Only one nurse commented that she would never enter data to the records during the ward round. All others (n=11) would like to enter immediately notes to the patient record, for instance information concerning discharge and organisational issues (n=5), tasks to be done (n=5) and notes to the next shift or questions to the physician (n=3).

The consultation meeting at the nurses' office took place in the course of the later afternoon. The physician and nurse sat together at the desk to discuss primarily medications and other prescriptions as well as questions the nurse had. In total, 102 accesses to the records have been observed during these meetings. Each record was taken out of the cart following the order of the patient rooms. Most frequently, medications (about 27.6% out of 102 accesses), questions to the physician (22.3%), other prescriptions (about 19%) and the chart (11.7 %) were used. Nurses also skimmed back and forth, for example switching between the list with medications and the questions to the physician. They also used the worksheet in order to better orient themselves and to coordinate their access to the records. About 10% out of 102 accesses concerned data entry to the record, in order to check off an item or to jot down a brief note, such as a sign when a catheter needed to be removed.

Communication during handovers (category C) happened verbally. All participating nurses met at the nurses' office and each summarized relevant information on the patient's current status or tasks and nursing activities that needed to be done next. They either retrieved the information from the worksheet, the patient record or from memory. Usually, nurses reported one patient after the other in the same order as the patient rooms.

We identified no specific interaction patterns for the other tasks (categories D-F), which means that patient record accesses happened rather spontaneously. For example, a nurse first looked up the ordered laboratory tests, and then completed all involved steps such that the blood sample and the order form could be dispatched to the lab. Or a nurse would look up the wound documentation form, prepare the required materials on a cart and then go to the patient to perform the wound care.

As the patient record was not available in the patient room, or because nurses did not have enough time during the ward rounds to enter information immediately, documentation usually took place at a later time.

Tasks under categories D and E were virtually all performed in the patient room. When nurses collected data at the bedside, as for example vital signs, they hardly could memorise several values and thus noted them on a memo. Later on, back to the nurses' office, they transcribed the values to the record. Other information, such as the psychosocial situation could not be recalled because of the amount of information. For this task, nurses took the corresponding form to the patient's room to fill it in on site. Afterwards, the form was brought back to the nurses' office and filed to the patient record. Other forms, such as the fluid balance or pain chart as well as vital sign assessment forms were kept in the patient room for a longer period, particularly when often needed, as it is the case when the patients are critical. Further, there was information displayed directly at the bedside as well, for instance when a patient had ordered bed rest, so that all healthcare professionals were informed.

In the majority of the cases, information concerning assistance of the patient could be recalled and thus was documented later on without any difficulties. However, information concerning patient's positioning was an exception. As the observations revealed, one nurse on an evening shift had difficulties remembering whether she had turned the patient from a dorsal position to the right or to the left side. On another ward, nurses kept the position information form at the bedside to prevent such a problem. When nurses had to complete tasks of the category F, they usually had to handle multiple forms. With the exception of family discussions and the processing of meal orders, tasks of this category were carried out in the nurses' office.

During the course of the interviews, participants were asked, what kind of information would be important to enter into the record while still in the patient room. The majority of the nurses (n=11) stated that all parameters such as pulse, temperature, blood pressure or aeration should be recorded immediately. Besides this, nurses would like to have on site access to medications and other prescriptions (n=4), pain assessment (n=3), daily progress notes (n=2), care plans (n=2) and open issues (n=2). However, it was not considered a good idea to have wound care documentation in the patient room due to hygienic reasons. Four nurses stated that the whole record should be available during their last round on the ward. Overall, nurses would like to have mobile access to care plans and all other information needed in relation to tasks of the categories A, B and D.

3.4 Completion of the Shift

Towards the end of their shift, nurses started to complete their entries to the record. To do this, all observed nurses from the morning shift (n=3) stayed at the nurses' office at the desk. During this phase, the oncoming shift nurses had already arrived, and since both groups stayed in the nurses' office, the room was temporarily rather crowded and the situation hectic.

All observed nurses from the evening shift (n=5) performed a so called "last round" towards the end of their shift. They took with them the cart with all the records, the worksheet and the prepared medications to go from one patient room to the next. The three observed surgery nurses completed all entries to the records during this round in the patient room. One nurse on a medical ward captured only part of the information on site, for instance administered medications or vital signs. After returning to the nurses' office, she completed the documentation by filling in the remaining forms such as the daily progress notes. Another nurse from internal medicine had a similar routine and completed the majority of the paperwork after the last round in the nurses' office.

All nurses used the worksheet when it came to finishing the documentation. Using the sheet, they checked whether they had addressed all the pending issues and documented completed tasks in the record. Interaction with the records was similar to the beginning of the shift. Each record was sequentially accessed according to the room order (interaction pattern of type 1). Afterwards, nurses went through the record from the front to the back and sometimes skimmed back. During this shift phase, we observed 159 accesses in total to the record, most frequently to the chart and the daily progress notes (each of them 14% out of 159 accesses). Medications and other prescriptions were accessed quite often as well (each of them 10%), as were entries about care delivered (11%) and the care plan (8%). For each patient, nurses wrote daily progress notes. The interviewees stated that overall, they most frequently recorded daily progress notes (n=11) and filled in the chart (n=9). To make sure, that important information was communicated to the next shift, nurses also wrote such information to the notes to the next shift. At both hospitals, these notes acted somewhat like a blackboard. Some nurses reported verbally to the oncoming nurse as well. Upon completion of these tasks, nurses ended their shift.

4 Discussion

To be aware of the actual nurses' working practices is an essential prerequisite for an ergonomic CPR design. Our study was undertaken in a real setting of two acute care hospitals in Switzerland and used both, objective and subjective methods. This method is well-established in the user-centred software development [14]. The analysis enabled us to gain a comprehensive insight into the nurses' practices and their interaction routines with the patient records during morning and evening shifts.

According to the participants who had work experience in hospital wards abroad, nurses' working practices are similar at Swiss, German and Austrian hospitals. As far as other studies allow a comparison, the nurses' work in other Western countries does not seem to differ much from our findings. Nurses manage a substantial part of the patient record, such as the chart, medications, prescriptions, questions to the physician, care plan or the daily progress notes [7, 15-19]. They also perform a diverse range of tasks, e.g. preparing and administering medications for all patients, participating in ward rounds or recording vital signs [9, 20, 21]. Since we could not find other studies focusing on the nurses' interaction with the patient record in different working contexts, we cannot assess to what extent our results are generalizable to other countries. But we feel that on the basis of our detailed description, it should be possible to identify corresponding interactions.

The study demonstrates the importance of temporarily used documentation tools, i.e. worksheets and memos, which confirms the findings of previous studies [22-25]. As our study shows, worksheets play a vital role in all three phases of the shift and they provide nurses with a concise overview of pending issues and other relevant information. In spite of this fact, to our knowledge, the concept of such worksheets is not yet supported by today's CPR systems. We outline later on, how this tool could be integrated in a CPR to meet the users' needs.

All worksheets consisted of a patient list, but on each ward, the sheet contained different columns, and nurses either worked with a personal or a shared worksheet. One of the advantages of personal sheets is their ubiquitous availability. In contrast, centrally managed worksheets allow for a quick access to vital patient information – such as the resuscitation status - by all nurses even when the responsible nurse is temporarily away. Both advantages could be easily consolidated with a mobile CPR, in which the implementation should allow for custom designed lists according to the individual needs.

Nurses used the patient record during all three phases of their shift, but did not access all type of information with equal frequency. During shift phases one and two, nurses accessed most frequently medications, other prescriptions and the chart either to retrieve or enter information. During shift phase three, the chart and the daily progress notes were accessed most frequently to document information. Today's organisation within the records reflect these access frequencies, as the chart, medications and other prescriptions are placed in the front of the folder and the daily progress notes in the back. Questions to the physician and care plans were also accessed frequently during all three shift phases. This fact should be used to reduce the interaction time with the CPR system.

We observed two types of characteristic interaction patterns during all three phases of the shift. The patterns emerged during the completion of tasks that called for access to all patient records – for example during the preparation of medications. With regards to tasks that were carried out individually depending on the patient status or organisational issues - as for instance wound care or discharge - accesses to the records were spontaneous. The design of a CPR system should facilitate nurses to learn and use it easily. In the following, we outline design suggestions and recommendations, which we consider suitable for all three phases of the shift and interaction modes with the patient record, as well as for stationary and mobile hardware.

To reflect the identified interaction patterns, the CPR should support sequential access to the patient records according to the room order as well as by going through a single record from the front to the back. A rough navigation functionality should allow nurses to jump easily from one information type, that is source – such as the chart, medications, other prescriptions - to another by just one click or a shortcut. Additionally, a fine navigation functionality should provide the option of skimming through the record forms back and forth in an efficient manner, with a single click or shortcut. As soon as the routine reaches the end of one patient record, the navigation mode should switch automatically to the first form of the next patient record (according to the room order). To minimize the interaction time with regards to sequential access to a single form or type of information, the system should additionally facilitate users to first choose a specific form – for example the medications list – and then to step through with a click from one patient record to the next, with the chosen form always at display.

Aside from a navigation streamlined to the access patterns, the CPR user interface should also allow for direct access to records and forms respectively. Hence, it should enable the user to access records by means of a patient list and forms via a menu, tab or other facility. In that way, the navigation should suit to both types of accesses – those with and those without interaction patterns.

To ease detection of new information, the CPR system should highlight information that has been added or changed by other healthcare professionals individually for each nurse, from the time on since they last accessed the record. This functionality could be implemented by using flags or - within a form - by highlighting added information with colour to improve the visual orientation.

At the beginning of the shift, nurses amended and updated the information on the worksheet for each assigned patient. To support this task adequately by a CPR system, for each individual patient the row of the worksheet list should be displayed together with the content of the retrieved record form. Nurses would then be able to either manually enter information to the row, or to copy information from the accessed form to the worksheet, for instance by means of a click or drag&drop interaction.

Since the worksheet is frequently used in all shift phases – for example during the ward rounds or completion of documentation at the end of the shift – it would make sense to permanently display a patient worksheet row together with the accessed forms. Furthermore, it would be useful to integrate the questions to the physician into the worksheet, to support quick access to these information when needed. Switching to the entire worksheet list should be feasible with one click or shortcut.

The morning ward round seems to be common in all Western countries [21, 15, 27]. During these rounds, the interaction with the patient record takes place in a mobile and hectic environment, and nurses usually consult the same forms as the physician to discuss the patient situation. Hence, it would be useful to find a CPR design that meets the needs of both professional groups. Streamlining the CPR user interface accordingly would lead to a significant reduction of interaction time [28]. The interviewed nurses stated to most frequently consult the forms with the questions to the physician during ward rounds. As already suggested, this information should be displayed permanently as part of the worksheet row. Other specific nursing information – such as nursing assessments, care plans or daily progress notes - were retrieved rather infrequently. For these information, a higher interaction time is acceptable and therefore, the navigation functionality could be primarily adapted to the physicians' interaction preferences during ward rounds. Aside from this particular working context, the CPR user interface should be streamlined specifically to the nurses' needs as described above.

In mobile contexts, nurses usually made no extensive data entries, i.e. they just jotted down a few words on the record form, worksheet or memo. This habit could be supported with a pen-based input facility. Berglund's study concluded that a PDA has the potential to be accepted as a supportive tool in healthcare organisations [4]. However, the display of the mobile device should not be too small, because small displays reduce the overview of displayed information and increase interaction time [29]. Another undesirable effect of a small displays is that such a device needs a different graphical user interface from that of a stationary PC. We therefore confirm Cole's recommendation [30] not to use PDA's to avoid such disadvantages. Tablet

PC's integrate both criteria – a pen-based input facility and a sufficiently large display – but they are still too heavy and not easily stowed in a lab coat pocket. Hence, we conclude that further studies are required to investigate more suitable devices, such as wearable computers or e-paper devices.

5 Conclusion

As our study demonstrated, nurses are extremely skilled in using their paper-based documents. They regularly use patient records and worksheets to enter or to consult information and to coordinate their activities. The observations showed that the use of these tools is closely interweaved. Therefore, worksheets should be carefully integrated into the CPR system, and the systems's navigation functionalities should be streamlined to the identified interaction patterns. As soon as suitable mobile devices are available, CPR's will provide an ubiquitous access to needed information, which makes the use of memos obsolete.

Only a system that reflects the professionals working practices will encounter their acceptance. Mobile CPR's with high usability as outlined here would make a substantial contribution to reach this aim.

Acknowledgements. We would like to express our sincere thanks to the staff of the hospitals Stadtspital Waid in Zurich and Kantonsspital Luzern for their generous support of this study. Special thanks to Susanne Büge for her organizational support, and to Niroshan Perera and Ela Hunt for their critical review of the manuscript. This work is granted by Datonal AG, Switzerland.

References

1. Ammenwerth, E., Buchauer, A., Bludau, B., Haux, R.: Mobile information and communication tools in the hospital, IMIA Yearbook pp.338–357 (2001)
2. Ragneskog, H., Gerdnert, L.: Competence in nursing informatics among nursing students and staff at a nursing institute in Sweden. Health Inf. Lib. J. 23, 126–132 (2006)
3. Brender, J., Ammenwerth, E., Nykänen, P., Talmon, J.: Factors Influencing Success and Failure of Health Informatics Systems. Meth. Inf. Med. 45, 125–136 (2006)
4. Berglund, M., Nilsson, Ch., Revay, P., Petersson, G., Nilsson, G.: Nurses' and nurse students' demands of functions and usability in a PDA. Int. J. Med. Inf. 76, 530–537 (2007)
5. Choi, J., et al.: MobileNurse: hand-held information system for point of nursing care. Computer Methods and Programs in Biomedicine 74, 245–254 (2004)
6. Wu, J., Wang, S., Lin, L.: Mobile computing acceptance factors in the healthcare industry: A structural equation model. Int. J. Med. Inf. 76(1), 66–77 (2007)
7. Potter, P., Boxerman, S., Sledge, J.A., Boxerman, S.B., Grayson, D., Evanoff, B.: Mapping the nursing process. J. of Nursing Administration 34, 101–109 (2004)
8. Wolf, L.D., Potter, P., Sledge, J.A., Boxerman, S.B.: Grayson, Describing nurses' work: combining quantitative and qualitative analysis. Human Factors 48(1), 5–14 (2006)
9. Ebright, P.R., Patterson, E.S., Chalko, B.A.: Understanding the complexity of registered nurse work in acute care settings. J. of Nursing Administration 33, 630–638 (2003)

10. Ammenwerth, E., Kutscha, U., Kutscha, A., Mahler, C., Eichstadter, R., Haux, R.: Nursing process documentation systems in clinical routine - prerequisites and experiences. Int. J. Med. Inf. 64, 187–200 (2001)
11. Poissant, L., Pereira, J., Tamblyn, R., Kawasumi, Y.: The impact of electronic health records on time efficiency of physicians and nurses: a systematic review. JAMIA 12(5), 505–516 (2005)
12. Darbyshire, Ph.: Rage against the machine?': nurses' and midwives' experiences of using Computerized Patient Information Systems for clinical information. J. Clin. Nurs. 13, 17–25 (2003)
13. Oulasvirta, A.: The Fragmentation of Attention in Mobile Interaction and What to Do with It. Interactions 12(6), 16–18 (2005)
14. Rauterberg, M., Spinas, Ph., Strohm, O., Ulich, E., Waeber, D.: Benutzerorientierte Software-Entwicklung, vdf (1994)
15. van der Meijden, M.J., Tange, H.J., Boiten, J., Troost, J., Hasman, A.: An experimental electronic patient record for stroke patients. Part 1: Situation analysis. Int. J. Med. Inf. 125, 58–59 (2000)
16. van der Meijden, M.J., Tange, H.J., Boiten, J., Troost, J., Hasman, A.: An experimental electronic patient record for stroke patients. Part 2: System description. Int. J. Med. Inf. 58-59, 127–140 (2000)
17. Ammenwerth, E., et al.: PIK-Studie 2000/2001, Evaluation rechnergestützter Pflege-dokumentation auf vier Pilotstationen, Forschungsbericht der Univers. Heidelberg (2001)
18. Parker, J., Brooker, Ch.: Everyday English for International Nurses. A guide to working in the UK, Churchill Livingstone (2004)
19. Martin, A., Hinds, C., Felix, M.: Documentation practices of nurses in long-term care. J. Clin. Nurs. 8, 345–352 (1999)
20. Manias, E., Aitken, R., Dunning, T.: How graduate nurses use protocols to manage patients' medications. J. Clin. Nurs. 14, 935–944 (2005)
21. Manias, E., Street, A.: Nurse-doctor interactions during critical care ward rounds. J. Clin. Nurs. 10, 442–450 (2001)
22. Hardey, M., Payne, S., Coleman, P.: Scraps: Hidden nursing information and its influence on the delivery of care. J. Adv. Nurs. 32, 208–214 (2000)
23. Kerr, M.P.: A qualitative study of shift handover practice and function from a socio-technical perspective. J. Adv. Nurs. 37(2), 125–134 (2002)
24. Strople, B., Ottani, P.: Can Technology improve intershift report? What the research reveals. J. of Professional Nursing 22(3), 197–204 (2006)
25. Allen, D.: Record-keeping and routine nursing practices: the view from the wards. J. Adv. Nurs. 27, 1223–1230 (1998)
26. Payne, S., Hardey, M., Coleman, P.: Interactions between nurses during handovers in elderly care. J. Adv. Nurs. 32(2), 277–285 (2000)
27. Tang, P., LaRosa, M., Gorden, S.: Use of Computer-based Records, Completeness of Documentation, and Appropriatness of Documented Clinical Decission. JAMIA 6, 245–251 (1999)
28. Reuss, E.: Visualisierungs- und Navigationskonzepte für das computerbasierte Patienten-dossier im Spital. Dissertationsschrift, ETH Zürich (2004)
29. Watters, C., Duffy, J., Duffy, K.: Using large tables on small display devices. Int. J. Human-Computer Studies 58, 21–37 (2003)
30. Cole, E., Pisano, E.D., Clary, G.J., Zeng, D., Koomen, M., Kuzmiak, C.M., Kyoung, B., Pavic, Y.D.: A comparative study of mobile electronic data entry systems for clinical trials data collection. Int. J. Med. Inf. 75(10-11), 722–729 (2006)

Organizational, Contextual and User-Centered Design in e-Health: Application in the Area of Telecardiology

Eva Patrícia Gil-Rodríguez[1], Ignacio Martínez Ruiz[2], Álvaro Alesanco Iglesias[2], José García Moros[2], and Francesc Saigí Rubió[3]

[1] Av. del Canal Olímpic s. n. 08860 Castelldefels (BCN) Spain
Internet Interdisciplinary Institute
Universitat Oberta de Catalunya
egilrod@uoc.edu
[2] c/ María de Luna 1, 50018 – Zaragoza (Spain)
Aragon Institute for Engineering Research (I3A)
University of Zaragoza
{imr, alesanco, jogarmo}@unizar.es
[3] Av. Tibidabo 45-47
Health Sciences Programe
Universitat Oberta de Catalunya
fsaigi@uoc.edu

Abstract. Currently, one of the main issues of Human Computer Interaction (HCI) in the area of e-Health is the design of telemedicine services based on the study of interactions established between people and new technologies. Nevertheless, the research in HCI has been mainly focused from a cognitive point of view, and most studies don't consider other variables related to the social impact of technology. This gap is extremely important in e-Health, because Information Systems (IS) changes qualitatively work processes in health organizations. In this paper we present an interdisciplinary perspective that includes the cognitive and organizational aspects to the technical design of new e-Health services in the area of telecardiology. The results obtained permit to accurately define not only technical requirements and network resources management, but also user requirements, contextual particularities of the scenarios where these services could be implemented and the IS impact in organizational processes. Moreover, the ethnographical methodology proposed has been applied to specific emergencies scenarios, obtaining conclusions that allow new user-centered designs that can be directly integrated in medical practice in an efficient way.

Keywords: Contextual inquiry, e-Health, ethnographical methodology, scenario analysis, telecardiology, User-Centered Design.

1 Introduction

It is well known that the purpose of Human Computer Interface (HCI) [1],[2] is to obtain theoretical knowledge about the interactions established between people and

A. Holzinger (Ed.): USAB 2007, LNCS 4799, pp. 69–82, 2007.

new technologies, and that this knowledge is reflected in the design of technological devices' functionalities and their interfaces, guaranteeing success in further implementations. In fields like e-Health, we can find a wide sphere in which to develop these kinds of study because it is common to find a context in which users are not able to take full advantage of the potential offered by these technological devices due to the design criteria [3].

Nevertheless, within this discipline, a great proportion of user studies have been carried out from a cognitive point of view, without considering other variables related to the social impact of technology in organizations. From our interdisciplinary perspective, we maintain that Information and Communication Technologies (ICT) produce qualitative changes in health organizations and establish a permeable relationship between the daily routine of health-related professionals and their working process.

Where Information Systems (IS) are concerned, any implementation of ICT which is aimed at increasing the coherence and continuity of health care, is directly proportional to organizational change. This is to be understood as the setting up of coordination mechanisms among health services together with the resulting adjustments of coordination at management level, and the creation of alliances, agreements, etc. between health organizations and related health staff [4]. This principle is also valid for any project implementing a telemedicine system, as it is well known that technology changes not only healthcare staff working systems but also working routines within the healthcare system. To achieve this goal, the participation of clinical staff in the development of new technological systems is mandatory [5]. It must be stated that any organizational change has to be undertaken in parallel with cultural change, described as a change in the behavior of all the staff involved, facultative and non-facultative. In order to succeed, any telemedicine system model is compelled to consider updating knowledge, processes and skills together with establishing protocols and guides, and the implication of these on faculty staff/workers. In this sense, health organizations/institutions are advised to encourage co-responsiveness with the project among the faculty.

Bearing all of the above in mind, one of the main handicaps affecting the implementation of a telemedicine system is management and organizational adaptation. Even so, IS must also be designed as inter-operable, and data to be transmitted must be compatible, standard and a result of consensus. In addition, the definition of a given telemedicine project must reflect the real situation of the institutions involved, both at the level of technological requirements and at an organizational level. In fact, three main ideas are identified as key factors for success: solving real healthcare problems, having a user-centered design, and viewing the designed solutions as services that can be offered.

In this sense, some studies have demonstrated the importance of contextual and user studies in reducing cultural resistance to technological innovation processes in organizations [6]. Because of this, we must take into account not only the cognitive requirements of users, but also contextual, practical and symbolic requirements. That means analyzing how technology is appropriated by people in their specific context, imbuing it with unpredicted meanings that do not follow developers' guidelines [7]. As a result of this, organizational variables are needed to study specific scenarios in real medical healthcare; cultural variables allow us to analyze the meanings that patients and doctors give to their daily routines and medical equipment; cognitive (the process of information with or without ICT devices), symbolic (cultural meanings of ICT) and practical variables of ICT uses (the ways of improving health-related professionals' tasks and patients' quality of life) will also be very useful in the design of telemedicine services. Knowledge of these variables serves to improve the design of Graphical User Interfaces (GUI) by minimizing, for example, the usual negative associations that patients make between medical technology and their illness.

To gather data about these kinds of variables, ethnographic methodology is required [8]. In our ethnographical research, we adopt the sociological perspective of Actor Network Theory (ANT) [9]. ANT studies the relations established between human actors (people and organizations: physicians, doctors, patients) and artifacts (Electronic Patient Record (EPR), medical equipment). From ANT, artifacts are understood as actors (like humans), because they have the ability to change relations in social organizations in a very meaningful way.

Hereafter we will show how this perspective could be applied in the specific context of telecardiology service design and which are its most relevant contributions and added values within this field in particular, but also in optimizing transversal research work that results in improved Quality of Life (QoL) for patients and health professionals. The methodology of our research is presented in Section 2; in our methodology perspective we take into account organizational, contextual and user variables, in order to translate this qualitative information into the service requirements found in telecardiology. The results obtained, including specific examples of implementation obtained from the methodology proposed, their technical requirements and some preliminary results from the contextual study of a specific scenario – telecardiology emergencies - are detailed in Section 3. Finally, discussion and conclusions are presented in Section 4.

2 Materials and Methods

Telecardiology is one of the most relevant areas in e-Health [10] and, in this study, the two main telecardiology services have been included:

- *Teleelectrocardiology.* This is a well-studied telemedicine topic as the electrocardiographic (ECG) signal is by far the most used cardiac signal in establishing an initial idea of what is going on in the heart. It deals with the transmission of ECG signals from a remote site (where the patient is located) to the expert site (where the cardiologist who interprets the signal is located). Teleelectrocardiology can be carried out in two ways: real-time (using lossy compression techniques) and 'store & forward' transmission (S&F, with lossless compression), depending on whether interaction between the remote site and the expert site is required [11].
- *Teleechocardiology.* The diagnostic technique of echocardiography is widespread due to its low cost and non-invasive ability to capture dynamic images of the heart in real time. It can be also be carried out in two ways, real-time and S&F but, unlike in teleelectrocardiology, real-time applications are preferred since it is very important to have interaction between the remote site and the expert site in order to complete a consultation successfully [12].

In this paper, we have designed a methodology to define potential scenarios in this key area of telecardiology, from user and health organizations points of view (or what technology must do, according to the needs of users and organizations). This proposed methodology could improve the efficiency and acceptability of the implementation of the service in this specific area. Thus, two methodology phases have been distinguished as described below.

2.1 Exploration of Telecardiology Scenarios in Health Assistance

First, we study the real needs of health organizations, identifying specific welfare levels where electrocardiography and echocardiography transmissions will help health-related professionals in their work. Thus, this first phase uses qualitative methodologies, based on interviews and focus groups, in order to define the telecardiology scenarios of interest. The design of interviews takes into account: organizational variables, needed to study specific scenarios in real medical healthcare; cultural variables, specific to the meanings that patients and doctors give to their daily routines and medical equipment; and ICT use variables: the observation of the use of ICTs by health-related professionals and patients, taking into account cognitive (the process of information with or without ICT devices), symbolic (cultural meanings of ICT) and practical aspects of this use (the ways of improving health-related professionals' tasks and patients QoL, these proving very useful in the design of the service). In order to gather information about these variables, we held 25 interviews with experts in telemedicine and nine interviews with relevant cardiologists in Catalonia (Spain). In addition, we set up a focus group with the coordinator and other relevant members of the Catalan Cardiology Association in order to identify specific characteristics of the proposed services. Focus groups [13] engage in discussion about a specific topic – in this case, ideal scenarios from a medical perspective involving the implementation of a telecardiology service. In these sessions, different points of view about the same topic can be raised and discussed to understand the specific characteristics of each scenario and the ideal services to be implemented in a telecardiology setting.

The next step is to analyze the content of the interviews and focus group. Content analysis allows us to extract and order different categories of information related to the variables considered in the study. The content analysis is supported by the qualitative analysis software ATLAS-ti [14], which helps to identify the most relevant subjects and reveals inter-relationships (see Fig. 1). As result of this analysis, the different implementation scenarios of the service and its potential users are established. Analyzing the scenarios allows us to study the implementation of new technologies in specific contexts and to consider applications not yet developed [15]. The scenarios are descriptive narratives based on real events in which selected people carry out specific actions, and these descriptions have shown themselves to be effective in managing interdisciplinary work between technology developers and user study experts, because they enable the technology being applied, its functionalities, the characteristics of its users and their interactions, to be clearly illustrated. By also taking into account the technical requirements of these scenarios, we can then select different but equally realistic proposals for implementation of the technology in particular environments and explore how services and equipment can respond to problems: improvement of the level of accessibility to medical services, the job of healthcare professionals, and the patient QoL.

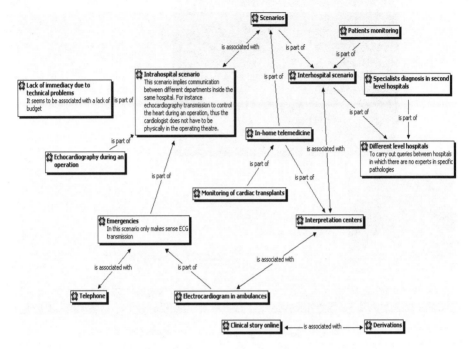

Fig. 1. Network of relationships obtained with ATLAS-ti

2.2 Contextual Inquiry of the Selected Scenarios

This second phase studies the variables previously defined in our theoretical framework in situ (through the scenarios already used) by means of ethnographic methodology [16], [17]. To achieve our project goals, it is especially pertinent that we use an ethnographic approach since it allows us to obtain an in-depth knowledge that takes into account diverse factors such as contextual and organizational factors, and of course, factors relating to the uses of the ICT that keep in mind not only users' cognitive aspects, but also the symbolic and practical dimensions of these uses.

In each specific case of telecardiology, we identify the socio-technical network related to the scenario involving not only the health-related professionals, technical staff and patients, but also the technological devices, transmission networks, and clinical information specific to each context. From this ethnographical approach, we gather the point of view of different actors involved in the scenario: cardiologists, health-related professionals, patients, etc. with specific roles in the environment under investigation. To gather this information, it is necessary to develop participant observations and customized interviews in each healthcare unit involved.

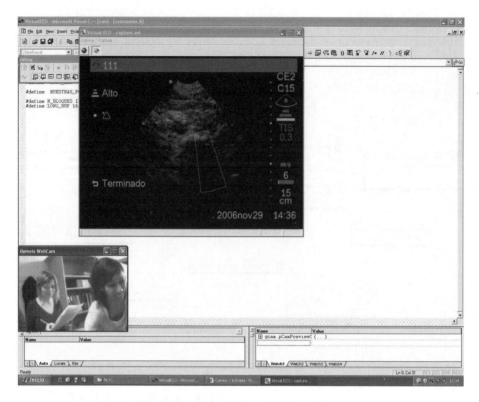

Fig. 2. Users' observation in a real scenario

In the case of an emergency telecardiology scenario, we observed participants for a period of 10 weeks and held eight focus groups and three individual interviews with the different actors in this scenario, including professionals from mobile units, the emergency coordination centre and a cardiologist from the hospital.

This methodological design allows us to build up a picture of the complete workflow of this specific scenario and the real needs of the users interacting with the technology, and therefore to design a telecardiology system with customized functionalities and GUI. Collecting these cognitive, symbolic and practical characteristics of ICT use on daily tasks in specific scenarios is a key-point for efficient GUI design.

3 Results

In this section, we show results about specific telecardiology scenarios and services as defined from an organizational and user's point of view. Moreover, the particular technical requirements of each scenario are detailed in order to guarantee the quality of the service from a technological point of view. Finally, we describe some preliminary results from the ethnographical study of the selected scenario of emergencies based on mobile teleelectrocardiology.

3.1 Optimal Selection of Telecardiology Scenarios and Services

From the first methodological phase, we define some specific telecardiology scenarios. Thus, the results identify three relevant contexts in which the implementation of a telemedicine system would be of value to the cardiology experts:

I.   *Home telemonitorization:* home telecardiology is only really useful for patients with heart transplants because it's a privileged scenario for monitoring this kind of patients. In order to avoid patient displacements, we can transmit ECG and echocardiographical signals, and it is also only necessary to control basic parameters such as blood pressure and patient weight increases, which do not suppose challenges for networked transmission.

II. *Remote specialist consultation of echocardiographies and ECGs, in order to avoid patient and cardiologist displacements.* It could be developed by means of an established protocol: for instance, a cardiologist from a rural hospital that periodically consults an expert from a tertiary hospital about doubts regarding the interpretation of echocardiographs. It could even have an occasional character, as in the case of specific pathologies consultation, in which the doctor that attends the patient is not as expert as another doctor located at a hospital in another city. And teleechocardiology can be very useful in a rural or inter-hospital environment, because in this scenario the displacement of expert cardiologists is often required for image diagnostics. In this case, it is important to note that echocardiography transmission requires the presence of a cardiologist or expert carrying out the test, in order to guarantee the quality of diagnostic images. On the other hand, the electrocardiography transmission seems very useful in an inter-hospital scenario, especially when there are no experts in interpretation of electrocardiographs attending patients. The outsourcing of diagnostic interpretation or cardiology test devices is included in this scenario. Telediagnosis service contracts involve image transmission between professionals from different countries. Outsourcing cardiology test devices also involves image transmission, in this case between health organizations. In both cases, it is necessary to transmit signals because the people who administer the test and the people who interpret it do not share either a temporal or physical space.

III. *ECG remote transmission in emergencies.* In the context of transmitting vital signs and information about patients from a remote location, a very relevant scenario is telecardiology emergencies, where ECG transmissions from mobile units enable fast diagnosis and treatment of some acute heart diseases.

3.2 Technical Requirements of the Selected Telecardiology Scenarios

From the previous contextual and sociological results, where the relevant telecardiology scenarios have been selected (see Fig.3), this subsection explains their related technical requirements. This technological analysis is supported by the multimedia technologies [18] on which the new e-Health services are usually based. Heterogeneous environments are differentiated according to type of service and differing quality of service degrees [19], and assessed over very diverse network topologies [20]. The main technical characteristics of each scenario are described below and summarized in Table 1:

- *Home scenario.* A patient at home transmits the medical signals to the hospital in order to be supervised remotely. The technical features of this communication usually correspond to fixed networks based on narrowband technologies (Public Switched Telephone Network, PSTN, or Digital Subscriber Line, DSL). Although both S&F and real-time transmissions modes are allowed, S&F is the most usual and it is often that transmitted signals are compressed with lossless techniques.
- *Rural scenario.* Its basic characteristics are associated to the communication between doctors of different locations (primary healthcare centres and hospitals). The technical features of this communication are very similar to home scenario, although real-time transmissions are now more usual in order to allow telediagnosis, inter-hospital work, etc. If the data volume is high, optimal codification and compression methods are necessary in order to select the best parameters (compression rate, digital images resolution, etc.) that guarantee the quality of service technical levels.
- *Emergencies scenario.* Emergency telemedicine is synonym of wireless telemedicine since mobile channels are the main way of transmitting medical information from a moving ambulance (mobile unit) to the hospital where the patient is transported. The communications are based on third generation (3G) Universal Mobile Telecommunication System (UMTS), and they often are RT and using lossy compression techniques.

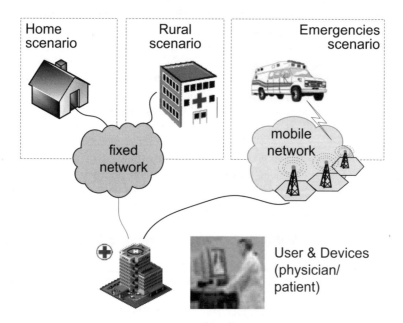

Fig. 3. Generic telecardiology scenarios considered in this study

Table 1. Technical description of the scenarios and services selected in this study

Scenario	technical parameters for teleelectrocardiology	technical parameters for teleechocardiology
Home	f_s = 250Hz/channel h = 8-12-16bits/channel s = 10 MB (original sizes)	f_s = 24-30frames/s h = 8-16-24bits/pixel s = 640x480/704x576pixels
Rural	f_s = 250Hz/channel h = 8-12-16bits/channel s = 1-4MB (2/3:1-*lossless*)	f_s =12-24-30frames /s h = 8-16-24bits/pixel s = 352x288/512x512pixels
Emergencies	f_s = 250Hz/channel h = 8bits/channel s = 500kB (10/15/20:1-*lossy*)	f_s = 1-5-15 frames /s h = 2-4-8bits/pixel s = 176x144/352x240pixels
Scenario	Operation modes	
Home	S&F: the signals are pre-acquired and further digitally transmitted to the hospital	
Rural	S&F, but there are RT projects based on portable devices for on-line telediagnosis	
Emergencies	RT: the communication between the mobile unit and the hospital is on-line.	

However, the analysis of the implementation of an efficient e-Health solution requires the integration not only of these technological requirements but also the user variables. Thus, a complete study of health organizations, leading to a detailed description of their environments, is presented in the next subsection and refers specifically to the case of mobile teleelectrocardiology for emergencies.

3.3 Organizational, Contextual and User Variables in Mobile Teleelectrocardiology

In order to design a specific e-Health service from this interdisciplinary point of view, we need to take into account organizational, contextual and user variables in a specific telecardiology scenario: teleelectrocardiology from mobile units to hospitals in an emergency scenario. The main objective of our ethnographic methodology is to study real user, contextual and organizational needs in this specific scenario [21]. At this point, we present some preliminary results about characteristics of teleelectrocardiology in a mobile scenario from a clinical, organizational, contextual and user point of view.

From a clinical point of view, transmission of electrocardiography from mobile units helps cardiologists to make decisions about the best available treatment for patients with acute disease. In some specific acute coronary syndromes, the efficacy of therapy depends on the time of administration, being of maximum benefit in the first hour following the beginning of the symptoms and diminishing progressively up to 12 hours after they began. Nevertheless, 50% of the patients arrive at the hospital within 4 hours of acute illness, and we must reduce the time required to assist them [22].

Some therapies can be applied in a mobile unit, but more efficient therapies should only be applied in a hospital setting. The transmission of an ECG from a mobile unit helps doctors who are not necessarily specialized in the area of cardiology to make a diagnosis and to take the best decision in the least time. It is especially interesting in rural zones is that they are able to administer the therapy from a mobile unit. On other occasions, they will be able to move patients to a nearer hospital, always taking into account the trade-off that exists between the time of administration and the efficacy of the therapy. The unit co-ordinating emergency care via these transmissions can also help to make decisions about which is the most appropriate hospital in the network to receive the patient. This coordination makes a direct admission into the cardiology unit of the hospital possible, and makes it possible to attend to patients as soon as they arrive.

General coordination of the emergency services is vital. Through this, technology facilitates a global organization of available resources, and enables the services to assist patients who would not otherwise be able to access help in such short timeframes. The complete workflow of transmission is shown in Fig. 4.

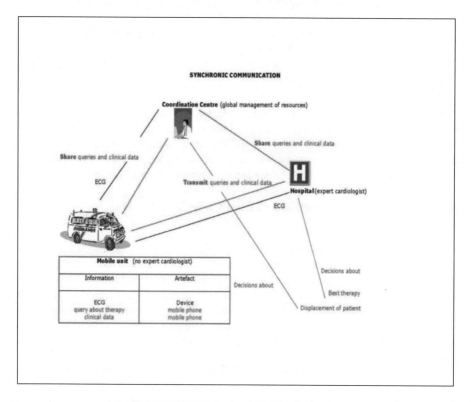

Fig. 4. Mobile Telectrocardiology Scenario

But what about health-related professionals from mobile units?

This kind of professional works under very stressful conditions, and their ability to communicate with experts is extremely important to enable necessary decisions to be made in a short time. However, some resistance exists to this kind of transmission system, due to the following reasons:

- It is very important to pay constant attention to the patient, and technological devices being used should not in any way interfere with this. Perceived problems are the number of staff, the inability to locate required professionals in the hospital, and the simple need to be holding a mobile telephone at all times.
- It is also necessary to transmit the patient's clinical data, since the transmission of an ECG by itself is not sufficient to arrive at a diagnosis. What is the best way to transmit all this data, without impacting negatively on the quality of care being delivered to the patient in an emergency situation?
- The resulting ability of non-expert personnel to make decisions as a result of these transmissions creates some resistance amongst health-care workers, having the potential to radically change perceptions of expertise and responsibility in the health-care arena.
- On the other hand, organizing across a network also creates another cultural change: that of assigning patients to different hospitals based not on their geographical proximity, but according to the suggestions of the emergency coordinating centre as to which is the best place for a particular patient.
- Finally, it is perceived that the new services could create more work without improving the effectiveness of the system in any way.

These types of resistance should be borne in mind as key points when designing a service and thinking about the user's interaction with devices in mobile units. Keeping these things in mind will improve efficacy in assistance to patients, and will also improve the quality of this care in emergency situations.

4 Discussion and Conclusions

Technological innovations in the field of telemedicine are extremely important, and a study of the organizational, contextual and user variables affecting their implementation is vital in order to guarantee that those innovations respond to existing problems in the healthcare system. In this paper, and following an interdisciplinary perspective, the real needs of health organizations have been analyzed in order to select the most appropriate environments in the area of telecardiology: home, rural an emergencies scenarios.

In the specific case of telecardiology, the transmission of both electrocardiograms and echocardiographs of the heart are promising fields. In this study, we have identified different levels of care at which the implementation of telemedicine could improve the quality of care and the efficacy of the health care system. For this, as well as taking into account the variables already mentioned, we have also included technical aspects in order to guarantee the viability of implementing the service in each specific scenario.

The results obtained about organizational, contextual, user and technical aspects permit to address key requirements in the further design of e-Health services:

- Organizational aspects: the implementation of e-Health services creates organizational transformations in work processes, as well as cultural transformations in professionals' daily routines. Organizational culture is one of the aspects that complicate the implementation of advanced communications technologies in the healthcare field, as the application of these technologies always supposes qualitative changes in work processes, values, forms of coordination etc. It seems that that inertia makes it difficult to adapt to these changes. Resistance may also emerge to the implementation of services based on advanced communications technologies because it may result in an increase in the volume of work, or at least that is the perception of many professionals who express resistance.
- Contextual aspects: it's therefore important to adjust the system to the specific scenario: e.g, taking into account the specific characteristics of emergency assistance of patients, designing audio interfaces that allow continuous assistance and transmission of clinical data, etc…
- User aspects: usability (the system must be easy to use and to learn, and it mustn't disturb professionals' work), user profiles and requirements (take into account different levels of users in design, e.g. experts and non-experts in cardiology, cardiologists and emergency coordination centre professionals).
- Technical aspects: take into account technical characteristics of every specific scenario (store-and-forward or real-time transmission mode, suitable compression techniques, recommended resolution for acquired signals, etc.).

From the ethnographic methodology proposed, a study in situ of the organizational, contextual, and user requirements related to the specific emergencies scenario based on mobile teleelectrocardiology has been carried out. The first results have allowed quantifying the maximum time of therapy administration regarding every syndrome, and deciding about the best application location: the mobile units or the hospital.

The further lines of this work will be focused on the translation of these requirements into recommendations for the new designs of e-Health services that include customized devices and GUIs, incorporating usability characteristics in order to be intuitive and easy to learn and use by health-related professionals with different profiles [23]. Furthermore, an evaluation of the designed service will be performed from both technological and user points of view in its specific context.

Acknowledgments. This work was supported by projects TSI2004-04940-C02-01, TSI2007-65219-C02-01, and FIT-350300-2007-56 from Science and Technology Ministry (MCYT) and European Regional Development Fund (ERDF), Pulsers II IP IST-27142 from VI Framework. We acknowledge also the Catalan Association of Cardiology and the Medical Emergency System in Catalonia (Spain), without which collaboration this research would not be possible.

References

1. Jacko, J.A., Sears, A.: The Human-Computer Interaction Handbook. New Jersey Lea Eds (2003)
2. Holzinger, A.: Usability Engineering for Software Developers. Communications of the ACM 48(1), 71–74 (2005)
3. Holzinger, A., Errath, M.: Designing Web-Applications for Mobile Computers: Experiences with Applications to Medicine. In: Stary, C., Stephanidis, C. (eds.) User-Centered Interaction Paradigms for Universal Access in the Information Society. LNCS, vol. 3196, pp. 262–267. Springer, Heidelberg (2004)
4. Saigí, F.: L'evolució de la Història Clínica Compartida. L'opinió dels experts. E-Salut. Revista de Sistemes d'Informació en Salut 1, 27–30 (2006)
5. Monteagudo, J.L., Serrano, L., Salvado, C.H.: La telemedicina:sciencia o ficción? An. Sist. Sanit. Navar. 28(3), 309 (2005)
6. Lanzi, P., Marti, P.: Innovate or preserve: When technology questions co-operative processes. In: Bagnara, S., Pozzi, A., Rizzo, A., Wright, P. (eds.) 11th European Conference on Cognitive Ergonomics ECCE11 - S (2002)
7. Silverstone, R., Hirsch, E. (eds.): Consuming technologies: media and information in domestic spaces, Barcelona (1996)
8. May, C., Mort, M., Williams, T., Mair, F., Gask, L.: Health technology assessment in its local contexts: Studies of telehealthcare. Social Science & Medicine 57(4), 697–710 (2003)
9. Latour, B.: Reassembling the Social: an Introduction to Actor-Network-Theory. Oxford Clarendon (2005)
10. Demiris, G., Tao, D.: An analysis of the specialized literature in the field of telemedicine. J. Telemed. Telecare 11(6), 316–319 (2005)
11. Red europea de investigación de robótica y telemedicina, OTELO. mObile Tele-Ecography using an ultra Light rObot, (Last access: 28/06/07) Available at: http://www.bourges.univ-orleans.fr/otelo/home.htm
12. Hjelm, N.M., Julius, H.W.: Centenary of teleelectrocardiography &telephonocardiology. J. Telemed. Telecare 11(7), 336–339 (2005)
13. Rosenbaum, S., Cockton, G., Coyne, K., Muller, M., Rauch, T.: Focus groups in HCI: wealth of information or waste of resources? In: Conference on Human Factors in Computing Systems (2002)
14. Muñoz, J.: Qualitative data analysis with ATLAS-ti (Last access:28/06/07) Available at: antalya.uab.es/jmunoz/indice/filetrax.asp?File=/jmunoz/cuali/Atlas5.pdf
15. Carroll, J.M.: Making use: Scenario-Based Design of Human-Computer Interactions. MIT Press, Massachussets (2000)
16. Latimer, J.: Distributing knowledge and accountability in medical work. An ethnography of multi-disciplinary interaction. Cardiff School of Social Sciences Working Papers Series, papers 1 – 10 (6), Last access:28/06/07, Available at: http://www.cardiff.ac.uk/socsi/research/publications/workingpapers/paper-6.html.
17. Savage, J.: Ethnography and health care. BMJ 321, 1400–1402 (2000)
18. Harnett, B.: Telemedicine systems and telecommunications. J.Telemed.Telecare 12(1), 4–15 (2006)
19. Hardy, W.C.: QoS measurement and evaluation of telecommunications Quality of Service. IEEE Communications Magazine 40(2), 30–32 (2002)
20. Gemmill, J.: Network basics for telemedicine. J.Telemed.Telecare 11(2), 71–76 (2005)

21. Beyer, H., Holtzblatt, K.: Contextual Design. Morgan Kauffman Publishers, San Francisco (1998)
22. Piqué, M.: Síndrome coronària aguda amb aixecament persistent de l'ST. Paper de la fibrinòlisi en l'estratègia de reperfusion. XIX Congrés de la Societat Catalana de Cardiologia (2007)
23. Lie, M., Sorensen, K.H.: Making technology our own? Domesticating technology into Everyday Life. Scandinavian University Press, Olso (1996)

The Effect of New Standards on the Global Movement Toward Usable Medical Devices

Torsten Gruchmann and Anfried Borgert

Use-Lab GmbH, Buergerkamp 3, D-48565 Steinfurt, Germany
torsten.gruchmann@use-lab.de,
anfried.borgert@use-lab.de

Abstract. Ergonomics and usability do not only pertain to computer work places. More and more they are becoming an important factor in the process of ensuring safety, efficacy, and a return on investment. Especially in the medical field, the increasing life expectancy requires among other things continually new methods of therapy and diagnosis. Triggered by competition and higher functionality of the devices medical products becomes more and more complex. The growing density of achievements in the public health care system such as DRG's, as well as the larger number of various medical devices imply for physicians and nursing staff a heavy burden that can deteriorate therapy quality and safety. The rapidly changing user requirements have a high pressure also on products' development. The critical questions refer equally to an efficient development combined with reduced expenses for development time and costs, as to an optimized use of the products by the customer with reduced process costs and an increased user satisfaction and safety for the welfare of patients and users. In order to guarantee the safe use of medical devices, a usability engineering process (UEP) should be integrated into a medical device development program. This article describes challenges and possible solutions for medical devices' manufacturers seeking to do so.

Keywords: Human Factors, Usability, EN 60601-1-6:2004, Use Error, Safety, Medical Product Development, ROI.

1 Benefits of Usability Engineering

In the past, usability or ergonomics have been predominantly associated with producing user friendly computer workplaces offering adaptable lightning conditions, back-friendly chairs and height adjustable desks. Nowadays, attention to usability and ergonomics extends beyond the workplaces, consumer products, and automotive products also medical devices and aims to develop and apply knowledge and techniques to optimize system performance, whilst protecting the safety, health, and well-being of all individuals involved.

Many companies have identified usability as an important factor for customer relation management and user satisfaction.

A. Holzinger (Ed.): USAB 2007, LNCS 4799, pp. 83–96, 2007.
© Springer-Verlag Berlin Heidelberg 2007

Because high-tech products have already become commodity and the trend is shifting toward simple while pleasing user interfaces, usability is becoming the distinguishing feature.

Some product developers still regard usability engineering as an elective activity that is subject to reduction or elimination when budgets are limited. They do not recognize that reducing or eliminating usability activities may negatively affect the company's future profitability.

While companies whose products offer greater usability can already profit from a positive sales trend, other companies spend millions to litigate and settle claims arising from medical device use errors, If they had the possibility to starting the development process again from the beginning, it is almost certain that they would invest heavily in human factors to identify and remedy usability problems before introducing a given device to market.

Meanwhile several manufacturers integrating usability into their development program have realized that usability engineering can prevent not only future usability problems, but also reduce time and costs by accelerating the time to market and increasing the return of investment.

As Wiklund projects in his analysis [1], opportunities for savings and return include:

Simplified learning tools
Faster time to market
Increased sales
Product liability protection
Reduced customer support
Longer design life

Wiklund estimates a significant return on investment in human factors. Here is a sample analysis of the potential return on investment based on his original projections. Estimating 2.000.000 Euro development costs for a product and usability engineering costs of about 300.000 Euro for two years of development time, the following savings can be calculated:

One time:
Time to market: 187.000 Euro
Design life: 400.000 Euro
Develop learning tools: 25.000 Euro

Annual:
Product liability: 220.000 Euro
Customer support: 100.000 Euro
Produce learning tools: 66.000 Euro
Increased sales: 120.000 Euro

Accordingly, for a period of 5 years, the total savings are 3.142.000 Euro, which means a return on the invested costs approaching 10- to 1.

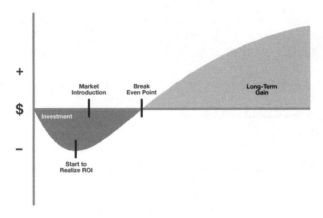

Fig. 1. Timeline of return of investment. Courtesy of Wiklund Research & Design (Concord, Massachusetts/USA).

Parallel to the medical device standards, which are already mandatory, currently there are activities administered by the IEA (International Ergonomics Association) to work on an international standard considering usability in the industrial design process. The IEA EQUID committee aims to develop and manage activities related to the use of ergonomics knowledge and methods in the design process of products, work systems and services. This objective is to be accomplished through the definition of process requirements for the design of ergonomic products, work systems and services, and the establishment of a certification for ergonomics quality in design (EQUID) program.[1]

2   Errors and Error Sources in Medicine

The report of the Institute of Medicine, "To Err Is Human" [2] shocked the public by estimating that 44.000 to 98.000 people die in any given year from medical errors that occur in hospitals. "To Err Is Human" asserts that the problem is not bad people in health care; it is that good people are working in compromised systems that must be made safer. The FDA receives some 100.000 reports a year; more than one third of them involving use errors. The FDA also commented that 44% of medical device recalls were identified as being related to design problems, and that use errors were often linked to device design. In addition more than one third of medical incident reports involve use error, and more than half of device recalls related to design problems involve user interface [3, 4, 5].

---

[1] The International Ergonomics Association or IEA is a federation of about forty individual ergonomics organizations from around the world. The mission of the IEA is to elaborate and advance ergonomics science and practice, and to improve the quality of life by expanding its scope of application and contribution to society For further information see: http://www.iea.cc/

Especially high stress levels can exceed users' abilities to operate a device properly. In those situations the user is acting very intuitively based on his rudimental knowledge, being not aware of all the details learned in introductions and instructions before.

This is why a device that can be used safely in low stress situations could be difficult or dangerous to use in high stress surroundings.

In addition manufacturers of medical products often seek to differentiate themselves from competitors by offering products with a higher degree of functionality not necessarily reflecting changes that address the needs of users. The result of the added functionality (i.e., increased complexity) is that some users might have greater difficulty accessing and correctly using the primary operating functions of the device especially when they are feeling stressed and fatigued.

This reveals another reason for a dysfunctional design given that medical product developers are often unfamiliar with typical working conditions of a hospital surrounding. Specifically, it is unlikely they have worked under typical working conditions or observed the interaction of caregivers with medical products. As a result, developers create a high-tech product in accordance with their personal mental models rather than the requirements of the typical users.

Such lack of information about typical users could reflect in part the lack of opportunities to obtain such information due to the reduction of nursing staff and increased workload necessitated by the push for dramatic savings in health care as well as the Diagnostic Related Groups (DRG's).

The clear implication is that improved design of medical devices and their interfaces can reduce errors.

3 Usability Engineering Process in Medical Product Development

To counter the development of medical products that do not meet the needs of users, efforts are underway for requiring that usability engineering is integrated into the product development process via the international standard EN 60601-1-6:2004, Medical Electrical Equipment – Part 1-6: General Requirements for Safety – Collateral Standard: Usability.

The focus of this standard, which is collateral to the safety standard for electro-medical devices EN 60601-1, Medical Electrical Equipment – Part 1: General Requirements for Safety is to identify use-related hazards and risks through usability engineering methods and risk control. The standard points to the importance of usability engineering not only for patient safety and satisfaction of the user, but also with regards to time and costs of product development.

In 2004 the standard became mandatory for the certification of medical devices and specifies requirements for a process of how to analyze, design, verify and validate the usability, as it relates to safety of medical electrical equipment addressing normal use and use errors but excluding abnormal use (Figure 2).

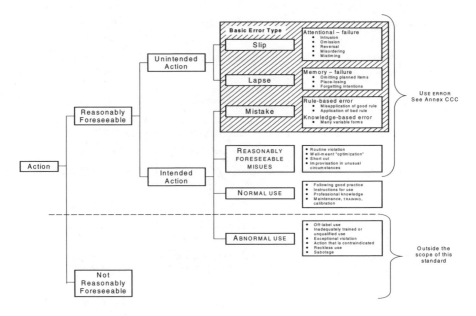

Fig. 2. Summary of the taxonomy of user action [6]

This was three years after AAMI released its human factors process guide AAMI HE74:2001 in the USA. The usability engineering process is intended to achieve reasonable usability, which in turn is intended to minimize use errors and to minimize use associated risks [6].

Manufactures must be familiar with these requirements if they want to prove to regulatory bodies that they have a usability program in place ensuring that they have produced a safe device via a user interface that meets user's needs and averts use errors that could lead to patient injury and harm. That is to say usability is no longer only in the companies' interest in terms of advantages in competition, better time to market or higher return of investment, however now they are legally obliged to fulfill the standard.

On first impression, the standard's advantage might not be clearly visible for those companies that are already integrating usability in their product development process since they already have to address several other standards such as quality- or risk management standards including partly usability activities. One (confusing) result is, that now too many parallel standards are treating usability and through the variety of regulations, it is not obvious how to take an integrated response to the posed requirements and options. The link between risk management, usability engineering and R&D process is shown below (Figure 3).

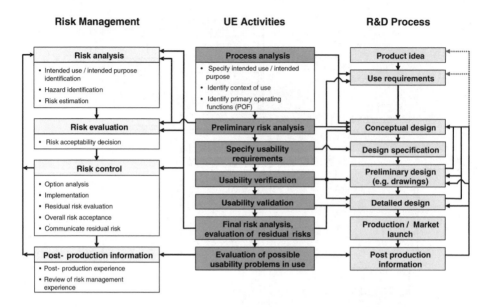

Fig. 3. Link between risk management, usability engineering and R&D process. Courtesy of Use-Lab GmbH (Steinfurt, Germany).

Often companies are focusing exclusively on meeting the "letter of the law" and are not willing to increase their expenditures to obtain a higher degree of usability, even if near-term spending is assured to bring a longer-term return on investment. But it is not only following a standard that leads to a high-quality user interface [7, 8, 9].

Medical device manufacturers whose interest is not only to meet the standard's requirements and make their device more user-friendly beyond all standards and regulations are sometimes asking for the "voice of the nurse", which means the requirements of typical users, before starting the development process. This always leads to a more user centered and intuitive design [10].

More often, however, manufacturers give an engineer the responsibility to access the HF of a design, even if that person is not formally trained in HF or they engage in-house resources for human factors. Some of them also seek external usability consulting services. But the problem is often the same. The requirements of human factors are considered late in the development process and still too often, usability engineering is done toward the end of a product development cycle, before market launch of the product. The manufacturers might feel that they can address the need to integrate human factors activities into their overall product development plans more superficially by only conducting a usability test near the end of the design process.

Usually at that point, the only revisions that can be made are rather superficial and major changes in the product design related to bad usability are infeasible. If usability testing demonstrates that there are fundamental deficits, developers try to make the product as "usable" as possible to verify that the product is acceptably easy and intuitive to use and their main goal is to comply with the standard.

This illustrates why it is important to start the usability engineering process and to involve typical users early in the needs definition phase of a development effort, and in iterative design evaluations.

3.1 Specification of User Requirements

The usability standard asks for the involvement of users ("representative intended operators") only in the validation phase. But the integration of users early on, preferably at the onset of the development process, is an effective and efficient method to improve the usability of medical devices.

Ideally, user requirements and needs should be identified through direct involvement of typical users by gathering data through interviews and observations of intended operators at work. This is important for determining the intended use and purpose of the product as well as the primary operating functions, including frequently used functions and safety-relevant functions.

Fig. 4. Identifying the requirements of the users in their typical environment

It is also important to analyze the context in which the device is used as well as user and task profiles to determine factors that influence its use.

Analysis should reflect on each significant category of users (e.g., nurses, physicians, patients or biomedical engineers) as well as different use environments (e.g., indoor or outdoor use of transport ventilation systems).

This goes as well for the consideration of cultural diversity and the challenges it presents to medical device usability. The question of where to focus the research efforts depends mainly on where the manufacturers have the largest market share and into which countries they wish to expand their business, although this approach involves the risk of not revealing nuances which might impact the use of a device.

For example, in France, CT or MRI contrast agents are prescribed in advance and patients take them with them to their appointment. If contrast agent injector manufacturers whose systems offer multi-dosing (e.g., several consecutive injections from one contrast agent container) were not aware of these differences, it would prevent them from developing customized solutions for the French market.

Although the standard does not directly address whether trials in multiple countries are necessary, it is a concern for many devices.

For example, in the United States, ventilators are typically used by respiratory therapists. However that specialization does not exist in Europe and the training and duties of typical users would therefore be different in those two markets.

Other reasons for the importance of performing data collection and analysis on an international basis are:

- different organizational and infrastructural systems
- different education of caregivers
- different user expectations and user experience
- different regulations as pertains reimbursement of medication or medical treatment from health insurance companies
- different learning behavior
- cultural differences regarding interpretation of colors, icons and symbols

Furthermore it has been observed that products are often carriers of culture [11], since users like to identify themselves and their culture with the product.

Some cultures consider learning to use complex devices a challenge since it gives them a sense of accomplishment whereas others might regard the same device as unnecessarily complex, absorbing too much learning time.

Or medical device consumers in Europe mostly value smaller products, whereas customers in China preferably seem to place a higher value on larger products. An example for that is a laboratory device (Figure 5) which was very successful on the German market thanks to its small and compact size but Chinese customers for example felt that such a small device should cost less [12]. Nevertheless these cultural differences must be rated very critically and individually from one product to another.

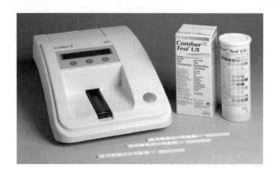 

Fig. 5. Product design for different international markets (Europe vs. China)

After all, conclusions regarding the usability specifications and requirements can be drawn and documented for the preliminary design from the whole collected data about the intended users, the use context and the primary operating functions.

Defining the usability specifications is the moment when the compliance with the standard implicates understanding problems of how to interpret the wording and how to fulfill the standard's requirements.

Unfortunately paragraphs are not clearly enough formulated. E.g. the differences between "primary operating functions" and "operator actions related to the primary operating functions" or "operator-equipment interface requirements for the primary operating functions" [13] are not self-explaining and may lead to confusion.

Furthermore the standard does not clearly enough differentiate between requirements describing qualitative aspects such as "the system shall be easy and intuitive to use", and specifications describing quantitative variables such as "the time to complete a special task is less than 15s" [14].

3.2 Usability Verification of Design Concepts

The usability specifications are now translated into design device ideas. These ideas can be presented in the form of narratives, sketches (Figure 6), animated power point presentations, or rough mock-ups.

To assess the strengths and weaknesses of various concepts, it is advisable to include typical users again in a concept evaluation exercise, such as a series of interviews, focus groups (Figure 7), or similar opinion gathering activities. Such user research usually reveals one preferred concept, or at least the positive characteristics of a few that can be melded into a hybrid design solution.

Verification is also necessary to determine if the design complies with the requirements of the usability specification.

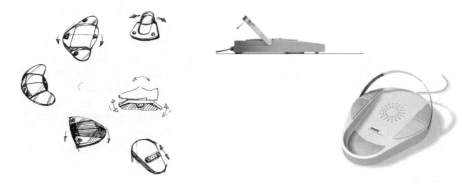

Fig. 6. Conceptual design in scribbles or 3D graphics

Fig. 7. Focus groups or expert reviews for usability verification

The verified concepts are the basis for the design and development of a prototype or functional model. Throughout the verification process, the design requirements will be validated against the current design.

Additional methods used for verification are heuristic analysis and expert reviews. Because it is often hard for designers to recognize usability trouble spots in their own designs an expert review is an important technique. When followed by usability testing, an expert review ensures that a manufacturer's usability investment is optimized.

3.3 Usability Validation

After the follow-up design, another optional verification and the detailed design, the prototype has to be validated for usability.

Unlike clinical trials the usability validation is conducted with a higher degree of concentration on the product. In contrast to clinical trials the validation test can use a prototype of the device and simulate a real-world environment but also including rare events.

The usability validation of the prototype is evaluating whether the actual design meets the specifications or not. Thus, this validation determines if the device can be used safely, with a minimum of risks and errors.

It is very important that the validation of the product or preferably the prototype of the product is conducted with actual potential users and that the use scenarios for the validation include typical situations of actual use (Figure 8).

It is also of great importance that worst case situations are included into the scenarios to ensure that the device can be used safely under stress conditions. Objective measures, such as the time it takes to complete tasks and a user's error rate, provide more-scientific validation of the user interface than questionnaires do.

To identify 80% - 90% of the critical problems in the design that are relevant for the patient's safety the involvement of approximately 10 users can be sufficient [15].

Test participants are usually not introduced into the device rather they have to use and rate the devices' primary operating actions intuitively based on their general knowledge on this topic not considering the lack of familiarity with the specific user interface.

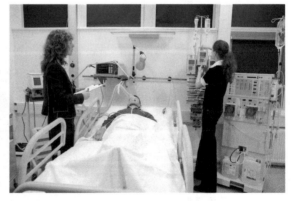

Fig. 8. Typical test scenarios

Each step of the test has to be rated following a special rating scheme and criteria (Figure 9) resulting in a list of design issues and use related hazards and risks. These results are prioritized. Finally an overview of which action failed the specification and which has passed under safety perspectives has to be made [16, 17, 18].

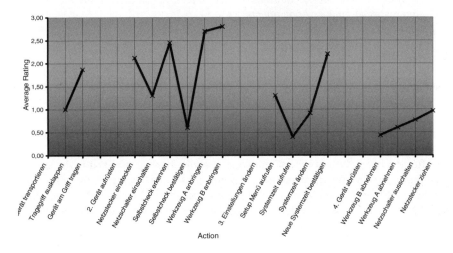

Fig. 9. Rating scheme of a usability validation test

4 Standard's Implementation Problems

It some how seems as if the standard might reduce human factors practices only to safety; and not totally includes efficacy and satisfaction.

Companies often do not see the need or the surplus value of including additional human factors efforts into the overall product development process especially if they have made all necessary efforts to comply with the standard. It is likely that confusion might arise from the difficulty of understanding of when the standard ends and at what point additional human factors engineering starts. Beyond the standard's requirements as it pertains to safety human factors specialists would provide suggestions for design improvements in order to make the product more user-friendly. However, the return of investment of additional human factors efforts is not quite evident to medical device manufacturers from the first outset.

For medical device developers the terms validation and verification might be a bit confusing since these terms cannot be brought clearly enough in line with typical terms of development processes. In several companies the meaning of validation and verification is completely different than these terms are used in the standard. Sometimes verification is used as an acronym for the whole product development; sometimes it is just one step within the whole process done by an external person. Contrary to the standard manufacturers call validation tests e.g. clinical trials or beta tests.

An easy way of recalling the difference between validation and verification is that validation is ensuring "you built the right product" and verification is ensuring "you built the product right." Verification testing is performed to confirm that a product complies with its requirements and specifications, whereas validation testing is performed to confirm that it satisfies stakeholder needs.

Several product developers are against this standard because it requires another set of documentation in addition to all the other risk- and quality standards and because almost all of the standard's requirements have already been included in the FDA and other international guidance documents.

5 Summary

Medical device companies want to get new products to market quickly and efficiently, all while controlling costs and fulfilling the standards necessary for certification.
A commitment to usability in medical product design and development offers enormous benefits, including greater user productivity, more competitive products, lower support costs, and a more efficient development process.

After the integration of the usability engineering process described in the standard, the requirements of the different users should be taken into consideration, factors that can negatively influence the success of the product should be reduced to a minimum, and the product should be safe, easy and intuitive to use. Time and budget necessary for redesign at the end of the development process should be a thing of the past and the probability of a use related hazard or risk for the patient and the user as well should be reduced.

However, a clear understanding of standards' compliance issues provides an important tool for success but it also seems to be a challenging endeavor. To produce design excellence medical product's developers and designers must still exercise high skill and creativity without neglecting the requirements of the usability standard.

In other words, good user interface design is no "cook book" and it requires the involvement of trained and experienced human factors specialists.

Several manufacturers are already investing considerable effort into applying the standard by performing a lot of predevelopment analysis, co-developmental user research and user centered validation tests on their own or with support of human factors specialists.

Nevertheless experience shows that some manufacturers still start implementing the usability engineering process late in the validation phase.

This is not because of their unwillingness; it is rather a question of the misinterpretation of the new standard and its influence to existing standards. Thus it might happen that manufacturers are convinced of having collected all necessary data for the usability engineering file, however the submitted documented is only the user manual. Then they learn about the challenge to implement serious modifications without running out of budget and time, which is always impossible.

Once the importance of such proceedings is known, in their next project almost all manufacturers integrate the usability engineering much earlier in the specification phase understanding the ROI calculation mentioned at the beginning of this article.

Companies that lack a dedicated human factors team will need help to meet the standard as well as remain competitive in a marketplace with increasingly user-friendly technologies.

To enable companies to do so the following ideas for improvement of the usability standard would be helpful:

- simplify the understanding of the standard's wording and make it fit to R&D terminology,
- generalize the standard covering requirements of all countries;
- provide suggestions about how to integrate usability engineering into running product development programs,
- provide suggestions about incorporating usability engineering documentation into existing documentation requirements, e.g. in ISO 13485, ISO 9001, ISO 14971 and
- adapt the standard in a way that it acts as a guidance document that describes how usability and human factors engineering can be efficiently incorporated into the existing design control requirements of ISO 13485.

The upcoming standard EN 62366 focusing on all medical products and not limited to electro medical products might consider some of these deficits. Alternatively the IEA initiative EQUID could be a source for help.

Nevertheless, once understood right the usability engineering process can be easily integrated into the development process and can help to identify and control use related hazards and risks offering a great change to ensure user friendly medical devices contributing greatly to enhancing patient safety [18, 19]. However it's also a challenging endeavor.

References

1. Wiklund, M.E.: Return on Investment in Human Factors. MD&DI Magazine , 48 (August 2005)
2. Institute of Medicine; To Err Is Human: Building a safer Health System. National Academic Press, Washingdon Dc (2000)
3. Carstensen, P.: Human factors and patient safety: FDA role and experience. In: Human Factors, Ergonomics and Patient Safety for Medical Devices; Association for the Advancement of Medical Instrumentation (2005)
4. FDA (Food and Drug Administration), Medical Device Use-Safety: Incorporating Human Factors Engineering into Risk Management, U.S. Department of Health and Human Services, Washington DC (2000)
5. Holzinger, A., Geierhofer, R., Errath, M.: Semantic Information in Medical Information Systems - from Data and Information to Knowledge: Facing Information Overload. In: Proc. of I-MEDIA 2007 and I-SEMANTICS 2007, pp. 323–330 (2007)
6. EN 60601-1-6:2004, Medical Electrical Equipment - Part 1-6: General Requirements for Safety - Collateral Standard: Usability; International Electro technical Commission (2005)
7. Holzinger, A.: Usability Engineering for Software Developers. Communications of the ACM 48(1), 71–74 (2005)
8. Holzinger, A.: Application of Rapid Prototyping to the User Interface Development for a Virtual Medical Campus. IEEE Software 21(1), 92–99 (2004)

9. Memmel, T., Reiterer, H., Holzinger, A.: Agile Methods and Visual Specification in Software Development: a chance to ensure Universal Access. In: Coping with Diversity in Universal Access, Research and Development Methods in Universal Access. LNCS, vol. 4554, pp. 453–462. Springer, Heidelberg (2007)
10. Holzinger, A., Sammer, P., Hofmann-Wellenhof, R.: Mobile Computing in Medicine: Designing Mobile Questionnaires for Elderly and Partially Sighted People. In: Miesenberger, K., Klaus, J., Zagler, W., Karshmer, A.I. (eds.) ICCHP 2006. LNCS, vol. 4061, pp. 732–739. Springer, Heidelberg (2006)
11. Hoelscher, U., Liu, L., Gruchmann, T., Pantiskas, C.: Cross-National and Cross-Cultural Design of Medical Devices; AAMI HE75 (2006)
12. Wiklund, M.E., Gruchmann, T., Barnes, S.: Developing User Requirements for Global Products; MD&DI Magazine (Medical Device & Diagnostics Industry Magazine) (April 2006)
13. Gruchmann, T.: The Usability of a Usability Standard. In: 1st EQUID Workshop, Berlin (2007)
14. Gruchmann, T.: Usability Specification; Gemeinsame Jahrestagung der deutschen, österreichischen und schweizerischen Gesellschaften für Biomedizinische Technik 2006 (2006)
15. Nielsen, J.: Usability Engineering. Academic Press, London (1993)
16. Gruchmann, T.: The Impact of Usability on Patient Safety. BI&T (AAMI, Biomedical Instrumentation & Technology) 39(6) (2005)
17. Gruchmann, T., Hoelscher, U., Liu, L.: Umsetzung von Usability Standards bei der Entwicklung medizinischer Produkte; Useware 2004, VDI-Bericht 1837, VDI-Verlag, pp.175–184 (2004)
18. Hoelscher, U., Gruchmann, T., Liu, L.: Usability of Medical Devices; International Encyclopedia of Ergonomics and Human Factors, 2nd edn. pp. 1717–1722. CRC Press/Taylor & Francis Ltd, Abington (2005)
19. Winters, J.M., Story, M.F.: Medical Instrumentation: Accessibility and Usability Considerations. CRC Press/Taylor & Francis Ltd, Abington (2007)

Usability of Radio-Frequency Devices in Surgery

Dirk Büchel, Thomas Baumann, and Ulrich Matern

University Hospital Tuebingen, 72072 Tuebingen, Germany
Experimental-OR and Ergonomics
dirk.buechel@experimental-op.de

Abstract. This paper describes the usability evaluation conducted at the Experimental-OR of the University Hospital Tuebingen. Subjects of the study were three radio-frequency (RF) devices for surgery. The aim of this study was to detect the differences in the usability among devices which have the same mode of operation and to announce the usability problems occurring with their use. Standard usability test methods for interactive devices have been adapted and verified for the use in the medical field, especially for validating RF devices. With the aid of these methods several usability problems and their consequences to user satisfaction have been be identified. The findings of these tests provide data for the improvement in further development of RF devices.

Keywords: HCI, human machine interface, usability, radio-frequency, surgery.

1 Introduction

As a result of the permanently increasing number of complex and complicated technological applications in medicine and health care, staff members are extremely challenged. This situation is tightened by the lack of usability. For economic reasons, the health care system is relying increasingly on lower-skilled personnel with limited health care education and training [20]. Many of the devices in this field are not intuitive, hard to learn and use [6]. User errors are inevitable. U.S. studies verify that medical errors can be found among the ten most common causes of death in medicine [5, 13]. The Institute of Medicine's (IOM) estimates that more than half of the adverse medical events occurring each year are due to preventable medical errors, causing the death of 44.000 to 98.000 people [13]. In a study of the Experimental-OR in Tuebingen, Montag et al. have researched 1.330 cases from the OR, reported to the BfArM (German federal institute for drugs and medical products) between the years 2000 and 2006, regarding their reasons. 37 % of the reported errors can be traced back to operator errors and, therefore, to a lack of communication between user and the device [17]. Another analysis of adverse events with RF devices in surgery from 1996 to 2002 within the Medizinprodukte-Bereiber-Verordnung [19] (German regulation for users of medical devices) was performed by the BfArM [3]. In this study 82 of the 113 listed adverse events (78% of those with harm to the patient) are not the results of technical or electrical defects but due to possible usability problems. Six adverse events of the category "labeling / attribution mix-up" have been listed (see figure 1).

A. Holzinger (Ed.): USAB 2007, LNCS 4799, pp. 97–104, 2007.

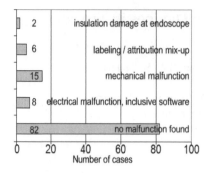
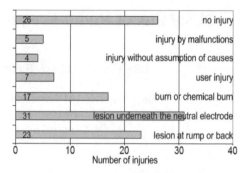

Fig. 1. Malfunctions and injuries with RF devices reported to the BfArM (adapted) [3]

Considering all these findings of the adverse events reported to the BfArM, it is distinguishable that attention to ergonomics and usability requirements is unavoidable for optimising the OR workplace. Findings of two surveys among the attendees of the German Surgical Society annual conferences 2004 and 2005 verify this: 70% of the queried surgeons and 50 % of the OR nurses reported are unable to operate

Fig. 2. RF-device without self- descriptiveness needs further information

the medical devices within the OR safety. More than 40 % of them were involved in situations of potential danger for staff or the patient [16]. The aim of this study was to figure out the usability problems of currently available RF devices.

2   Materials and Methods

2.1   Preliminary Study for Definition of User Goals and Tasks

To generate tasks for the usability test, as similar as possible to the actual tasks within the real OR environment, manufacturer and OR employees have been surveyed with a structured questionnaire about their typical goals and tasks while using RF devices. Three manufacturers, seven OR employees and 29 trainees for technical surgical assistants (Operations-Technische Angestellte (OTAs)) of the University Hospital Tuebingen participated as representatives of the user group. User goals and tasks identified in this survey are shown in figures 3 and 4.

Clustering the findings for the rating of importance of the specific tasks according to manufacturers, OR qualified employees and trainees for technical surgical assistants, differences between users (technical surgical assistants and OR employees) and manufacturers could be determined (for example, see figure 5).

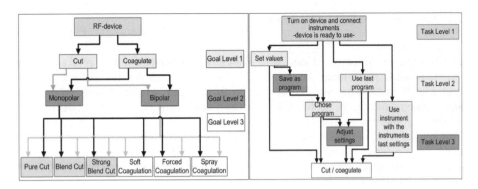

Fig. 3. User Goals   **Fig. 4.** User Tasks

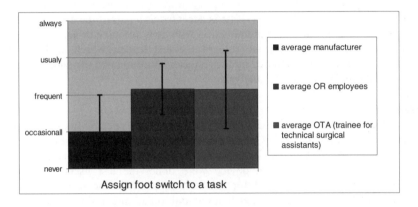

Fig. 5. Adapted example for importance rating of the tasks

2.2 Usability Test

The usability test was performed, conforming to the applicable standards such as ISO 9241-110, ISO 9241-11 and EN 60601-1-6 [6, 8, 9] at the Experimental-OR in Tuebingen. The aim of the usability test was to evaluate: Overall usability, functional design, safety, adequacy for professionals, adequacy for rare use and adequacy for untrained use. Furthermore, usability will be evaluated with regard to: start-up, setting values, choosing programs, saving programs and managing error notifications. Measures for usability are, according to the standards mentioned above, "effectiveness", "efficiency" and "user satisfaction" [8].

2.3 Task for the Test

Considering the user goals and actions (results of the preliminary study) test tasks were selected and transformed into test scenarios, such as setting values and saving a program or reacting on error notifications. Within the usability test the test users have to perform twelve tasks with each of the three devices. The devices are tested in randomised order, to prevent learning effects.

2.4 Methods Used for the Usability Test

Thinking Aloud

The method of "thinking aloud" is a method to collect test users' cognitions and emotions while performing the task [4, 12, 14, 18]. Exercising this method, the test leader urges the test users to loudly express their emotions, expectations, cogitations and problems during the test [2]. In this study test teams consisting of two persons where used to enhance the verbalisation of their cognitions by arising a discussion. This special form of thinking aloud is called "*constructive interaction*". Owing to the much more natural situation, more user comments could be expected [11].

Live Observation and Video Documentation

Live observation was chosen as observation method. The advantages of this method are, that the test leader is able to motivate the test user for *thinking aloud*, that he can act as a communication partner and that, he is able to achieve insights which are likely to be invisible to an outsider, if he is accepted as a part of the events [1, 2]. The disadvantage of interfering the natural action with own initiative or activity of the test leader, was avoided by a passive role of the test leader, which he left only in exceptional cases or when initiated by the test user. Additionally, the entire action was audio-visually recorded for later analysis.

Questionnaires

After completing each task the test users filled out a task specific questionnaire. For this purpose the After-Scenario-Questionnaire (ASQ) [15] was adapted for this study. Additional positive and negative impressions for performing the task had to be stated. When the twelve tasks for one device were done, the test users answered the IsoMetrics questionnaire adapted for this test to evaluate the tested device. The IsoMetrics [10] questionnaire is an evaluation tool to measure usability according ISO 9241-10 [7]. After the end of the entire test the test users filled out a comparison questionnaire. With this deductive method subjective comparative data of the devices have been achieved. For this the test users evaluated the three devices with a ranking from 1 (very bad) to 10 (very good) for the following criteria: overall impression, design, operability and satisfaction.

2.5 Samples

17 test users from the medical sector attended this study. OR employees with daily contact to the objects of investigation were not considered as test users for this study, as they might have preferences or repugnances for the devices they already used. Therefore a surgeon and two OR nurses, with no more contact to RF devices have been selected as representatives for the real users, as well as 12 medical students and two technical surgical assistant trainees. For this user group it was assumable that they have the necessary knowledge, but that they are not prejudiced to one of the devices. Furthermore the sample varies in age and sex.

3 Findings and Analysis

For each device several usability problems were found - overall 25 main problems.

Table 1. Some examples of usability problems

Error	Error analysis	Usability problem
Instrument activation not possible via food switch	Food switch inserted in wrong connector	Connectors not labelled clearly – confusions possible
Scrolling direction for menu chosen wrong	Scrolling direction of jog dial not clear	Turning direction of jog dial not conform to intuitively chosen direction of scrolling.
Confirmation via „Press any Key" not possible	Only buttons at device intended for confirmation but not buttons at instrument's handle	If confirmation call occurs users think they can confirm it via the buttons at the handle

These and other usability problems let to a prolonged procedure time of 10+ minutes. For example, some test users needed, for a typical task ("cut" and "coagulation"– see figure 6) up to 13:46 minutes, whereas other test users treated the patient within only 1:24 minutes. Reasons for this problem are unclear labelled connectors for the food switch as well as missing user information at the device and/or within the manual.

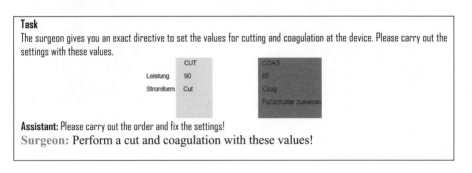

Task
The surgeon gives you an exact directive to set the values for cutting and coagulation at the device. Please carry out the settings with these values.

	CUT	COAG
Leistung	90	85
Stromform	Cut	Coag
		Fußschalter zuweisen

Assistant: Please carry out the order and fix the settings!
Surgeon: Perform a cut and coagulation with these values!

Fig. 6. Adapted test task

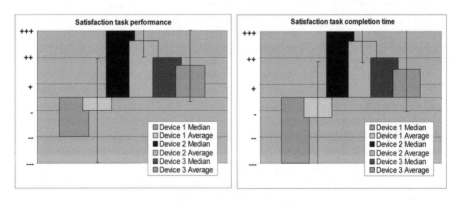

Fig. 7. User satisfaction of the task described in Fig. 6

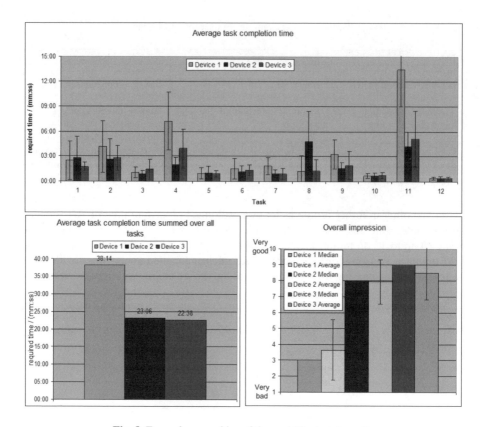

Fig. 8. Exemplary graphics of the usability tests' results

The average time for the completion of the task with device 1 was 7:12 minutes, whereas the average time was only 3:55 and 1:57 with the other devices. Figure 7 show that such problems directly affect user satisfaction.

It is obvious that one of the three tested devices (device 1) severely differs from the others regarding usability. This was not only detectable by the 1.6 times longer average task completion time summed over all tasks, but also by the subjective user impressions (see figure 8).

4 Discussion

This study shows that devices which are already on the market differ drastically regarding their usability. The deficiencies found, lead to longer procedure time and cost intensive failures with a high potential of hazards to patients and employees. Therefore, it seems to be absolutely necessary for surgical departments to implement the knowledge of usability testing into their purchasing process.

Looking at the studies mentioned in chapter 1, it can be assumed, that unsatisfying usability bears a high risk in health care. Therefore, it is necessary to investigate medical devices regarding usability and to implement these findings into the product design. For a comparable usability evaluation, standard usability test and evaluation methods have to be developed.

Different opinions about main tasks to be performed RF devices have been identified between manufacturers and users. Therefore it is necessary that manufacturers focus more detailed on the wishes but also on the behaviour of the users during the procedure.

The study design was adequate to test the usability of RF devices. Especially constructive interaction and video observation provide a lot of high quality and useful data. A comparison of the average results of the comparison questionnaires and the IsoMetrics show a validity of 91%. Vantages of the comparison questionnaire are subjective comparing data date between the devices, whereas the IsoMetrics shaped up as a useful tool for validating user opinion regarding the dialog principles in ISO 9241-10 and 9241-110.

The ASQ can be used to evaluate a special task and draw conclusions about the impact of special usability problems to users' satisfaction. Therefore the combined usage of ASQ and IsoMetrics or other standard questionnaires improve sensibility of the Usability Test. For a final conclusion which test setup and methods of measurement are best for medical systems, further studies have to be performed.

References

1. Bortz, J., Döring, N.: Forschungsmethoden und Evaluation für Sozialwissenschaftler. Springer, Heidelberg (1995)
2. Büchel, D., Spreckelsen v., H.: Usability Testing von interaktiven Fernsehanwendungen in England. TU Ilmenau, Ilmenau (2005)
3. Bundesinstitut für Arzneimittel und Medizinprodukte. Online at, http://www.bfarm.de(last access:2007 -07-30)
4. Carrol, J.M., Mack, R.L.: Learning to use a word processor: By doing, by thinking, and by knowing. In: Thomas, J.C., Schneider, M.L. (eds.) Human factors in computer systems, pp. 13–51. Ablex Publishing, Norwood (1984)
5. Centers for Disease Control and Prevention Deaths: Leading Causes for 2003. National Center for Health Statistics, National Vital Statistic Reports 55(10) 7 (2007)
6. DIN EN 60601-1-6 Medizinische elektrische Geräte; Teil 1-6: Allgemeine Festlegungen für die Sicherheit - Ergänzungsnorm: Gebrauchstauglichkeit, 1. Ausgabe, Beuth-Verlag (2004)
7. DIN EN ISO 9241-10: Ergonomische Anforderungen für Bürotätigkeiten mit Bildschirmgeräten, Teil 10: Grundsätze der Dialoggestaltung - Leitsätze, CEN - Europäisches Komitee für Normung, Brüssel (1998)
8. DIN EN ISO 9241-11: Ergonomische Anforderungen für Bürotätigkeiten mit Bildschirmgeräten, Teil 11: Anforderungen an die Gebrauchstauglichkeit - Leitsätze, CEN - Europäisches Komitee für Normung, Brüssel (1998)
9. DIN EN ISO 9241-110: Ergonomische Anforderungen der Mensch-System-Interaktion, Teil 110: Grundsätze der Dialoggestaltung. CEN - Europäisches Komitee für Normung, Brüssel (2004)
10. Gediga, G., Hamborg, K.-C.: IsoMetrics: Ein Verfahren zur Evaluation von Software nach ISO 9241-10. In: Holling, H., Gediga, G. (eds.) Evaluationsforschung. Göttingen: Hogrefe, pp. 195–234 (1999)

11. Holzinger, A.: Usability Engineering Methods for software developers. Communications of the ACM 48(1), 71–74 (2005)

12. Jøorgenson, A.H.: Using the thinking-aloud method in system development. In: Salvendy, G., Smith, M.J. (eds.) Designing and using human-computer interfaces and knowledge based systems, Proceedings of HCI International 89, Volume 2. Advance in Human Factors/Ergonomics, vol. 12B., pp. 742–750. Elsevier, Amsterdam (1989)

13. Kohn, L.T., Corrigan, J.M., Donaldson, M.S.: To err is human. National Academy Press, Washington, DC (1999)

14. Lewis, C.: Using the Thinking-aloud method in cognitive interface design. IBM Research Report RC 9265. Yorktown Heights, NY: IBM T.J. Watson Research Center (1982)

15. Lewis, J.R: An after-scenario questionnaire for usability studies Psychometric evaluation over three trials. ACM SIGCHI Bulletin 23(4), 79 (1991)

16. Matern, U., Koneczny, S., Scherrer, M., Gerlings, T.: Arbeitsbedingungen und Sicherheit am Arbeitsplatz OP. In: Deutsches Ärzteblatt, vol. 103, pp.B2775–B2780 (2006)

17. Montag, K., RÖlleke, T., Matern, U.: Untersuchung zur Gebrauchstauglichkeit von Medizinprodukten im Anwendungsbereich OP. In: Montag, K. (ed.) Proceedings of ECHE2007 (in print)

18. Nielson, J.: Usability Engineering. Academic Press, Boston (1993)

19. Verordnung über das Errichten, Betreiben und Anwenden von Medizinprodukten (Medizinprodukte-Betreiberverordnung MPBetreibV). Fassung vom 21. August 2002, BGBl. I 3396 (2002)

20. Weinger, M.B.: A Clinician's Perspective on Designing Better Medical Devices. In: Wiklund, M., Wilcox, S. (eds.) Designing Usability into Medical Products, Taylor & Francies, London (2005) (Foreword)

BadIdeas for Usability and Design of Medicine and Healthcare Sensors

Paula Alexandra Silva[1] and Kristof Van Laerhoven[2]

[1] Lancaster University, Computing/Infolab21
Lancaster, LA1 4YR, UK
palexa@gmail.com
[2] Darmstadt University of Technology, Department of Computer Science
64289 Darmstadt, Germany
kristof@mis.tu-darmstadt.de

Abstract. This paper describes the use of a technique to improve design and to develop new uses and improve usability of user interfaces. As a case study, we focus on the design and usability of a research prototype of an actigraph - electronic activity and sleep study device - the Porcupine. The proposed BadIdeas technique was introduced to a team of students who work with this sensor and the existing design was analysed using this technique. The study found that the BadIdeas technique has promising characteristics that might make it an ideal tool in the prototyping and design of usability-critical appliances.

Keywords: Human–Computer Interaction, Usability Engineering, Medical Informatics, Human Factors, Innovation, BadIdeas.

1 Introduction

Wearable biomedical sensors still have many unsolved challenges to tackle. Along with reliability, security or communication infrastructure issues, usability comes first in several studies and projects [4, 6, 7]. When searching the literature for usability studies of wearable sensors, we realise that they are scarce to inexistent; this leads us to believe that this is still a largely unexplored domain. As fields such as ubiquitous computing evolve, usability for wearable sensors will gain importance, as it may be the safest way to facilitate the entrance of sensors into our daily routines.

At present, there are many ongoing projects (e.g., [6] or [4]) exploring eHealth and sensors technology potential. This will improve not only doctors' and care takers' professional work but also, and more important, patients' and citizens' quality of life in general. Certainly these will necessarily cause an impact in sensors usability investigation, promoting the development of this domain.

2 Background

The BadIdeas technique claims to favour divergent and critical thinking [3]. These are two fundamental characteristics to develop new uses and applications for a given technological component (as frequently done in ubiquitous computing), or to assess and improve the quality of solutions (as demanded in the process of any technology based

A. Holzinger (Ed.): USAB 2007, LNCS 4799, pp. 105–112, 2007.

project). The BadIdeas technique asks a certain group of participants to think of bad, impractical, even silly ideas within a specific domain and then uses a series of prompts to explore the domain, while directing participants into transforming their initial wacky thoughts into a good practical idea. As detailed in [3], this technique obeys four phases: i) generation of (bad) ideas; ii) analysis: what, why and when not; iii) turning things around; and iv) making it good. The second step that elicits a series of prompts (see Table 1) is crucial, as it allows and induces us to a deep and elaborated analysis of the problem domain. Prompt questions for BadIdeas (adapted from [3])

The bad	The good
1. What is bad about this idea?	1. What is good about this idea?
2. Why is this a bad thing?	2. Why is this a good thing?
3. Anything sharing this feature that is not bad?	3. Anything sharing this feature that is not good?
4. If so what is the difference?	4. If so what is the difference?
5. Is there a different context where this would be good?	5. Is there a different context where this would be bad?

By aiming at bad ideas, this technique reduces subjects' personal attachment towards their 'good ideas' and fosters 'out-of-the-box' thinking, thereby bringing out new ideas. Additionally, by stimulating critique and interrogation it largely raises a subject understanding of almost any domain.

Our ongoing study of the BadIdeas technique shows that we can use this method not only to explore a general problem, as we ought to do when thinking about new uses for a certain technological component (see 3.2.), but also to solve a particular issue of a problem, such as a usability flaw (see 3.3). Moreover, BadIdeas have the advantage of being potentially used by anybody, facilitating various types of user, such as doctors, care-givers or patients, to be involved in the development process.

The Porcupine (Fig. 1) [8] is a wearable sensing platform that monitors motion (acceleration and posture), ambient light, and skin temperature. It is specifically designed to operate over long periods of time (from days to weeks), and to memorise the user's data locally so that it can be uploaded at a later stage and be analysed by a physician (or in general, a domain expert).

It is currently used in three healthcare-related projects: A decisive project from the early design stages of the Porcupine project that involved the analysis of activity levels of bipolar patients over days. Psychiatrists used and use this type of long-term actigraphy to detect changes in the mood of the patient and predict phases of depression or manic behaviour. Actigraphs are usually worn like a wrist watch by the patients.

A similar project focuses on monitoring activities of elderly users, to automatically asses their independence and detect whether they are still fully able to perform tasks (e.g. household activities). A third healthcare-related application focuses solely on sleep patterns and detection of sleep phases during the user's sleep. This is again based on the observation that certain sleep patterns have different activities and activity intensities associated with them.

In a philosophy of participatory design [2], current users, activity researchers and stakeholders from the previous projects, identified areas where improvements were desirable in terms of usability. One problem is the scalability of deploying the Porcupines in a large trial involving dozens of patients: The feedback from psychiatrists in

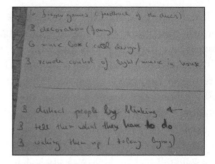

Fig. 1. The current version of the Porcupine prototype

Fig. 2. Subject A ideas; Good and bad ideas signed by *G* and *B*, respectively

this matter resulted in the replacement of the memory chips by a memory card slot: patients would just replace the card and send it over, while the worn Porcupine would keep on logging. A second problem was the need to change the battery: this would include stocking up on disposable batteries and losing time while changing them. The subsequent Porcupine version used a battery that is rechargeable via the existing USB connector. Experiments after the first versions similarly resulted in buttons and LEDs being cut down in the next version as well.

We analysed the Porcupine performance under five possible usability metrics:

Adaptability – The applications in which the Porcupine sensor unit is used require minimum adaptability. They are worn continuously by the same person, at the same place on the body. Given these, it can be used in almost any context or by any user.

Comfort of use –The Porcupine is worn continuously by users, in a location on the body that is subject to motion. A common place so far was the user's dominant wrist, which tends to be indicative of actions taken. The disadvantage of using this location, however, is that it is in plain view, and that its weight and size are critical.

Reliability –The data is only valid under the assumption that the device is not taken off, unless its data needs to be uploaded, and is thus linked with comfort.

Robustness –The electronics are coated by epoxy, therefore robust to impacts or drops. However, the wrist location does expose the device to splashes of water, which can be problematic if they end up in the battery or USB connectors.

Ease of use –The Porcupine's use is straightforward: most of the time it needs to be worn without maintaining any user interface. Only when configuring or uploading the data from memory to a host computer via USB, some user interaction is required.

From the above analysis, the fact that the Porcupine is worn on a wrist strap still presents usability issues, so we looked for alternatives, by applying the BadIdeas.

3   Experiment

The experiment involved seven post-graduate students from Darmstadt University of Technology; the first author moderated the study as a BadIdeas facilitator. Apart from the BadIdeas facilitator, all participants had very strong engineering backgrounds and were familiar with the sensor. Ages varied from 24 to 32.

The experiment occurred in three stages: first, a simple illustrative session with the purpose of exploring new uses for Porcupines inside dice; second and focal part, to identify alternative uses for the wrist-worn Porcupine; and third, a post-hoc analysis of the results obtained with the second. The first two phases were carried out during a meeting of roughly 60 minutes. After explaining the experiment goal and task, the two BadIdeas exercises took place, both following the same organization: brief presentation, individual bad ideas writing, verbal sharing ideas with the group and solving in group of the BadIdeas.

An informal environment and structure was preferred as these affect and favour idea generation [1]. This was present in the friendly atmosphere amongst the group and the way information was conveyed and gathered. No formal printed text was used for the brief, neither for the participants to write their ideas; verbal communication and regular blank sheets of paper were used instead (see Fig. 2).

We intentionally omitted the analytic parts of the technique, typical of the second phase, as our ongoing experience shows that people tend to get inhibited, too attached, or even stuck to the structure of the technique. We wanted participants to place all their effort into the problem understanding and generation of solutions, so we removed any tempting rigid structures. No restrictions were given respecting the number of ideas or levels of badness. There were also no domain restrictions.

3.2   Part I – Porcupine in a Die

In the first part of the meeting, we wanted to demonstrate to the   participants how the BadIdeas technique works and what kind of results it allows one to achieve. Accordingly, we used a playful example and aimed at finding new uses for an existing 'smart die' which has the Porcupine's components embedded in, as detailed in [9].

Roughly the problem brief was: *"Suggest us new contexts of use for the Porcupine die that does not include using it in games"* (its most obvious application). After explaining this apparently simple task and goal, students still needed "a kick-start" from the facilitator that needed to provide some silly, bad and, apparently unfeasible, examples (e.g. *"what if you could eat them?"*, or *"make them out of water"*) in order to inspire the participants. As can be seen in Table 1, each participant provided on average 2.88 ideas. But, more good than bad ideas, as they were asked to. This evidence is not only observed by us, but also confirmed by the participants, who when asked to go through their written ideas and sign the good and bad ones (see Fig. 2) realised that in fact they wrote more good than bad ideas, even though they were purposely aiming at bad ones. In total, and after eliminating the repetitions, thirteen ideas were supplied. From these, the majority selected the "to use as ice cubes" as the bad idea to transform into good. Embedding the unit in an ice cube was bad because the melted water would damage the circuits of the sensor. But, once one of the participants said the ice was good to cool down drinks and cocktails, someone else said it would be great if that ice cube could tell us when to stop drinking, in case we were driving. Then, someone suggested the sensor was coated with some special material to allow its integration in an ice cube. And, without notice, seconds after, the students were already discussing marketing measures to promote the smart ice cube.

Table 1. Gathered ideas: good, bad, average and total per participant

Subject	Good Ideas PI	Bad Ideas PI	Total Ideas PI	Total Ideas PII
A	2	2	4	3
B	2	0	2	0
C	2	1	3	4
D	2	0	2	3
E	0	1	1	0
F	1	0	1	1
G	1	2	3	0
H	5	2	7	3
		Average:	**2,88**	**1,75**

3.3 Part II – Porcupine Wrist Worn Sensor

The second part of the experiment had a clear usability problem to be solved in respect to a well-defined and stable sensor. Based on the feedback of its users (from patients to doctors), a yet unsolved usability issue of the Porcupine was related to the fact, that it was worn on a wrist strap, which was not always appreciated by the patients. So alternatives were asked from the participants: *"Think of a different way of wearing/using the Porcupine. Exclude its current wrist-worn possibility from your suggestions"*. On average, the participants had now only 1.75 ideas. The first session turned out positive, as participants were already aware of their persistent tendency to switch into generating good ideas instead of pure bad ideas, as suggested and aimed for; no good ideas were written, but also some of the participants were not able to have any ideas.

3.4 Part III – Post-Hoc Analysis and Results

After gathering all the ideas and reorganising them by removing the repeated ones, we rated them from five to one according to their inventive potential; five represented the most challenging and one the less challenging ideas.

Ideas	Rating
Hide them in unexpected places to annoy people (in sofa: frighten/surprise people, in door: trigger bucket of water, in phone: phone turns off when picked up)	5
To use in a wig	5
Recording and erasing data	5
Ring to punch	4
Records for only one second	4
Running out of power	4
Ball and chain for ankle	3
Distract people by blinking	3
To display cryptographic text	2
Annoying	2
Waking people up (too long lying)	2
Blink randomly	1
Tell people what to do	1

Accordingly, the most unexpected, odd and surprising ideas were considered as more challenging and the most boring and obvious as the least challenging. The three most inventive ideas were further analysed. We reorganised the experiment's participants into three groups and redistributed the selected ideas to the groups, ensuring that the idea would not be given for transformation to its creator. An email was sent to all participants of the experiment where each group/student was given one idea to solve: i) A Porcupine that records and erases data; ii) A Porcupine to use in a wig; and iii) A Porcupine to hide in unexpected places (sofa, door, phone) to annoy people. To each of the ideas, the students were asked to answer two questions that somehow summarized the remaining three phases of the BadIdeas:

i) *What is/are the property/ies that make it that bad? and ii) Is there a context/object/situation in which that property or one of those properties is not bad?*

Finally, they were asked to aim at (re)designing the porcupine in order to integrate that feature, by explaining how (by writing), by making sketches or by developing simulations. They were also advised to keep as many records of things as possible.

Six of the participants replied and answered their challenges. From our point-of-view, they succeeded as they were all able to complete the exercise and find new uses for the Porcupine. But, surprisingly, they forgot the healthcare and elderly domain and

Table 2. Results of post-hoc session

Idea	Subject	What are the properties that make it bad?	Is there a context/object/situation in which the property(ies) is not bad?
i)	A	Monitoring might invade **privacy**, **erasing** important data	Erasing is good for **solving privacy** issues (e.g., clearing up criminal cases); **Porcupine** can **decide** what to delete/record
	H	**Erasing** data that is valuable	Erasing data when it is not valuable anymore is good, the **erasing** would then happen **after** a data **abstraction** phase
ii)	C	**Big** and **heavy** to use on the head, **uncomfortable**	The location is good **to detect** certain **head-related actions**, which is great for e.g., pilots, drivers, policeman
	E	/	/
	G	Wig is an unusual location, **user acceptance**	Wig-worn unit could **warn** if the wig is shifting out of place, it might also be good as a **party-gadget**
iii)	B	/	/
	D	**Embedding** it makes it **hard to use** them	Hiding them in objects might be good **for users that are hostile towards the unit**, such as prisoners that need to be monitored
	F	To annoy is **not a valid purpose**	For **security**, e.g., **hiding and annoying burglars** instead of the user, or **espionage**: **monitoring people** unaware of the unit

went to completely different areas when exploring and solving their bad ideas. From this, we learned that when using the BadIdeas method, we always need to ensure that we remind the participants to go back to the problem domain area, when turning things around and making them good. In the next sections we report the participants' answers also synthesised in Table 2.

Porcupine that records and erases data. As bad features the participants stated that a Porcupine would be problematic for privacy reasons and because it could eliminate important data. When analysed more carefully those properties appeared good, if referring to the elimination of criminal cases or if the Porcupine could decide what to delete. For this, a new algorithm was needed, to allow this evolution. In this case, the bad idea was not related with a different place/object of use, so participants did not provide new contexts of use, but they found a way of saving memory use.

Porcupine to use in a wig. Concerning the use of the Porcupine in a wig, participants indicated that it was big, heavy, uncomfortable and even embarrassing to use. Nonetheless, they affirmed that location would change to an advantage if it was used by some specific type of public, e.g. policemen, drivers or pilots.

Porcupine to hide in unexpected places (sofa, door, phone) to annoy people. Finally, participants envisioned possible uses and advantages in the fact of having Porcupines hiding everywhere. The obvious problems were its lack of purpose and also the difficulty of embedding them in the potential hiding places. Once these were resolved, the hidden Porcupine appeared as positive to monitor people that were not aware or were hostile towards the unit, such as burglars and prisoners or patients.

4   Conclusion and Future Work

The experiment was experienced as funny and productive. Although vague, these adjectives transmit a positive opinion about the technique. Participants also showed interest about results in the weeks after the study, and kept on mentioning the bad ideas they had. This makes us believe that participants would use the technique again.

Considering the whole process, the use of constraints was important to force the participants to think out of the box, in the first example to keep them away of the most obvious application, and the second, to keep them away of the traditional wrist-worn Porcupine. At the end, novel alternatives were achieved.

Besides being not as natural or obvious, it was still possible to communicate and continue the BadIdeas process by email. Although, we probably lost the synergetic energy generated by live group discussions. To answer this and other potential open questions, further experiments are needed and should be implemented.

Finally, as referred in other papers [10] and concerning the domain of this paper, as well as other unexplored or relatively new domains, it is crucial to keep the necessity of designing for purpose present. To these domains, guidelines lists are harder to apply and follow and probably should serve more as questioning and thinking starting points than as rules of thumb. So a good practice for evaluation and (re)design is to keep the purpose of the design present and as the main goal to serve.

One of the most important phases in the (re)design of any new appliance or user interface is the one where its core ideas are generated. This paper argues for a novel technique, named BadIdeas, to generate new design ideas and analyse existing products. The generation of ideas is then followed by a process of understanding and modification of ideas. We have applied this technique in a monitored session for the analysis of a research prototype used for actigraphy in several healthcare-related applications, with a team of students that work on and with the prototype.

The two purposes we had in mind at the beginning of this study were fulfilled as the students were able to conclude the exercise successfully by finding novel user for the Porcupine dice and to identify design alternatives to the wrist-worn Porcupine. In the study, we found that participants require some adjustment to the generation of bad or silly ideas, rather than the usual goal of coming up with good ideas in a certain direction. Especially helpful to this end was the 'warming up' with a mock-up problem, such as the dice in our study, as well as exemplary bad ideas from the facilitator or moderator. Once this was done, the remainder of the study proved to be intuitive, with the participants effortlessly coming up with bad ideas.

One important aspect that stood out was the need for reminding participants to return back to the original area where a usability issue was targeted – as this was not explicitly done in our study, the resulting ideas were often deviating a lot from the intended application.

Acknowledgements. We are grateful to the students involved in the experiment that happily shared their time and actively participated during and after this paper's study.

References

1. Csikszentmihalyi, M.: Creativity: Flow and the Psychology of Discovery and Invention. Harper Perennial, New York (1996)
2. Dix, A., Finlay, J., Abowd, G D, Beale, R.: Human Computer Interaction. Prentice-Hall, Englewood Cliffs
3. Dix, A., Ormerod, T., Twidale, M., Sas, C., da Silva, P.G.: Why bad ideas are a good idea. In: Proceedings of HCIEd.2006-1 inventivity, Ireland, pp. 23–24 (March 2006)
4. FOBIS - Nordic Foresight Biomedical Sensors (26/6/2007), http://www.sintef.no/fobis/
5. IEEE/EMBS Technical Committee on Wearable Biomedical Sensors and Systems, http://embs.gsbme.unsw.edu.au/ (accessed on the 26/6/2007)
6. Korhonen, I.: IEEE EMBS Techical Committee on Wearable Biomedical Sensors and Systems. Pervasive Health Technologies. VTT Information Technology http://embs.gsbme.unsw.edu.au/wbss/docs/rep_ambience05.pdf (accessed on 26/6/2007)
7. Svagård, I.S.: FOBIS. Foresight Biomedical Sensors. WORKSHOP 1. Park Inn Copenhagen Airport SINTEF ICT(6th October 2005) (accessed on 26/6/2007) http://www.sintef.no/project/FOBIS/WS1/Background%20for%20the%20foresight%20FOBIS_Svagard.pdf
8. Van Laerhoven, K., Gellersen, H.-W., Malliaris, Y.: Long-Term Activity Monitoring with a Wearable Sensor Node. In: Proc. of the third International Workshop on Body Sensor Nodes, pp. 171–174. IEEE Press, Los Alamitos (2006)
9. Van Laerhoven, K., Gellersen, H-W.: Fair Dice: A Tilt and Motion-Aware Cube with a Conscience. In: Proc. of IWSAWC 2006, IEEE Press, Lisbon (2006)
10. Silva, P.A., Dix, A.: Usability: Not as we know it! British HCI, September, Lancaster, UK. 2007, vol. II, pp.103-106 (2007)

Physicians' and Nurses' Documenting Practices and Implications for Electronic Patient Record Design

Elke Reuss[1], Rahel Naef[2], Rochus Keller[3], and Moira Norrie[1]

[1] Institute for Information Systems, Department of Computer Science,
Swiss Federal Institute of Technology ETH, Haldeneggsteig 4, 8092 Zurich, Switzerland
[2] Departement of Nursing and Social Services, Kantonsspital Luzern,
Spitalstrasse, 6016 Lucerne, Switzerland
[3] Datonal AG, Medical Information Systems Research,
Lettenstrasse 7, 6343 Rotkreuz, Switzerland
reuss@inf.ethz.ch,
rahel.naef@ksl.ch,
rkeller@datonal.com,
norrie@ inf.ethz.ch

Abstract. Data entry is still one of the most challenging bottlenecks of electronic patient record (EPR) use. Today's systems obviously neglect the professionals' practices due to the rigidity of the electronic forms. For example, adding annotations is not supported in a suitable way. The aim of our study was to understand the physicians' and nurses' practices when recording information in a patient record. Based on the findings, we outline use cases under which all identified practices can be subsumed. A system that implements these use cases should show a considerably improved acceptance among professionals.

Keywords: EPR data entry; User-centred system design; Usability.

1 Introduction

In a hospital, the patient record serves as a central repository of information. It represents a pivotal platform for different groups of professionals and is an important management and control tool, by which all involved parties coordinate their activities. Because of the high degree of work division and the large number of involved persons, it is an essential prerequisite that all relevant information is documented in the record. Primarily, nurses and physicians are responsible for maintaining the patient records.

Today's electronic patient record (EPR) systems adopt the paper paradigm virtually as it stands and hence extensively ignore human factors [1, 2]. There are a variety of ways for users to overcome the fixed structure of paper-based forms, for example by adding annotations. However, the rigidity of electronic forms and the limited facilities to present information might restrict users when it comes to data entry in the EPR. Although some systems already allow the user to add annotations to clinical images such as X-rays, this does not really seem to solve the problem.

Heath and Luff [3] investigated the contents of medical records in general practices and found that doctors need to be able to make a variety of marks and annotations to

A. Holzinger (Ed.): USAB 2007, LNCS 4799, pp. 113–118, 2007.

documents. Bringay et al. [4] had similar findings in a paediatric hospital unit and in addition realized that professionals also communicate messages by adding post-its to the record. They implemented a solution which provides textual annotations that are edited and displayed in separate windows and which are linked to other documents of the EPR. Their prototype additionally supports sending messages.

Only a few references can be found in the literature about the documenting practices of healthcare professionals in hospitals [4, 5]. These investigations focused on physicians and participants were merely observed and not interviewed. Nurses manage a substantial part of the patient record including the chart, all prescriptions and the care plan so it is essential to analyse their practices as well. Furthermore, understanding the user requirements is a pivotal step in user-centred design [6]. Hence, we concluded that further analysis was needed. To investigate the documenting practices of both physicians and nurses in hospitals, we addressed the following questions: (a) What documenting practices actually exist, and (b) Do professionals always find an adequate means to record information? We start with a brief description of our study design and then present an analysis of results and the implications for EPR design that should reflect the professionals' routines and needs when entering data. We then discuss the findings and present concluding remarks.

2 Study Design

Structured interviews were carried out with 20 physicians and 12 nurses at three different Swiss hospitals in units of internal medicine, surgery and geriatrics. The questions dealt with the professionals' practices when recording information in the patient record. In particular, we asked participants, what kind of different documenting practices they use (such as adding marginal notes and sketches, or encircling text), whether they were always able to adequately record information and, if not, what strategies of evasion they use.

Each interview was tape recorded and, for the purpose of analysis, later on transcribed to a detailed report. Additionally, we asked participants to give us typical examples of forms that illustrate their practices.

3 Results

We first describe the results of the analysis and then outline the implications for the EPR design.

3.1 Documenting Practices

The majority of the physicians (n=13) and nurses (n=8) expressed the need for flexible ways of capturing information in the patient record, such as facilities to draw sketches, to make annotations or to individually highlight relevant information. Physicians add annotations to comment on sketches, or they sketch a finding, for instance the lymph node status, to supplement a textual description. Nurses often make notes to the medication list, other prescriptions or the chart. For instance, they add marginal comments if the administering of a drug does not fit the given time pattern on the paper form (see figure 1), or when the drug is located at the patient room because the patient takes it themself.

Furthermore, professionals frequently record questions that should be answered by colleagues. Nurses quite often have questions for the physicians. If the question is addressed to the attending physician of the ward, it is written on the corresponding record form (question to the physician). This form is consulted the next time that the nurse meets the physician, which is usually during the daily round or in the afternoon meeting, to clarify the questions. But some questions concern professionals that are rather infrequently on the ward, for example the anaesthesiologist. Hence, nurses attach a post-it to the anaesthetic chart to ask, for example, something about the pre-medications. In that way, it is ensured that the anaesthesiologist will notice the question the next time that they access the record.

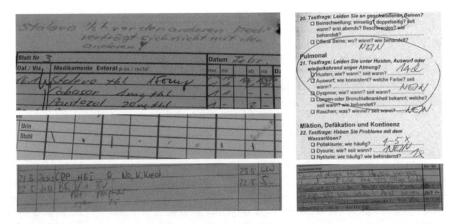

Fig. 1. Examples of annotations and marks on paper-based patient record forms: (on the left, from top to bottom) annotation about an incompatibility of drugs, annotation to an elimination pattern and to a lab test order; (on the right, from top to bottom) highlighting on a patient status form and items that are ticked off using coloured marks.

Highlighting facilities were rated to be very important as marks improve the visual orientation and hence ease the task of finding relevant information. Some physicians argued that bold text could easily be overlooked, whereas coloured marks or encircled areas would attract attention. Nurses usually tick off an item in the record by marking it with a colour.

Most of the physicians (n=15) and nurses (n=10) were not always able to adequately record information. This happened due to the absence of suitable forms and form elements such as fields, areas etc., or because elements were too small. Users either omitted such information or wrote it in generic fields such as the progress notes or the notes to the next shift. For example, physicians did not record personal notes or a sketch indicating the exact location of the patient's pain.

Or they wrote information about social aspects or results from a discussion with the general practitioner in the progress notes. Nurses frequently used the notes to the next shift when they were uncertain where to record information. For instance, one nurse stated that they used the notes to the next shift in order to enter information about the patient's fear of surgery. Participants judged all these strategies to be critical, as information might get lost or its retrieval could be hindered.

3.2 Implications for EPR Design

Based on our findings, we outline in the following the identified use cases (UC)s. When information is entered in the patient record, there seem to exist five additional UCs together with the basic UC (see table 1). The basic UC applies when professionals access one of the specified EPR forms and fill it in entirely or partially. If none of the given forms or elements fits, users should be able to either add an annotation or an independent entity.

Table 1. Documenting practices and identified use cases (UC)

UC ID	Name (brief description)
UC01	Add annotation (users annotate an EPR form)
UC02	Insert entity (users insert an entity into the record on document level)
UC03	Feed question (users feed a question)
UC04	Add note (users feed a note)
UC05	Add mark (users add marks that are overlayed to the form and its contents, respectively)

Annotating should be provided by a shortcut and a context menu. This functionality opens a floating text/sketch frame that is automatically placed by the system to an empty area on the form. Additionally, the user should be able to move the frame to fix it in the desired position (as is the case on paper as shown in figure 1) and to anchor it in order to define to which element the annotation belongs. If users would like to record information separately - such as personal notes or notes about a discussion with the general practitioner - the system should offer them the possibility to enter such information to an independent entity on the document level. Each entity should be given a suitable name by the user that will be used for retrieval. Furthermore, the system should provide a facility to easily file the entity to one of the record's folders.

Feeding a question constitutes another UC. The user opens by means of a shortcut or context menu a dialog window, in which the question, the receiver's role and, if necessary, a deadline is entered. By default, the receiver is set to the role which is most often asked by the questioner. For example, nurses most often ask a question to the attending physician on the ward. This question and answer process can be realized with a workflow management system. The task list should be managed on the level of the role, not the person itself. If a task was assigned directly to a person, the assignment would still apply even if the context had changed, thus not tapping the full potential of dynamic assignment.

It is therefore necessary for one part to do a dynamic specialisation of a role, for example a specialisation of the general role "attending physician" to "attending physician on the ward X", and for the other part to dynamically infer the assignment of a person to this specialized role, for example, assignment of the person to the specialized role "attending physician on ward X" for the concrete ward at a given date/time.

If the task list were managed on the level of the specialized role, the task responsibility would go along with the assignment. In that way, the handling of questions is independent of personnel changes due to changes of shift, cases of emergency or holidays. After completion, the question is dispatched and receivers are given notice. In addition, questions and answers are listed in a separate question/answer form to provide users with an overview and a sorting facility. A similar use case applies, when users add a note that should be essentially perceived by one or several colleagues. The user captures the note and defines, if required, one or more roles that are notified. Additionally, the notes are journalized to the notes form in the EPR.

Finally, the system should allow the user to mark information in the record such as text, sketches and images with colour, to encircle something or to add an arrow. This overlayed information should be shown by default only to the author. But the system should also offer a functionality that switches on the overlay of a colleague.

4 Discussion

Our study demonstrated a variety of documenting practices that should be reflected in future EPR system design. Other works have already shown that physicians need to annotate, to mark and to add notes [4, 5]. As our investigation revealed, these features should be supported for nurses as well. In addition, healthcare professionals in hospital also need a facility to add entities and to feed questions.

As Cimino et al. [6] showed, users tend to type less than they intended to record when using a coded data entry system. This seems to confirm our finding that the rigidity of today's EPR forms implicates the risk of misfiling or even omitting important data. To avoid such problems, the system should allow the user to capture all information adequately. Users either add information to forms or they would like to capture data separately. Hence, insertion of annotations in the form of text or sketches and also independent entities should provide for the required flexibility of the system.

In contrast to Bringay et al. [4], we suggest to implement the annotation functionality so that the EPR form and the annotation are shown at the same time. We assess this to be important for the system's usability. On the one hand, we believe that they should be displayed together because both pieces of information are connected. On the other hand, healthcare professionals prefer to have quick access to required information and even admit to not consulting information if there is an additional effort required to access it [2]. This means that if users first have to click on a link – as proposed by Bringay et al. – there is a risk that users will miss important facts. Adding annotations to images is already offered by some commercial EPR systems (see for example www.medcomsoft.com), where the image and its annotation are displayed simultaneously. But it is not clear whether these solutions show sufficient usability, i.e. if time and effort would be too high to capture data. We reckon that further user studies with prototypes are needed to find a usable solution.

When users need to add an independent entity, this suggests that the EPR has not been designed properly due to poor requirements analysis. But on the other hand, it is quite difficult to completely gather all requirements in such a complex field of application. Hence, it is assumed that the system should provide the insertion of entities to prevent misfiling or loss of information.

Messaging, i.e. adding notes, has already been realized [4]. But when messages are addressed to a specific person, this might cause problems because personnel changes are common practice in a hospital. Therefore, we propose to use a role-based concept to manage messages and questions. Chandramouli [8] describes in his work a concept that allows privileges to be assigned dynamically to roles considering contextual information. We suggest the use of a similar concept to realize the dynamic responsibility assignment as described in this paper.

5 Conclusion

As our study shows, healthcare professionals in hospitals use documenting practices that are not yet reflected in commercial and prototype EPR systems. In a next step, we aim to develop a prototype that implements the identified UC's in order to evaluate the suggested concepts with user studies. This will potentially provide further input to a best possible ergonomic EPR design in the demanding hospital environment.

References

1. Nygren, E.: From Paper to Computer Screen. Human Information Processing and Interfaces to Patient Data. In: Proceedings IMIA WG6 (1997)
2. Reuss, E.: Visualisierungs- und Navigationskonzepte für das computerbasierte Patientendossier im Spital, Dissertationsschrift, ETH Zürich (2004)
3. Heath, C., Luff, P.: Technology in action, p. 237. Cambridge University Press, Cambridge (2000)
4. Bringay, S., Barry, C., Charlet, J.: A specific tool of annotations for the electronic health record. In: International Workshop of Annotations for Collaboration, Paris (2005)
5. Bricon-Souf, N., Bringay, S., Hamek, S., Anceaux, F., Barry, C., Charlet, J.: Informal notes to support the asynchronous collaborative activities. Int. J. Med. Inf. (2007)
6. Cimino, J., Patel, V., Kushniruk, W.: Studying the Human–Computer–Terminology Interface. JAMIA 8(2), 163–173 (2001)
7. http://www.medcomsoft.com, accessed on 2007/07/29
8. Chandramouli, R.: A framework for multiple authorization types in a healthcare application system. In: Computer Security Applications Conference Proceedings, pp. 137–148 (2001)

Design and Development of a Mobile Medical Application for the Management of Chronic Diseases: Methods of Improved Data Input for Older People

Alexander Nischelwitzer[1], Klaus Pintoffl[1], Christina Loss[1], and Andreas Holzinger[2]

[1] University of Applied Sciences FH JOANNEUM, A-8020 Graz, Austria
School of Information Management
Digital Media Technologies Laboratory
alexander.nischelwitzer@fh-joanneum.at
[2] Medical University Graz, 8036 Graz, Austria
Institute for Medical Informatics, Statistics & Documentation (IMI)
Research Unit HCI4MED
andreas.holzinger@meduni-graz.at

Abstract. The application of already widely available mobile phones would provide medical professionals with an additional possibility of outpatient care, which may reduce medical cost at the same time as providing support to elderly people suffering from chronic diseases, such as diabetes and hypertension. To facilitate this, it is essential to apply user centered development methodologies to counteract opposition due to the technological inexperience of the elderly. In this paper, we describe the design and development of a mobile medical application to help deal with chronic diseases in a home environment. The application is called *MyMobileDoc* and includes a graphical user interface for patients to enter medical data including blood pressure; blood glucose levels; etc. Although we are aware that sensor devices are being marketed to measure this data, subjective data, for example, pain intensity and contentment level must be manually input. We included 15 patients aged from 36 to 84 (mean age 65) and 4 nurses aged from 20 to 33 (mean age 26) in several of our user centered development cycles. We concentrated on three different possibilities for number input. We demonstrate the function of this interface, its applicability and the importance of patient education. Our aim is to stimulate incidental learning, enhance motivation, increase comprehension and thus acceptance.

Keywords: User centered development, Mobile usability, Patient compliance, Patient education, Mobile learning, Elderly people.

1 Introduction and Motivation

The demographical structure in many industrial countries tends towards an increasing population of elderly people. Within the next 20 years, around 25% of the population in European countries will be aged 65 and more [1]. Consequently, whatever progress in health conditions we expect, an increase in care within the next decades is required. This increasing average age of the total population inclines towards a subsequent rise

A. Holzinger (Ed.): USAB 2007, LNCS 4799, pp. 119–132, 2007.

of chronic diseases such as diabetes and hypertension, and in conjunction with today's life expectancy, a dramatic raise in medical costs.

CODE-2 Study. According to the results of the Costs Of Diabetes in Europe Type 2 study [2], which analyzed the financial expenditures for managing specific diabetes-related complications and long-term effects, the annual expense incurred due to type 2 diabetes in Germany averages 4600 Euros per patient. Only seven percent of these costs are spent on medication, and over 50 percent account for the treatment of complications. At the same time, only 26 percent of patients have an acceptable value of HbA1c, a long term indicator for diabetes control. The study concludes that diabetes control is inadequate in most patients and in order to reduce total costs, the focus should be turned to early prevention of complications and that an initial increase in treatment costs due to preventive measures can be more than compensated by savings occurring from prevention of complications [2], [3]. To date, there are no concepts for care continuity in diabetes control to be used by all patients; moreover health care providers do not even require patients to record their values using a diabetes diary. Another common example for the lack of clinical concepts in public health is hypertension, which is a common disease amongst the elderly: Untreated hypertension can lead to cardiovascular events, thus elderly have significantly higher expenditures per capita for hypertension and per hypertensive condition [4]. In Austria, for example, only 20 percent of people with hypertension are on proper medication to control this condition [5].

Older people and new technologies are currently one of the important research and development areas, where accessibility, usability and most of all life-long learning play a major role [6]. Interestingly, health education for the elderly is a still neglected area in the aging societies [7], although computer based patient education can be regarded as an effective strategy for improving healthcare knowledge [8] and the widespread use of mobile phones would enable mobile learning at any place, at any time. [9]. However, elderly patients form a very specific end user group and there is a discrepancy between the expected growth of health problems amongst the elderly and the availability of patient education possibilities for this end user group, therefore a lot of questions remain open and need to be studied [7]. However, although technology is now available (at least in Europe) at low cost, the facilitation of usage is only one aspect which must be dealt with, it is necessary to understand the uncertainties and difficulties of this particular end user group. Research is therefore also aimed at investigating ways to increase motivation and improve acceptance in order to make the technology more *elder user friendly* [10].

MyMobileDoc. In order to address some of the above mentioned issues we have developed MyMobileDoc providing a user centered and user friendly interface with no automatic electronic data exchange by sensors, since our focus is on personal awareness. The main functions include a daily medication remainder service and the provision of immediate feedback to the patient, providing educational information about relevant medical details, i.e. chronic diseases. Our main goal was to provide a personal tool designed to increase compliance and, most of all, the possibility of increasing personal responsibility and personal awareness, since we found that patient compliance, acceptance and patient education is vital. Technology should enable the patients to actively become involved in their management of chronic diseases.

2 Technical Background

Worldwide, there are a lot of various groups working on the use of mobile phones and health applications, patient monitoring, automatic data transmission and sensors [11], [12]. However, only a few are dealing with manual data entry [13], and issues of patient education and learning, motivation and acceptance of these technologies amongst elderly people are most often neglected – although elderly people are a completely different end user group. Their motivation is different, their frustration level is lower and they may have to overcome previous, negative experience [10].

2.1 Communication Concept of MyMobileDoc

The most important part in the structure of MyMobileDoc is **the patients themselves**, who regularly send their data (blood pressure; blood glucose levels, pain intensity, contentment level, etc.) via their mobile phone, or by using a standard web client on any device (Game Boy, Surf Pad etc.) to the Medical Data Centre (MDC) on the server (figure 1). The received data can then be evaluated automatically or by medical staff. On the basis of this data, appropriate feedback can be immediately forwarded to the patient. It is also possible to identify possible emergency states from the received data and on that basis, help can be called. Technologically, the patient's mobile could also be located by an Assisted Global Positioning System (AGPS). The patient data can be stored in the MDC and the patient can view his medical diary at any time. Also, the medical professionals can have access to the data, administer the patient's user information and check medical data by using the MyMobileDoc web front end for medical professionals. Patient's relatives or his nurses can gain access to the saved medical values to observe the patient's health status. However, in all these cases, data protection measures and secure data transfer must be taken into account [14]. In this paper, we concentrate on the user interface (solid lines within figure 1).

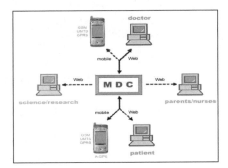

Fig. 1. The basic communication structure of the MyMobileDoc system (MDC = Medical Data Center); In this paper we concentrate only on the user interface development (solid lines)

Before creating a working prototype of MyMobileDoc, some decisions about suitable hardware for the clients and the server, as well as the programming tools to be used, had to be made.

3.1 Clients

Within the technical concept of MyMobileDoc we distinguish between two types of clients: one interface for the medical professional which is basically a standard Internet Browser; and the second type for patients, which should provide them with a maximum of flexibility. This means every terminal would be possible, however, we have chosen mobile phones due to their high availability. For our prototype, we used a Motorola A920 (figure 2) which included an integrated Flash-Player (version Flash Player 5), since we already had experience with this technology [15].

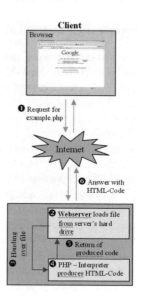

Fig. 2. A view of the user interface of MyMobileDoc: the login screen on a Motorola A920 mobile phone (left); PHP connects the client with the database server (right)

4 End User Group Analysis and Medical Indications

A suitable instrument for diagnosing and treating certain medical conditions usually has to be chosen carefully by the medical professional. If, for example, a physician decides to use the wrong tool, the treatment might not be as effective, and a patient's condition may even deteriorate. If, as in our case with MyMobileDoc, the patient is the end user and uses the tool, the situation is even more critical: It is obvious that the better the usability of the device, the more patients will be able and willing to use it. We followed the general guideline that our technological model should match the mental models of the end users [16].

Consequently, for successful everyday use of such a particular application, the following issues must be carefully considered during the design and development: the profile of the patient (compliance) and his underlying clinical condition (disease).

4.1 Patient Compliance

Identifying suitable patients is a crucial factor for successful treatment. This task is performed by medical doctors after a physical examination. Based on clinical experience, we suggest dividing patients into three groups in order to determine whether or not they will be reliable supporters in their own therapy (compliant) and benefit from MyMobileDoc.

Basically, the compliance level of a certain patient is generally hard to predict from age, gender or social factors alone. The lack of a general accepted definition of compliance makes it difficult to operationalize and consequently evaluate the psychological concept of compliance [17]. However, taken as a behavioral concept, compliance involves actions, intentions, emotions, etc. which are also known from general usability research. Therefore, indirect methods such as interviews are generally used because they have the advantage of revealing the individual's own assessment of their compliance. Assessment by nurses and physicians has usually been based on either the outcome of compliance or information obtained in interviews. Self report measures are the most commonly used to evaluate the compliance of adolescents with chronic disease. Reasons for their popularity can be seen in easy applicability and low costs [17]. During our study we identified the following three groups listed in Table 1, who reflect three commonly observed circumstances that can influence the patient's level of compliance.

Table 1. Three commonly observed groups of patients and their estimated likelihood of their compliance

	Will use MyMobileDoc	Level of education	Motorical skills	Ability to understand risk factors
Group 1	likely	high	sufficient	given
Group 2	unlikely	low to medium	sufficient	usually limited
Group 3	unpredictable	varying	insufficient	varying

A typical group 1 patient could be a 45 year old teacher who seeks medical advice for recently diagnosed Type 2 diabetes. He tries hard to get back to work as soon as possible. We can expect him to use MyMobileDoc for treatment, but he might not really need it as he soon knows a lot about his condition and treatment options.

A patient out of group 2 could be a 50 year old woman who has been living on a farm since the age of 20. She does not have social contacts other than her family. She is overweight and has never heard about cholesterol, nor has she ever used a mobile phone. Using MyMobileDoc will be a big challenge to her. She will have to learn about the interactions of lifestyle and diabetes, and compared to group 1 she will need much more assistance and control when using MyMobileDoc. A member of group 3, a 70 year old man who suffered a stroke that left his right arm and leg paralyzed, is physically unable to use MyMobileDoc for diabetes control, irrespective of his educational background. However, this group usually has help from relatives or (mobile) nurses available. All three groups would benefit from MyMobileDoc. Patients from group 2 need more control and education, whereas group 3 patients are in need of physical help. What we are responsible for is usability.

3.2 Clinical Conditions

Basically, there are a lot of medical conditions that can be monitored with MyMobileDoc, some are discussed here.

Blood pressure. Diagnoses of high blood pressure (hypertension) can only be established after multiple tests. Having these done by patients, avoids higher than normal values due to anxiety at the physicians office. Therapy usually lasts many years, so a mobile application can here save time and consultation costs. After introducing new medication, a patient should monitor blood pressure several times a day, also during pregnancy. Immediate feedback and information on such issues can help to raise awareness and compliance.

Blood glucose levels. Type 2 diabetes control is easy for physically fit patients. From a message generated by the database server, patients can learn about the impact of diet on their condition. If registered values are too high over a longer period, the attending physician can contact them and advise. Moreover, it is possible to include fast acting insulin into therapy, an option that only a few patients use outside the hospital. Type 1 diabetes is not an indication: MyMobileDoc depends on network connection, and hypoglycemia is, if not treated rapidly, a life threatening condition. There are diabetes diaries available on paper and on PDA's. However, compared with MyMobileDoc, none of them offer their end users spontaneous feedback and the opportunity to get educational information.

Peak-Flow. In asthmatics, acute exacerbation can be prevented by regularly measuring the peak expiratory flow, and adjusting medication accordingly. As Peak-Flow values often decrease days before the patient suffers clinical deterioration, a database server could alert the patient. Follow-up examinations are also based on evaluating Peak-Flow values written down by the patient but this is only possible retrospectively. Here a mobile application can optimize therapy and reduce the number of consultations.

Anticoagulation therapy. After cardiac valve prosthesis, heart surgery, certain forms of arrhythmia, deep vein thrombosis, pulmonary embolism and other conditions, patients might have to take medications to avoid blood clotting. Serious bleeding can be a complication if too much anticoagulant is taken, which means that therapy has to be constantly monitored. Currently, the vast majority of patients have the tests for International Normalized Ratio (INR) done by a doctor every five weeks, and the ESCAT (Early Self Controlled Anti Coagulation Trial) study showed that only 54 % of patients had acceptable INR values [18]. If patients performed these easy tests, which involve a finger prick themselves, the treatment quality would be much better, as the goal is to check the INR once a week.

Pain control. Chronic pain can be controlled ideally, if treated *before* the pain raises, thus, immediate reaction in essential. Through an electronic pain scale at the touch screen display the patient can inform the medical professional.

Subjective data. Besides the possibility of enter objectively measured data (body weight, circumference of a swollen ankle, number of steps a patient can walk, oxygen saturation, pulse rate etc.) the most interesting possibility is to record subjective parameters (depression, nervousness, contentment level etc.).

4   Some Lessons Learned During User Centered Development

4.1   Card Sorting and Paper Prototyping: Developing the Navigational Structure

We were aware of the requirement of keeping the navigation within MyMobileDoc as intuitive as possible in order to encourage, especially elderly, to use the system, even if they had no computer literacy. We applied both card sorting [19], [20] and rapid paper prototyping (paper mock-ups) [21], [22] for the design of these structures.

Card Sorting. During the Card Sorting session (see Fig. 4 left side), we prepared potential functions and menu entries on individual cards and asked people from the target end user group (N=15) to sort them in order to cluster what could belong together. At first we used Open Card Sorting, in order to exploratory analyze our application domain and afterwards we applied Closed Card Sorting, in order to comparatively test the suitability of several structures and to select the most suitable one. This helped to understand how end users categorize. With cluster analysis (dendograms) we got hierarchical structures, thus, we gained some understanding of the *mental models* of our end users. However, the limitations of this methods are well known, consequently we did not rely on any one method alone, but combined methods in order to minimize conceptual and practical errors [23]. The results reflected the logical structure for our end user group and elicited insight into semi-tacit knowledge. Unfortunately, Card Sorting did not deliver feedback whether the initial choice of options was reasonable.

Paper Prototyping. This method again proved an effective and inexpensive way to evaluate both content and navigation structure. Our test persons used their fingers to reach a specified goal, for example launching an emergency call, on a paper-simulated interface. Along with paper prototyping, we also used thinking aloud in order to gain insight into end user behavior. However, we experienced that to perform thinking aloud studies with elderly people was more difficult, possibly due to the fact that thinking aloud requires much more load on the short-term memory.

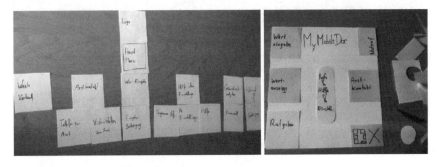

Fig. 4. A picture from the card sorting experiment (left); paper prototype (right)

4.2   The Numeric Input Interface Evaluation

Since the input interface for bio-parameters is the part of our application most often used, we evaluated three different layouts: calculator style, cursor style and slider style (see Fig. 4).

All of our 15 patients aged from 36 to 84 (mean age 65) and 4 nurses aged from 20 to 33 (mean age 26) took part in this evaluation study. None of them had any problems with the touch screen; similar to previous studies the touch screen proved again to be both usable and useful for elderly people [24], [25].

The task consisted of inputting a blood pressure value of 145 over 96, the results are as follows:

Calculator Style. 13 out of the 15 persons preferred to use the calculator-like method. This turned out to be most intuitive, although we experienced that the digit 0 has to be placed below the digit 8, which enabled visually impaired people to use the system.

Cursor Style. The second method involved using two buttons to select a digit, and another two to increase or decrease it. Contrary to our expectations, only a few patients understood this method generally, and only 2 patients under the age of 40 preferred it, in comparison to the calculator method.

Slider Style. The third layout consists of two sliders that can be moved over the display when pressed. This was very easy to understand, however, it involves a lot of practice in handling the touch screen and was definitely not suitable for the elderly, possibly due to the fact that there are far more possible blood pressure values, than pixels on the screen. However, when using a pain scale with 10 positions only, the slider method may have proved preferable.

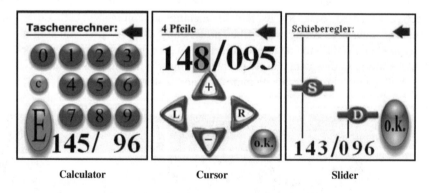

Calculator   Cursor   Slider

Fig. 4. Three different input possibilities were tested (from left to right): calculator style, cursor style and slider style

4.3   Reading Text from Mobile Phone Screens

During our studies we found that the older users were unable to understand certain terminology. Wording, which constitutes obvious usage amongst developers, is absolutely unclear to the elderly, e.g. user interface, front end, click, touch, etc.

We experienced that our end users preferred to *page* through the text, displayed on the mobile device, instead of scrolling through it, because the handling of elements that support scrolling are far more difficult than the handling of pagination-elements, and scrolling easily can lead to a loss of orientation between the lines [26]. Therefore on mobiles, the amount of data that can be viewed at a time should not exceed a certain number of lines. In our user studies we found that just five information blocks – call them chunks – are the maximum that patients found to be still manageable (see figure 5). The commonly known limits on human short-term memory make it impossible to remember everything given an abundance of information [27], [28]. According to Nielsen [29], humans are poor at remembering exact information, and minimizing the users' memory load has long been one of the top-ten usability heuristics. Facts should be restated when and where these are needed, rather than requiring users to remember things from one screen to the next.

Further, we saw in our study, that our end users preferred the use of sans serif fonts. It is generally known – although often neglected in mobile design – that serif fonts should be avoided, as serifs tend to impair readability at small sizes [30].

Out of this group, some fonts are recommended for use on mobile devices, for example Verdana. Embedded fonts have to be rendered as vectors on the mobile's flash player. The third group, pixel fonts, provides exceptionally crisp screen text at small sizes. Their coordinates have to be set to non fraction values und their height is specified for best results.

Similar to the study of [31], we found that *text size* did not affect our end users as much as we previously expected. Interestingly, not to scroll is more important than the text size. This is possibly, because the as awkward perceived additional scrolling caused by larger text compensates any beneficial effect of the larger text size. Also interesting was that many older users associated bullets with buttons, they just want to click on any round objects.

5   MyMobileDoc and Patient Education

Patient education has emerged as a result of health promotion and disease management programs in response to increased pressure to provide more relevant and concrete information at less cost [32], [8], [7]. Several studies reported that elderly patients having very little computer literacy had successfully learned computer based information about disease-related self care and have also reported satisfaction with computer based learning technologies [33]. In the following sections of this chapter, we describe some of the functions and our experiences during our end user studies.

5.1   Main Menu

After entering the user name and password, a logically designed start screen appears. The house symbol in the right upper corner is consistent throughout the application and designed to reload this start page (home page, see Fig. 5, left side).

This screen is the *patient's cockpit* to all the functions and displays essential information: The end user is presented with his name, and a text field for a comment, which can be changed by the attending doctor for each patient. It could for instance contain something like *"just added new medication"*. There is a second text field located below, containing the heading of medical news. It can also be changed by the doctor, but is the same for all patients. In the right upper corner of the screen, a *traffic light symbol* is displayed, which indicates how a patient is currently doing. This uses a mathematical algorithm that takes into account the last seven entries of values, with the last three being more heavily weighted. The values considered to be out of range can be set by the doctor individually for each patient. If, for example, the last blood pressure entered was 220 over 120, which is way too high, the traffic light shows red, irrespective of preceding entries. Icons along the top and left side of the text block are for launching further functions.

Fig. 5. Left: The main menu acts as the starting point to all functions and serves as the "point of information"; Centre and Right: The *adviser* acts as a point for unobtrusive patient education

5.2 Entering Blood Pressure

Pressing the notebook and pen icon on the top left of the main menu launches the input screen (Figure 6), which is described above, is similar to a calculator. There is no need to enter a preceding 0, as a three-digit blood pressure higher than 500 would not be compatible with life. So if the first digit is higher than four, the software

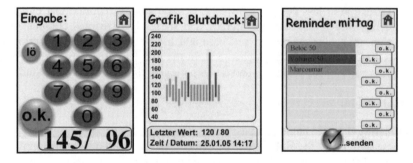

Fig. 6. The patient's input interface for entering self obtained blood pressure values (left), und the graphical representation of the last values entered (right); Pushed from the server, the reminder prevents patients from forgetting to take their medication.

concludes that the value contains two digits only. After pressing the OK-button, a confirmation display appears, followed by a feedback to the patient, whether transmission was completed. Also, a comment is generated, which depends on the value entered. For alarmingly high values, this could be *"take Nitrolingual Pumpspray immediately, consult doctor!"* Pressing OK again loads the main menu, with the updated traffic light.

Pressing the graphic-icon loads an easy to understand graphical interpretation of the last 25 entries.

High values are highlighted red and the last entry is given in detail (Figure 6). Pressing the newspaper icon loads closer information to the news heading displayed in the bottom text field of the main menu.

The book symbol, left of the main menu, loads an e-learning facility for patients. Only basic information is provided but this is what the patients often desperately need as most of them, for example, do not know the normal value for blood pressure.

Emergency Call. The Red Cross icon is to be used in emergencies only, and that's why there is a confirmation screen after pressing it. An emergency number is being dialed by the Smartphone, and Assisted GPS (AGPS) technology can be used to localize a patient.

Medical Consultation. An icon with a seated medical doctor/medical doctor sitting down/ can be pressed in order to establish a connection to the attending doctor, either by telephone call or by video telephony, which the Motorola A920 also supports. The doctor's number does not have to be looked up by the patient. It can be changed by the doctor over the web interface, and the patient just has to press a dial symbol. Additionally, he has his doctor's name and number on his display.

Health Check. When pressing the traffic light icon, the patient receives more information on his status over the last couple of days. A comment is generated by the central database, stating why the traffic light has a certain color. Also, the percentage of values out of range is displayed.

Reminder Service. The reminder (Figure 6) differs from the other functions in that it is pushed by the server and loads automatically when the patient has to take his medication, which is usually four times a day. The user can mark each pill as having been taken or not. If he just presses o.k., all medications are marked as being recognized by the patient in the database. This is in order to not inconvenience compliant patients by requiring them to press all the buttons four times a day. It is clear that the reminder service can not control patient compliance, as users can mark tablets as being taken without doing so. But there are conditions, such as chronic pain or high blood pressure, where medications may be taken only when needed, which could deliver valuable information for further treatment to the doctor. But in most situations, the service simply prevents patients from forgetting to take the tablets.

6 Conclusion and Future Work

MyMobile Doc is a simple and flexible system providing *no* automatic data exchange between the measuring device (biosensor) and the mobile phone. This has, of course, the disadvantage of less comfort, since there are automatic data transmission sensors available; however, we neglected this intentionally, as we found that the patients received more motivation and a sense of responsibility when entering the data by themselves and, most of all, the patients can enter subjective personal data, for example pain intensity and contentment level – where no sensors are available. Consequently, MyMobileDoc must be seen as a mobile patient education tool, which supports patients with chronic diseases and increases patient compliance and raises acceptance. However, it must be pointed out that asking patients to enter numerical data manually is likely to lead to errors – especially with elderly patients or patients who are partially incapacitated by the effects of their chronic disease. It is also important to note that our intention was primarily to motivate cognitive involvement and the **interaction** of the patients with the device in order to stimulate awareness. Following this approach, of course, is only possible for certain patients and after a training phase. MyMobileDoc offers the end user – independent of their location – aid in the electronic documentation of the health status and supports them in the **interpretation, comprehension** and **awareness** of measured medical data.

Every design and development of mobile applications for elderly must support the end users in overcoming their fears and enable them to accept technological aids and mobile devices without reservations; i.e. the design must reflect this acceptance and not be the cause of new biases.

However, much further and deeper research is necessary in order to understand the behavior of elderly end users and in order to design and develop better usable applications.

Acknowledgements. We are grateful for the valuable comments provided by the anonymous reviewers.

References

1. Grundy, E., Tomassini, C., Festy, P.: Demographic change and the care of older people: introduction. European Journal of Population-Revue Europeenne De Demographie 22(3), 215–218 (2006)
2. Liebl, A., Spannheimer, A., Reitberger, U., Gortz, A.: Costs of long-term complications in type 2 diabetes patients in Germany. Results of the CODE-2 (R) study. Medizinische Klinik 97(12), 713–719 (2002)
3. Liebl, A., Neiss, A., Spannheimer, A., Reitberger, U., Wieseler, B., Stammer, H., Goertz, A.: Complications co-morbidity, and blood glucose control in type 2 diabetes mellitus patients in Germany - results from the CODE-2 (TM) study. Experimental and Clinical Endocrinology & Diabetes 110(1), 10–16 (2002)
4. Roberts, R.L., Small, R.E.: Cost of treating hypertension in the elderly. Current Hypertension Reports 4(6), 420–423 (2002)
5. Silberbauer, K.: Hypertonie - die Lawine rollt. Oesterreichische Aerztezeitung 59(12), 40–43 (2004)

6. Kleinberger, T., Becker, M., Ras, E., Holzinger, A., Müller, P.: Ambient Intelligence in Assisted Living: Enable Elderly People to Handle Future Interfaces. In: Universal Access to Ambient Interaction. LNCS, vol. 4555, pp. 103–112 (2007)
7. Visser, A.: Health and patient education for the elderly. Patient Education and Counseling 34(1), 1–3 (1998)
8. Lewis, D.: Computers in patient education. Cin-Computers Informatics Nursing 21(2), 88–96 (2003)
9. Holzinger, A., Nischelwitzer, A., Meisenberger, M.: Mobile Phones as a Challenge for m-Learning: Examples for Mobile Interactive Learning Objects (MILOs). In: Tavangarian, D. (ed.) 3rd IEEE PerCom, pp. 307–311. IEEE Computer Society Press, Los Alamitos (2005)
10. Holzinger, A., Searle, G., Nischelwitzer, A.: On some Aspects of Improving Mobile Applications for the Elderly. In: Stephanidis, C. (ed.) Coping with Diversity in Universal Access, Research and Development Methods in Universal Access. LNCS, vol. 4554, pp. 923–932 (2007)
11. Pinnock, H., Slack, R., Pagliari, C., Price, D., Sheikh, A.: Understanding the potential role of mobile phone-based monitoring on asthma self-management: qualitative study. Clinical and Experimental Allergy 37(5), 794–802 (2007)
12. Scherr, D., Zweiker, R., Kollmann, A., Kastner, P., Schreier, G., Fruhwald, F.M.: Mobile phone-based surveillance of cardiac patients at home. Journal of Telemedicine and Telecare 12(5), 255–261 (2006)
13. Ichimura, T., Suka, M., Sugihara, A., Harada, K.: Health support intelligent system for diabetic patient by mobile phone. In: Khosla, R., Howlett, R.J., Jain, L.C. (eds.) KES 2005. LNCS (LNAI), vol. 3683, pp. 1260–1265. Springer, Heidelberg (2005)
14. Weippl, E., Holzinger, A., Tjoa, A.M.: Security aspects of ubiquitous computing in health care. Springer Elektrotechnik & Informationstechnik, e&i 123(4), 156–162 (2006)
15. Holzinger, A., Ebner, M.: Interaction and Usability of Simulations & Animations: A case study of the Flash Technology. In: Rauterberg, M., Menozzi, M., Wesson, J. (eds.) Human-Computer Interaction Interact, pp. 777–780 (2003)
16. Adams, R., Langdon, P.: Assessment, insight and awareness in design for users with special needs. In: Keates, S., Clarkson, J., Langdon, P., Robinson, P. (eds.) Designing for a more inclusive world, pp. 49–58. Springer, London (2004)
17. Kyngas, H.A., Kroll, T., Duffy, M.E.: Compliance in adolescents with chronic diseases: A review. Journal of Adolescent Health 26(6), 379–388 (2000)
18. Kortke, H., Korfer, R., Kirchberger, I., Bullinger, M.: Quality of life of patients after heart valve surgery - First results of the early self-controlled anticoagulation trial (ESCAT). Quality of Life Research 6(7-8), 200–200 (1997)
19. Zaphiris, P., Ghiawadwala, M., Mughal, S.: Age-centered research-based web design guidelines CHI 2005 extended abstracts on Human factors in computing systems, pp. 1897–1900 (2005)
20. Sinha, R., Boutelle, J.: Rapid information architecture prototyping Symposium on Designing Interactive Systems, pp. 349–352 (2004)
21. Holzinger, A.: Application of Rapid Prototyping to the User Interface Development for a Virtual Medical Campus. IEEE Software 21(1), 92–99 (2004)
22. Holzinger, A.: Usability Engineering for Software Developers. Communications of the ACM 48(1), 71–74 (2005)
23. Adams, R.: Universal access through client-centred cognitive assessment and personality profiling. In: Stary, C., Stephanidis, C. (eds.) User-Centered Interaction Paradigms for Universal Access in the Information Society. LNCS, vol. 3196, pp. 3–15. Springer, Heidelberg (2004)

24. Holzinger, A.: User-Centered Interface Design for disabled and elderly people: First experiences with designing a patient communication system (PACOSY). In: Miesenberger, K., Klaus, J., Zagler, W. (eds.) ICCHP 2002. LNCS, vol. 2398, pp. 34–41. Springer, Heidelberg (2002)

25. Holzinger, A.: Finger Instead of Mouse: Touch Screens as a means of enhancing Universal Access. In: Carbonell, N., Stephanidis, C. (eds.) Universal Access, Theoretical Perspectives, Practice, and Experience. LNCS, vol. 2615, pp. 387–397. Springer, Heidelberg (2003)

26. Giller, V., Melcher, R., Schrammel, J., Sefelin, R., Tscheligi, M.: Usability evaluations for multi-device application development three example studies. In: Human-Computer Interaction with Mobile Devices and Services, pp. 302–316. Springer, Heidelberg (2003)

27. Miller, G.A.: The magical number seven, plus or minus two: Some limits of our capacity for processing information. Psychological Review 63, 81–97 (1956)

28. Baddeley, A.: The concept of working memory: A view of its current state and probable future development. Cognition 10(1-3), 17–23 (1981)

29. Nielsen, J.: Medical Usability: How to Kill Patients Through Bad Design. Jakob Nielsen's Alertbox, April 11, http://www.useit.com/alertbox/20050411.html(last access: 2007-08-16)

30. Holzinger, A.: Multimedia Basics,Design. In: Developmental Fundamentals of multimedial Information Systems, vol.3, Laxmi Publications, New Delhi (2002)

31. Chadwick-Dias, A., McNulty, M., Tullis, T.: Web usability and age: how design changes can improve performance ACM Conference on Universal Usability: The ageing user, pp.30–37 (2002)

32. Rippey, R.M., Bill, D., Abeles, M., Day, J., Downing, D.S., Pfeiffer, C.A., Thal, S.E., Wetstone, S.L.: Computer-Based Patient Education for Older Persons with Osteoarthritis. Arthritis and Rheumatism 30(8), 932–935 (1987)

33. Stoop, A.P., Van't Riet, A., Berg, M.: Using information technology for patient education: realizing surplus value? Patient Education and Counseling 54(2), 187–195 (2004)

Technology in Old Age
from a Psychological Point of View

Claudia Oppenauer, Barbara Preschl, Karin Kalteis, and Ilse Kryspin-Exner

University Vienna, 1010 Vienna, Austria
Institute for Clinical, Biological and Differential Psychology
claudia.oppenauer@univie.ac.at,
barbara.preschl@gmx.at,
karin.kalteis@wuk.at,
ilse.kryspin-exner@univie.ac.at

Abstract. The aim of this paper is to foster interdisciplinary research on technology use in old age by including psychological theories and dimensions such as cognitive, motivational and emotional factors which referring to recent research studies highly influence acceptability and usability of technical devices in old age. Therefore, this paper will focus on psychological theories in the context of ageing with regard to attitudes towards technology use, acceptability and the importance of user involvement at the very beginning of technological product development and design.

Keywords: Gerontopsychology; Assistive Technology; Usability; Accessibility; Home-Care.

1 Introduction

Internationally, the number of old people and consequently of people needing health care is increasing. Nevertheless, so far it is unsure how to deal with this problem especially because financial costs for traditional health care systems become more and more unaffordable. At the same time there is a trend towards single households, and private care systems within the families will eclipse even more in the future. These facts promote technical support and compensation for financial and personal resources of our societies in order to prolong an independent life in old age [20, 33].

Because of age related changes in functional abilities such as psychomotor, perceptive and cognitive skills older people often have difficulties in using "general" technology [35]. Multidisciplinary research on the technology-gerontology interface is of prime importance for a better understanding why technology is difficult to use, how to adapt technology to the needs of older people, and how to train the elderly to use technology [10]. This paper focuses on psychological research and theories which contribute to a better understanding of the needs and difficulties of older people due to technology use.

A. Holzinger (Ed.): USAB 2007, LNCS 4799, pp. 133–142, 2007.

2 Psychological Theories in the Context of Aging and Technology

2.1 Cognitive Theory of Aging

A main domain in gerontopsychology is the discussion about cognitive functions and performances in old age. Over the years research studies generated a different picture of intelligence and cognitive functions and found that *Crystallized Intelligence*, which refers to the knowledge and life experience of a person, takes until old age and can be well trained. On the opposite *Fluid Intelligence* referring to reasoning and concentration decreases with old age and simultaneously the ability to cope with new and complex situations. To conclude, cognitive performances can not be generalized and differ from person to person and are connected with mental activity resources, education and genetic factors [23].

Certainly the use of technical systems requires not only Crystallized but also Fluid Intelligence. According to the facts mentioned above cognitive functions in old age show a high *cognitive plasticity:* by cognitive training it is possible to increase or maintain cognitive functions as memory, attention, concentration or problem solving [20]. Moreover, the motivation to use a technical device can be enhanced by training of special skills which are necessary for use of a technical system [18].

As a result technical solutions should be designed and developed in cooperation with the end-users and their needs to assure that the product fits with the cognitive level of the end-users and to achieve high motivation for use [18].

2.2 Successful Aging with Technology

Concerning psychological theories about abilities basic competences and expanded competences can be distinguished [3, 4]. Basic competences are *Activities of Daily Living (ADLs)* such as personal hygiene, food preparation etc. *Instrumental Activities of Daily Living (IADLs)* are important for independent living as well as for social functioning and include for example preparing meals, money transfers, shopping, doing light or heavy housework or using a telephone. Old people have to deal with a decline in these capabilities. Plenty of technical devices support independent living of the elderly. Mollenkopf et al. [30] distinguish t*echnology of every day life* from technical systems which should compensate for impaired (psycho)motor and cognitive skills, so called *Assistive Technology (AT)*. Furthermore, these technological devices can be divided into *high and low technology* [30]. Security handle bars or ramps are examples for the latter. Technology based on microelectronics (e.g. safety alarm systems, and monitoring devices) is called High Technology.

Referring to Baltes and Baltes [2, 3] successful ageing is a process of *Selective Optimization with Compensation*. Because of age related losses people select areas of life to optimize these domains and to increase capacity, e.g. by cycling in stead of taking the bus to train psychomotor abilities. Furthermore, when optimization is no longer feasible older people face the challenge to find compensatory alternatives. The consequence is a restricted but effective way of life.

According to this model old people could use the help of AT in order to optimize or maintain activities of daily living and leisure activities [32].

2.3 Personal Competences and Environmental Demands

Psychological theories of environment and ageing are relevant in this context because a technological product is always embedded in a person's environment [30]. In the *Competence –Press Model* by Lawton [22] for instance, personal competences face challenges by aspects of the environment on the individual (environmental press). Depending on whether individual competences and environmental press fit together coping with new situations in old age is more or less successful.

2.4 Theories of Control

Theories of Control as for example the Locus of Control Theory by Rotter [7] have a high impact on many areas of human life. Ideally humans should have a high internal locus of control in their life in order to interpret life events as a consequence of their own behaviour. Research studies show that people with a high internal locus of control are more successful in dealing with technologies [7].

Behaviour, needs and capabilities can be enhanced or ignored by the environment Furthermore the environment sometimes leads to dependency and loss of capacity in case of overprotective care or disrespect towards a person's autonomy. According to Baltes, there are two main models dealing with loss of independency in old age: *Learned Dependence* and *Learned Helplessness* [1, 39]: If assistance is provided in situations with which a person would be able to cope on his/her own, this ability will vanish by and by. Because of the overprotection (e.g. by a caretaker) it becomes redundant for the person to carry out an activity anymore and finally due to missing training the person will unlearn it – the person is loosing independency in this situation. The phenomenon of learned dependency is often observed in health care and family systems because claims of independency and autonomy of the elderly are ignored by caregivers and family members. A reason for this is that caregivers' perception focus on deficits and ignore competences [1, 21, 33].

These concepts are especially relevant in the context of monitoring technologies in health care systems. Monitoring technologies bear the danger of fostering external control beliefs and could in last resort lead to learned dependency and helplessness [32].

Another theory according to perceived control is the concept of *self-efficacy* by Bandura [6]. Perceived self-efficacy subsumes people's beliefs about their capabilities to influence situations and life events. These beliefs strongly determine emotional, cognitive and motivational structures as well as psychological well-being. People with strong self-efficacy perception interpret difficult or new tasks rather as challenging than threatening. Failures are more likely attributed to insufficient effort or lacking knowledge. Consequently persons with high self-efficacy are less vulnerable for depression or other affective disorders. For the impact of self-efficacy on technology use see 3.3 Research Studies.

2.5 A Multidisciplinary Approach

A multidisciplinary approach (following the Competence-Press Model by Lawton), including gerontology, technology and psychology is adopted by the Center for Research on Aging and Technology Enhancement (CREATE) [10]. The so-called *Human*

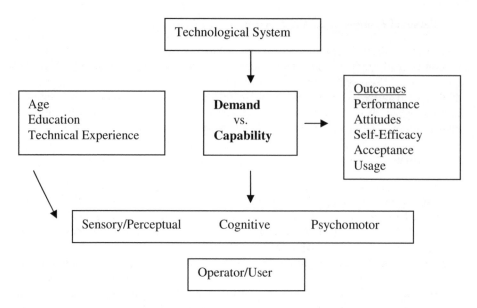

Fig. 1. Human Factors Approach by Czaja et al. [10]

Factors Approach (see Fig.1) helps study ageing and technology while examining the relationships between demands of the technical system on the one hand and cognitive, psychomotor and perceptual capabilities of the user on the other hand.

The degree of fit between these components determines performance on the technical system, attitudes and self-efficacy beliefs about using technical devices, acceptance and usability of the system. A main goal of this research program is developing theoretically and empirically driven design guidelines for technical systems including aspects of the user-system interface (hardware, software and training) [10]. In the following, results of the most recent research including these components are illustrated.

3   Older People and Technology Use

3.1   Assisted Living in Old Age

According to Cowan and Turner-Smith [9] Assisted Technology can be defined as 'any device or system that allows an individual to perform a task that they would otherwise be unable to do, or increases the ease and safety with which the task can be performed'. Consequently AT can prolong independence and autonomy of the elderly in order to stay in their own home. Of course, this is only possible if AT is developed according to the needs of old people and if the technical device supports their feeling of being independent and does not stigmatize. In fact, AT has the potential to partially substitute various social and health-care interventions or at least assist them. Findings of interviews in England and Scotland show that the acceptability of AT depends on the felt need for assistance (individual need and home environment lead to a felt need for assistance), the recognition of the product quality (efficiency, reliability, simplicity and safety of the device) and on availability and financial costs [25].

Kleinberger et al. [19] postulate three factors which should be met if AT is used by old people and support them staying longer independent: 1. Assisted technologies should be ambient, which means that they offer their service in a very sensitive way and are integrated in daily environment (e.g. movement sensors in the wall, which are invisible). 2. They have to be adaptive to the individual needs and capabilities of the elderly. 3. AT services have to be accessible.

3.2 Usability

In recent projects and research articles usability of technical devices or systems is a very common discussed term. Although it is obvious that use and acceptance of a system are major conditions for the success of a new product, user involvement has not always been the current method to guarantee this. To avoid additional costs and technical problems evaluation of usability should take place at the very beginning of product development. Seal, McCreadie, Turner-Smith and Tinker [38] postulate three main reasons for involving older users in assistive technology research: First, applications and further problems may be avoided. Second, stereotypes of technology minimizing older peoples' handicaps are reduced. Third, cost-effective technology, "Design for all" products, can be developed with the participation of the elderly.

There are many ways to evaluate usability of a technical system. One distinction can be made in *inspection* and *test methods*. Inspection methods do not involve end users themselves but the judgment of usability specialists and work with cognitive techniques like heuristic evaluation, cognitive walkthroughs and action analysis. Despite the possibility of usability evaluation at an early stage and a very analytical style of this method, needs of end users are only anticipated. On the contrary, test methods like thinking loud, field observation and questionnaires gain information directly from the user and provide needs, preferences but also individual problems and concerns. Finally, both approaches should be combined when usability is evaluated to guarantee both subjective view of the user and expert opinion [14].

Another possibility for user involvement is focus groups. In focus groups interviews about a particular topic are co-ordinated by a moderator. This method is especially relevant if interactions within a group are relevant for the product development. Focus groups can be used at different project phases: for exploration in identifying creative new solutions to well known problems at an early stage of a project or for detailed information at late-time proceeded stages [38].

In the UTOPIA project (Usable Technology for Older People – Inclusive and Appropriate) needs for new ICT products were identified and concepts relating to these needs were developed. Information about needs of elderly people was gained by questionnaires, interviews, focus groups and workshops. In-home interviews showed how technical devices matched with the user's life and in which situations these were used. The stereotype that older people less likely than younger people use technology was approved. However this result was not true for all technologies and even can not be reduced to the complexity of a technology. Concerning the fear of technologies, old users regard their fear of a product as the fault of themselves and not of the product design and development [11].

Involving users in the design and development process of technologies has become state of the art in recent European projects dealing with Assistive Technology. Within the FORTUNE Project even a curriculum framework was developed for teaching users about the project principles in order to increase involvement in future research activities [38]

User involvement in the development of patient oriented systems e.g for touch screen panel PCs in hospitals are essential and lead to high acceptance. Practice with touch screen technology is compared to other input devices easy to learn even for people with little knowledge about computers. One of the advantages of touch screens is that the input device is output device too. Because of the direct eye-hand co-ordination users are able to experience their sense of touch, sense of sight and sense of hearing which encourages a sense of immersion [40]. According to the motto *less is more* design of such systems should be simple and easy [15, 16]. Mobile computers in medicine provide the opportunity for economic working for medical staff as well as economic time for patients.

Patients with handicaps in their visual or motor functions often experience problems in filling out personal data related questionnaires. In the MoCoMed-Graz project (Melanoma Pre-care Prevention Documentation) patients could login with a code at a touch based Tablet Pc and complete the questionnaires for the clinical information system and for a scientific database for research in skin cancer. The project acted on a User-Centered Design (UCD) approach [31] including four levels: paper mock-up studies, low-fi prototypes, hi-fi prototypes and the system in real life. Low-fi level was conducted with the paper mock-up which meant that screen designs and dialogues based on paper elements. The high-fi level already worked with a full functional prototype of the touch screen and in the end the final version was tested in real-life. Of course this procedure is very time-consuming but a precondition for user acceptance [17].

3.3 Interdisciplinary Research Studies

In the last ten years research concentrated on the difficulties of older adults when using technologies [5, 27, 29, 35, 36, 43]. For age-related decline in vision and hearing and compensational strategies in conjunction with product guidelines see for instance Fozard [12] and Schieber [37].

In Japan, a quantitative study about computer attitudes, cognitive abilities and technology usage among older adults showed that higher *cognitive abilities* were related to the use of products whose usage ratio was high (e.g. computer, copier, fac-similes and video recorder) [42]. The European Mobilate survey also exhibited a correlation between technology use (ATM) and cognitive functioning [41].

Mollenkopf regards technical systems as socio-culturally-shaped artefacts [26, 27]. Societal stereotypes influence the development and design of technologies as well as the acceptance or rejection by potential user groups. The findings of a qualitative study in Germany indicated that the fear about what is new, motivation to use a technical device, the ease of use and advice and training are linked to acceptance or rejection of a technical system (e.g. household technology, safety alarm systems, wheelchairs and medical technology) [27].

According to Rogers et al. older people are – contrary to stereotypes – willing to use new technologies [36]. Their level of acceptance even increases if older adults receive adequate training and if the benefits of the technical device are clearly understood. The results of another German representative survey show that some devices (e.g. microwave, washing machine, hearing devices, blood pressure meter, video recorder and computer) are associated with fear, bad experiences and the need for easier use [28]. The findings of the Japanese study indicated that positive computer *attitudes* were related to greater usage of computers as well as to greater usage of computerized products (e.g. car navigation system, ATM and mobile phone) [42]. A quantitative study carried out by the authors indicated that *motivational aspects* heavily influenced Safety Alarm System's use, i.e. higher motivation was related to higher system use [34].

Finally, *self-efficacy beliefs* about capabilities to use technical devices have to be mentioned. Results of a German study on digital wristwatch use showed that higher internal locus of control significantly coincided with higher coping strategies relating to technical problems [7]. Findings of the already mentioned European Mobilate survey indicated that higher technology use is related to higher internal locus of control and lower external locus of control [41]. Correlations showed a slight tendency that technology users feel less controlled by external circumstances and that they generally feel empowered concerning their own life.

4 Outlook and Discussion

Technical support and compensation for financial and personal resources in health care systems do not only require new technical solutions but also inclusion of the elderly in the design and development process. From a psychological point of view it is significant to consider age-related losses in cognitive and psychomotor functions as well as resources and capabilities of the elderly [21]. In gerontechnological research user participation in design process is a fundamental premise to develop products that are suitable for people of all ages (design for all) [13]. Nevertheless, population of the elderly is as heterogeneous as other age groups and in the end only consideration of individual needs will lead to acceptability and motivation to use. From a technical point of view, it is particularly important to focus on the user to tap the full technological, potential and to avoid that this potential remains unrealized [24]. Ideally user involvement and therefore usability evaluation should start as soon as possible in product development to avoid financial and technical disadvantages [38]. Special training programs, information giving and counselling following cognitive theories of aging may increase acceptance and reduce anxiety towards a technical product [18, 21, 25]. According to the Competence-Press Model by Lawton [22] family members and significant others also play a major role whether a person will use technology. Thus, it is essential to provide information not only to the person concerned but also to the social environment.

Further research is needed in the context of motivational aspects that lead to acceptance or decline of a technical product. Additionally, ethical aspects - especially in the context of monitoring devices and old people with dementia- have to be taken into account. In this regard, elderly people have to be entitled to their own decision, either in favour of a technical product or against it. Therefore it is significant that a person is well informed about the product and has the ability for informed choice making and consent [8]. According to Mollenkopf et al. this "informed consent" must be guaranteed [30].

Technical support can make a vital contribution to professional health care systems. But cost analyses including quality of life and evaluation of technical devices should be taken into account [32]. Technical devices always have to be seen in relationship to human care and will never replace human support [25]. This offers the opportunities to future investigations of the interactions between technical and human support in health care systems.

References

1. Baltes, M.M.: Verlust der Selbständigkeit im Alter: Theoretische überlegungen und empirische Befunde. Psychologische Rundschau (46), 159–170 (1995)
2. Baltes, P.B., Baltes, M.M. (eds.): Successful aging: Perspectives from the behavioral sciences. Cambridge University Press, New York (1990)
3. Baltes, P.B., Baltes, M.M.: Erfolgreiches Altern: Mehr Jahre und mehr Leben. In: Baltes, M.M., Kohli, M., Sames, K. (eds.) Erfolgreiches Altern, Bedingungen und Variationen, Hans Huber, Bern, pp. 5–10 (1989)
4. Baltes, M.M., Wilms, H.U.: Alltagskompetenz im Alter. In: Oerter, R., Montada, L. (eds.) Entwicklungspsychologie, Ein Lehrbuch, Belz, Weinheim, pp. 1127–1130 (1995)
5. Baltes, P.B., Lindenberger, U., Staudinger, U.M.: Life-Span Theory in Developmental Psychology. In: Damon, W., Lerner, R.M. (eds.) Handbook of Child Psychology, Theoretical Models of Human Development, vol.1, pp. 1029–1043. Wiley & Sons, New York (1998)
6. Bandura, A.: Self-efficacy. Freemann, New York (1997)
7. Beier, G.: Kontrollüberzeugungen im Umgang mit Technik. Ein Persönlichkeitsmerkmal mit Relevanz für die Gestaltung technischer Systeme, dissertation. de - Verlag im Internet GmbH, Berlin (2004)
8. van Berlo, A., Eng, M.: Ethics in Domotics. Gerontechnology 3(3), 170–171 (2005)
9. Cowan, D., Turner-Smith, A.: The role of assistive technology in alternative models of care for older people. In: Royal Commission on Long Term Care, Research, Appendix 4, Stationery Office, London, vol. 2, pp.325–346 (1999)
10. Czaja, S., Sharit, J., Charness, N., Fisk, A.D., Rogers, W.: The Center for Research and Education on Aging and Technology Enhancement (CREATE): A program to enhance technology for older adults. Gerontechnology 1(1), 50–59 (2001)
11. Eisma, R., Dickinson, A., Goodmann, J., Syme, A., Tiwari, L., Newell, A.F.: Early User Involvement in the Development of Information Technology-Related Products for Older People. Universal Access in the Information Society (UAIS) 3(2), 131–140 (2004)
12. Fozard, J.L.: Using Technology to Lower the Perceptual and Cognitive Hurdles of Aging. In: Charness, N., Schaie, K.W. (eds.) Impact of Technology on Successful Aging, pp. 100–112. Springer, Heidelberg (2003)
13. Fozard, J.L., Rietsema, J., Bouma, H., Graafmans, J.A.M.: Gerontechnology: Creating enabling environments for the challenges and opportunities of aging. Educational Gerontology 26(4), 331–344 (2000)
14. Holzinger, A.: Usability Engineering for Software Developers. Communications of the ACM 48(1), 71–74 (2005)

15. Holzinger, A.: Finger instead of Mouse: Touch Screens as a Means of Enhancing Universal Access. In: Carbonell, N., Stephanidis, C. (eds.) Universal Access. Theoretical Perspectives, Practice, and Experience. LNCS, vol. 2615, pp. 387–397. Springer, Heidelberg (2003)
16. Holzinger, A.: User-Centered Interface Design for Disabled and Elderly People: First Experiences with Designing a Patient Communication System (PACOSY). In: Miesenberger, K., Klaus, J., Zagler, W. (eds.) ICCHP 2002. LNCS, vol. 2398, pp. 34–41. Springer, Heidelberg (2002)
17. Holzinger, A., Sammer, P., Hofmann-Wellenhof, R.: Mobile Computing in Medicine: Designing Mobile Questionnaires for Elderly and Partially Sighted People. In: Miesenberger, K., Klaus, J., Zagler, W., Karshmer, A.I. (eds.) ICCHP 2006. LNCS, vol. 4061, pp. 732–739. Springer, Heidelberg (2006)
18. Holzinger, A., Searle, G., Nischelwitzer, A.: On Some Aspects of Improving Mobile Applications for the Elderly. In: Stephanidis, C. (ed.) Coping with Diversity in Universal Access, Research and Development Methods in Universal Access. LNCS, vol. 4554, pp. 923–932 Springer, Heidelberg (2007)
19. Kleinberger, T., Becker, M., Ras, E., Holzinger, A., MÜller, P.: Ambient Intelligence in Assisted Living: Enable Elderly People to Handle Future Interfaces. Universal Access to Ambient Interaction, Lecture Notes in Computer Science vol. 4555 (200/), pp.103–112
20. Krämer, S.: Technik und Wohnen im Alter - Eine Einführung. In: Stiftung, W. (ed.) Technik und Wohnen im Alter. Dokumentation eines internationalen Wettbewerbes der Wüstenrot Stiftung, Guttmann, Stuttgart, pp. 7–26 (2000)
21. Kryspin-Exner, I., Oppenauer, C.: Gerontotechnik: Ein innovatives Gebiet für die Psychologie? Psychologie in Österreich 26(3), 161–169 (2006)
22. Lawton, M.P., Nahemow, L.: Ecology and the aging process. In: Eisdorfer, C., Lawton, M.P. (eds.) Psychology of adult development and aging, pp. 619–674. American Psychological Association, Washington (1973)
23. Lehr, U.: Psychologie des Alters. Quelle & Meyer, Wiebelsheim (2003)
24. Mayhorn, C.B, Rogers, W.A., Fisk, A.D.: Designing Technology Based on Cognitive Aging Principles. In: Burdick, D.C., Kwon, S. (eds.) Gerontechnology. Research and Practice in Technology and Aging, pp. 42–53. Springer, New York (2004)
25. McCreadie, C., Tinker, A.: The acceptability of assistive technology to older people. Ageing & Society 25, 91–110 (2005)
26. Mollenkopf, H.: Assistive Technology: Potential and Preconditions of Useful Applications. In: Charness, N., Schaie, K.W. (eds.) Impact of Technology on Successful Aging, pp. 203–214. Springer, Heidelberg (2003)
27. Mollenkopf, H.: Technical Aids in Old Age - Between Acceptance and Rejection. In: Wild, C., Kirschner, A. (eds.) Safety- Alarm Systems, Technical Aids and Smart Homes, Akontes, Knegsel, pp. 81–100 (1994)
28. Mollenkopf, H., Meyer, S., Schulze, E., Wurm, S., Friesdorf, W.: Technik im Haushalt zur Unterstützung einer selbstbestimmten Lebensführung im Alter - Das Forschungsprojekt Sentha und erste Ergebnisse des Sozialwissenschaftlichen Teilprojekts. Zeitschrift für Gerontologie und Geriatrie 33, 155–168 (2000)
29. Mollenkopf, H., Kaspar, R.: Elderly People's Use and Acceptance of Information and Communication Technologies. In: Jaeger, B. (ed.) Young Technologies in old Hands. An International View on Senior Citizen's Utilization of ICT, DJØF, pp. 41–58 (2005)
30. Mollenkopf, H., Schabik-Ekbatan, K., Oswald, F., Langer, N.: Technische Unterstützung zur Erhaltung von Lebensqualität im Wohnbereich bei Demenz. Ergebnisse einer Literatur-Recherche (last access: 19.11.2005), http://www.dzfa.uni-heidelberg.de/pdf/Forschungsberichte/fb19.pdf

31. Norman, D.A., Draper, S.: User Centered System Design. Erlbaum, Hillsdale (1986)
32. Oppenauer, C., Kryspin-Exner, I.: Gerontotechnik - Hilfsmittel der Zukunft. Geriatrie Praxis 2, 22–23 (2007)
33. Poulaki, S.: Kompetenz im Alter: Möglichkeiten und Einschränkungen der Technik. Verhaltenspsychologie und Psychosoziale Praxis 36(4), 747–755 (2004)
34. Preschl, B.: Aktive und passive Nutzung von Hausnotrufgeräten - Einflüsse auf Trageverhalten und Nutzungsverhalten. Eine gerontotechnologische Untersuchung aus psychologischer Sicht. Unveröffentlichte Diplomarbeit, Universität Wien (2007)
35. Rogers, W.A., Fisk, A.D.: Technology Design, Usability and Aging: Human Factors Techniques and Considerations. In: Charness, N., Schaie, K.W. (eds.) Impact of Technology on Successful Aging, pp. 1–14. Springer, New York (2003)
36. Rogers, W.A., Mayhorn, C.B., Fisk, A.D.: Technology in Everyday Life for Older Adults. In: Burdick, D.C., Kwon, S. (eds.) Gerontechnology. Research and Practice in Technology and Aging, pp. 3–18. Springer, New York (2004)
37. Schieber, F.: Human Factors and Aging: Identifying and Compensating for Age-related Deficits in Sensory and Cognitive Function. In: Charness, N., Schaie, K.W. (eds.) Impact of Technology on Successful Aging, pp. 42–84. Springer, New York (2003)
38. Seale, J., McCreadie, C., Turner-Smith, A., Tinker, A.: Older people as Partners in Assistive Technology Research: The Use of Focus Groups in the Design Process. Technology and Disability 14, 21–29 (2002)
39. Seligman, M.E.P.: Helplessness. Freemann, San Francisco (1975)
40. Srinivasan, M.A., Basdogan, C.: Haptics in Virtual Environments: Taxonomy, research, status, and challenges. Computer and Graphics 21(4), 393–404
41. Tacken, M., Marcellini, F., Mollenkopf, H., Ruoppila, I., Szeman, Z.: Use and Acceptance of New Technology by Older People. Findings of the International MOBILATE Survey: Enhancing Mobility in Later Life. Gerontechnology 3(3), 126–137 (2005)
42. Umemuro, H.: Computer Attitudes, Cognitive Abilities, and Technology Usage among Older Japanese Adults. Gerontechnology 3(2), 64–76 (2004)
43. Vanderheiden, G.C.: Design for People with Functional Limitations Resulting from Disability, Aging, or Circumstance. In: Salvendy, G. (ed.) Handbook of Human Factors and Ergonomics, pp. 1543–1568. Wiley, Chichester (1997)

Movement Coordination in Applied Human-Human and Human-Robot Interaction

Anna Schubö[1], Cordula Vesper[1], Mathey Wiesbeck[2], and Sonja Stork[1]

[1] Department of Psychology, Experimental Psychology, Ludwig-Maximilians-Universität München, Leopoldstr. 13, 80802 Munich, Germany
[2] Institute for Machine Tools and Industrial Management, Technische Universität München, Boltzmannstr. 15, 85747, Garching, Germany
anna.schuboe@lmu.de

Abstract. The present paper describes a scenario for examining mechanisms of movement coordination in humans and robots. It is assumed that coordination can best be achieved when behavioral rules that shape movement execution in humans are also considered for human-robot interaction. Investigating and describing human-human interaction in terms of goal-oriented movement coordination is considered an important and necessary step for designing and describing human-robot interaction. In the present scenario, trajectories of hand and finger movements were recorded while two human participants performed a simple construction task either alone or with a partner. Different parameters of reaching and grasping were measured and compared in situations with and without workspace overlap. Results showed a strong impact of task demands on coordination behavior; especially the temporal parameters of movement coordination were affected. Implications for human-robot interaction are discussed.

Keywords: Movement Coordination, Joint Action, Human-Human Interaction.

1 Introduction

Coordinating our movements and actions with other people is an important ability both in everyday social life and in professional work environments. For example, we are able to coordinate our arm and hand movements when we shake hands with a visitor, we avoid collisions when we walk through a crowded shopping mall or we coordinate the timing of our actions, e.g., when we play in a soccer team or interact with our colleagues. In all these apparently easy situations, our human motor system faces a variety of challenges related to action control: movement sequences have to be planned, perceptual information has to be transferred to motor commands, and the single movement steps have to be coordinated in space and time. These processes are performed by complex internal control mechanisms. To perform a single movement (such as shaking the visitor's hand), the acting person has to select the right motor program together with the parameters needed for its execution. A single, simple movement can, in principle, be executed in an infinite number of possible ways.

The need to specify the correct parameters is referred to as the "degrees-of-freedom" problem [1]. Once a motor program has been selected and the movement

A. Holzinger (Ed.): USAB 2007, LNCS 4799, pp. 143–154, 2007.

parameters have been specified, the actor still has to adapt his movement "online" to an environment that may have changed either during movement planning or may still change during movement execution. For example, moving obstacles (such as another person also reaching for the visitor's hand) may obstruct the originally planned movement path.

Movement coordination, however, is not restricted to situations in which humans interact: The fast progress in robotic science will lead to more and more interactions with robots and computers, for example, with household robots or in professional domains such as industrial or medical applications [2]. As robots normally are capable of even more ways to perform a specific movement compared to humans, the amount of possible solutions to the degree-of-freedom problem is even more complex [3]. Consequently, a useful approach is to investigate principles in human-human coordination and to transfer them to human-robot coordination, as this strategy helps, first, to reduce the degrees-of-freedom in robots and, second, to improve the adaptation of robots to the specific needs of a human interaction partner.

The present paper suggests that human-robot interactions can be facilitated when principles found in human-human interaction scenarios are applied. For example, the finding that performance is smoother when interacting partners refer to a common frame of reference has strongly influenced human-machine communication [4]. In the following, we will describe recent findings from human motor control and interpersonal action coordination, before we present a scenario for investigating human-human and human-robot interaction. Finally, we will discuss its relevance for applied scenarios in working domains such as industrial or medical human-robot interactions.

1.1 Human Motor Control

Research in cognitive psychology and neuroscience has attempted to detect the underlying mechanisms of a variety of human movement abilities such as catching a ball or grasping a cup [5]. A diversity of general principles in the motor system has been found that constrain the amount of possible movement patterns. For example, when humans grasp an object, the relation between the grasp and the transport phase of the movement is such that the size of the maximal grip aperture of the actor's hand (i.e. the distance between thumb and index finger) is determined by the speed of the reaching movement [6].

Generally, the human motor system attempts to minimize movement costs; a specific cost function determines which type of movement requires least resources and demands the smallest coordination effort [1]. For example, one such principle is to minimize torque change during an arm movement [7] as this provides the most efficient use of energy resources. To further illustrate this point, Rosenbaum and colleagues [8] put forward the notion of a constraint hierarchy. Such a hierarchy is set up for each motor task and determines the priorities of single task steps in overall task hierarchy. For example, when carrying a tray of filled glasses keeping the balance is more important than moving quickly.

Dependent on the priorities that are specific for a certain task, the motor system can determine the most efficient way to perform a movement so that costs are minimized.

Several additional constraining mechanisms have been specified that reduce the space of solutions to the degree-of-freedom problem of human movement planning.

First, constraints arise from the movement kinematics themselves, e.g. trajectories are normally smooth and have a bell-shaped velocity profile. Second, different parameters determine the exact form of a movement, including shape and material of a to-be-grasped object [9], movement direction [10], the relation of object size and movement speed or the timing of the maximal grip aperture [11]. Third, another important mechanism is to plan movements according to their end-goal [8]: That means, particular movements are chosen with respect to the actor's comfort at the goal posture ("end-state comfort effect") that may as well be dependent on the subsequent movement. A variety of studies have shown the end-state comfort effect for different kinds of end-goals and movement parameters (e.g. [11], [12], [13]).

Besides determining such constraining parameters of movement control, researchers have attempted to clarify the internal mechanisms responsible for planning, executing, and monitoring movements. Generally, three different types of mechanisms are assumed to work together [5]. First, motor commands are sent out via reafferences from the motor system to the periphery (arms, legs, etc.) where they are executed and transformed into consequent sensory feedback. Second, the same motor commands are fed into an internal forward model that uses these efference copies of the motor commands to predict the sensory consequences of possible movements before they actually happen. This is crucial for monitoring and online adaptation of movements because it allows the system to calculate in advance what the most likely outcome of a certain movement will be. The predictions can subsequently be compared with the sensory feedback. The resulting prediction error, i.e. the difference between simulated and real sensory consequences, can be used to train the forward model and improve its performance online. Third, an inverse internal model works in the opposite direction by computing motor commands with respect to the goal movement, i.e. the motor system determines what groups of muscles need to be used in order to reach a certain sensory effect. Thus, with these three internal processes a regulatory circuit is formed that plans, controls and executes movements and corrects online for possible errors.

1.2 Human-Human Interaction

Many situations afford people to work together in order to achieve a common goal. However, although the internal model approach described above works well on an individual's level, motor control is even more complicated in interaction situations. Compared to intrapersonal movement planning, the degree of coordination difficulty increases when two or more people act together because they have to infer the other person's next movements and action goals based on what they observe [14].

Recent research on action observation has revealed first ideas how humans can use sensory (mainly visual) information to understand and predict another person's actions [15].

For example, Meulenbroek et al. [16] investigated whether movement kinematics are adopted from another person by simple observation. The task for pairs of participants was to one after the other grasp and transport objects of variable weight to a different location while the other one observed the movement. Results showed that the person lifting the object secondly was systematically less surprised by the weight of the object than the first person, thus showing that people seem to be able to integrate observed actions into expectations about object features.

However, mere observation of actions might in some cases not be sufficient for interpersonal coordination because, for example, the timing of actions might be critical. Thus, it is necessary to plan and execute own actions based on predictions of what the other person will do instead of only responding to observed actions [15]. Knoblich and Jordan [14], for example, examined the human capability to anticipate others' actions. In their study, pairs of participants performed a tracking task with complementary actions. Results show that participants learned to predict another person's actions and to use this information in an anticipatory control strategy to improve overall tracking performance.

With respect to interpersonal movement coordination, Wolpert and colleagues assumed that the same general mechanisms of internal control that are responsible for planning, executing, and monitoring movements in an individual may also be applied to group situations [17]. That means the same three types of mechanisms are used to understand and predict others' actions: In that case, the motor commands from a person A will provide communicative signals (e.g. gestures or body signals) that the interaction partner B can perceive. This enables B to understand A's movement and, accordingly, use this information to coordinate his or her own motor actions with the partner. Thus, the theory of internal models can account both for intra- and interpersonal motor control.

1.3 Human-Robot Interaction

Also in many applied scenarios like in industry or health care domain, people work together with other humans or with robots in order to achieve common goals. Manifold robots are applied in industry, health care or other work environments. Robots can be defined as machines that can manipulate the physical world by using sensors (e.g. cameras, sound systems, or touch sensors) to extract information about their environment, and effectors (e.g. legs, wheels, joints, or grippers) that have the purpose to assert physical forces on the environment [3].

Although different areas of application have special needs and demands on robot complexity and their ability to interact with humans, there are some requirements that are similar across different professional domains such as industrial manufacturing or health care applications. Such requirements include safety aspects (especially in human-robot interaction), and cost efficiency. Robots are mainly used to improve and extent human capabilities in performing complex tasks, e.g., in situations where humans fail to perform a task due to their physiological constitution (e.g. when strong forces are needed in combination with high precision performance as, e.g., in fracture repositioning [18]) or when tasks have to be performed in dangerous or hostile environments (e.g. when dealing with radiation or with extreme temperatures).

An additional requirement that may be more prominent in medical than in industrial robots is high flexibility. Assistive robots in surgery, for example, need to be adjusted to the specific patient, surgical procedure and surgical staff whenever used, although general aspects and settings of the situation may remain similar. High flexibility is even more important for robots that are used in rehabilitation or in daily-living assistance [19]. Also the human-machine interface has to be of low complexity, as robot adjustment may be done by people who are in the main not trained to interact with robots. Especially assistive robots in rehabilitation or in daily-living assistance have to be designed in a way so that they are accepted and can be handled by their human users.

Thrun [2] suggested classifying robots that are used in work environments and social situations dependent on the amount of interaction robots have and the workspace they share with humans: industrial robots are considered to mainly work in factory assembly tasks, separated from humans in special areas where they normally do not directly interact with humans. Professional service robots, however, may work in the same physical space as humans (e.g. robots assisting in hospitals), and have an intermediate amount of interaction usually limited to few and specific actions. Personal service robots have the closest contact to humans as they interact with them directly and in manifold ways, e.g., by assisting elderly people or people with disabilities [2]. A second type of classification referred to the role robots are assigned in assisting or supporting humans, e.g., in robot-assisted surgery, ranging from a passive, limited and low-risk role in the common task to an active role with strong involvement, high responsibility and high risk [20]. State-of-the-art robots aim at combining both these attributes, that is, close interaction with humans, and an active role and high responsibility in task performance. An important trend in current robotics is therefore the attempt to build highly autonomous robots that can navigate freely, take own decisions, and adapt to the user [2], [3], [4]. The advantage of such autonomous robots is that they are capable of accommodating to changing environments without the need of explicit reprogramming, what makes them especially suitable for the interaction with humans.

Yet human-robot interaction is an even more complicated task than intra- or interpersonal movement coordination. Besides safety (and many other) aspects, the robot would – similar to a human interaction partner - have to build internal models of other people's actions. Nevertheless, many new routes to build social robots are taken: One interesting approach is to construct robots that can learn from humans by observation [21], [22] and by autonomously finding a way to imitate the action of a human model. In order to do so, the robot needs to observe the human worker and acquire representations of the environment and the interaction partner. In this sense, the model by Wolpert, Doya and Kawato [17] for intra- and interpersonal movement coordination can also account for human-robot interactions: The robot uses observations about the human actor to form a set of predictions that can be compared to the real movements and that enables the robot to infer the human's current state and predict subsequent actions.

Following this line of reasoning, robot performance in human-robot interaction will benefit from investigations on human-human interaction in many ways. First, the same principles that control interpersonal movement coordination may apply to human-robot interaction. These principles, when isolated, may be learned and taken over by the robot. Second, robot interaction partners that learn from humans may be able to operate autonomously in direct human-robot interaction. Third, such robots will be easier to interact and pose fewer problems with respect to safety aspects. Building such robots that can predict and adapt to human movements (and other human behavior) is an important future goal.

Our research aims at providing methods and data to examine human movements during normal actions. In the remaining part of this paper, we introduce a scenario that examines general mechanisms of human-human movement coordination in a relatively natural environment in order to transfer results to situations in which humans and robots work together. An important aspect will be the optimization of the workspace humans share and what consequence workspace overlap has on coordination performance.

2 The Ball Track Scenario

The present scenario uses a wooden ball track (i.e. a children's toy to let marbles run down) consisting of five building blocks of different form and size. Participants either perform alone or with a partner in joint action. Although the task as such is relatively simple, it offers a very flexible design for examining a large amount of possible parameters of motor planning and control. In the present paper, we focus on movement coordination and the influence of workspace overlap. Coordination performance was compared in a condition with and without workspace overlap, of special interest were the actors' temporal coordination parameters.

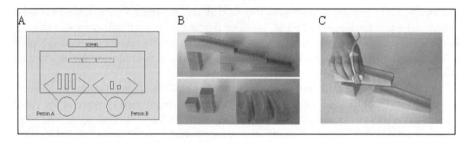

Fig. 1. A: Seating arrangement during joint performance. B: Example of an initial block position (bottom) and goal ball track (top); C: Marker of Polhemus Fastrak.

2.1 Method

In our scenario, participants built a ball track from wooden blocks (height 4 cm, width 4 cm, length between 4 and 16 cm) while their arm and hand movement parameters were assessed. They sat alone or next to a second person at a table with the ball track blocks and a screen in front of them (Fig. 1A). In each trial, the participants' task was to grab the blocks from a predetermined initial position and to stack them according to a visual instruction showing the goal ball track (Fig. 1B, top). As our main focus was on the temporal aspects of movement coordination, we analyzed movement onset time, latency and length of individual performance steps (e.g. grasp and transport components), velocity, and acceleration patterns and overall variances of movements. A Polhemus Fastrak [23], a magnetic motion tracker with a six degree-of-freedom range (X, Y and Z coordinates and azimuth, elevation and roll), was used for movement recording (sampling rate: 120 Hz). The receiver markers were mounted on the back of the hand or on the index finger and thumb (Fig. 1C). Later on, we plan to additionally use eyetracking as well as the electroencephalogram (EEG) to measure electrical brain potentials while the block stacking task is performed.

2.2 Parameters

The scenario allows the examination of a variety of different aspects of human movement coordination. As our main interest concerns parameters in human-human interaction that can be transferred to human-robot interaction, our main focus is put on the mechanisms during joint action situations. Thus, by varying the goal position and orientation of the ball track, we can construct situations in which *obstacle avoidance* becomes necessary. In specific, this is the case when the workspaces of the interaction partners overlap because one person needs to reach over to the other's workspace in order to place a block at the correct position. Consequently, we expect to find timing delays in the movement pattern if the other person's actions have to be carried out before the subsequent movement can be started. Moreover, movement trajectories are hypothesized to differ in the overlap compared to the non-overlap situation.

A second parameter we examine is *individual versus joint performance*. This approach has already been widely used [15] because it allows to directly observe the influence of a second person on an individual's performance. In our scenario, we hypothesized that, when people work in a group, the kinematics and temporal aspects of the interaction partner's movements will be taken into account and used to adapt the own motor actions. For example, if the tempo in which both participants stack their blocks varies strongly at the beginning of the experiment, we expect that over time they accommodate such that they find an intermediate tempo during movement execution or that one person takes over the other's tempo.

Besides these parameters that address the interaction performance of both participants, we are interested in parameters that determine the individual's movement. Thus, a further experimental variation concerns the relation of *initial and goal block orientation*. Based on the above described findings on the dependence of end-goal on movement planning, as e.g. the end-state-comfort effect [8], [11], we want to examine hand orientation at pick up and put down sites. We expect individuals to use different initial hand orientations that depend both on the initial and on the goal position of the blocks.

Moreover, we will examine *uni- versus bimanual working,* i.e. we compare the condition in which participants use only one hand to the condition in which they work with both hands. We also intend to analyze "two persons, one hand" and "one person, two hands" setups. Results showing similar movement patterns in these conditions would support the idea that the own motor system is used to understand and predict other people's actions [24].

Finally, we will vary *timing constraints* imposed to our participants as this is a usual feature of many applied tasks and *task complexity*, for example by using more natural stimuli like normal household objects or medical equipment.

2.3 Results

In a recent experiment, we recorded movement parameters of pairs of participants while they worked on building a simple ball track. In each trial, five wooden blocks were positioned in a fixed arrangement (as shown e.g. in Fig. 1B, bottom). As dependent variables, we focused on the temporal coordination during interference from the partner's movements (avoidance of a dynamic obstacle). Thus, we contrasted trials in which the first person had to start either directly in front of his or her own body (no workspace overlap) with those trials in which the movement had to start in the work area of the other person (workspace overlap). We hypothesized a delay in the case that a workspace had to be used by both persons at the same time compared to the no workspace overlap condition.

Temporal Coordination. Results provided first evidence for some of the hypothesized effects. As can be seen in Fig. 2, people systematically coordinated the timing of their responses. The onset of a movement both at pick up (bottom) and put down positions (top) is strongly coupled with the partner's movement pattern.

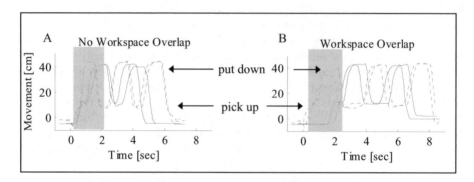

Fig. 2. Temporal coordination in joint performance: vertical movements of one pair of participants [smooth lines: left-sitting person, dotted lines: right-sitting person]. A: No workspace overlap. B: Workspace overlap between first and second person which leads to a relative delay of the movement onset of the second person [transparent area: coordination of first movement onsets].

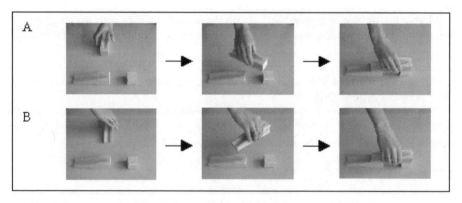

Fig. 3. Different initial block orientations lead to different grip orientations, even if the goal position is the same. A: A comfortable grip is possible also during grasping. B. Here, a rather uncomfortable grip has to be used during pick up in order to secure maximum comfort and stability while placing the block to its end position.

Workspace Overlap. However, the exact timing of their coordinative behavior was modulated by the task demands. In case of overlapping workspaces (Fig. 2B) the second person's movement onset was more delayed compared to the no overlap condition (Fig. 2A). In fact the movement only started with the end of the first person's movement because the second person had to wait until the first person had left the mutual workspace.

Hand Orientation. Finally, we could also observe that hand orientation during grasping depended on the relation of block orientation at the initial and the goal position. That means that participants grasped the blocks differently when the blocks were rotated at their initial or goal position in order to acquire a comfortable and safe hand orientation during placing the blocks (Fig. 3).

3 Discussion

The aim of this paper was to introduce a scenario that allows experimental investigation of a variety of questions related to movement coordination in humans and robots. We presented the theoretical background, the methodology and an overview of relevant variables. In addition, results of first experiments indicate that the timing of interpersonal coordination strongly depends on the task context. Specifically, partners have to delay their movements in the case that the mutual workspace overlaps. Our long-term goal is to extract movement parameters in human-human interaction in order to transfer results to applied situations in which robot and humans work together. We consider this a useful approach to the examination of ways in which the growing use of robots in all major areas of society can be supported, e.g. in applications like household assistance systems or in professional domains such as industry or health care. Adapting robot performance to principles observed in human-human interaction will make human-robot interaction safer and will improve the robots' acceptance by human interaction partners.

3.1 Implications for Human-Robot Interaction in Health Care

Generally, robots have proven to be useful in manifold ways [2]. For example, industrial robots support work in factory assembly and perform high-precision tasks or operate in highly dangerous environments. Service robots directly interact and operate in the same physical space as humans. In health care, robots assist during surgery provide helpful information for diagnosis and facilitate rehabilitation [20], [25]. In many of these scenarios, the interaction between robot and human is indirect, that is, the human user controls the robot; and usually both operate from spatially separated locations [2]. More challenging, however, and of future relevance in human-robot interaction are robot applications that involve direct interaction where humans and robots act in the same physical space. In health care, such applications may be robots assisting elderly people, people with disabilities or robots in rehabilitation. Rehabilitation is a wide field that reaches from half-automatic wheelchairs and other systems providing physical assistance [19], [25], [26] to training robots up to approaches of psychological enrichment for depressed or elderly people [26], [27]. These applications require special attention to safety issues as highly autonomous and adaptive robots are used. It has to be guaranteed that humans and robots act without posing a risk to each other, without hindering or disrupting each other's actions. One promising way to meet these requirements is to design robots that are capable of predicting human movement behavior as well as of planning and executing their own effector movements accordingly. In particular, robots are needed that "know" to a certain degree in advance whether a human movement is planned, where the movement will terminate, towards what goal it is directed and which movement path may be taken. Such mechanisms would enable the robot to stop or modify an own movement in case the human comes close, is likely to interfere or may be hurt by the actual robot movement. In such a case, the robot may autonomously calculate a new trajectory that does not interfere with the human movement path.

Besides the safety aspect, these applications impose additional challenges. First, systems must be relatively easy to operate because most users are not specifically trained with them. Second, robots interacting directly with patients have to be designed in a way that they are accepted by their human user [28].

Our work provides basic steps towards building such robots. By examining human ways of dealing with a shared workspace in human-human interaction we aim at extracting general principles that facilitate coordination. If we come closer to answering the question how one person can estimate and predict the movement parameters of another person, we might be able to use this knowledge for professional applications such as in assistive robots in health care. Additionally, transferring human-human interaction principles to human-robot interaction may facilitate robot handling and increase the acceptance of the human user.

3.2 Future Directions

State-of-the-art robots can nowadays perform many sophisticated tasks. Direct robot-human interaction in shared physical space poses one of the biggest challenges to robotics because human movements are hard to predict and highly variable. Breaking down these complex biological movements into general principles and constraints and applying similar rules as in human-human interaction will help to train robots in natural tasks.

A future aspect of human-human and human-robot interaction may lie in the combination and integration of different operators' actions into the same frame of reference. Similar to intrapersonal control of goal-directed behavior, movement goals and end-states may play an important role in human-human and human-robot action coordination, as they provide a common external frame of reference that may be taken to program common actions [4] as well as for establishing a shared conceptualization in verbal communication [29]. Referring to a common reference may also become relevant in surgery when high-expert teams interact from separate spatial locations with restricted direct visual contact.

Finally, as our scenario consists of predefined single steps only, it lacks the necessity of verbal communication between the interaction partners to establish a common ground. The investigation of more complex cooperation tasks with uncertainty about the next performance step will give insights into the role of verbal communication in action coordination [29].

To conclude, the value of understanding the parameters which determine the outcome of internal forward and inverse models in humans is twofold: First, mechanisms observed in human intrapersonal movement coordination can be used in order to solve the degree of freedom problem in movement trajectory planning of robots. Second, transferring results from interpersonal movement coordination will enhance adaptive behaviors in machines and allow the safe and efficient collaboration of human workers and robots.

Acknowledgments. This work was funded in the Excellence Cluster "Cognition for Technical Systems" (CoTeSys) by the German Research Foundation (DFG). The authors would like to thank their research partners Florian Engstler, Christian Stößel and Frank Wallhoff for fruitful discussions and contributions in the CoTeSys project "Adaptive Cognitive Interaction in Production Environments" (ACIPE).

References

1. Rosenbaum, D.A.: Human Motor Control. Academic Press, San Diego (1991)
2. Thrun, S.: Toward a framework for human-robot interaction. Human-Computer Interaction 19, 9–24 (2004)
3. Russel, S., Norvig, P.: Artificial Intelligence: A Modern Approach, 2nd edn. Prentice Hall, Englewood Cliffs (2003)
4. Stubbs, K., Wettergreen, D., Hinds, P.J.: Autonomy and common ground in human-robot interaction: a field study. IEEE Intelligent Systems, 42–50 (2007)
5. Wolpert, D.M., Kawato, M.: Multiple paired forward and inverse models for motor control. Neural Networks 11, 1317–1329 (1998)
6. Wing, A.M., Turton, A., Frazer, C.: Grasp size and accuracy of approach in reaching. Journal of Motor Behavior 18, 245–260 (1986)
7. Uno, Y., Kawato, M., Suzuki, R.: Formation and control of optimal trajectory in human multijoint arm movement. Biological Cybernetics 61, 89–101 (1989)
8. Rosenbaum, D.A., Meulenbroek, R.G.J., Vaughan, J.: Planning reaching and grasping movements: Theoretical premises and practical implications. Motor Control 2, 99–115 (2001)
9. Cuijpers, R.H., Smeets, J.B.J., Brenner, E.: On the relation between object shape and grasping kinematics. Journal of Neurophysiology 91, 2598–2606 (2004)
10. Bennis, N., Roby-Brami, A.: Coupling between reaching movement direction and hand orientation for grasping. Brain Research 952, 257–267 (2002)
11. Rosenbaum, D.A., van Heugten, C.M., Caldwell, G.E.: From cognition to biomechanics and back: The end-state comfort effect and the middle-is-faster effect. Acta Psychologica 94, 59–85 (1996)
12. Ansuini, C., Santello, M., Massaccesi, S., Castiello, U.: Effects of end-goal on hand shaping. Journal of Neurophysiology 95, 2456–2465 (2006)
13. Cohen, R.G., Rosenbaum, D.A.: Where grasps are made reveals how grasps are planned: generation and recall of motor plans. Experimental Brain Research 157, 486–495 (2004)
14. Knoblich, G., Jordan, J.S.: Action coordination in groups and individuals: learning anticipatory control. Journal of Experimental Psychology: Learning, Memory, and Cognition 29, 1006–1016 (2003)
15. Sebanz, N., Bekkering, H., Knoblich, G.: Joint action: bodies and minds moving together. Trends in Cognitive Sciences 10, 70–76 (2006)

16. Meulenbroek, R.G.J., Bosga, J., Hulstijn, M., Miedl, S.: Joint action coordination in transfering objects. Experimental Brain Research 180, 333–343 (2007)
17. Wolpert, D.M., Doya, K., Kawato, M.: A unifying computational framework for motor control and social interaction. Philosophical Transactions of the Royal Society of London 358, 539–602 (2003)
18. Egersdörfer, S., Dragoi, D., Monkman, G.J., Füchtmeier, B., Nerlich, M.: Heavy duty robotic precision fracture repositioning. Industrial Robot: An International Journal 31, 488–492 (2004)
19. Saint-Bauzel, L., Pasqui, V., Gas, B., Zarader, J.: Pathological sit-to-stand predictive models for control of a rehabilitation robotic device. In: Proceedings of the International Symposium on Robot and Human Interactive Communication, pp. 1173–1178 (2007)
20. Camarillo, D.B., Krummel, T.M., Salisbury, J.K.: Robot technology in surgery: past, present, and future. The American Journal of Surgery 188, 2S–15S (2004)
21. Schaal, S.: Is imitation learning the route to humanoid robots? Trends in Cognitive Sciences 3, 233–242 (1999)
22. Breazeal, C., Scassellati, B.: Robots that imitate humans. Trends in Cognitive Sciences 6, 481–487 (2002)
23. Polhemus: http://www.polhemus.com
24. Rizzolatti, G., Fogassi, L., Gallese, V.: Neurophysiological mechanisms underlying the understanding and imitation of action. Nature Reviews 2, 661–670 (2001)
25. Taylor, R.H.: A perspective on medical robotics. Proceedings of the IEEE 94, 1652–1664 (2006)
26. Shibata, T.: An overview of human interactive robots for psychological enrichment. Proceedings of the IEEE 92, 1749–1758 (2004)
27. Kanade, T.: A perspective on medical robotics. In: International Advanced Robotics Program Workshop on Medical Robotics (2004)
28. Breazeal, C.: Social interactions in HRI: the robot view. IEEE Transactions on Systems, Man, and Cybernetics – Part. C: Applications and Reviews 34, 181–186 (2004)
29. Brennan, S.E., Clark, H.H.: Conceptual pacts and lexical choice in conversation. Journal of Experimental Psychology: Learning, Memory, and Cognition 22, 1482–1493 (1996)

An Orientation Service for Dependent People Based on an Open Service Architecture

A. Fernández-Montes[1], J. A. Álvarez[1], J. A. Ortega[1],
Natividad Martínez Madrid[2], and Ralf Seepold[2]

[1] Universidad de Sevilla, 41012 Sevilla, Spain
Escuela Técnica Superior de Ingeniería Informática,
alejandro.fdez@gmail.com, {juan,ortega}@lsi.us.es
[2] Universidad Carlos III de Madrid, 28911 Leganés, Madrid, Spain
Departamento de Ingeniería Telemática
{natividad.martinez, ralf.seepold}@uc3m.es

Abstract. This article describes a service architecture for ambient assisted living and in particular an orientation navigation service in open places for persons with memory problems such as those patients suffering from Alzheimer's in its early stages. The service has the following characteristics: one-day system autonomy; self-adjusting interfaces for simple interaction with patients, based on behavioural patterns to predict routes and destinations and to detect lost situations; easy browsing through simple spoken commands and use of photographs for reorientation, and independence of GISs (Geographic Information Systems) to reduce costs and increase accessibility. Initial testing results of the destination prediction algorithm are very positive. This system is integrated in a global e-health/e-care home service architecture platform (OSGi) that enables remote management of services and devices and seamless integration with other home service domains.

Keywords: Health care, dependent people, Alzheimer, service platform, OSGi, orientation service.

1 Introduction

The World Health Organization uses the term e-health to explain the relations between institutions, public health, e-learning, remote monitoring, telephone assistance, domiciliary care and any other system of remote medicine care. Each aspect of this very wide spectrum has undergone major technical improvement in recent years, however health care systems often lack adequate integration among the key actors, and also commonly fail to take certain social aspects into account which slow down the acceptance and usage of the system.

The social groups addressed by the work presented in this paper are made up of elderly or disabled people. Elderly people especially need to interact with health care services in a transparent and non-intrusive way, since their technical abilities are limited in many cases. Currently, some initiatives specifically address the training of elderly people to handle modern interfaces for assisted living [1], and elderly people

A. Holzinger (Ed.): USAB 2007, LNCS 4799, pp. 155–164, 2007.

are also the target of a EU project called SENIORITY [2], to improve the quality of assistance to elder people living in European residences or at home by means of integrating a quality model with new telemonitoring and telecommunications devices. Several design aspects need to be specially taken into consideration for elderly users, considering for instance physical [3] or visual [4] accuracy. Therefore, one of the objectives of the service architecture presented here is to offer a Human Computer Interface (HCI) which avoids technological barriers to elderly or disabled people.

Furthermore, there is another factor influencing the market penetration of health care services. Daily care for dependent people is often organized in two unconnected, parallel ways. On the one hand, dependent people always prefer to contact first of all their relatives and friends if they need anything. According to several studies, dependent people are reluctant to use many health care services because they do not personally know the operator or contact person in the service centre, and hence only use these services in emergency cases. Therefore, another objective of this work is to integrate these relatives and friends into the health care service provision, in an effort to increase the usability of the system.

A first scenario for the proposed service architecture addresses the mobility support for Alzheimer patients. For both these and for people suffering orientation problems or mild cognitive impairment (MCI), daily activities that require leaving the home and moving within the city or town present an important challenge, a high risk of getting lost and a certain possibility of accidents. In these situations, this group of people would benefit from personal navigation systems with simple human interfaces which would help them find the appropriate route, guiding them if necessary to their goal without configuring the system. The concrete target group in the study are the members of the Alzheimer Sevilla association (http://www.alzheimersevilla.com) that have provided the requirements and supported the tests.

The main objective of the service designed for this scenario is to develop a system that enables, in open areas, the detection of lost or disorientated persons suffering from Alzheimer's in its early or intermediate stages, or from similar psychical problems. Potentially dangerous situations can be prevented with the assistance of a system with a follow-up and intelligent navigation functionality which is able to distinguish the moment at which a patient loses orientation and can therefore help him reach his destination.

The next section presents the state of the art and introduces the proposal. The general service architecture of the system is then presented, and the orientation service is described in detail. Finally, some results and conclusions are presented.

2   State of the Art

The capacity of orientation and path-following of both humans and animals has been thoroughly studied [5] [6] [7]. Navigation or "wayfinding" systems have evolved from textual descriptions to 3-D interfaces with audio [8]. For persons with reduced or no sight, these types of systems have advanced considerably, with work such as [9], although these systems are commonly centred on navigation within interior areas.

Furthermore, the consequences of mental disabilities in learning and following paths have been studied [10], as well as the problems people have when trying to use public transport [11]. These orientation problems affect the independence and social life of these people. Considerable medical research suggests that one of the best ways to prolong their independence is to help them complete their daily routines. Several papers address the reorientation of people with mental disabilities in open [12] and in closed areas [13].

In this paper, a reorientation system for open areas is presented with the following characteristics: System autonomy of approximately 24 hours; self-adjusting interfaces for simple interaction with patients based on predictions of destinations to support decision-making; easy browsing through simple spoken commands and photographs, and independence of GISs (Geographic Information Systems), due to their high cost which would reduce accessibility. Instead, web planners such as Google Maps or Yahoo Maps are used, and are complemented with information about public services such as buses, trams or trains.

Moreover, this system is going to be integrated in a global e-health/e-care home service architecture. Several approaches address the integration of home-collected/monitored patient data and the use of mobile devices to view the information or receive alerts, like [14]. In the related work, several monitoring sensors form a body network on the patient or user and communicate with a base station to store the information, which can be visualized by the patient or by the medical personnel with a PDA. The approach has some commonalities with our work, in the sense that several monitoring data can be recorded and stored, but we use a service platform (OSGi) that allows remote management of services and devices (relevant if the users have no special technical knowledge) and seamless integration with other home service domains (like communication and audiovisual), with the purpose of allowing a direct participation of relatives and friends in the e-care services.

3   E-Care Service Architecture

The proposed service is part of a general e-health and e-care service platform, whose architecture is depicted in Fig 1. The general scheme is divided into three environments: the home of the dependent person, the home of one (or several) relatives or friends, and the medical/care centre. One of the main goals of this service architecture is to involve trusted people as well as professional people in the care activities.

In the home of a dependent person, several networks and devices can exist: medical, audiovisual and automation networks with different devices connected by wire or wireless to a residential gateway (RGW) with an embedded OSGi framework. (de facto standard for services in home, personal or automotive gateways). Heart-rate and blood-pressure monitors, and blood glucometers are examples of integrated devices in the medical network. The automation network includes sensors and actuators, for example light sensors and heating actuators. The audiovisual network typically includes a television, an IP camera or a webcam with microphone, necessary for the dependent person to communicate with relatives and carers. The residential gateway is able to physically interconnect all networks and devices, and to host the different services which can be managed remotely by the care service provider and/or by relatives or friends. The environment in the home of the relative only needs the audiovisual network for the communication, while the Medical/Assistance Centre requires servers to store the medical data. Medical staff and carers access information and communicate with the dependent person via PCs with audiovisual capabilities.

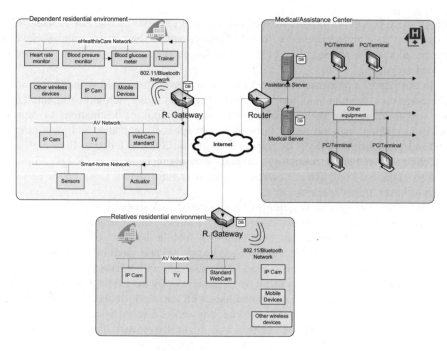

Fig. 1. General architecture of the e-care system

If the dependent person is outside the home, the same architecture applies, the gateways being a mobile phone or PDA. External communication takes place through GPRS/UMTS/Wi-Fi, and communication with the PAN (Personal Area Network) is carried out via Bluetooth or infrared (for instance, with a pacemaker). The services may use the audio and video facilities of the mobile device.

4 Orientation Service: *InMyOneWay*

The following scenario illustrates the orientation service that is presented in this paper. Pedro is 68 and suffers from Alzheimer in an early stage. Although he still has enough cognitive faculties to live alone, his family is afraid that he may get lost on the street and be unable to find his way home, so he is using the *InMyOneWay* system.

Unfortunately, one day Pedro loses his bearings and gets lost. Although he is on one of his usual routes, he is not able to recognize where he is. In his anxiety, he cannot remember that he is wearing an intelligent device, but after some instants of wandering, the device detects a strange behaviour pattern inside a known route. It automatically sends SMSs to relatives, indicating the exact position and a description of the behaviour. After one minute, no relative has called Pedro, therefore the device itself vibrates, and Pedro picks up the terminal. It shows four pictures: two of destinations where the system predicts that he is going to, and other two of his closest reference places, and tells him with text and voice to choose where he wants to go to. Once Pedro has chosen one of them, the orientation service specifically indicates how to reach it, using simple instructions and pictures of the places where he is walking. Shortly after reaching the reference place following the indications, his daughter phones asking whether he wants to be picked up, or if he knows where he is and does not need help. After the fright, Pedro prefers to be picked up, which is quite easy, as his daughter can check the street and house number closest to his position.

The functionality and data are distributed between the mobile device itself, the residential gateway, and the Internet (Fig. 2). The business logic and the information about the routes are in the mobile device. Connections to the network are sporadic to update information about the places along the usual routes of the user. These periodical updates are carried out using the residential gateway, which uses information coming from the Internet about public transport and services, traffic information, street repairs affecting the routes and new pictures of the area. Furthermore, the residential gateway processes the data to make it accessible to the terminal via Web services.

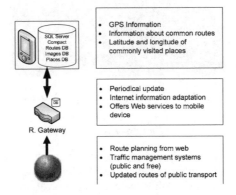

Fig. 2. Service function and data distribution

This architecture places most processing in the mobile device which needs to periodically recover the GPS position, calculate distances with the routes in the database, predict the destination and use pictures for navigation. Nevertheless, there are many devices allowing these computation and communication capabilities for a reasonable price, since the trend in the mobile phone market is heading towards a fusion between PDA and phone devices, handled with one or two hands, with high resolution screens and several communication interfaces.

We interviewed the people responsible for the Alzheimer Sevilla association (http://www.alzheimersevilla.com) to obtain the requirements for the devices. These devices should (1) be small and light enough not to annoy the carrier, (2) not attract attention to prevent theft, and (3) provide interaction that is simple and almost without buttons since people with Alzheimer find complex devices difficult to handle.

An HTC P3300 terminal was used, integrating the GPS, phone and PDA functionalities. This device fulfils the first two requirements. To improve the HCI, the picture-based interface previously described was used, as shown in Fig. 3. The photographs were not taken with the intention of being a part of a subsequence of images to give route directions, like in the studies of [15].

Fig. 3. HTC P3300 with Adaptive Interface

The complete functionality is based on several sub-processes:

1. Positioning. Through the integrated GPS chip, the receiver obtains the latitude, length and height once every second (configurable period).
2. Destination prediction. With the cumulated positioning data and the device databases, the most probable destinations of the user are calculated.
3. Lost detection. With the information of the detected destinations and current route, several patterns indicating disorientation situations are examined.
4. Navigation. Once a disorientation pattern has been detected, a communication with the user is established in order to offer pictures and instructions that allow him to regain his orientation and reach the desired destination.
5. Communication. When risk situations are detected, and taking into account that a fundamental requirement is to use simple interfaces and even to avoid interfaces for certain users, the communication capability of the device (GSM/GPRS) is used to make calls or send SMSs to relatives or carers without the need for the user to write anything.

5 Previous Experience Modelling

In order to retrieve information on the daily activity of the Alzheimer patients when they are outside it is necessary to build a model for previous experiences of their journeys. To this end, two techniques are used: (1) Retrieval of previous routes based on a GPS device, and (2) Route generation for frequently visited places using several points of reference.

The first technique makes it possible to record past routes which accurately represent the daily dynamic of the patient. However, since this method provides no information on where the dependant can get lost, this situation is counteracted by generating routes from information supplied by the user such as data about his residential location, the stores where the dependant goes shopping, his relatives' residential locations and the medical centre where he goes if he suffers any illness.

The data about streets and numbers where these places are located was linked to the corresponding geographical coordinates of latitude and longitude (geocoding process) using existing web tools such as Google Maps and Yahoo Maps APIS.

6 Destination Prediction

A journey (or path) can be defined as a set of points with information on sequentially ordered latitudes and longitudes. Additional information can be included, such as height or speed relative to the previous point.

When adaptive interfaces based on the activity performed are to be generated, understanding where the user is heading is of prime importance. If this goal is accomplished, the man-machine interaction is successful and image selection of possible destinations is more accurate. Software must compare points from the current journey to points of stored journeys to determine the possible destinations. To study the similarity of two paths, some distance functions have been defined in order to disregard irrelevant details such as shortcuts or short deviations due to roadworks.

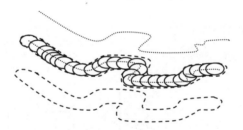

Fig. 4. Generation of the scope of a path

Scope of a path: Given a labelled path $X_{A-B} = \{ p_0, p_1, ..., p_n \}$ and a point q, this point belongs to the scope of the path if there exists some point of the path whose distance to q is less than a given value of δ:

$$q \in scope(X_{A-B}), \; if \; \exists j \mid dist(p_j, q) \leq \delta.$$

In this paper it is determined that $\delta = 85$ meters is an acceptable distance to consider. A point is in the scope of a path if it belongs to the generated region (see Fig. 4).

Similarity of paths: One path matches another if it has a certain percentage of points from their total that belong to the scope of the other. The similarity level is expressed in the following manner:

Given $X_{A-B} = \{ p_0, p_1, ..., p_n \} \wedge Y_{C-D} = \{ q_0, q_1, ..., q_m \}$, we define:

Similarity($X_{A-B} Y_{C-D}$)=100*(Number of qi ∈ scope(X_{A-B}))/m)

Hence, given X_{A-B}={ $p_0, p_1, ..., p_n$ } ^ Y={ $q_0, q_1, ..., q_m$},

Y is identical to X_{A-B} if ∀i, q_i ∈ scope(X_{A-B}) then Y is labelled with the class A-B and therefore becomes a labelled path: Y_{A-B}

Since the number of points of the two paths is not necessarily the same, the features are not commutative. Given X={ $p_0, p_1, ..., p_n$ } ^ Y={ $q_0, q_1, ..., q_m$ }

similarity(X,Y)= similarity(Y,X) if and only if n=m.

Having checked the similarity index with all the representative (canonical) paths over all the points of the non-canonical path, the predicted destination in each point was obtained. The point when the real destination was predicted not being altered until the end of the path, was called **detection point**. Results were coherent with common sense: Until the itinerary did not enter in a non common area with other paths, it was impossible to distinguish where we go. Nonetheless, we also observed that although there were overstrike paths, frequently small changes taken on the paths allowed the system to distinguish the correct destination. The results (shown in Fig. 5) were very hopeful because the remaining distance to the destination measured in a straight line, on average was 69,65% of the total path and the remaining time to reach the destination was more than the 70% of the total (when the journey finally stops).

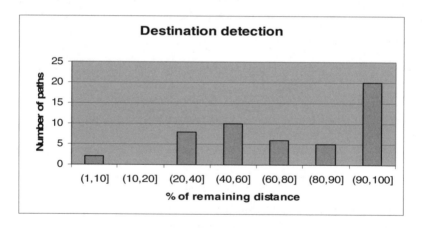

Fig. 5. On-line results of the destination detection algorithm

It was considered that a new detection point that remained stable for some meters, could be defined as a **decision point**. The spatial pertinent value seemed to be 120 meters. This value adds this amount of time and space to the decision point, being specially harmful for the paths in which the decision arrived very late (cases on the left of the graphic where the remaining distance to reach the destination is less than the 10% of the total distance path), as it causes that the arrival to the destination comes before we can predict it. In summary, the results were very positive: from the data collected about paths travelled during a month and five days, we got the actual destination in 98% of cases, after having made only 30,35% of the total path.

7 Lost Detection

In the same manner, the detection of disoriented dependants is essential to the system in order to be useful. The following situations were detected as a risk for the dependant.

- **Lost during known journeys or at known places**. When the dependant gets lost at a known place, similarity-of-path techniques are not applicable because the dependant has not left the route. Hence patterns which can show us that the dependant has got lost are required and these are defined as long delays at intermediate points which are not public transport elements (waiting for a bus is something normal).
- **Lost in new places**. When the system detects that the user is leaving a frequently used path can ask the dependant if he knows where he is going. If the answer is affirmative the system will be quiet the rest of the journey. Otherwise the dependant is offered some images of his frequently visited places.

8 Results and Conclusions

The *InMyOneWay* system is initially configured for the city of Seville and its public bus transportation service, but can be easily adapted to other cities. The system is currently in the testing phase. The pure functionality can be tested by the development team, but not the usability. The field evaluation is complicated, since the number of targeted users is reduced. Many families only realize that one of its members is suffering from Alzheimer's when he/she has already got lost and from this point do not trust them alone in the city. The ideal tester would be Alzheimer patients whose illness is detected in the early stages, or elderly people with (or without) memory problems who can make use of and evaluate the system, and obtain the extra benefits of using the scheduling capabilities of the system, such as referring to bus timetables, or exploring new routes for certain destinations.

Moreover, the cost of obtaining adequate pictures (for orientation and not for artistic purposes) to cover a complete city is very high. A Web portal is under preparation to allow the sending of pictures and video sequences of routes as a basis for a wider navigation system. Regarding the improvements in the detection of disorientation patterns, two options are being considered. One is the introduction of accelerometers and gyroscopes to detect fine scale movements and recognize strange movements which can be considered disorientation symptoms, such as turning around several times. The other option is the introduction of heart pulse sensors to help detect states of nervousness.

Acknowledgments. This research is supported by the Spanish Ministry of Education and Science under the +D project InCare (Ref: TSI2006-13390-C02-02).

References

1. Kleinberger, T., Becker, M., Ras, E., Holzinger, A., Müller, P.: Ambient Intelligence in Assisted Living: Enable Elderly People to Handle Future Interfaces. In: Stephanidis, C. (ed.) Universal Access in HCI, Part II, HCII 2007. LNCS, vol. 4555, pp. 103–112. Springer, Heidelberg (2007)
2. The SENIORITY EU project. Online at: http://www.eu-seniority.com/ (last access: 2007-09-01)
3. Holzinger, A., Searle, G., Nischelwitzer, A.: On some Aspects of Improving Mobile Applications for the Elderly. LNCS, vol. 4554, pp. 923–932. Springer, Heidelberg (2007)
4. Holzinger, A., Sammer, P., Hofmann-Wellenhof, R.: Mobile Computing in Medicine: Designing Mobile Questionnaires for Elderly and Partially Sighted People. In: Miesenberger, K., Klaus, J., Zagler, W., Karshmer, A.I. (eds.) ICCHP 2006. LNCS, vol. 4061, pp. 732–739. Springer, Heidelberg (2006)
5. Álvarez, J.A., Ortega, J.A., González, L., Velasco, F., Cuberos, F.J.: Ontheway: a prediction system for spatial locations. Winsys (August 2006)
6. Golledge, R.G: Wayfinding behaviour: Cognitive mapping and other spatial processes. John Hopkins University press, Baltimore (1999)
7. Golledge, R.G., Stimson, R.J.: Spatial Behavior: A Geographic Perspective. In: Spatial and Temporal Reasoning in Geographic Information Systems, Guildford Press, New York (2004)
8. Smith, S.P., Hart, J.: Evaluating distributed cognitive resources for wayfinding in a desktop virtual environment. In: 3dui, vol. 00, pp.3-10 (2006)
9. Tjan, B.S, Beckmann, P.J., Roy, R., Giudice, N., Legge, G.E.: Digital Sign System for Indoor Wayfinding for the Visually Paired. In: Cvprw, vol. 0, pp.30 (2005)
10. Moore, M., Todis, B., Fickas, S., Hung, P., Lemocello, R.: A Profile of Community Navigation in Adults with Chronic Cognitive Impairments. Brain Injury (2005)
11. Carmien, S., Dawe, M., Fischer, G., Gorman, A., Kintsch, A., Sullivan, J.: Socio-technical environments supporting people with cognitive disabilities with using public transportation. ACM Trans. Comput-Hum. Interact. 12(2), 233–262 (2005)
12. Patterson, D.J., Liao, L., Gajos, K., Collier, M., Livic, N., Olson, K., Wang, S., Fox, D., Kautz, H.: Opportunity knocks: A system to provide cognitive assistance with transportation services. Ubiquitous Computing (2004)
13. Liu, A.L., Hile, H., Kautz, H., Borriello, G.: Indoor wayfinding: Developing a functional interface for individuals with cognitive impairments. In: Assets 2006. Proceedings of the 8th International ACM SIGACCESS Conference on Computing and Accesibility, pp. 95–102. ACM Press, New York (2006)
14. Ahamed, S.I, Haque, M., Stamm, K., Khan, A.J.: Wellness Assistant: A Virtual Wellness Assistant on a Handheld Device using Pervasive Computing. In: ACM SAC 2007. Proceedings of the 22nd Annual ACM Symposium on Applied Computing, Seoul, Korea, pp. 782–787 (2007)
15. Beeharee, K., Steed, A.: A natural wayfinding exploiting photos in pedestrian navigation systems. In: Mobile HCI 2006. Proceedings of the 8th conference on Human-computer interaction with mobile devices and services, pp. 81–88 (2006). ISBN:1-59593-390-5

Competence Assessment for Spinal Anaesthesia

Dietrich Albert[1], Cord Hockemeyer[1], Zsuzsanna Kulcsar[2], and George Shorten[2]

[1] University of Graz, 8010 Graz, Austria
Department of Psychology
Cognitive Science Section
dietrich.albert@uni-graz.at,
cord.hockemeyer@uni-graz.at
[2] HSE South, Cork University Hospital, Cork, Ireland
Department of Anaesthesia and Intensive Care Medicine
george.shorten@mailp.hse.ie,
zsuzsanna.kulcsar@gmail.com

Abstract. The authors describe a new approach towards assessing skills of medical trainees. Based on experiences from previous projects with (i) applying virtual environments for medical training and (ii) competence assessment and personalisation in technology enhanced learning environments, a system for personalised medical training with virtual environments is built. Thus, the practical training of motor skills is connected with the user-oriented view of personalised computer-based testing and training. The results of this integration will be tested using a haptic device for training spinal anaesthesia.

Keywords: Competence Assessment, Competence-based Knowledge Space Theory, Spinal Anaesthesia, Usability in Ambient Assisted Living and Life Long Learning, Simulations in Medicine.

1 Introduction

The current methods used for training in medical procedural (or technical) skills are inefficient and may jeopardise patient safety as medical trainees are required to practice procedures on patients. The resultant worldwide move towards competence-based training programmes has necessitated the search for valid and reliable competence assessment procedures (CAPs). The challenges in developing such CAPs lie in defining each competence and taking account of the many factors which influence learning and performance of medical procedures. Such determinants include cognitive, motor, communication, and emotional (e.g. fatigue, anxiety, or fear) factors. In other domains, competence-based knowledge space theory (CbKST) has been successfully applied to enhance learning, assess competence and facilitate personalised learning [see, e.g., 1, 2, 3]. The objective of our starting project is to transfer this approach to the medical domain in order to develop a valid, reliable and practical CAP for one medical procedural (and motor) skill, spinal anaesthesia. In order to do so, we will comprehensively describe the competences, generate algorithms necessary to assess individual performance, implement the CAP in a user-friendly, web-based format and test it in simulated and real clinical settings for construct validity and reliability.

A. Holzinger (Ed.): USAB 2007, LNCS 4799, pp. 165–170, 2007.

In the following sections, we give short introductions to spinal anaesthesia and to competence-based knowledge space theory. Section 4 finally describes the work approach for the new project.

2 The Spinal Anaesthesia Technique

Regional anaesthesia is a major component of modern anaesthetic practice. It means blocking the nerve supply to an area of the body so the patient cannot feel pain in that region. Many procedures can be performed on patients who are awake, using regional anaesthesia. This may avoid the risk and unpleasantness sometimes associated with general anaesthesia and may provide specific benefits.

The most common type of regional anaesthesia is *spinal anaesthesia*. This involves putting local anaesthetic through a needle near the spinal cord into the fluid, which surrounds it to anaesthetize the lower region of the body (see Fig. 1).

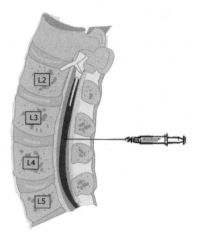

Fig. 1. Spinal anaesthesia

Spinal anaesthesia quickly blocks pain with a small amount of local anaesthetic, it is easy to perform and has the potential to provide excellent operating conditions for surgery on the lower part of the abdomen and legs. The local anaesthetic agents block all types of nerve conduction and prevent pain and may also prevent movement in the area until the block wears off. The effects of different local anaesthetics last for different lengths of time, so the length of the anaesthetic can be tailored to the operation.

Teaching spinal anaesthesia to medical trainees traditionally follows two steps, (i) teaching declarative knowledge, e.g., on anatomy, and (ii) supervised practical training on patients. Due to cost factors and to recent changes in European work laws, the second step of this education has to be strongly reduced. The resulting gap in medical training shall be compensated through computer supported competence assessment and (later on) competence teaching.

The DBMT project (Design Based Medical Training, see http://www.dbmt.eu/] has been developing a haptic device supporting a first-stage practical training in spinal anaesthesia in a virtual environment [4]. Figure 2 shows a photo of the device in its current stage of development. This device allows the trainee to obtain the sensory experience connected of performing spinal anaesthesia without putting patients to the risk that it is applied by practical novices.

Fig. 2. Training device for spinal anaesthesia

3   Competence-based Knowledge Space Theory (CbKST)

Knowledge space theory (KST) was originally developed for the adaptive assessment of knowledge in a behaviouristic paradigm [5, 6]. The research groups around Albert in Heidelberg and Graz extended this approach by connecting the observed behaviour to its underlying competences [7]. These extensions then built the basis for developing approaches for personalised technology enhanced learning [8, 2, 3].

The basic idea underlying KST and CbKST is to structure a domain of knowledge by prerequisite relationships. The key idea is to assign to each competence a set of prerequisite competences, i.e. a set of other competences which have to be acquired first[1]. If we define the *competence state* of a trainee as the subset of competences within the regarded domain of knowledge that this trainee has acquired then the set of possible competence states, the *competence space,* is restricted by the prerequisite relation. The prerequisite relation then provides a good basis for personalised assessment and training of competences thus imitating a private teacher.

For the personalised assessment [6, Chapter 9-11], we start testing a competence of medium difficulty. Depending on the correctness of the trainee's answer, we can draw conclusions not only on the mastery of the actually tested competence. In case of a correct answer, we have also obtained evidence for the mastery of the prerequisites. Vice versa, in case of a wrong answer, we have also obtained evidence for the

---

[1] In a more advanced approach, the prerequisites of a competence are modelled as a family of prerequisite competence sets, the so-called clauses. Each clause can represent, e.g., a different approach to teach the respective competence or a different path to solve a respective problem.

non-mastery of those competences for which the actually tested competence is a pre-requisite. Modelling the trainee's competence state through a likelihood distribution on the set of all possible competence states, each piece of evidence leads to an update of this likelihood distribution. Within the update, the likelihood of states consistent with the observed behaviour is increased while the likelihood of inconsistent states is decreased. Subsequent competences to be tested are selected on the basis of this up-dated distribution. The assessment procedure selects always competences of medium difficulty to be tested where the difficulty of competences depends on the current likelihood distribution.

Previous experiences and simulations show that this procedure leads to a drasti-cally reduced number of competences to be actually tested [1]. Furthermore, this assessment delivers a non-numeric result, i.e. we do not only know the percentage of competences mastered by the trainee but we also know exactly which competences are mastered and which competencies still need some training.

This leads to the application of the competence space for personalised learning. Based on the non-numerical assessment result one can easily determine which compe-tences need to be trained, and based on the prerequisite relation one can derive in which order the missing competences should be acquired. Thus we avoid frustrating our train-ees by trying to teach them competences for which they lack some prerequisites.

The techniques described herein have been applied successfully in several research prototypes [see, e.g., 8, 2], and in the commercial ALEKS system (Assessment and Learning in Knowledge Spaces, see http://www.aleks.com/), a system for the adaptive testing and training of mathematics in and beyond the K12 area.

More recently, CbKST has been applied for assessing learners' competences im-plicitly, i.e. based on their ongoing actions instead of explicitly posing problems to the learner [9, 10, 11]. Especially the situation in game-based learning bears many similarities to virtual environments for practical medical training.

4 Competence Assessment for Spinal Anaesthesia

The project "Competence Assessment for Spinal Anaesthesia" is a collaborative pro-ject involving five partners from four countries. A course is built based on the device and contents developed within the DBMT project. After defining the overall set of competences involved in this course, specific competences are assigned to each ele-ment of the course. A training system is implemented in a modular, service oriented architecture using existing parts where possible. Figure 3 shows a sketch of the ele-ments of this system.

The experiences from applying CbKST to game-based learning [10,12] show its high suitability for the non-numeric assessment of competences in complex task solu-tion processes. Observing the trainees every step in the virtual reality training envi-ronment, the system cannot only judge whether the spinal anaesthesia would have been applied successfully but it will also determine which competences need more training or are missing at all. In a further development to a full-fledged e-learning system it would also offer appropriate teaching and training material on these compe-tences. In the sequel, the architecture of such a full-fledged e-learning system on spi-nal anaesthesia is briefly sketched.

The learner communicates through the Web with the core learning system. This learning system channels content of classical nature to the learner while communication to the virtual environment is, of course, to take place directly through the respective devices. Any actions of the learner, e.g. activities in the virtual environment or answers to test problems, are passed on to the assessment module which updates its learner model according to the observed action and to the underlying competences. This learner model is then used by the learning system for selecting new contents and can also be used by the virtual environment for adapting current situations.

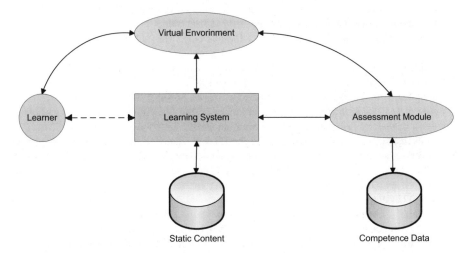

Fig. 3. Architecture of the medical training system

Developing this medical training system involves several aspects of usability and usability research. First, there is, of course, the consideration of results of usability research, on a general level [see, e.g., 13] as well as for the specific area of medical training systems based on virtual reality [14]. A special focus will be on the interplay and integration of the virtual reality and the classical e-learning parts of the system.

A second aspect is the adaptivity of the system which is an ambivalent feature with respect to usability. While adaptivity to the individual user generally is regarded positive, in e-learning it includes some change of the visible content based on the trainee's learning progress leading to possible unwanted confusion. Solving this ambivalence still remains an open challenge for research.

Acknowledgements

The authors wish to express their gratitude for the helpful input of Liam Bannon, University of Limerick, Ireland, and for the constructive remarks of the anonymous reviewers to an earlier version of this paper.

References

1. Hockemeyer, C.: A Comparison of non-deterministic procedures for the adaptive assessment of knowledge. Psychologische Beiträge 44, 495–503 (2002)
2. Conlan, O., Hockemeyer, C., Wade, V., Albert, D.: Metadata driven approaches to facilitate adaptivity in personalized eLearning systems. The Journal of Information and Systems in Education 1, 38–44 (2002)
3. Conlan, O., O'Keeffe, I., Hampson, C., Heller, J.: Using knowledge space theory to support learner modeling and personalization. In: Reeves, T., Yamashita, S. (eds.) Proceedings of World Conference on E-Learning in Corporate, Government, Healthcare, and Higher Education, AACE, Chesapeake, VA, pp. 1912–1919 (2006)
4. Kulcsár, Z., Lövquist, E., Aboulafia, A., Shorten, G.D.: The development of a mixed interface simulator for learning spinal anaesthesia (in progress)
5. Doignon, J.-P., Falmagne, J.-Cl.: Spaces for the assessment of knowledge. International Journal of Man.-Machine Studies 23, 175–196 (1985)
6. Doignon, J.-P., Falmange, J.-Cl.: Knowledge Spaces. Springer, Berlin (1999)
7. Albert, D., Lukas, J. (eds.): Knowledge Spaces: Theories, Empirical Research, Applications. Lawrence: Erlbaum Associates, Mahwah (1999)
8. Hockemeyer, C., Held, T., Albert, D.: RATH - a relational adaptive tutoring hypertext WWW-environment based on knowledge space theory. In: Alvegård, C. (ed.) CALISCE 1998. Proceedings of the Fourth International Conference on Computer Aided Learning in Science and Engineering, Chalmers University of Technology, Göteborg, Sweden, pp. 417–423 (1998)
9. Heller, J., Levene, M., Keenoy, K., Hockemeyer, C., Albert, D.: Cognitive aspects of trails: A Stochastic Model Linking Navigation Behaviour to the Learner's Cognitive State. In: Schoonenboom, J., Heller, J., Keenoy, K., Levene, M., Turcsanyi-Szabo, M. (eds.) Trails in Education: Technologies that Support Navigational Learning, Sense Publishers (2007)
10. Albert, D., Hockemeyer, C., Kickmeier-Rust, M.D., Peirce, N., Conlan, O.: Microadaptivity within Complex Learning Situations – a Personalized Approach based on Competence Structures and Problem Spaces. In: ICCE 2007. Paper accepted for the International Conference on Computers in Education (November 2007)
11. Kickmeier-Rust, M.D., Schwarz, D., Albert, D., Verpoorten, D., Castaigne, J.L., Bopp, M.: The ELEKTRA project: Towards a new learning experience. In: Pohl, M., Holzinger, A., Motschnig, R., Swertz, C. (eds.) M3 – Interdisciplinary aspects on digital media & education, Vienna: Österreichische Computer Gesellschaft, pp. 19–48 (2006)
12. Hockemeyer, C.: Implicit and Continuous Skill Assessment in Game–based Learning. In: EMPG 2007. Manuscript presented at the 38th European Mathematical Psychology Group Meeting, Luxembourg (September 2007)
13. Holzinger, A.: Usability engineering for software developers. Communications of the ACM 48(1), 71–74 (2005)
14. Arthur, J., Wynn, H., McCarthy, A., Harley, P.: Beyond haptic feedback: human factors and risk as design mediators in a virtual knee arthroscopy training system SKATSI. In: Harris, D. (ed.) Engineering Psychology and Cognitive Ergonomics, Transportation Systems, Medical Ergonomics and Training, Ashgate, Aldersho, UK, vol. 3, pp. 387–396 (1999)

Usability and Transferability of a Visualization Methodology for Medical Data

Margit Pohl, Markus Rester, and Sylvia Wiltner

Vienna University of Technology, 1040 Vienna, Austria
Institute of Design & Assessment of Technology
{margit, markus}@igw.tuwien.ac.at

Abstract. Information Visualization (InfoVis) techniques can offer a valuable contribution for the examination of medical data. We successfully developed an InfoVis application – Gravi – for the analysis of questionnaire data stemming from the therapy of anorectic young women. During the development process, we carefully evaluated Gravi in several stages. In this paper, we describe selected results from the usability evaluation, especially results from qualitative studies. The results indicate that Gravi is easy to use and intuitive. The subjects of the two studies described here especially liked the presentation of time-oriented data and the interactivity of the system. In the second study, we also found indication that Gravi can be used in other areas than the one it was developed for.

Keywords: Information Visualization, Usability, Utility, Transferability, Focus Groups, Interviews.

1 Introduction

Computers are used increasingly in a medical context. The usability of such systems is especially important as the consequences of mistakes made when using such systems can be critical. Computers are used for a wide variety of purposes in medicine. In this paper, we describe an InfoVis tool supposed to support psychotherapists in their work with anorectic young women. The aim of information visualization is to represent large amounts of abstract data (e.g. results from questionnaires, data about the development on financial markets, etc.) in a visual form to make this kind of information more comprehensible. Medicine is a very important application area for InfoVis methods [1], especially because of their flexibility and the possibility of representing time oriented data. Chittaro also points out the significance of applying design guidelines offered by Human-Computer Interaction in such applications. For abstract data, there is usually no natural mapping of data on the one hand and visual representation on the other hand (in contrast to e.g. geographical information systems where there is a natural mapping between maps and physical space). Therefore, the design and testing of InfoVis methodologies is especially important. A more general description of the significance of usability research in information visualization can be found in [2] and [3]. A description of different

A. Holzinger (Ed.): USAB 2007, LNCS 4799, pp. 171–184, 2007.
© Springer-Verlag Berlin Heidelberg 2007

methods of usability testing can be found in [4] and [5]. For the application of such methods in a study evaluating the feasibility and usability of digital image fusion see [6] and for the importance of user centered development in the medical domain [7].

InfoVis methodologies are supposed to support humans in analyzing large volumes of data. In many cases, these processes can be seen as activities of exploration. [8] describes the process of seeing as a series of visual queries. [9] investigates user strategies for information exploration and uses the concept of information foraging to explain such behavior. [10] and [11] use the term sensemaking to describe the cognitive amplification of human cognitive performance by InfoVis methodologies. These methodologies help to acquire and reorganize information in such a way that something new can be created. [12] pointed out the importance of information visualization for „creating creativity". [13] distinguishes between two different forms of search – simple lookup tasks and searching to learn. Searching to learn (which is similar to exploration) is more complex and iterative and includes activities like scanning/viewing, comparing and making qualitative judgments. This distinction seems to be important as different forms of evaluation methods are appropriate for these two forms of search. Testing the usability of exploratory systems is more difficult than testing systems for simple lookup (see e.g. [14]). We developed a report system (described in [15]) to investigate the usability of an exploratory InfoVis methodology in medicine.

The following study describes an investigation in how best to support psychotherapists in their work. The aim of these therapists is to analyze the development of anorectic young women taking part in a psychotherapy. During this process a large amount of data is collected. Statistical methods are not suitable to analyze these data because of the small sample size, the high number of variables and the time oriented character of the data. Only a small number of anorectic young women attend a therapy at any one time. The young women and their parents have to fill in numerous questionnaires before, during and after the therapy. In addition, progress in therapy is often not a linear process but a development with ups and downs. All of this indicates that InfoVis techniques might be a better method of analysis of these data. The aim of the therapists is to predict success or failure of the therapy depending on the results of the questionnaires, and, more generally, to analyze the factors influencing anorexia nervosa in more detail. This process is explorative in nature as there exists no predefined "correct" solution for the therapists' problems but several plausible insights might be got during the analysis of the data.

The study presented here is part of a larger project called in2vis. The aim of the project was the development of an appropriate InfoVis methodology to support the therapists in their work with anorectic young women. This methodology is called Gravi (for a more detailed description see below). This methodology was evaluated extensively. In addition, it was compared to other methods of data analysis (Machine Learning and Exploratory Data Analysis). The evaluation process took place in two stages – the usability study and the utility study (insight study and case study). In a last stage of the utility study, we tested whether the visualization methodology developed for a certain area of application could be useful for other areas as well. We called this transferability study. For an overview and short description of the different stages see Fig. 1.

We distinguish usability and utility aspects of the system to avoid mixing up usability problems and problems with the InfoVis methodology and its mapping of data and visualization as such. In this paper, we present results from the usability study and from the so-called transferability study which also investigated usability aspects. Machine Learning and Exploratory Data Analysis were not considered in the usability and the transferability study. The results of the utility study have been presented elsewhere [16, 17]. A preliminary description of parts of the usability study were described in [15].

Stage	Method	Subjects	Aim	Outcome
usability study	usability inspection	1 usability expert	spot most obvious glitches	31 severe usability problems
	heuristic evaluation	27 semi-experts in usability (students)	in depth testing	447 reports ⇒ 576 problems 221 different problems
	focus groups		additional usability assessment	no new problems BUT different perspective
insight study	insight reports	32 domain novices (students)	patterns of insight & cognitive strategies	876 reports ⇒ 2166 insights 668 different insights
	log files		used vis. options & exploration strategies	56055 log file entries
	focus groups		relativize findings & aid correct interpretation	transcription of 3x 100min
case study	interviews	2 real users (clinicians)	feasibility & usefulness in real life	transcription of 1x 60min
	observation & thinking aloud			notes on 1x 180min
transferability study	interviews	14 experts of other domains	usefulness in other domains	transcription of 14x 60min

Fig. 1. Overall evaluation study design. At four different stages diverse evaluation methods are utilized. Quantitative and qualitative methods complement each other. Appropriate subjects are tested according to the various aims of the respective stages. The two parts which results are discussed in this paper are highlighted (red and blue).

2 Description of the Visualization Methodology

The human perceptual system shows great ability to locate and organize things spatially, perceive motion, etc. This is utilized in Gravi [18] by positioning icons on the screen. These represent the patients and questionnaires they answered. According to the answer a patient gave to a question, the patient's icon is attracted by the question's icon. A spring-based system model is used to depict this so that every patient is connected to every question (see Fig. 2 left side). This leads to the formation of clusters of patients who gave similar answers (see Fig. 2 right side). The therapists are especially interested in those variables which predict the outcome of the therapy (successful or not successful). By analyzing clusters of "positive" and "negative" cases they can identify those variables.

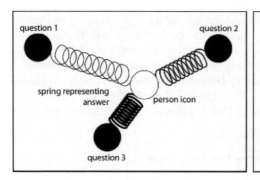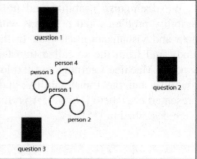

Fig. 2. Concept of spring-based positioning (left), leading to formation of clusters (right)

Gravi provides various interaction possibilities to explore the data and generate new insights. The icons and visual elements can be moved, deleted, highlighted and emphasized by the user. Each change leads to an instant update of the visualization. For details on mental model, visualization options, user interactions, and implementation see [19].

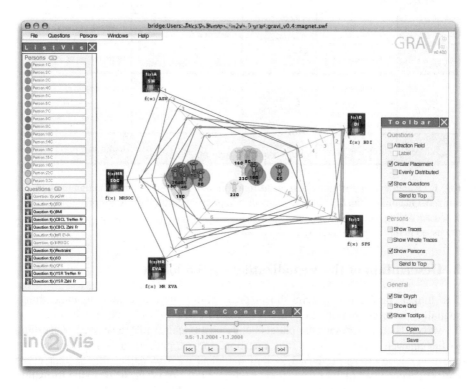

Fig. 3. Typical Screenshot of Gravi – Two clusters show similar answering behavior of patients with positive therapy outcome (green icons) and those with negative outcome (red icons). Star Glyphs – a connected line of the answer scores on each questionnaire for every patient – communicate exact values of all patients on showed questionnaires.

Many different visualization options are available, like Star Glyphs (see Fig. 3). To represent the dynamic, time dependent data Gravi uses animation. The position of the patients' icons change over time. This allows for analyzing and comparing the changing values. The therapists need this feature to visualize information recorded at different points in time. The development in time is a very important aspect of the analysis of the progress of the therapy. To visualize the change over time of patients in one view there is the option to show the paths the patients' icons will follow in the animation. These paths are called traces (see Fig. 4)

Fig. 4. Traces allow for the analysis of the changing answering behavior of multiple patients over all five time steps (i.e. the whole therapy which lasts for about one year). Shown here are the traces of four patients who start at almost identical positions according to the same five questionnaires of Fig. 3. Already at time step "2" we see divergent changes in answering behavior of those with positive and those with negative therapy outcome.

3   Usability Study (Focus Groups Following Heuristic Evaluation)

We utilized three different evaluation methods in our usability study: an informal inspection, heuristic evaluation, and focus groups (see Fig. 1). The first method revealed severe usability issues which have been corrected before any further evaluation took place. This should increase the quality of following assessments. Heuristic evaluation [20] and focus groups complemented each other with their different outcomes. For results of the heuristic evaluation see [15].

"Focus groups are structured group interviews that quickly and inexpensively reveal a target audience's desires, experiences, and priorities" [21].   They are a method of qualitative research which involve group discussion and group interviewing. Participants discuss and comment on the topic that is under investigation. This relies heavily on their subjective view and experience [22]. Although one major strength of focus groups is that they may reveal issues not anticipated so far, due to the interactive setting, they have to be prepared well. So a guideline for the group moderator is very advisable to ensure that the discussion stays focused and covers predefined questions [cf. 23].

Whereas the outcome of heuristic evaluation was an extensive list of usability problems, focus groups allowed the subjects to give detailed arguments for their views on the usability of the assessed visualization methodology. Also a form of severity rating was compiled when the subjects discussed their views on the most important usability problems.

The subjects were computer science students who attended a course where they not only received lectures on usability but also had to work out several assignments. Therefore we can describe them as semi-experts in usability. Furthermore, they also received an introduction to the application domain in order to ensure basic understanding of real users' data and tasks.

The focus groups lasted about an hour each. The same 27 subjects who also took part in the heuristic evaluation were split into two groups of similar size for the focus groups. This is necessary because otherwise the discussion would be too chaotic or, and this would be even worse for an interesting outcome, the participants would not start a lively discussion referring to each other at all due to a too big group size. Another important argument for holding two different groups is that there may be some group members who are rather dominant regarding discussed topics. By holding at least two focus groups the overall outcome of this evaluation method will probably more fruitful.

The guideline for the discussion started with an introduction including an explanation of the focus group methodology and the showing of some screenshots of Gravi, so the subjects remember all the main features of the visualization. First, the overall impression of the usability was discussed. Most important here were the following questions: *Can Gravi be operated intuitively? What was the most severe usability problem? What was the best designed feature from a usability perspective?* Thereafter various parts of the user interface were subject to discussion. Understandability was the basic question for all these parts, but also structure or positioning were discussed for main menu, windows and toolbars, context menus, and workplace.

3.1 Intuitive Operation of Gravi

In focus group A the participants made 16 statements, 11 of which were on the positive or even very positive side. Nevertheless there were some constraints mentioned. The basic interactions and the mental model were assessed very understandable. Complex functionalities were rated more confusing.

In focus group B 11 statements are documented. 8 of these were – similar to group A – on the positive side. But in contrast to group A, the mentioned restrictions in group B have in common that at the beginning Gravi cannot be used intuitively but some time is needed to learn how to operate the system.

3.2 Most Severe Usability Problem

There were 46 statements concerning the most important usability problem. These 46 statements documented 27 different problems, 10 of which were mentioned more than once. With 7 mentions the problem "undo/redo is missing" was the most frequently documented one. For more details on this result see [15].

An evident characteristic of the gathered material is that, except for a few statements, the problems mentioned by the different focus groups were quite divergent. That confirms the claim for the necessity of conducting more than one focus group.

3.3 Best Designed Feature from a Usability Perspective

Members of group A stated 7 good features, including drag & drop interactions (two times), instant updates of visualization after any interaction, time control window to navigate through the time-dependent data, and the idea of only one workbench where all interaction and visualization takes place. Two more statements were concerned with positive aspects of the visualization methodology and are not usability issues as such.

Participants of group B stated similar positive features like drag & drop interaction but also some additional good elements: the possibility to quickly switch between much and less detail of the visualized data, the neutral design of the user interface that does not distract from reading the visualization, the beneficial color coding of the data, the existing alternative ways to accomplish some desired interaction (see Fig. 5), and direct interaction which gives the feeling of touching the visualization and therefore eases understanding.

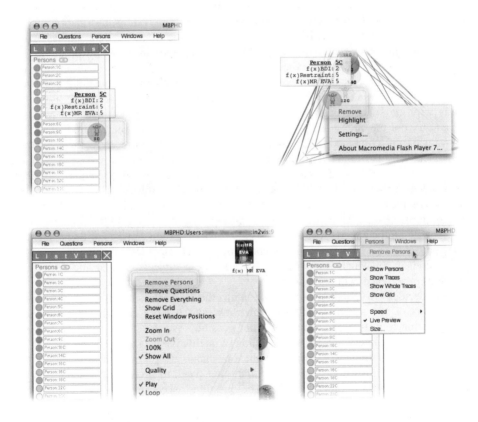

Fig. 5. Alternative ways to remove icons of patients from the workbench: drag & drop interaction with a single icon: moving from workbench to ListVis window (top left), context menu of a single icon (top right), context menu of workbench to remove all icons (bottom left), and main menu entry to remove all icons from workbench (bottom right)

4 Usability and Transferability (Interviews with Experts)

The visualization methodology Gravi has been developed for a very specific purpose – the analysis of the therapy process of anorectic young women. As the development process for such methodologies is costly and time consuming, it is interesting to assess whether this tool could be used in other contexts as well. To achieve this, we conducted an investigation about the transferability of Gravi to other content areas (as e.g. other forms of psychotherapy, history, teaching, etc.). We especially asked content experts from these other areas whether they could think of concrete examples how to use Gravi in their area of expertise and discussed these examples with them. As far as we know, no other study of this kind has been conducted. This study was especially interesting because we could also test Gravi with content experts in similar domains as the one it was originally intended for (that is, other forms of psychotherapy).

An important aspect of this transferability study was also to assess the usability of Gravi with content experts (as opposed to the original usability study which had been conducted with students). Such studies can yield very interesting information about the usability of InfoVis tools gained in more realistic contexts than studies with students as subjects. But there are problems as well. One of the main problems of usability testing with content experts is their lack of time. This was also obvious in this study. The experts did not have enough time to really get acquainted with the system, and the results are, therefore, not as detailed as the ones from the usability study with the students who got a long introduction to the system and had enough time to try out the system on their own. Nevertheless, we think that such studies can be valuable if combined with other usability studies because they increase the ecological validity of the overall evaluation of the system.

The transferability study was not only restricted to experts in medicine but also included experts from other areas with similar problems (historians, teachers, etc), that is, complex, sometimes time oriented data and ill-structured knowledge domains. All in all, there were 14 content experts participating in this study. In this paper, we concentrate on the experts from the medical area. This group consists of five subjects, all of whom have a background in psychotherapy and have at least passing knowledge of the domain for which Gravi was developed. These subjects had varying degree of previous experience with computers.

We decided to conduct interviews with the subjects. Because of the time constraints of the subjects, a longer interview and a more rigorous procedure of investigation (e.g. thinking aloud) was not possible. We developed a framework of questions for the interview (for a description of qualitative data analysis see e.g. [24]). The aim of this framework was, on the one hand, to get information about the usability of the finished system and, on the other hand, to investigate whether Gravi could be transferred to other domains. The questions asked are described in the next paragraph. The questions asked in the interviews were also used as categories for the interpretation of the interviews. In addition, we also looked for indications for major usability problems. An indication for a usability problem might, e.g., be if users could not understand a feature of the system after two or three explanations and made the same mistake again and again.

Table 1. Guideline for transferability interviews

	Topic	Time Target
1.	Introducing interviewers, research project, research question, and interview procedure. Get permission to record interviews.	5 minutes
2.	Brief presentation of application domain:	10 minutes
	• Anorexia nervosa: definition, causes, consequences, and treatment of disease	
	• Abbreviations and meanings of used questionnaires (e.g., BDI – Beck's Depression Index)	
	Initial training (i.e. presentation of most basic features and interactions with Gravi):	
	• Drag & Drop	
	• Highlight	
	• Add and remove of patient icons and questionnaire icons	
	• Color-coding of icons (e.g., green – positive therapy outcome)	
	• Transparency of icons (all data is missing)	
	• Tooltips for details on patient and questionnaires	
	• Spring metaphor of visualization for positioning the patients' icons	
	• Time-dependent data and navigation through time steps	
3.	Scenario 1: What is the influence of 5 given questionnaires (i.e. the answering behavior of the patients on these) on the therapy outcome? Visualization is already operated by subjects, although assistance is provided whenever needed.	15 minutes
4.	Scenario 2: What statements can be made about patient 18? What is the predicted therapy outcome of this so far unclassified patient? (See Fig. 6) Visualization is operated by subject without any support.	10 minutes
5.	Presentation of further features and visualization options of Gravi (e.g., traces for visualizing paths of patients' icons over all time steps)	5 minutes
6.	Questioning of subjects on usability and utility of Gravi:	15 minutes
	• Was the general impression of Gravi positive or negative? (Give reasons)	
	• Was the system understandable? (Give reasons)	
	• Are there any concerns about shortcomings of the visualization?	
	• What was the best feature of Gravi?	
	• How was the learnability of Gravi?	
	Questioning of subjects on transferabiltiy of Gravi:	
	• What are possible fields of application in the subjects' domains? Sketch concrete examples if possible.	
	• Is the visualization technique suitable for the data of the subjects' domains or are improvements and/or modifications necessary?	

The experiment for the transferability study lasted approx. one hour. In the beginning, the subjects were introduced to Gravi. This usually took about 30 minutes. It should be mentioned that this gave subjects only a superficial impression of the system. Then, the subjects had to solve a very easy example on their own to get a better understanding of how the system worked. Then the subjects were asked how confident they were about the solution they found, which positive and negative features of the system they could identify and whether the system was easy to use. The last question was whether they could think of a concrete application area for Gravi in their own domain. In the guideline for the interviews (see Table 1), usability aspects were explicitly addressed. In addition, usability issues could also be derived from the interaction of the subjects with the system.

Subjects were tested in their own offices (which meant that they were sometimes disturbed by phone calls etc). The same laptop was used for all tests. The interviews were recorded and then transcribed.

The simple task the subjects had to solve during the experiment was to predict success or failure of a patient from the data of two time-steps (instead of five time-steps for the patients who had finished the therapy). See Fig. 6 for a prototypical visualization configuration to answer this question.

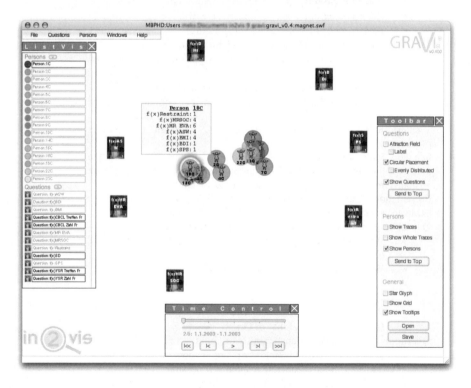

Fig. 6. Prototypical visualization configuration to predict the therapy outcome of patient 18 (highlighted). The position of the icon of patient 18 amongst patients with positive outcome allows for the hypothesis that patient 18 will also have an positive therapy outcome.

Four of the five subjects could accomplish this task. One subject felt that s/he had not enough data. This subject also felt insecure about the use of computer technology in general. This fact influenced her/his understanding of the system negatively. This subject also had tight time constraints which made the situation of the interview fairly stressful. It should be mentioned that, despite this, the subject formulated a very interesting example from her own domain. It is probably also characteristic that this subject preferred more static features of the system to the more dynamic ones (that is, representation of dynamic data).

Of the four subjects who solved the example, only three gave a confidence rating (two of them high confidence and one medium confidence). The fourth subject mentioned that a reliable confidence rating could only be given when extensive information about the subject domains and especially the behavior of a large enough sample of other patients was available.

The answers relating to the most positive feature of the system were rather heterogeneous. Two subjects thought that the dynamic presentation of time oriented data were very advantageous. Two other subjects thought that the circular positioning of the icons for the questions was very positive. Apparently, this gives a compact impression of the data. One subject mentioned that s/he found that the visual presentation of data in general was very positive.

Several negative features were mentioned. The system quickly becomes confusing when many questions and persons are added. A good introduction to the system is necessary to avoid technical problems. One subject found it difficult to remember information from one time-step to the next. Problems of a more epistemological nature were also mentioned. One subject found that positioning the questions in an arbitrary manner on a circle is problematic. S/he thought that the system should suggest an appropriate position. Another subject thought that the level of aggregation of data might be too high so that important information might get lost. Yet another subject indicated that the system only works well if one can be sure to have identified all necessary variables for predicting success or failure of the therapy.

In general, the subjects found the system quite easy to use. Two subjects said that a good introduction was necessary. As mentioned before, one subject had difficulties to use the system (probably because of the lack of previous experience with computers and the lack of time). The protocols of the interviews which also contain the process of the introduction show no other serious difficulties in the usage of the system which indicates that the subjects' subjective impressions and their actual behavior correspond, which is not always the case in usability research.

The last question was that subjects should find an example from their own work experience which could be supported by Gravi. These examples were very heterogeneous in level of detail. To a certain extent, most of the examples were very similar to the original application domain as all subjects had a background in psychotherapy. In general, all subjects suggested the application of Gravi for the support and analysis of psychotherapy but some of them indicated that their data might not be appropriate (e.g. too many subjects, too few variables; too heterogeneous samples). One subject thought that Gravi could only be used for the analysis of group psychotherapy but another subject thought it possible to compare all her/his patients using Gravi. One subject suggested that the source of the data do not have to be questionnaires but could also be ratings made by the therapists themselves. Two subjects also suggested an application of Gravi to support university teaching. One of these examples was quite elaborate. Students should be rated according to their ability in conducting the first interview with prospective patients. The data resulting from this rating process should be analyzed by Gravi. This analysis was supposed to improve the teaching process and give teachers more insights into what could go wrong.

5 Conclusion

InfoVis techniques can offer a valuable contribution for the examination of medical data. We successfully developed an InfoVis application – Gravi – for the analysis of questionnaire data stemming from the therapy of anorectic young women. During the development process, we carefully evaluated Gravi in several stages. In this paper, we describe selected results from the usability evaluation, especially results from qualitative studies. In a first stage, usability was evaluated using students as subjects. These students were semi-experts in the area of usability. Therefore, they could give more precise feedback concerning usability issues. Based on the results of this study, Gravi was improved considerably. The results of the transferability study, using content experts as subjects, is an indication of this. The content experts all found the system easy to use, and of a total of 14 all but one could solve a small example with Gravi on their own after a short introduction.

It is slightly difficult to compare the results of the focus groups and the interviews with the content experts because the students concentrated on usability issues whereas the answers of the content experts also cover other issues apart from usability aspects. There is some indication that the presentation of time-oriented data and the interactivity of the system (drag & drop, instant updates of the screen etc.) is seen as positive by members of both groups. The answers concerning the negative features of the system are very heterogeneous. The content experts often mentioned problems of a more epistemological nature which cannot be called usability problems.

It seems that Gravi can easily be used for other application areas in medicine. The subjects especially mentioned other uses for the analysis of therapy processes. One problem which has to be solved in this context is the question whether Gravi can only be used for group therapy or also for individual therapy.

Acknowledgments. The project "Interactive Information Visualization: Exploring and Supporting Human Reasoning Processes" is financed by the Vienna Science and Technology Fund (WWTF) [Grant WWTF CI038]. Thanks to Bernhard Meyer for the collaboration in conducting the transferability interviews.

References

1. Chittaro, L.: Information Visualization and its Application to Medicine. Artificial Intelligence in Medicine 22(2), 81–88 (2001)
2. Plaisant, C.: The Challenge of Information Visualization Evaluation. In: Proc. AVI 2004, pp. 109–116. ACM Press, New York (2004)
3. Tory, M., Möller, T.: Human Factors in Visualization Research. IEEE Transactions on Visualization and Computer Graphics 10(1), 72–84 (2004)

4. Holzinger, A.: Usability Engineering Methods for Software Developers. Communications of the ACM 48(1), 71–74 (2005)
5. Holzinger, A.: Application of Rapid Prototyping to the User Interface Development for a Virtual Medical Campus. IEEE Software 21(1), 92–99 (2004)
6. Wiltgen, M., Holzinger, A., Groell, R., Wolf, G., Habermann, W.: Usability of image fusion: optimal opacification of vessels and squamous cell carcinoma in CT scans. Springer Elektrotechnik & Informationstechnik, e&i 123(4), 156–162 (2006)
7. Holzinger, A., Geierhofer, R., Ackerl, S., Searle, G.: CARDIAC at VIEW: The User Centered Development of a new Medical Image Viewer. In: Zara, J., Sloup, J. (eds.) Central European Multimedia and Virtual Reality Conference, pp. 63–68, Czech Technical University, Prague (2005)
8. Ware, C.: The Foundation of Visual Thinking. In: Tergan, S.-O., Keller, T. (eds.) Knowledge and Information Visualization. Searching for Synergies, pp. 27–35. Springer, Heidelberg (2005)
9. Pirolli, P.: Exploring and Finding Information. In: Carroll, J.M. (ed.) HCI Models, Theories, and Frameworks. Toward a Multidisciplinary Science, pp. 157–191. Morgan Kaufmann, San Francisco (2003)
10. Russell, D.M., Stefik, M.J., Pirolli, P., Card, S.: The cost structure of sensemaking. In: Proceedings of the INTERCHI 93, ACM Conference on Human Factors in Computing Systems, ACM Press, New York (1993)
11. Card, S.: Information Visualization. In: Jacko, J.A., Sears, A. (eds.) The Human-Computer Interaction Handbook. Fundaments, Evolving Technologies and Emerging Applications, pp. 544–582. Lawrence Erlbaum, Mahwah (2003)
12. Shneiderman, B.: Creating Creativity: User Interfaces for Supporting Innovation. In: Carroll, J.M. (ed.) Human-Computer Interaction in the New Millenium, pp. 235–258. Addison-Wesley, San Francisco (2002)
13. Marchionini, G.: Exploratory Search: From Finding to Understanding. Communications of the ACM 48(4), 41–46 (2006)
14. Bertini, E., Plaisant, C., Santucci, G. (eds.) BELIV 2006.Proceedings of the 2006 AVI workshop on BEyond time and errors: novel evaLuation methods for Information Visualization, p. 3. ACM Press, New York (2006)
15. Rester, M., Pohl, M., Hinum, K., Miksch, S., Popow, C., Ohmann, S., Banovic, S.: Assessing the Usability of an Interactive Information Visualization Method as the First Step of a Sustainable Evaluation. In: Holzinger, A., Weidmann, K.-H. (eds.) Empowering Software Quality: How can Usability Engineering reach this goal? Wien: Oesterreichische Computergesellschaft, pp. 31–43 (2005)
16. Rester, M., Pohl, M., Wiltner, S., Hinum, K., Miksch, S., Popow, C., Ohmann, S.: Evaluating an InfoVis Technique Using Insight Reports. In: Proceedings of the IV 2007 Conference, pp. 693–700 (2007)
17. Rester, M., Pohl, M., Wiltner, S., Hinum, K., Miksch, S., Popow, C., Ohmann, S.: Mixing Evaluation Methods for Assessing the Utility of an Interactive InfoVis Technique. In: Jacko, J. (ed.) HCII 2007. Human-Computer Interaction. Interaction Design and Usability. LNCS, vol. 4550, pp. 604–613. Springer, Heidelberg (2007)
18. Hinum, K., Miksch, S., Aigner, W., Ohmann, S., Popow, C., Pohl, M., Rester, M.: Gravi++: Interactive information visualization to explore highly structured temporal data. Journal of Universal Comp. Science 11(11), 1792–1805 (2005)
19. Hinum, K.: Gravi++ – An Interactive Information Visualization for High Dimensional, Temporal Data. PhD thesis, Institute of Software Technology & Interactive Systems, Vienna University of Technology (2006)

20. Nielsen. Heuristic Evaluation. In: Usability Inspection Methods, vol. ch.2, pp. 25–62. John Wiley & Sons, Chichester (1994)
21. Kuniavsky, M.: User Experience: A Practitioner's Guide for User Research, p. 201. Morgan Kaufmann, San Francisco (2003)
22. Mazza, R.: Evaluating information visualization applications with focus groups: the coursevis experience. In: BELIV 2006. Proc. of the 2006 AVI workshop on BEyond time and errors: novel evaLuation methods for InfoVis, ACM Press, New York (2006)
23. Mazza, R., Berre, A.: Focus group methodology for evaluating information visualization techniques and tools. In: IV 2007. Proc. of the 11th International Conference Information Visualization, pp. 74–80. IEEE Computer Society, Los Alamitos (2007)
24. Bortz, J., Döring, N.: Forschungsmethoden und Evaluation, 4th edn. Springer, Heidelberg (2006)

Refining the Usability Engineering Toolbox: Lessons Learned from a User Study on a Visualization Tool

Homa Javahery and Ahmed Seffah

Human-Centered Software Engineering Group
Department of Computer Science and Software Engineering
Faculty of Engineering and Computer Science
Concordia University
Tel.: 514-848-2424, ext. 3024
Fax: 514-848-2830
{h_javahe,seffah}@cs.concordia.ca

Abstract. This paper details a usability study on a bioinformatics visualization tool. The tool was redesigned based on a usability engineering framework called UX-P (User Experiences to Patterns) that leverages personas and patterns as primary design directives, and encourages on-going usability testing throughout the design lifecycle. The goals were to carry out a design project using the UX-P framework, to assess the usability of the resulting prototype, and to mitigate the test results into useful recommendations.

Keywords: Usability Engineering Framework, Usability Testing, User Experiences, Conceptual Design, Personas, Patterns.

1 Introduction

Designing a software tool or a web site in biomedical sciences or health care so that they are effective in achieving their purpose of helping users (biomedical practitioners, physicians, research scientists, etc.) in performing a task such as searching through patient records, interacting with a large set of data for imaging, or organizing a body of information into useful knowledge, is not a trivial matter. Even for everyday business, the discipline of usability engineering – which studies system usability, how this is assessed, and how one develops systems so that they bridge the conceptual gap between user experiences and the tool's functionality – is still maturing.

Most often usability engineering (UE) methods are applicable in testing after the development and deployment of a system, rather than in providing guidance during the early development phase. You can think of it as being able to diagnose the disease (of poor usability) but not being able to prevent it. In comparison to the traditional software engineering approach, UE begins by getting to know the intended users, their tasks, and the working context in which the system will be used.

Task analysis and scenario generation are performed, followed by low-fidelity prototyping and rough usability studies, ending with a high-fidelity prototype that can

A. Holzinger (Ed.): USAB 2007, LNCS 4799, pp. 185–198, 2007.

be tested more rigorously with end-users before the deployment of the final application.

The need to build a tighter fit between user experiences and design concepts is described as one of the main challenges in UE [1]. To advance the state-of-the art and narrow this existing gap, we require frameworks and processes that support designers in deriving designs which reflect users, their needs and experiences. We have proposed a set of UE activities, within the context of a design framework that we call UX-P (User Experiences to Patterns), to support UI designers in building more user-centered designs based on user experiences.

In this paper, we will present a study we carried out with a bioinformatics visualization tool. The objectives of the study were to redesign the tool using our framework, and then assess its usability in comparison to the original version. We will first overview UX-P, its core activities, and its key principles. We will then demonstrate how these activities were carried out with the Protein Explorer application. Finally, we will discuss the experiments we carried out to assess the usability of our new prototype: Testing methods, the study design, and results from quantitative and qualitative usability assessment.

2   A Learning Study: The UX-P Framework

User experience descriptions and UI conceptual designs are two major artifacts of user-centered design (UCD) [2]. User experiences is an umbrella term referring to a collection of information covering a user behavior (observed when the user is in action), expectations, and perceptions – influenced by user characteristics and application characteristics. In current practice, user experiences are captured in narrative form, making them difficult to understand and apply by designers. In our framework, we use *personas* as a technique to allow for more practical representations of user experiences. Personas [3] were first proposed in software design as a communication tool to redirect the focus of the development process towards end users. Each persona should have a name, an occupation, personal characteristics and specific goals related to the project. We extend the persona concept to include details about interaction behaviors, needs and scenarios of use, which can be applied directly to design decisions.

A conceptual design is an early design of the system that abstracts away from presentation details and can be created by using *patterns* as design blocks. Similar to software design patterns [4], HCI design patterns and their associated languages [5] [6] are used to capture essential details of design knowledge. The presented information is organized within a set of pre-defined attributes, allowing designers, for example, to search rapidly through different design solutions while assessing the relevance of each pattern to their design. Every pattern has three necessary elements, usually presented as separate attributes, which are: A context, a problem, and a solution. Other attributes that may be included are design rationale, specific examples, and related patterns. Furthermore, we have enhanced the patterns in our library to include specific information about user needs and usability criteria.

The UX-P Framework facilitates the creation of a conceptual design based on user experiences, and encourages usability testing early on in product development. In essence, the main design directives in our framework are personas and patterns; appropriate design patterns are selected based on persona specifications. The framework is based on a set of core UE principles, enriched with "engineering-like" concepts such as reuse and traceability. Reuse is promoted by using design patterns. Furthermore, since we have a logical link between persona specifications and the conceptual design, traceability to design decisions is facilitated. Our framework consists of the following phases:

- Modeling users as personas, where designers create personas based on real data and empirical studies.
- Selecting patterns based on persona specifications, where certain attributes, needs, and behaviors drive the selection of candidate patterns for the desired context of use.
- Composing patterns into a conceptual design, where designers use a subset of the candidate patterns as building blocks to compose a conceptual design.

Designers are free to repeatedly refine the artifacts produced at each phase before proceeding to the next phase. Two additional sources of information contribute to the above phases: Empirical studies and other UCD artifacts. First, empirical studies using techniques such as task-based or psychometric evaluations provide the groundwork for a better understanding of users and their needs, resulting in a first set of personas. Usability inquiries are useful for eliciting information about interaction behaviors and usability-related design issues. In particular, heuristics and user observations with a prototype or similar application can help gather additional information for persona creation. This information feeds directly into our design decisions. Secondly, other UCD artifacts besides personas and patterns are necessary in creating an overall design. User-task, context of use, and interaction models provide essential information during any design process and are important guides for establishing UI structure and system-related behavioral details.

With the goal to understand, adapt and refine the UX-P Framework to the context of biomedical applications, we redesigned a tool called Protein Explorer [7]. Protein Explorer is a web-based software application for biomedical research, used for the visual prediction and analysis of bimolecular structure. Users explore various macromolecule structures such as Protein and DNA in 3D, using a web browser. The Protein Explorer browser interface (see Figure 1) is split into four windows, organized as panes. The window on the right is the visualization interface, containing the 3D macromolecule. The structural data for this molecule comes from a Protein Data Bank and users can view molecules by entering their Protein Data Bank ID. The upper left window provides information about the molecule and includes display options. Furthermore, it splits into two windows with tips and hints in a "child" frame. There are links on this window that lead you to other resources. The lower left window of Protein Explorer is a message box detailing the atoms clicked in the visualization, and allows users to type commands in a scripting language called chime.

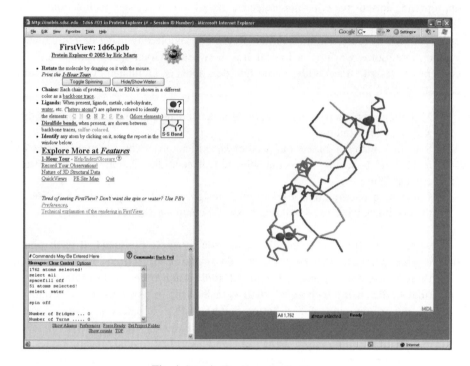

Fig. 1. Protein Explorer Application

3   Modeling Users as Personas

We used a sample of 22 users from the biomedical-related fields to create personas. We carried out usability inquiries in the form of field studies and user observations on two bioinformatics visualization tools, Cn3D [8] and ADN-Viewer [9]. We did not use the Protein Explorer for any pre-design user observation to ensure that the new design was developed based on our framework, without any possible positive "side-effects" due to the re-iteration and/or the discovery of usability issues with the original version of the tool.

For each participant, a complete set of user attributes was recorded. The average age range was 20-34 years old, although our sample also included adults that were older (45+ years old). We had 8 males and 14 females in our sample. In terms of education level, all of our users had a university degree, with a large subset having advanced degrees. Some user attributes, such as education level, were recorded based on an initial questionnaire administered to participants. Others, such as learning speed, were recorded during user observations. Furthermore, we noted information about goals and interaction details for each user, as well as typical scenarios for a subset of the most representative users.

We noted a number of distinguishing interaction behaviors and attribute dependencies between users. Attributes that caused notable differences in interaction behavior were domain experience (in Bioinformatics), background (its values were defined as being either "biology" or "computer science"), and age. The following are some examples of our observations:

- Users with low product experience were often confused when interacting with either tool; features were not sorted according to their mental model.
- Users with significant product experience were feature-keen and reluctant to learn a new design paradigm.
- The biologists needed more control when interacting with the tool. They were extremely dissatisfied when processes were automated, wanting to understand how the automation worked. They had an experimental problem-solving strategy, where they followed a scientific process and were repeat users of specific features.
- Users from computational backgrounds had a linear problem-solving strategy where they performed tasks sequentially, and exhibited comfort with new features.
- Older adults (45+) were more anxious when interacting with the system, and were less comfortable in manipulating the visualization.
- Older adults had a high need for validation of decisions. They would ask others for help in performing more complex tasks.
- As age increased, the expectation of tool support increased. This is due in part to a decrease in learnability with older users.

Table 1. Persona for Protein Explorer

Persona 1: Marta Aviles	
	Quote: "I want to be able to access specific features without having to worry about the details" **General Profile [shortened]:** Marta is a 27 year-old computational biologist in a pharmaceutical company. She has a Masters in Computer Science. She works in a highly-competitive industrial setting, over 50 hours/week. She uses various bioinformatics tools on a daily basis, working mostly with predictive models for RNA. She uses 3D visualization features to assess the predictive power of specific molecular interactions. She needs to manipulate the visualization easily, and see the inner workings of the tool (computations). She has experience programming in database and object-oriented languages. She is often in a rush and struggling to keep up with deadlines.
Goals	Application: Creates, manipulates and analyzes predictive models. Professional: Move up to middle-management. Personal: Get married and settle down.
Demographics	Marta is a 27-year old female, with an average income. She has no dependents.
Knowledge & Experience	Marta is fluent in English and Spanish. She has a high education level and average domain experience in Bioinformatics. She is an expert computer user, with advanced programming experience.
Psychological profile & needs	Marta has little need for validation of decisions and guidance. She is a proactive user, and has somewhat of a high learning speed. She has an average need for control and behavior to features.
Attitude & Motivation	Marta has a positive attitude to IT, and a somewhat high level of motivation to use the system.
Special Needs	Marta has no disabilities. She acts at times like an expert user.

We constructed the following set of personas by using the original user descriptions as a guide: (1) Marta Aviles, a young bioinformatics professional working in industry. She has medium domain experience, is from a computer background and is 27 years old. Her needs are average control and behavior towards features; she requires a balance between efficiency and simplicity. (2) Zhang Hui, a senior Parasitology professor. He has high domain experience, comes from a Biology background and is 61 years old. He is feature-keen, needs lots of control and has a high need for validation of decisions. (3) Sue Blachford, a mature adult who is a medical practitioner with limited experience in Bioinformatics. She has low domain experience, comes from a biology background, and is 42 years old. She is feature-shy, needs simplicity, and needs guidance.

Each persona included a set of scenarios, or user stories. We gathered these scenarios during our observations. Scenarios help limit the assumptions designers make about users and their tasks. They depict stories about users, behaviors and actions and include characteristics such as setting, actors, and goals. The key is to include enough detail to make significant design issues apparent. One scenario for Marta is illustrated in Table 2.

Table 2. Scenario for Persona Marta

Description	Marta is working on a new predictive model for determining protein structure. She just read a paper on a new method, and wants to examine some ideas she has with her work on the Hemoglobin molecule. She is sitting in her office, munching away on her lunch, and interacting with the 3D visualization tool. She searches for the option to view multiple surfaces concurrently, but has some trouble setting the initial parameters for determining molecular electrostatic potential. She only needs access to this feature, and is not interested in setting other constraints on the molecule. She gets slightly confused with all the advanced biomedical terminology, but after spending 20 minutes trying different values, she gets one that she is happy with.
Specific needs	Control and behavior to features (average). Balance between efficiency and simplicity.
Features	Advanced options, customizing parameters, tracing inner workings of tool,
Interaction details	Marta does not give up easily if she can't manipulate the molecule in the way she would like or if she can't find the information she needs. She wants to get the job done, without worrying about all the details. She is comfortable with computing technology and terminology.

4   Task Analysis and Modeling

To better understand the user-tasks involved for 3D bioinformatics visualization tools, we built a task model. Figure 2 illustrates a subset of the model. Note that we created it using the ConcurTaskTree notation and its associated tool [10], although we could have used any other relevant tool. The task model was used as a guiding artifact during design pattern selection and composition.

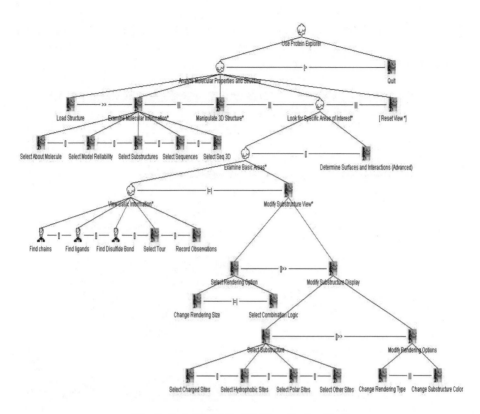

Fig. 2. Task Model for 3D Visualization Tool

5   Selecting and Composing Patterns

The personas led us in choosing appropriate design patterns for the prototype. The design patterns were chosen from a pattern library which consisted of web, visualization and graphical user interface patterns from several sources [5] [6] [11]. Each persona resulted in a different set of design patterns which would fulfill their particular needs. For example, Details on Demand [5] suggests the display of item details in a separate window so as to minimize clutter. The usability criteria for this pattern are minimalist designs and feedback. Since Sue has low domain experience, as well as a high need for simplicity and guidance, it was a highly recommended pattern for her. To facilitate the retrieval of appropriate patterns, we have built a recommender-type system called P2P Mapper which, based on a scoring system, proposes candidate patterns based on particular user types, their needs and attributes.

The patterns for Zhang and Sue varied the most, whereas Marta contained patterns from both personas. We decided that a compromise solution would best fit our purpose. We therefore redesigned the tool using the following recommended patterns: Button Groups (1), Card Stack (2), Good Defaults (3), Legend (4), Multi-level help (5), Details on Demand (6), Tool Tips (7), Convenient Toolbar (8), Action Panel (9),

Command History (10), Filter (11), and Reduction Filter (12). Figure 3 illustrates a subset of the patterns used; the numbers in brackets and in the figure are corresponding.

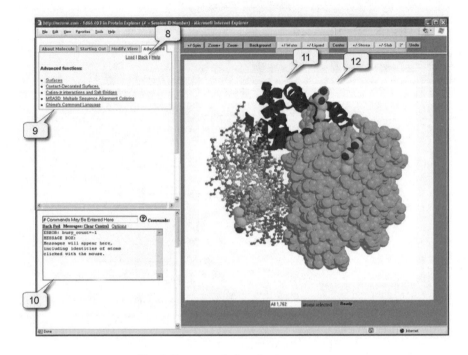

Fig. 3. Prototype designed using patterns

The selected patterns were used as "building blocks" and composed into a pattern-oriented design. The only requirement we had was to keep the same "look and feel" for the design, including the use of several panes. By using personas, we had a better understanding of the user and context of use. We composed patterns by using a pattern-oriented design model which addresses the logical organization of the entire application, as well as the use of key pattern relationships. The latter exploits the *generative* nature of patterns, resulting in a network of patterns which can be combined in a design. As an example, certain patterns are sub-ordinate to others, meaning that they can be composed within their super-ordinate pattern. A simple example is a *home page pattern*, which consists of page element patterns such as the *toolbar*. The reader is referred to [12] for further details.

6   Testing the Prototype with End-Users

We conducted a series of usability tests after redesigning the tool using our framework. In this section, we discuss the experimental methods used, the study design, and the results.

6.1 Experimental Methods

In usability engineering, user-oriented studies are required to motivate the research as well to assess the validation and accuracy of the proposals. Such studies need to draw on principles from both Human-Computer Interaction and empirical software engineering. Fenton and Pfleeger [13] discuss three main types of experimental assessment techniques to evaluate an application or method in software engineering: Surveys, case studies, and formal experiments. Surveys are retrospective studies, aiming to document relationships and trends after a particular event has occurred. Variables cannot be manipulated in this type of study, and large groups of projects and data sets are analyzed. Case studies are planned experiments where factors affecting a particular outcome are identified and documented. A typical project or situation is followed and analyzed, but the goal is not to capture information about all possible cases. In contrast to case studies, formal experiments are rigorously controlled investigations of a particular activity. Key factors are identified and manipulated, and the resulting outcome is documented. Since a great deal of control is required, the project scale is usually small.

Given that our study was not retrospective and the difficulty of control was high, case studies were the logical choice. The following are the planning steps we followed: (1) Conception, where the objectives and goals of the case study are clearly stated, to allow for evaluation. (2) Design, where the objectives are translated into a formal hypothesis and components of the experiment are clearly defined. This includes the hypothesis, treatment, dependent variables and independent variables. (3) Preparation, where subjects are prepared for application of the treatment. Examples include training staff and writing out instructions. (4) Execution, where the experiment is executed as described in the design. (5) Analysis, where all measurements are reviewed to ensure validity and where data is analyzed statistically. (6) Decision making, where a conclusion is drawn based on the analysis results.

The usability testing methods used for our case study consisted of well-established qualitative and quantitative assessment techniques from the HCI community, and results were reported according to the Common Industry Format (CIF) [14]. In large part, testing methods in HCI draw foundations and expertise from psychology experimentation. In particular, we used Testing, Inspection, and Inquiry methods [15] [16] for both our pre-design and post-design evaluations. Table 3 lists and defines the techniques applicable for each method; an asterisk (*) indicates the techniques we used.

6.2 Study Design

The objective of our case study was to assess the product usability of our prototype in comparison to the original Protein Explorer tool. According to [7], the original tool was designed with a focus on "functionality and usability." 15 end-users participated in usability testing. Our sample was a subset of the users observed during our pre-design evaluation. It is important to note that although some of our users had experience with bioinformatics visualization tools, none of them had any experience with the Protein Explorer. This was advantageous for us, since there was no transfer of learning effects from expert users. Furthermore, they were unaware of which

Table 3. Usability Testing Methods [15]

Method Technique	Description
Testing	
Thinking-Aloud Protocol *	user talks during test
Question-Asking Protocol	tester asks user questions
Shadowing Method	expert explains user actions to tester
Coaching Method *	user can ask an expert questions
Teaching Method	expert user teaches novice user
Codiscovery Learning	two users collaborate
Performance Measurement *	tester records usage data during test
Log File Analysis *	tester analyzes usage data
Retrospective Testing	tester reviews videotape with user
Remote Testing	tester and user are not colocated during test
Inspection	
Guideline Review	expert checks guideline conformance
Cognitive Walkthrough	expert simulates user's problem solving
Pluralistic Walkthrough	multiple people conduct cognitive walkthrough
Heuristic Evaluation *	expert identifies violations of heuristics
Perspective-Based Inspection	expert conducts focused heuristic evaluation
Feature Inspection	expert evaluates product features
Formal Usability Inspection	expert conducts formal heuristic evaluation
Consistency Inspection	expert checks consistency across products
Standards Inspection	expert checks for standards compliance
Inquiry	
Contextual Inquiry *	interviewer questions users in their environment
Field Observation *	interviewer observes use in user's environment
Focus Groups	multiple users participate in a discussion session
Interviews *	one user participates in a discussion session
Surveys	interviewer asks user specific questions
Questionnaires *	user provides answers to specific questions
Self-Reporting Logs	user records UI operations
Screen Snapshots	user captures UI screens
User Feedback	user submits comments

version of the tool (original vs. new) they were interacting with during the sessions. We performed task-based evaluations and open-ended interviews to compare the original design with the new design. Open-ended interviews included general questions about impressions of both versions of the tool (any differences, likes and dislikes) and specific questions about the user interface (navigation, etc.). Tasks were designed in conjunction with a biomedical expert. End-users of the tool typically follow a scientific process when performing tasks; i.e., the exploration of a particular molecule. We therefore designed each task as part of a scientific process. One example is presented below.

Exploring the Hemoglobin molecule
- Load Hemoglobin structure, with the PDB code 1HGA.
- Stop the molecule from spinning.

- Remove the ligands from the molecule.
- Modify view to "spacefill" (from "backbone") for all atoms of the molecule.
- Find out more about this molecule. For example, title and taxonomic source.
- In advanced options, find out how to view surfaces (multiple surfaces concurrently). This will allow you to use options such as "molecular electrostatic potential".

We used a *within-subjects* protocol, where each user performs under each condition; in our case, each user tested both designs, the Original Design (Design O) and the New Design (Design N). The advantage of this protocol is that there is less of a chance of variation effects between users, and we can obtain a large data set even with a smaller number of participants. In order to reduce the effect of learning, we varied the order of the designs [17] per participant – some users started with Design N, others with Design O. Furthermore, we varied each of the two scientific processes per design type. We logged task times, failure rates, and recorded the entire user experience with both designs.

6.3   Results

First, we present the **quantitative results**. Our independent variables were: (1) Variation of the design type and (2) variation of the design order used. Dependant variables were: (1) Task duration and (2) failure rate. However, we expected that the second independent variable has no effect on the results. More precisely, we expected that by effectively varying the starting type of the design we were able to reduce any effect of knowledge transfer between the designs to a minimum.

For **task duration**, we used the ANOVA test in order to compare task times of Designs O and N. Our hypothesis for the test was that we would have a statistically significant improvement of time required to complete a task in Design N compared to Design O. We performed an ANOVA two- factor test with replication in order to prove our hypothesis. The two factors selected where: (1) the order in which the user tested the designs (O/N or N/O) and (2) the design type tested (O or N). The goal of this test was to see whether each factor separately has an influence on the results, and at the same time to see if both factors combined have an influence on the design.

Table 4. Results for task duration and failure rates

Source of Variation	F	P-value	F crit	η^2
Task duration				
Factor 1	2.024175	0.167682	4.259675	0.03
Factor 2	35.70645	3.62E-06	4.259675	0.55
Interaction	3.182445	0.087084	4.259675	0.05
Failure rates				
Factor 1	4.033333	0.055991	4.259675	0.07
Factor 2	28.03333	1.97E-05	4.259675	0.49
Interaction	0.833333	0.37039	4.259675	0.01

The results (see Table 4) demonstrate that variation of the order in which the user has tested the design has no influence on the task times (p > 0.05). This means that the users were unaffected by transfer of knowledge from one design to another. Moreover, the test demonstrates that the combined effect of both variables has no statistically significant impact on the task times (0.05 < p < 0.10). Finally, the second factor is the only one that has a statistically significant effect on the task times: F = 35.71, p = 3.62 E-06, η^2 = 0.55. This demonstrates that there was a statistically significant improvement in task time in Design N when compared to Design O. We noted an average improvement of **52 %.**

For **failure rates**, our hypothesis was that there should be a significant improvement in failure rates with Design N versus O. Similar to task duration, we performed a two factor ANOVA test with replication in order to test our hypothesis, where the factors were the same as described above. The test results (see Table 4) demonstrated that there was a statistically significant improvement in failure rates in Design N when compared to Design O: Factor 2 has F = 28.03, p < 0.05 and η^2 = 0.49. Moreover, the test demonstrated that there is no statistically significant interaction between the two factors when considering their effect on failure rates (p > 0.05). Similarly, the test has demonstrated that the order under which the users have tested the designs has no statistically significant effect on the failure rates (p > 0.05).

The **qualitative results** were obtained from open-ended interviews with all users, carried out after task-based evaluations with both versions of the tool. The most common comments about the usability of the original version from end-users were as follows: (1) it is overloaded with content in the control pane; (2) the provided information is not filtered adequately, requiring users to spend lots of time reading irrelevant information, (3) navigation between pages is difficult, resulting in confusion when trying to reach the load page; and (4) manipulation of the visualization pane is difficult since it is unclear where the features for the visualization are located. Furthermore, we recorded user exploration sessions and used the think-aloud protocol. Our observations indicated a high level of frustration with users during their interaction with the original version of the tool.

The most common comments about the usability of the new prototype from end-users were as follows: (1) easier to locate information because of the structure; (2) organization of features and tools follows more closely with the scientific process in bioinformatics, (3) the interface is simpler and users feel more in control when interacting with it, (4) the use of tabs made navigation easier. Furthermore, during the recorded sessions, users seemed calmer and more comfortable during their interaction with the prototype.

13 out of 15 users indicated that they prefer the design of the new prototype compared to the design of the original tool. Simplicity and "feeling more in control" were cited as the most important reasons. Interestingly enough, one of the two users who indicated his preference for the original tool also cited "simplicity" as a reason, but in terms of the new prototype as being too simple, and the original version having all the information "handy." The other user indicated that the fonts were too small and the colors a bit confusing on the new prototype.

7 Concluding Remarks

In this paper, we presented a study we carried out with a bioinformatics visualization tool. The first objective of the study was to redesign the Protein Explorer using a UE framework which we developed, called UX-P. The framework leverages personas and patterns as the major directives in the design process, and encourages on-going usability testing. By doing so, it adheres to a few notable principles: A focus on varied user groups and their needs; the incorporation of behavioral rationale; the systematic and traceable application of gathered knowledge; an embracing of reuse; and taking a lightweight and pragmatic approach. We found that applying the framework facilitated our design activities, allowed us to incorporate sound UCD principles into our design, and afforded guidance to an often ad-hoc process. In other words, it provided us with some structure. Since the starting point was creating personas, the focus of the design activity was directed to the users early on. Furthermore, personas are a relatively lightweight user model, and we did not require a user or cognitive modeling specialist for their creation. By developing personas iteratively using empirical evidence (observations with 22 users), it allowed us to determine interaction behavior and determine usability issues with similar tools; these points were essential in selecting design patterns. In this vein, the framework follows the reuse paradigm through the use of these patterns, enabling us to make design decisions based on best practices. Notably, in current practice, there exists no commonly agreed upon UI design process that employs patterns and their languages as first class tools. Our second objective was to assess the usability of our prototype (designed following UX-P) in comparison to the original version (built with a focus on "functionality and usability" [7]). We performed usability testing within the context of a case study. The results were encouraging: The framework led to a more usable system as indicated by a number of usability indicators. We used a within-subjects protocol, and our sample size consisted of 15 end-users. Quantitative results indicated both a statistically significant improvement in task duration (a measure of user performance) and failure rates (a measure of effectiveness) with the prototype. Furthermore, qualitative results indicated a greater degree of satisfaction with the prototype for 13 out of the 15 users. From a design point of view, the UX-P Framework provided structure to our activities. It also ensured that user experiences act as a primary directive in the design process, which is essential in providing user-centered design solutions. From our case study, quantitative and qualitative usability indicators demonstrated a significant improvement in global usability for the prototype designed using UX-P; resulting in a more human-centered visualization tool. Feedback, as a result from this study, will be used to further refine and validate the framework.

References

1. Seffah, A., Gulliksen, J., Desmarais, M.C.: Human-Centered Software Engineering - Integrating Usability in the Software Development Lifecycle. Springer, Netherlands (2005)
2. ISO 13407: Standard on Human-Centered Development Processes for Interactive Systems. ISO Standards (1998)

3. Cooper, A.: The inmates are running the asylum: Why high-tech products drive us crazy and how to restore the sanity. SAMS Publishing, Indianapolis (1999)
4. Gamma, E., Helm, R., Johnson, R., Vlissides, J.: Design Patterns: Elements of Reusable Object-Oriented Software. Addison-Wesley, Reading (1995)
5. Tidwell, J.: UI Patterns and Techniques [online]. [Accessed March, 2003]. Available from, http://time-tripper.com/uipatterns/index.php (2002)
6. Welie, M.V.: Interaction Design Patterns [online]. [Accessed March, 2003](2003) Available from http://www.welie.com
7. Protein Explorer: FrontDoor to Protein Explorer [online]. [Accessed October 2006]. (2005) Available from, http://molvis.sdsc.edu/protexpl/frntdoor.htm
8. Cn3D: Cn3D 4.1 Download and Homepage [online]. [Accessed October 2006]. Available from, http://130.14.29.110/Structure/CN3D/cn3d.shtml (2005)
9. Gros, P.E., Férey, N., Hérisson, J., Gherbi, R., Seffah, A.: Multiple User Interface for Exploring Genomics Databases. In: Proceedings of HCI International (July 22-27, 2005) (2005)
10. Paternò, F.: Model-Based Design and Evaluation of Interactive Applications. Springer, New York (2000)
11. Wilkins, B.: MELD: A Pattern Supported Methodology for Visualization Design. PhD Thesis, The University of Birmingham (2003)
12. Javahery, H., Sinnig, D., Seffah, A., Forbrig, P., Radhakrishnan, T.: Pattern-Based UI Design: Adding Rigor with User and Context Variables. In: Coninx, K., Luyten, K., Schneider, K.A. (eds.) TAMODIA 2006. LNCS, vol. 4385, pp. 97–108. Springer, Heidelberg (2007)
13. Fenton, N.E., Pfleeger, S.L.: Software Metrics: a Rigorous and Practical Approach, 2nd edn. PWS Publishing Co. Boston (1998)
14. ISO/IEC 25062: Software Product Quality Requirements and Evaluation (SQuaRE), Common Industry Format (CIF) for usability test reports. ISO Standards (2006)
15. Ivory, M.Y., Hearst, M.A.: The state of the art in automating usability evaluation of user interfaces. ACM Computing Survey 33(4), 470–516 (2001)
16. Holzinger, A.: Usability Engineering for Software Developers. Communications of the ACM 48(1), 71–74 (2005)
17. Dix, A., Finlay, J.E., Abowd, G.D., Beale, R.: Human-Computer Interaction, 3rd edn. Prentice Hall, Englewood Cliffs (2003)

Interactive Analysis and Visualization of Macromolecular Interfaces between Proteins

Marco Wiltgen[1], Andreas Holzinger[1], and Gernot P. Tilz[2]

[1] Institute for Medical Informatics, Statistics and Documentation
Medical University of Graz, A-8036 Graz, Austria
[2] Centre d'Etude du Polymorphysme Humain, 27 rue Juliette Dadu, F-75010 Paris, France
marco.wiltgen@meduni-graz.at

Abstract. Molecular interfaces between proteins are of high importance for understanding their interactions and functions. In this paper protein complexes in the PDB database are used as input to calculate an interface contact matrix between two proteins, based on the distance between individual residues and atoms of each protein. The interface contact matrix is linked to a 3D visualization of the macromolecular structures in that way, that mouse clicking on the appropriate part of the interface contact matrix highlights the corresponding residues in the 3D structure. Additionally, the identified residues in the interface contact matrix are used to define the molecular surface at the interface. The interface contact matrix allows the end user to overview the distribution of the involved residues and an evaluation of interfacial binding *hot spots*. The interactive visualization of the selected residues in a 3D view via interacting windows allows realistic analysis of the macromolecular interface.

Keywords: Interface Contact Matrix, Bioinformatics, Macromolecular Interfaces, Human–Computer Interaction, Tumour Necrosis Factor.

1 Introduction

Proteins are the molecules of life used by the cell to read and translate the genomic information into other proteins for performing and controlling cellular processes: metabolism (decomposition and biosynthesis of molecules), physiological signalling, energy storage and conversion, formation of cellular structures etc. Processes inside and outside cells can be described as networks of interacting proteins. A protein molecule is build up as a linear chain of amino acids. Up to 20 different amino acids are involved as elements in protein sequences which contain 50-2000 residues. The functional specificity of a protein is linked to its structure where the shape is of special importance for the intermolecular interactions. These interactions are described in terms of locks and keys. To enable an interaction, the shape of the lock (for example the enzyme) must be complimentary to the shape of the key (the substrate). Examples are the antibody-antigen complexes in the immune system, complexes of growing factors and receptors and especially the tumour necrosis factor – receptor complex.

A. Holzinger (Ed.): USAB 2007, LNCS 4799, pp. 199–212, 2007.

The analysis of such interactions are of special importance for modern clinical diagnosis and therapy e.g. in the case of infectious diseases, disturbed metabolic situations, incompatibilities in pharmacology etc.

A protein can not be seen, for example by a microscope (with x-rays focussing lenses). It exists no real image, like a microscopic view from cell, of a protein. Instead a *model,* resulting from (among other methods) X-ray crystallography must be used [1]. The model is derived from a regular diffraction pattern; build up from X-ray reflections scattered from the protein crystal. This structural model is a three-dimensional representation of a molecule containing information about the spatial arrangements of the atoms. The model reflects the experimental data in a consistent way. Protein structural information is publicly available, as atomic coordinate files, at the protein database (PDB), an international repository [2].

Bioinformatics is an interdisciplinary discipline between information science and molecular biology [3-6]. It applies information processing methods and tools to understand biological and medical facts and processes at a molecular level. A growing section of bioinformatics deals with the computation and visualization of 3D protein structures [7-12]. In the world of protein molecules the application of visualization techniques is particularly rewarding, for it allows us to depict phenomena that cannot be seen by any other means. The behaviour of atoms and molecules are beyond the realm of our everyday experience and they are not directly accessible for us.

Modern computers give us the possibility to visualize these phenomena and let us observe events that cannot be witnessed by any other means.

In this paper we concentrate on the analysis of macromolecular interfaces between interacting proteins. The high complexity of the protein-protein interface makes it necessary to choose appropriate (and as simple as possible) representations, allowing the end user to concentrate on the specific features of their current interest without being confused by the wealth of information. As in all medical areas, the amount of information is growing enormously and the end users are faced with the problem of too much rather than too little information; consequently the problem of information overload is rapidly approaching [13]. To apply end user centered methods is one possibility to design and develop the applications exactly suited to the needs of clinicians and biologists in their daily workflows. Usability engineering methods [14] ensure to understand the biologists' interactions, which is necessary to determine how the tools are ideally developed to meet the end users experience, tasks, and work contexts. Several research groups have developed software to assist the analysis of macromolecular interfaces such as: coupled display of planar maps of the interfaces and three dimensional views of the structures or visualization of protein-protein interactions at domains and residue resolutions [15, 16].

We present the development of a tool (the interface contact matrix) for the representation and analysis of the residue distribution at the macromolecular interface, connected interactively with a 3D visualization. Atomic coordinate files of protein complexes (containing two interacting proteins) from the PDB structure database are used as input. Because real experimental data are used, measurements and calculations can be done on the protein structures. First the distances between the residues of the two chains of a complex are calculated. The residues, at a given distance, between the two chains are identified and the interface contact matrix, a plot of adjacent residues between the two amino acid chains, is constructed. The residues name and number within each chain are plotted on the respective axis (horizontal and vertical axis) and a corresponding entry is done at the appropriate place in the matrix wherever two residues of each chain come into contact. These matrix elements are annotated with several physicochemical properties. The identification of spatial narrow residues between the chains in the complex 3D structure and the additional representation of their pair wise interaction in an interface contact matrix, allows a suitable and easy to survey representation of the interface information. The interface contact matrix enables the end user an overview about the distribution of the involved residues in the macromolecular interface and their properties, an evaluation of interfacial binding *hot spots* and an easy detection of common or similar patterns in different macromolecular interfaces. Of course for a realistic and adequate visualization of a macromolecular interface a 3D representation is necessary. Therefore the elements in the interface contact matrix are linked with the 3D structure in that way that mouse manipulations on the matrix elements display the corresponding region in the 3D structure (the interface between proteins emphasis normally only a limited set of residues). This connection of the two representations, the interface contact matrix and the 3D structure, reveal from a wealth of information the context and connections without overwhelming the end user. The relationship between the two representations was technically realized by interactive windows. Additionally the identified residues in the interface contact matrix are used to define the molecular surface at the interface.

The resulting interface surface enables exploration of molecular complementarities, where physicochemical properties of the residues are projected on the surface. To illustrate the procedure, the complex between the tumour necrosis factor (TNF) and the TNF-receptor was used [17].

The TNF is a dominant proinflammatory protein causing destruction of cells, blood vessels and various tissues and plays an important role in the immune system by activation of immune cells. The TNF-receptor is located at the cell membrane and the TNF molecule is interacting with its extra cellular domain. The interface contact matrix is a tool which enables a better understanding of the interaction between receptor and acceptor molecules which is the precondition for the necessary treatment of excessive inflammation.

2   Materials and Methods

A large number of proteins and protein complexes are deposited in the PDB structure database. At the moment PDB contains more than 44.000 protein structures. The structural information, experimentally determined, is stored in the PDB[1] data files as a collection of Cartesian coordinates of the involved atoms (figure 1, figure 2).

2.1   Software

Based on the experimental data, representations of the protein structures are generated by the computer. The calculations of the macromolecular interface were done by a

---

[1] The PDB data base can be accessed through Internet and the PDB data files downloaded.

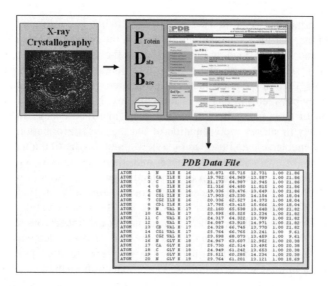

Fig. 1. Structures of protein complexes, determined by X-ray crystallography, are deposited in the PDB structure database. The structural information is stored in the PDB data files. These files contain: a running number, the atom type, the residue name, the chain identification, the number of the residue in the chain, the triplet of coordinates. The PDB data files are downloaded from the database as input files for protein analysis and visualization.

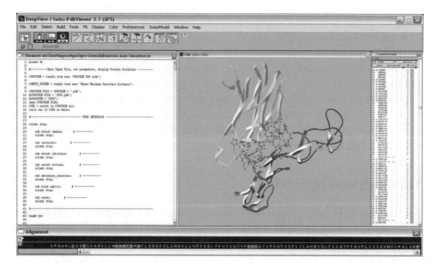

Fig. 2. The Swiss-PDB Viewer provides: main window (top), control panel (right), sequence alignment window (bottom) and display window (middle). In the script window (left) the program for the analysis of the protein-protein interface and the calculation of the interface contact matrix is read in. The calculations are done on the atomic coordinates in the PDB file. The routines for computation/visualization are initiated by mouse clicking in the script window.

special developed computer program for Windows PC, and the results from the analysis were visualized with the Swiss-Pdb Viewer [18], which consists of several windows (figure 2). The main window is used for manipulations (rotations, zooming etc.), representations and measurements of the displayed protein structures. The control panel window enables the selection of single residues for display. At the display window, the protein structures are visualized. The Swiss-Pdb viewer includes a parser of a scripting language. For the analysis of the protein interfaces, the proprietary program (written in the Deep View scripting language, a PERL derivate) was read in the script window. The program enables:

- identification of the residues involved in the interface between two chains,
- the determination of hydrogen bonds between appropriate atoms of the interface residues,
- the calculation and printing of the interface contact matrix,
- the annotation of physicochemical properties to the matrix elements.

The calculations for identifying the residues at the protein-protein interface are done automatically by use of the atomic coordinates from the PDB file (figure 3).

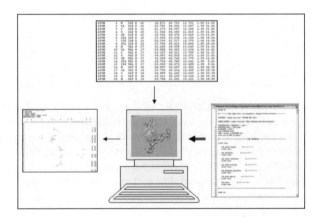

Fig. 3. The structural information stored in the PDB data file is used as input for computing and visualization. The calculations are done in a scripting language, and the output is represented as pair wise interaction of adjacent residues in the interface contact matrix.

At the end of the program run, routines are called by the program and displayed in the script window, allowing further manipulation of the results interactively by the user. The routines are initiated by mouse clicking in the script window and enables:

- the calculation of the molecular surfaces at the interface,
- the interactive visualization and analysis of the interface residues,
- the visualization of interfacial geometries in 3D

Interacting windows enables the relationship between the output data of the distance calculations and the corresponding parts of the protein 3D structure. This offers the possibility to interactively connect the elements in the interface contact matrix with structural properties. By mouse clicking on the appropriate output data the corresponding residue pairs (one on each chain) are highlighted on the display window allowing the evaluation of the structural properties.

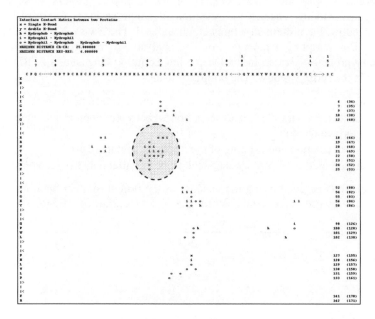

Fig. 4. The interface contact matrix is a plot of pair wise interactions between interfacial residues in the protein complex (vertical axis: TNF, horizontal axis: TNF-receptor). The residue name (one letter code) and number within each chain are plotted on the corresponding axis and an entry is made wherever two residues of each chain come into close contact. The contact is defined by a predefined threshold distance within the distance between at least two atoms (one of each residue) must fall in. Physicochemical properties are annotated to the matrix elements.

2.2 Computational Method

The macromolecular interface between proteins is defined by complementary surfaces with adjacent residues which are close enough to interact. First, the interatomic distances d_{ij} between each residue (n_i and m_j) of the two polypeptide chains A (TNF chain) and R (TNF-receptor chain) are determined automatically.

$$\forall_{n_i \in A} \forall_{m_j \in R} : (n_i, m_j) \to d_{ij}, \tag{1}$$

Then the residues, where at least two adjacent atoms are separated by a distance less than a predefined threshold distance, are identified as interface residues.

2.2.1 Interface Contact Matrix

These residues are used as input to construct the interface contact matrix:

$$I_l = \left\{ (n_i, m_j) \mid d_{ij} \leq d_l \right\}, \tag{2}$$

The two dimensional interface contact matrix is a plot of pair wise interactions between adjacent residues of the two chains in the protein complex (figure 4). The residue names and numbers within each chain are plotted on the respective axis (horizontal and vertical axis) and a corresponding entry is done at the appropriate place in the matrix wherever two residues of each chain interact. The values of the threshold distance d_1 (4-6 angstroms) are motivated by the range of hydrogen bonds and van der Waals bonds (figure 5). The matrix elements define then the macromolecular interface and they are used for further definitions and calculations. Combinations of physicochemical properties p_c are annotated to the matrix elements in the interface contact matrix:

$$\forall_{(n_i,m_j)\in I_1} \exists_{p_c\in P} : (n_i,m_j) \to p_c , \tag{3}$$

This indicates complementarities and correlations between the properties of the two residues across the interface like: electrostatic complementarities, correlation of hydrophobic/hydrophilic values, correlation of proton donator and acceptor across the interface resulting in hydrogen bonds.

2.2.2 Visualization of Macromolecular Interfaces

The interface contact matrix is interactively linked to a 3D visualization of the macromolecular structures, making the involved residues easily detectable in the wealth of information provided by the complex structures (figure 6). By mouse clicking on selected parts of the residue distribution in the matrix, highlights the corresponding residues in the 3D structure.

Molecular surfaces are suitable for the description of molecular shapes and the visualization of their complementary properties [19]. The residue distribution in the interface contact matrix is used to define the molecular surfaces (one for each protein in the complex) at the macromolecular interface. The following is done for the residues of each chain, identified in the interface contact matrix (components of the residue pairs $(n_i,m_j)\in I_1$), in the protein complex separately. To each atom A_k of a residue n_i a function ρ, depending on the size σ_k of the atom (with centre at r_k) and a weighting factor, is assigned:

$$n_i(A_k) \to \rho(w_k, r-r_k, \sigma_k) , \tag{4}$$

The weighting factor w_k, describes the contribution of the considered atom to the molecular surface. Then the functions of all the atoms of the residues in the interface contact matrix, contributing to the set of surface atoms S of the molecule are considered and summed up.

A common description of the molecular surface uses Gaussian functions, centred on each atom A_k [19]:

$$\rho(r) = \sum_{k\in S} w_k \, e^{-|r-r_k|^2 / \sigma_k^2} , \tag{5}$$

This defines the approximate electron density distribution (other functions are also used in literature).

Fig. 5. The macromolecular interface can be defined by residues on both polypeptide chains which are close enough to form interactions. The most important interactions are hydrogen bonds (range 3.6 angstroms) and van der Waals bonds. Usually the van der Waals bonds arise between two atoms with van der Waals spheres (1.4-1.9 angstroms) within a distance of 1.5 angstrom of each other with no overlapping. Hydrogen bonds (dotted lines) arise from the interaction of two polar groups containing a proton donor and proton acceptor. Van der Waals bonds (interacting spheres): Due to its electronic properties, an atom acts as a dipole which polarises the neighbour atom, resulting in a transient attractive force. The threshold distance d_1 in the interface contact matrix is motivated by the range of these interactions.

In the protein complex, the molecular surfaces of both proteins show complementary areas at the interface (because the calculations of the molecular surfaces at the interface are done by use of adjacent residues: $(n_i, m_j) \in I_1$ at close distances).

3 Results

The presented method allows the analysis of the protein-protein interactions at the level of the sequences (interface contact matrix) and on the level of the 3D structure. Both analysis levels are interconnected, enabling a relation between different kinds of exploited information.

The analysis of the interaction properties in the interface contact matrix facilitates an identification and evaluation of interfacial binding "hot spots", that means locally strong binding forces. Important binding forces are hydrogen bonds which arise from

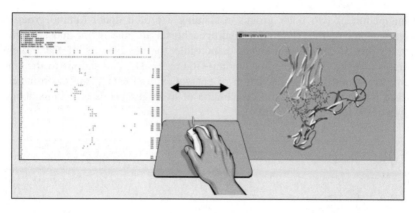

Fig. 6. Interactive windows enable a relationship between the elements of the interface contact matrix and the appropriate part of the 3D structure. Mouse clicking on the data of interest in the output window results in a highlighting of the corresponding parts of the protein structure in the display window. This enables an easy detection and structural analysis of the involved residues in the wealth of information provided by the complex interface structures.

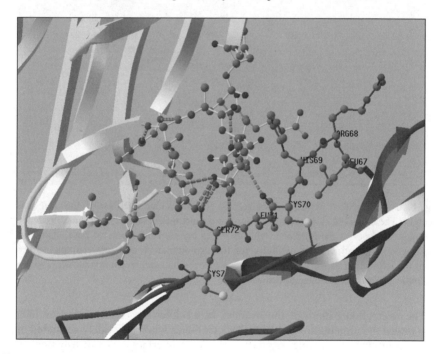

Fig. 7. This figure shows the 3D interface structure of the highlighted part of the interface contact matrix in figure 4. The TNF (upper part) is interacting with the extra cellular domain of its receptor (lower part). The residues at the macromolecular interface are visualized in a "ball-and-stick" representation. The covalent bonds are represented as sticks between atoms, which are represented as balls. The rest of the two chains are represented as ribbons. Residue names and numbers of the TNF receptor are labelled. The hydrogen bonds are represented by dotted lines.

the interaction of two polar groups containing a proton donor (amino group) and proton acceptor (carboxyl group). Hydrogen bonds are one of the most stabilizing forces in protein structure and are responsible for the binding of the protein to substrates and the factor-receptor binding (figure 5). The annotations to the contact patterns are useful for the evaluation of energetically effects at the macromolecular interface where the number and distribution of hydrogen bonds give hints about the strength of the interaction (figure 4).

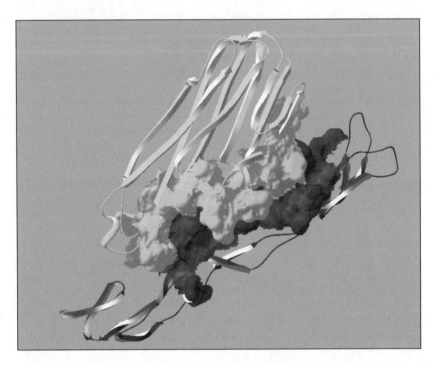

Fig. 8. Complementary molecular surfaces at the molecular interface of the TNF and its receptor. The identified residues in the interface contact matrix (figure 4) are used to define the molecular surfaces at the interface (with distances between the atoms of both chains till 7.5 angstroms). The surfaces are helpfully for the exploration of molecular complementarities. From the interaction of the two amino acid chains results complementary surfaces, where the two molecular surfaces come into contact.

The interaction pattern of the residues in the interface contact matrices indicate how folded the proteins are at the interface (residues witch are widely separated in the amino acid chain are acting together at the local interface). The distribution of the matrix elements along the columns in the matrix shows how embedded the interface residues of the receptor (horizontal axis) in the environment of the TNF residues are. This gives some hints for the exposure of the receptor residues to the adjacent residues in the macromolecular interface.

The visualization of the selected residues in a 3D view via interacting windows allows a realistic analysis of the macromolecular interface (figure 6). Due to the interactive windows, the selection of the residues in the interface contact matrix and the highlighting in the 3D structure allows an easy retrieval of the desired information out of the wealth of structural information without overwhelming the end user. This is of special importance for the exploration of highly embedded residues in the macromolecular interface, as well for matrix element distributions showing a high number of hydrogen bonds. It allows a fast and easy overview about the involved residues in their structural context. Complementary properties (for example: electrostatic, hydrophobic/hydrophilic values) of adjacent residues across the protein-protein interfaces can be detected in the interface contact matrix and studied in a 3D view. The relative orientations of the side chains of opposed residues and their reciprocal exposure are visualized (figure 7).

The visualization of the molecular surfaces of the residues at the interface of the two chains, where the two surfaces come into contact, allows insight into the interfacial geometries in 3D and the identification of molecular complementarities (figure 8).

4 Discussion

Exploring complementary protein surfaces provides an important route for discovering unrecognized or novel functional relationships between proteins. This is of special importance for the planning of an individual drug treatment. Better medication can be developed once the structures of binding sites are known.

There exist several approaches and methods for studying macromolecular interfaces like: the web resource iPfam, allowing the investigation of protein interactions in the PDB structures at the level of (Pfam) domains and amino acid residues and MolSurfer, which establish a relation between a 2D Map (for navigation) and 3D the molecular surface [15, 16]. The advantage of the presented method for the analysis of macromolecular interfaces is the representation at sequence level and at structural level and the connection between both views. That means the connection of statistical properties (distribution of the residues together with their annotated physicochemical values in the interface contact matrix) and the structural properties (reciprocal exposure of side chains, atomic interactions).

The selection of the residues in the interface contact matrix and the highlighting of the corresponding 3D structure, by aid of the interactive windows, enable an easy identification and structural analysis of the interfacial residues in the wealth of information provided by the complex interface structures. Common patterns in the interface contact matrices allow a fast comparison of similar structures in different macromolecular interfaces.

The analysis of the patterns in the interface contact matrices of slightly different protein complexes allows an easy detection of structural changes. Complementary properties or even possible mismatches of adjacent residues across the protein-protein interfaces can be detected in the interface contact matrix and studied in a 3D view. The representation with molecular surfaces shows complementary shapes.

To demonstrate the applicability of the method in clinical medicine we choose the analysis of the interface between the tumour necrosis factor (TNF) and its receptor [17]. The tumour necrosis factor molecule is particularly interesting. It is responsible for various immune responses such as inflammation and cytokine activation. TNF is interacting with the extra cellular domain of its receptor, which is located on the cell membrane. It plays a considerable role in the bodies' defence in inflammation, tumour pathology and immunology [20-22].

For this reason many forms of treatments have been building up to reduce the excessive TNF activity. Ethanercept as a receptor molecule (Enbrel) or Infliximab (a humanized antibody) are examples. On account of these molecules modern treatment has become a tremendous success based on the knowledge of the antigenity of TNF. Our macromolecular interface analysis and visualization system may help us to define better receptor and acceptor molecules for the neutralisation and excretion of the tumour necrosis factor. By the analysis of the residue distributions in the interface contact matrix and the associated visualisation of the macromolecular interface, the active sites of the reciprocal molecules can be studied and the concept of neutralisation and inactivation followed.

The interface contact matrix is a suitable frame work for further investigations of the macromolecular interfaces, providing the matrix elements with additional data. Further studies may investigate the macromolecular interfaces in more details by determining the most exposed interfacial residues and, in an additional step, by calculating and visualization of details at the atomic and electronic levels of these residues. The topic of future work will be the quantification of the contact between adjacent residues across the interface.

This can be done by Voronoi tessalation, a partition of the protein into cells with polygonal surfaces defining the neighbourhood of an amino acid and its contact area with adjacent residues. Of special importance is the interfacial accessibility, the area of common faces of the cells of adjacent residues on both chains.

The annotation of these values to the interface contact matrix allows the evaluation of the distribution of the contact areas of the matrix elements and the determination of the most exposed residues involved in the macromolecular interface.

These annotations result from structural analysis, which are nowadays advanced and high-throughput methods for the determination of the structure of factor-receptor complexes providing the information need to build the structure-activity relationships. Details at the atomic and electronic levels of the macromolecular interface needed for a deeper understanding of the processes, that remain unrevealed after structural elucidation, may be provided additionally by quantum theoretical calculations.

In every case, filling the "frame work" interface contact matrix with information means: connecting the matrix elements with physicochemical annotations, showing different and successively more properties of the macromolecular interface.

5 Conclusion

Our approach offers the advantage of connecting the interface contact matrix with a 3D visualization of the complex interfaces. In the interface contact matrix the involved residues of the macromolecular interface can be determined, (complementary) physicochemical properties be annotated and common pattern of different interfaces detected. The visualization of the selected residues in a 3D view via interacting windows allows a realistic analysis of the macromolecular interface.

We have used for demonstration of the method, the complex of TNF and its – Receptor representing a most rewarding concept of modern therapy. By computer visualisation, the macromolecular interface of the reciprocal molecules can be shown and the concept of neutralisation and inactivation followed.

This procedure of inactivation and neutralisation of detrimental molecules can barely figured out without the optical advices obtained with such methods. Molecular medicine means the understanding of diseases at a molecular level.

Hence thanks to these attempts of analysis and visualisation the construction and synthesis of reciprocal acceptor-, blocking- and neutralisation molecules may be very much enhanced and helpful for end users in diagnosis and treatment of inflammations and other diseases.

Acknowledgments. We thank the anonymous reviewers for their valuable comments.

References

1. Perutz, M.F., Muirhead, H., Cox, J.M., Goaman, L.C.: Three-dimensional Fourier synthesis of horse oxyhaemoglobin at 2.8 angstrom resolution: the atomic model. Nature 219, 131–139 (1968)
2. Berman, H.M., Westbrook, J., Feng, Z., Gilliland, G., Bhat, T.N., Weissig, H., Shindyalov, I.N., Bourne, P.E.: The Protein Data Bank. Nucleic Acids Research, 28, 235–243 (2000)
3. Lesk, A.M.: Introduction to Bioinformatics. Oxford University Press, Oxford (2002)
4. Gibas, C., Jambeck, P.: Developing Bioinformatics Computer Skills, O'Reilly (2001)
5. Chang, P.L.: Clinical bioinformatics. Chang. Gung. Med. J. 28(4), 201–211 (2005)
6. Mount, D.W.: Bioinformatics: sequence and genome analysis. Cold Spring Harbor laboratory Press, New York (2001)
7. Hogue, C.W.: Cn3D: A new generation of three-dimensional molecular structure viewer. Trends Biochemical Science 22, 314–316 (1997)
8. Walther, D.: WebMol- a Java based PDB viewer. Trends Biochem Science 22, 274–275 (1997)
9. Gabdoulline, R.R., Hoffmann, R., Leitner, F., Wade, R.C.: ProSAT: functional annotation of protein 3D structures. Bioinformatics 1,19(13), 1723–1725 (2003)
10. Neshich, G., Rocchia, W., Mancini, A.L., et al.: Java Protein Dossier: a novel web-based data visualization tool for comprehensive analysis of protein structure. Nucleic Acids Res. 1,32, 595–601 (2004)
11. Oldfield, T.J.: A Java applet for multiple linked visualization of protein structure andsequence. J.Comput. Aided. Mol. Des. 18(4), 225–234 (2004)
12. Wiltgen, M., Holzinger, A.: Visualization in Bioinformatics: Protein Structures with Physicochemical and Biological Annotations. In: Zara, J., Sloup, J. (eds.) CEMVRC 2005. Central European Multimedia and Virtual Reality Conference, pp. 69–74. Eurographics Library (2005)
13. Holzinger, A., Geierhofer, R., Errath, M.: Semantic Information in Medical Information Systems - from Data and Information to Knowledge: Facing Information Overload. In: Proceedings of I-MEDIA 2007 and I-SEMANTICS 2007, pp. 323–330 (2007)
14. Holzinger, A.: Usability Engineering for Software Developers. Communications of the ACM 48(1), 71–74 (2005)
15. Gabdoulline, R.R., Wade, R.C., Walther, D.: MolSurfer: A macromolecular interface navigator. Nucleid Acids Res. 1,31(13), 3349–3351 (2003)

16. Finn, R.D., Marshall, M., Bateman, A.: iPfam: visualization of protein-protein interactions in PDB at domain and amino acid resolutions. Bioinformatics 21(3), 410–412 (2005)
17. Banner, D.W., D'Arcy, A., Janes, W., Gentz, R., Schoenfeld, H., Broger, C., Loetscher, H., Lesslauer, W.: Crystal structure of the soluble human 55 kd TNF receptor-human TNF beta complex: implications for TNF receptor activation. Cell 73, 431–445 (1993)
18. Guex, N., Peitsch, M.C.: SWISS-MODEL and the Swiss-Pdb Viewer: An environment for comparative protein modelling. Electrophoresis 18, 2714–2723 (1997)
19. Duncan, B., Olson, A.J.: Approximation and characterization of molecular surfaces. Biopolymers 33, 219–229 (1993)
20. Kim, Y.S., Morgan, M.J., Choksi, S., Liu, Z.G.: TNF-Induced Activation of the Nox1 NADPH Oxidase and Its Role in the Induction of Necrotic Cell Death. Mol. Cell. 8, 26(5), 675–687 (2007)
21. Vielhauer, V., Mayadas, T.N.: Functions of TNF and its Receptors in Renal Disease: Distinct Roles in Inflammatory Tissue Injury and Immune Regulation. Semin Nephrol. 27(3), 286–308 (2007)
22. Assi, L.k., Wong, S.H., Ludwig, A., Raga, K., Gordon, C., Salmon, M., Lord, J.M., Scheel-Toellner, D.: Tumor necrosis factor alpha activates release of B lymphocyte stimulator by neutrophils infiltrating the rheumatoid joint. Arthritis Rheum. 56(6), 1776–1786 (2007)

Modeling Elastic Vessels with the LBGK Method in Three Dimensions

Daniel Leitner[1,2], Siegfried Wassertheurer[1], Michael Hessinger[3]
Andreas Holzinger[4], and Felix Breitenecker[2]

[1] Austrian Research Centers GmbH - ARC,
Donau-City-Straße 1, A-1220 Vienna, Austria
[2] Vienna University of Technology, Institute for Analysis and Scientific Computing,
Wiedner Hauptstraße 8-10, A-1040 Vienna, Austria
[3] Medical University of Graz, Clinical Department for vascular surgery,
Auenbruggerplatz 29, 8036 Graz
[4] Medical University of Graz, Institute for Medical Informatics, Statistics and
Documentation, Auenbruggerplatz 2/V, A-8036 Graz, Austria
dleitner@osiris.tuwien.ac.at

Abstract. The Lattice Bhatnagar Gross and Krook (LBGK) method is widely used to solve fluid mechanical problems in engineering applications. In this work a brief introduction of the LBGK method is given and a new boundary condition is proposed for the cardiovascular domain. This enables the method to support elastic walls in two and three spatial dimensions for simulating blood flow in elastic vessels. The method is designed to be used on geometric data obtained from magnetic resonance angiography without the need of generating parameterized surfaces. The flow field is calculated in an arbitrary geometry revealing characteristic flow patterns and geometrical changes of the arterial walls for different time dependent input contours of pressure and flow. For steady flow the results are compared to the predictions of the model proposed by Y. C. Fung which is an extension of Poiseuille's theory. The results are very promising for relevant Reynolds and Womersley numbers, consequently very useful in medical simulation applications.

Keywords: Simulation, Lattice Boltzmann Model, Haemodynamics, Elasticity, Computer Fluid Dynamics.

1 Introduction

In the western industrial countries cardiovascular diseases are the most frequent cause of death. Therefore a lot of research is done to get a better understanding of the cardiovascular system. Of special interest is the simulation of blood flow in three spatial dimensions using the vessel geometry that is obtained from magnetic resonance angiography. This enables an investigation of pressure and flow profiles and shear stress at the vessel wall. The appearing shear stress is important for the risk estimation of arteriosclerosis [1].

A. Holzinger (Ed.): USAB 2007, LNCS 4799, pp. 213–226, 2007.
© Springer-Verlag Berlin Heidelberg 2007

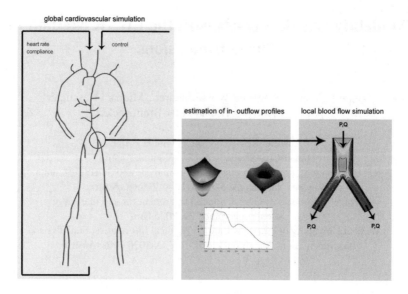

Fig. 1. From a one dimensional model boundary conditions for the more detailed three dimensional model are obtained

In this work a LBGK is used to simulate the blood flow in three spatial dimensions and to solve the incompressible Navier-Stokes equations with the LBGK method [2]. The LBGK method working as a hemodynamical solver on tomographic data has been presented in [3]. For the treatment of elasticity of the vessel walls boundary conditions where proposed by [4] where the vessel wall is represented as a surface. When the vessel walls are represented as voxels, a simpler approach has been proposed in [5], which does not need a parameterized representation of the vessel wall. In this work this approach will be extended to three spatial dimensions.

Blood flow simulation in three dimensions is mostly restricted to a region of interest, where geometrical data are obtained from tomographic angiography. At the in- and outlets appropriate boundary conditions must be applied. One possibility is to obtain the in- and outflow profiles from one-dimensional simulation of the cardiovascular system [6], see figure 1.

2 The LBGK Method

For simulating the flow field we use the LBGK D3Q15 model. A detailed description can be found in [7] and [2]. The LBGK method has proved to be capable of dealing with pulsative flow within the range of Reynolds and Womersley number existing in large arteries. The LBGK method has been successfully applied to the cardiovascular domain by A.M.M Artoli in [8] and [3]. In the following a short overview of the method shall be given.

2.1 The LBGK D3Q15 Method

LBGK Models are based on a statistical description of a fluid in terms of the Boltzmann equation. Thus it is a bottom up approach in developing a numerical scheme for solving the Navier-Stokes equations. Starting point is the Boltzmann equation with the BGK approximation of the collision integral with single relaxation time is given by

$$\frac{\partial f}{\partial t} + \xi \cdot \nabla f = -\frac{1}{\lambda}(f - f^{eq}) \tag{1}$$

where $f(\mathbf{x}, \mathbf{v}, t)$ is the probability distribution depending on the spatial coordinate $\mathbf{x}$, the velocity $\mathbf{v}$ and the time t. The value f^{eq} is the Maxwell distribution function and ξ is the macroscopic velocity.

When the Boltzmann is discretised in the spatial domain, in phase space and in time it yields

$$f_i(x + c \cdot \mathbf{c}_i \Delta t, t + \Delta t) - f_i(\mathbf{x}, t) =$$
$$-\frac{1}{\lambda}(f_i(\mathbf{x}, t) - f_i^{eq}(\mathbf{x}, t)) \tag{2}$$

where $c = \Delta x / \Delta t$, Δx is the lattice grid spacing and Δt the time step. The speed c couples the spatial and temporal resolution and therefore ensures Lagrangian behavior.

The particle distribution functions f_i evolve on a regular grid and represent particle densities traveling on the links $\mathbf{c}_i$, see figure 2. The term

$$f_i(\mathbf{x}, t) = f(\mathbf{x}, \mathbf{v}, t) \tag{3}$$

refers to the particle distribution on the lattice node $\mathbf{x}$ at the time t with the velocity $\mathbf{c}_i$.

The equilibrium density distribution $f^{eq}(\mathbf{x}, t)$ depends solely on the density $\rho(\mathbf{x}, t)$ and the velocity $\mathbf{u}(\mathbf{x}, t)$ of a lattice node $\mathbf{x}$. The density ρ and the velocity $\mathbf{u}$ are obtained from the density distribution function f_i. The density is given by

$$\rho(\mathbf{x}, t) = \sum_i f_i(\mathbf{x}, t). \tag{4}$$

Moment and velocity are given by

$$\mathbf{j}(\mathbf{x}, t) = \rho(\mathbf{x}, t)\mathbf{u}(\mathbf{x}, t) = \sum_i \mathbf{c}_i f_i(\mathbf{x}, t) \tag{5}$$

The discrete equilibrium distribution function is chosen as

$$f_i^{eq}(\rho,\mathbf{u}) =$$
$$\omega_i \frac{\rho}{\rho_0}(1+3(\mathbf{c}_i \cdot \mathbf{u})+\frac{9}{2}(\mathbf{c}_i \cdot \mathbf{u})^2 - \frac{3}{2}(\mathbf{u} \cdot \mathbf{u})) \tag{6}$$

with the weight coefficients chosen in a way that the zeros to fourth moments of the equilibrium distribution function equals the Maxwell distribution function. The chosen weights depend on the chosen velocities $\mathbf{c}_i$. For the D3Q15 method with the 15 velocities given in figure 2 the corresponding weights are:

$$\omega_0 = \frac{2}{9},$$
$$\omega_{1,...,6} = \frac{1}{9}, \tag{7}$$
$$\omega_{7,...,15} = \frac{1}{36}.$$

The mass and momentum equations can be derived from the model via multiscale expansion resulting in

$$\frac{\delta\rho}{\delta t}+\nabla \cdot (\rho\mathbf{u}) = 0$$
$$\frac{\delta(\rho\mathbf{u})}{\delta t}+\nabla \cdot (\rho\mathbf{u}\mathbf{u}) = \tag{8}$$
$$-\nabla p + v(\nabla^2(\rho\mathbf{u})+\nabla(\nabla \cdot (\rho\mathbf{u})))$$

where

$$p = c_s^2 \rho \tag{9}$$

is the pressure,

$$c_s = \frac{c}{\sqrt{3}} \tag{10}$$

is the speed of sound and

$$v = \frac{(2\tau-1)c^2}{6}\Delta t \tag{11}$$

is the kinematic viscosity.

The mass and momentum equations are exactly the same as the compressible Navier- Stokes equation if the density variations are small enough. Thus the compressible Navier- Stokes equation is recovered in the incompressible low Mach number limit.

2.2 Boundary Conditions

The boundary conditions for the LBGK method can be described in a heuristically and have a physical interpretation. The two most important boundary conditions are:

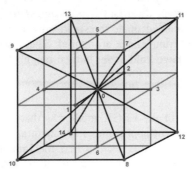

Fig. 2. The velocity directions $\mathbf{c}_i$ in the D3Q15 LBGK model

No-Slip: The no-slip condition describes a rigid wall and uses the fact that near a wall there is a no-slip situation, thus direct at the wall the fluid's velocity is zero. This effect is achieved by simply reflecting every incoming distribution function f_i, thus

$$f_i(\mathbf{x}, t+1) = f_{-i}(\mathbf{x}, t) \qquad (12)$$

where $-i$ denotes the index in the opposite direction of index i.

Pressure: In LBGK simulations density and pressure have to be handled carefully. The problem is that the pressure p is linearly related to the density ρ with $p = \rho \dfrac{k_B T}{m}$. Since the flow is incompressible density fluctuations must stay negligible. The pressure gradient is a result of small density fluctuation on a density ρ_0. Nevertheless it can be useful to change the density of certain nodes to achieve a predetermined pressure. The nodes distribution functions can be updated in two ways, one is to conserve the nodes momentum:

$$f_i^{new}(\mathbf{x}_p, t) = f_i^{eq}(\rho_0(\mathbf{x}, t), 0), \qquad (13)$$

the other one is to conserve the nodes velocity:

$$f_i^{new}(\mathbf{x}_p, t) = f_i^{eq}(\rho_0(\mathbf{x}, t), u(\mathbf{x}, t)). \qquad (14)$$

The boundary condition must be used in a way so that mass conservation is ensured. Over a certain time span the added density must equal the density that is removed from the lattice nodes.

In three dimensional blood flow simulation geometrical boundaries can be obtained from magnetic resonance angiography. The geometric structure can be revealed by the use of paramagnetic contrast agents. The acquired data are cross section images and are normally stored in the DICOM format, which is widely used for medical applications.

After the volume is cropped to the region of interest and a binary segmentation is performed. Every voxel is assigned to be solid or fluid in dependence of its density. If the density is above a certain threshold it is assigned to be 'fluid' otherwise it is assigned to be 'solid'. When an adequate threshold value is chosen a cartesian lattice is created, in which only relevant nodes are stored. These nodes are fluid nodes and solid no-slip nodes, which have fluid nodes in their neighborhood.

2.3 Implementation

One of the big advantages of the LBGK method is that its implementation is easy. Nevertheless there are different approaches implementing the method, each of them having different advantages. In this section first the implementation of the kinetic equation and the splitting of its operator is discussed. Further the methods ability of parallelization is examined.

LBGK schemes can be implemented very efficiently because of their explicit nature. The pseudo code for the LBGK method can be formulated as

```
while(running) {
for each node {
calculate kinetic equation
}
for each node {
calculate local equilibria
} }
```

First the structure of the kinetic equation will be discussed. The kinetic equation (in analogy to equation (2)) is given by

$$f_i(\mathbf{x}+\mathbf{c}_i, t+1) - f_i(\mathbf{x}, t) = -\frac{1}{\tau}(f_i(\mathbf{x}, t) - f_i^{eq}) \tag{15}$$

The operator can be split into a collision step and a streaming step in the following way:

$$
\begin{aligned}
f^*(\mathbf{x}, t) &= (1 - \frac{1}{\tau})f_i(\mathbf{x}, t) + \frac{1}{\tau}f_i^{eq} \\
f_i(\mathbf{x}+\mathbf{c}_i, t+1) &= f^*(\mathbf{x}, t).
\end{aligned}
\tag{16}
$$

This splitting of the operator is called collide-and-stream update order. The operator could as well be split into stream-and-collide update order.

For the knowledge of the equilibrium density distribution $f_i^{eq}(\rho, \mathbf{j})$ only the node itself is required, no neighboring nodes are needed. The density ρ and moment $\mathbf{j}$ can be calculated with the aid of the equations (4) and (5), the equilibrium is given in equation (6) with the corresponding weighting from equation (7).

Every step in the LBGK algorithm is very simple. Nevertheless the optimal low level implementation on certain CPUs is a hard task because the CPU cache must be used in an optimal way. A detailed work about optimization of computer codes for LBGK schemes can be found in [9] and with particulary regard to parallelization of LBGK in [10].

A great advantage of the LBGK method is its simple parallelization, which is possible due to the strictly local nature of the method. Considering CPUs with multiple cores this property is of increasing importance.

To adjust the method for multiple threads the set of nodes must be simply distributed on the processors. In each calculated time step the threads must wait for each others two times:

```
while(running) {
for each thread {
   calculate kinetic equation for all nodes
}
wait for all threads
for each thread {
   calculate equilibrium for all nodes
}
wait for all threads }.
```

3 Elasticity

In blood flow simulation it is important to consider the compliance of vessels. Therefore a boundary condition must be developed that describes the movement of the vessel wall in dependence of pressure. Fang et al. [4] have proposed a method which parameterizes the walls and uses a special treatment for curved boundaries. The method has been tested and successfully applied to pulsatile flow in two spatial dimensions [11].

The problem of this method is that the description of the vessel walls with the aid of surfaces is very complicated in three dimensions. The problem is comparable to the creation of feasible grids for the Finite Element Method (FEM) or the Finite Volume Method (FVM) from tomographic images, which is avoided using the LBGK method. Thus using this method the simplicity and advantages of the LBGK method are partly lost. Therefore in this work a simpler approach is chosen, which does not require parameterized walls but works on the voxel representation of the geometrical domain.

3.1 Introduction

In the following method elasticity is basically described by displacement of voxels, which represents the boundary as a solid wall. The displacement is dependent of the local pressure in the surrounding nodes. Every boundary node has a fixed threshold. When the surrounding pressure exceeds this threshold the node is replaced by a fluid node. On the other hand if the surrounding pressure goes below a certain threshold the fluid node is replaced by a solid node. The boundary conditions are updated in every time step. Thus the method is a realization of the hemoelastic feedback system described by Fung in [12], see figure 3.

To avoid a rupture of the vessel wall a cellular automaton (CA) is used to update the walls in every time step. For more information about CA the reader may refer to [13]. The proposed method offers some advantages compared to the classical approach:

- It avoids a surface representation of the geometrical domain but works directly on the voxel representation.
- The method does not increase the complexity of the algorithm because it works strictly local like the LBGK method.
- The approach enables a simple implementation in two and three dimensions.

In the following the steps of the algorithm are explained in more detail. First the representation of the volume with voxels is explained, next in which way the threshold values for the displacement are chosen and finally how the CA works which prevents rupture of the vessel wall.

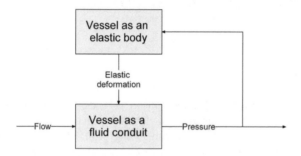

Fig. 3. A hemoelastic system analyzed as a feedback system of two functional units an elastic body and a fluid mechanism [12]

3.2 Voxel Representation

The geometrical data in the LBGK method are represented with the aid of voxels. The data structure that is used for this representation depends on the chosen implementation. For simplicity it is assumed that the geometrical data are stored in a two or three dimensional array. This array works as a look up table for the fluid dynamical computation. Note that the CA will work on the geometrical data array containing the boundary conditions, while the LBGK method will use the look up array to check if boundary conditions have to be applied.

When the vessel walls are considered to be rigid the layer of solid nodes that surrounds the fluid domain has a thickness of one. When elasticity is considered the layer of elastic nodes has to have the thickness of maximal circumference plus maximal narrowing. To set up the geometrical data in two dimensions is an easy task. Every solid node simply has to be replaced by a column of elastic nodes. In three dimensions more care must be taken when establishing the elastic layer.

In three dimensions the volume is created by binary segmentation from tomographic images. The nodes inside the vessel are fluid nodes, the nodes describing the tissue are no-slip nodes. The elastic layer is built from this rigid layer in every cross section image, see figure 4. This is done in a recursive way. The nodes of the first layer are exactly the no-slip nodes which are neighbors to fluid nodes. The following layers consist of the no-slip nodes next to elastic nodes. When the cross section images are put together the resulting geometry has the desired elastic layer in three dimensions. The result is comparable to an onion skin. The outside layers have large thresholds, which are descending moving inside the vessel. In the following section the choice of threshold values will be examined.

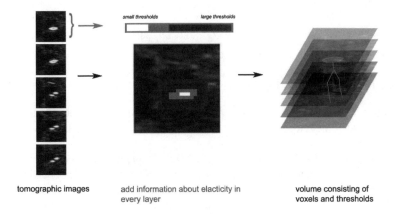

tomographic images   add information about elasticity in   volume consisting of
                           every layer   voxels and thresholds

Fig. 4. Elastic boundaries are created from tomographic images

3.3 Identification of the Threshold Values

When the pressure is higher than a certain value the no-slip boundary condition shall be replaced by a normal fluid node and vice versa. In this section it is described in which way the threshold values can be assigned.

For the simulation a linear pressure radius relationship is assumed:

$$r(z) = r_0 + \alpha \frac{p(z)}{2}, \tag{17}$$

where r_0 is the radius when the transmural pressure $p(z)$ is zero. The parameter α is a compliance constant, thus the threshold values are set to:

$$p(z) = \frac{2}{\alpha}(r(z) - r_0). \tag{18}$$

The parameters r_0 and α must be chosen carefully. In two spatial dimensions the radius r_0 can be set to the distance of the wall to the center line. The compliance constant α can be calculated from the maximal extension of the vessel.

In three spatial dimensions the situation is more complicated because the center line is not known in advance. In the voxel representation of the geometrical data the elastic boundary layer has a certain predetermined thickness, which predefines the maximal expansion and maximal contraction of the vessel. Two values for pressure must be chosen, p_{max}, the pressure where the maximal expansion occurs and p_{min}, the pressure where the maximal contraction occurs, thus

$$
\begin{aligned}
r_{max}(\mathbf{x}) - r_0 &= \alpha \frac{p_{max}(\mathbf{x})}{2}, \\
r_{min}(\mathbf{x}) - r_0 &= \alpha \frac{p_{min}(\mathbf{x})}{2}.
\end{aligned}
\tag{19}
$$

From the two equations the values r_0 and α can be easily calculated and the thresholds for the elastic layer can be chosen accordingly.

3.4 Cellular Automaton for Coherence of the Vessel Wall

The solid nodes are displaced when the pressure exceeds its threshold value. In this section a CA is developed to avoid rupture of the vessel wall introduced by this displacement process.

Note that the LBGK method and CAs are closely related. Historically LBGK schemes have even been developed from LGCA. The main difference between the two approaches is that the LBGK method has continuous state variables on its lattice nodes, while CA have discrete state variables in its cells. Appropriate update rules for the elastic walls boundary condition should therefore be strictly local and should have the same discretization in time and in the spatial domain as the LBGK model. The boundary conditions used by the LBGK method are normally defined in a separate lattice. This lattice can be interpreted as a CA with its own update rules which interact with the fluid dynamical model.

The CA has two different states, one is representing the fluid node and one is representing a no-slip boundary node. The update rules of these states are divided into two steps:

In the first step the pressure $p_{ca}(\mathbf{x},t)$ is compared to the threshold value t_p. If the nodes state is 'fluid' its pressure is used. If the node state is 'solid' the pressure of the neighboring fluid nodes are averaged, thus

$$
p_{ca}(\mathbf{x},t) = \begin{cases} p(\mathbf{x},t) & \textit{fluidnode} \\ \dfrac{1}{\# f} \displaystyle\sum_{i=1}^{N} p(\mathbf{x}+\mathbf{c}_i, t) & \textit{solidnode}, \end{cases}
\tag{20}
$$

where value $\# f$ is the number of fluid nodes surrounding the solid node $\mathbf{x}$. The pressure $p(\mathbf{x},t)$ of a solid node is defined to be zero. The value N is four in two dimensions and six in three dimensions corresponding to the four and six neighbors of the Neumann neighborhood. Every node $\mathbf{x}$ in the CA has a certain threshold value

t_p which is chosen according to the previous section. The boolean value P_{ca} is introduced:

$$P_{ca}(\mathbf{x},t) = \begin{cases} 1 & p_{ca}(\mathbf{x},t) \geq t_p \\ 0 & p_{ca}(\mathbf{x},t) < t_p. \end{cases} \tag{21}$$

The cells state is set to 'fluid node' if $P_{ca}(x,t) = 1$ and set to 'solid node' if $P_{ca}(x,t) = 0$.

In the second step the following rules are applied, which are chosen in a way that holes are closed, thus solid nodes diminish in fluid nodes and the other way around. In three spatial dimensions the rules are chosen in a similar way as in two dimensions. In general the method can be formulated as change of states when the following conditions are met:

$$change: \_fluid\_to\_solid : \sum_i^N P_{ca}(\mathbf{x},t) < t_{fs}$$

$$change: \_solid\_to\_fluid : \sum_i^N P_{ca}(\mathbf{x},t) > t_{sf} \tag{22}$$

with $t_{fs} = 2$ and $t_{sf} = 2$ in two dimensions ($N = 4$) and $t_{fs} = 2$ and $t_{sf} = 4$ in three dimensions ($N = 6$).

When the LBM switches from the state 'solid node' to 'fluid node' the fluid node is set to $f^{eq}(\rho,\mathbf{u})$, where $\mathbf{u}$ is zero and ρ is determined on the basis of the threshold value t_x. This simple numerical scheme leads to the analytical behavior representing linear pressure radius relationship.

4 Results

In this section Poiseuille flow in an elastic tube will be investigated and the numerical results are compared with the analytical solution.

The Poiseuille theory of laminar flows can be easily extended to elastic tubes. Normally Hook's law is used for the description of elasticity. But vessel walls do not obey Hook's law and therefor it is a good choice to assume linear pressure radius relationship [14], thus

$$r(z) = r_0 + \alpha \frac{p(z)}{2}, \tag{23}$$

where r_0 is the tube radius when transmural pressure $p(z)$ is zero and α is the compliance constant. When the derivative from this relationship in respect to z is taken, thus

$$\frac{\delta r(z)}{\delta z} = \frac{\alpha}{2}\frac{\delta p}{\delta z}, \tag{24}$$

the time dependent pressure gradient can be inserted into the Poiseuille velocity profile, which is given by

(a) Velocity field in an elastic tube

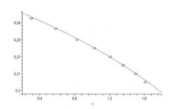

(b) Analytical values of r(z) (line) and calculated values (circles)

Fig. 5. An elastic tube in two dimensions

$$u_z(r) = \frac{(R^2 - r^2)}{4\mu}\frac{P_1 - P_2}{L} \tag{25}$$

for a tube in three dimensions. The flow Q is equal in every subsection of the tube, thus it is independent of z. The pressure flow relationship obtained by integrating over one cross section. In three dimension this yields

$$\frac{\delta p}{\delta z} = \frac{8\mu}{\pi a^4}Q \tag{26}$$

and in two dimensions

$$\frac{\delta p}{\delta z} = \frac{3\mu}{2a^3}Q. \tag{27}$$

Using equation (23) the following radius flow relationship is obtained in three dimensions

$$r(z) = \sqrt[5]{\frac{20\mu\alpha}{\pi}Qz + c_1} \tag{28}$$

and in two dimensions

$$r(z) = \sqrt[4]{12\mu\alpha Qz + c_2}. \tag{29}$$

The integration constant c can be calculated from the boundary condition, thus $c_1 = r(0)^5$ and $c_2 = r(0)^4$. These two equations are used to validate the numerical scheme.

In two dimensions the flow field has been calculated in a tube of 2 cm length and a radius of 0.225 cm at a transmural pressure of 0 mmHg. A resolution of 400*70 nodes is used, thus one lattice node equals 0.01 mm^2. Between the inlet and the outlet a pressure gradient of 1 mmHg is applied. The elastic boundary conditions evolve to a steady state which is represented in figure 5a. The radius $r(z)$ is plotted in figure 5b. The simulated result is in good accordance with the analytical results given by equation (29).

In three dimensions a rigid tube with 20 cm length and 2 cm radius at maximal expansion and a radius of 1.25 cm at transmural pressure of 0 mmHg is under investigation. This has been realized with a lattice of 40*40*200 nodes with a predetermined pressure gradient of 1 mmHg. Again a steady state evolves after a certain time. The three dimensional pressure and velocity is given in figure 6. The radius $r(z)$ behave according to equation (28).

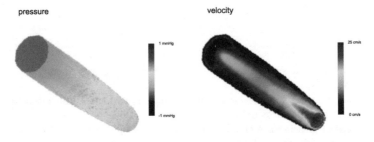

Fig. 6. Maximum intensity projection of velocity and pressure field in an elastic tube

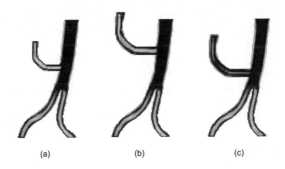

Fig. 7. Different examples of size and positioning of shunts

An important application is the simulation of blood flow in shunts. Different sizes of shunts can be examined and the position and alignment can be freely chosen in the simulation, see figure 7. The aim is to find an optimal design of a shunt in a patient specific way. Different configuration of the shunts lead to different pressure and velocity profiles during a cardiac cycle. The shunt should be created in a way to avoid turbulence and flow separation under the expected pulsatile pressure and flow fluctuations. Computer fluid dynamics is used as a tool for preoperation planning

5 Discussion and Conclusion

In this paper a simulation environment has been developed which is able to simulate blood flow through arbitrary patient specific geometries. For calculating the fluid flow numerically the LBGK method has been extended for a new type of boundary node supporting elastic vessel walls. A solver for two and three dimensional flow based on the LBGK method has been developed in Java working parallel on arbitrary multi-processor machines.

The LBGK method has proven to be feasible for hemodynamic calculations. The accuracy of the method is in good accordance to the accuracy of available boundary data like geometry, which is obtained from magnetic angiography, or pressure and velocities, which are obtained from in vivo measurements or global cardiovascular simulation. At the current resolution the method works very fast, in two spatial dimension it is even possible to calculate the fluid flow in realtime.

References

1. Suo, J., Ferrara, D.E., Sorescu, D., Guldberg, R.E., Taylor, W.R., Giddens, D.P.: Hemodynamic shear stresses in mouse aortas: Implications for atherogenesis. Thromb. Vasc. Biol. 27, 346–351 (2007)
2. Wolf-Gladrow, D.A.: Lattice-Gas Cellular Automata and Lattice Boltzmann Models- An Introduction. Lecture Notes in Mathematics. Springer, Heidelberg (2000)
3. Artoli, A.M.M., Hoekstra, A.G., Sloot, P.M.A.: Mesoscopic simulations of systolic flow in the human abdominal aorta. Journal of Biomechanics 39(5), 873–884 (2006)
4. Fang, H., Wang, Z., Lin, Z., Liu, M.: Lattice boltzmann method for simulating he viscous flow in large distensible blood vessels. Phys. Rev. E. (2001)
5. Leitner, S., Wasssertheurer, M., Hessinger, M., Holzinger, A.: Lattice Boltzmann model for pulsative blood flow in elastic vessels. Springer Elektronik und Informationstechnik e&i 4, 152–155 (2006)
6. Leitner, D., Kropf, J., Wassertheurer, S., Breitenecker, F.: ASIM 2005. In: F, H. (ed.) 18thSymposium on Simulationtechnique (2005)
7. Succi, S.: The Lattice Boltzmann Equation for Fluid Dynamics and Beyond. Oxford University Press, Oxford (2001)
8. Artoli, A.M.M., Kandhai, B.D., Hoefsloot, H.C.J., Hoekstra, A.G., Sloot, P.M.A.: Lattice bgk simulations of flow in a symmetric bifurcation. Future Generation Computer Systems 20(6), 909–916 (2004)
9. Kowarschik, M.: Data locality optimizations for iterative numerical algorithms and cellular automata on hierarchical memory architectures. SCS Publishing House (2004)
10. Wilke, J., Pohl, T., Kowarschik, M.: Cache performance optimizations for parallel lattice boltzmann codes. In: Kosch, H., Böszörményi, L., Hellwagner, H. (eds.) Euro-Par 2003. LNCS, vol. 2790, pp. 441–450. Springer, Heidelberg (2003)
11. Hoeksta, A.G., van Hoff, J., Artoli, A.M.M., Sloot, P.M.A.: Unsteady flow in a 2d elastic tube with the lbgk method. Future Generation Computer Systems 20(6), 917–924 (2004)
12. Fung, Y.C.: Biomechanics, Mechanical Properties of Living Tissues, 2nd edn. Springer, Heidelberg (1993)
13. Wolfram, S.: Cellular Automata and Complexity. Westview, Boulder (1994)
14. Fung, Y.C.: Biodynamics. In: Circulation, Springer, Heidelberg (1984)

Usability of Mobile Computing Technologies
to Assist Cancer Patients

Rezwan Islam[1], Sheikh I. Ahamed[2], Nilothpal Talukder[2], and Ian Obermiller[2]

[1] Marshfield Clinic, 3501 Cranberry Blvd, Weston, WI 54476, Wisconsin, USA
islam.rezwan@marshfieldclinic.org
[2] MSCS Dept., 1313 W Wisconsin Ave., Marquette University, Milwaukee, Wisconsin, USA
{iq, ntalukde, iobermil}@mscs.mu.edu

Abstract. Medical researchers are constantly looking for new methods for early detection and treatment of incurable diseases. Cancer can severely hinder the lives of patients if they are not constantly attended to. Cancer patients can be assisted with the aid of constant monitoring by a support group and a continual sense of self-awareness through monitoring, which can be enabled through pervasive technologies. As human life expectancy rises, incidents of cancer also increase, which most often affects the elderly. Cancer patients need continuous follow-up because of the state of their disease and the intensity of treatment. Patients have often restricted mobility, thus it is helpful to provide them access to their health status without the need to travel. There has been much effort towards wireless and internet based health care services, but they are not widely accepted due to the lack of reliability and usability. In this paper, we present a software called Wellness Monitor (WM). The purpose of WM is to utilize the portability and ubiquity of small handheld devices such as PDAs, cell phones, and wrist watches to ensure secured data availability, customized representation, and privacy of the data collected through small wearable sensors. WM explores how the social and psychological contexts that encompass the patients could be enhanced by utilizing the same technology, an aspect which is mostly unexplored. A further goal was to provide continuous psychological assistance.

Keywords: Pervasive health care, Wellness Monitor, mote, TinyOS, MARKS, cancer, chemotherapy, Tmote Sky.

1 Introduction

The aim of pervasive computing is to combine the world of computation and human environment in an effective and natural way so that the data can be easily accessible from anywhere, at anytime by the users [1]. The potential for pervasive computing is evident in almost every aspect of our lives, including hospitals, emergency and critical situations, industry, education, or the hostile battlefield. The use of this technology in the field of health and wellness has been termed *Pervasive Health Care*. When pervasive computing is introduced, cancer can be treated in a more organized and better scheduled fashion. With pervasive health care, cancer patients

A. Holzinger (Ed.): USAB 2007, LNCS 4799, pp. 227–240, 2007.
© Springer-Verlag Berlin Heidelberg 2007

can actively participate in their health status monitoring and take proactive steps to prevent or combat the deterioration of their bodies. Without such technology, cancer patients may not have the resources to handle emergency situations. The American Cancer Society estimates that 1,399,790 men and women will be diagnosed with cancer and 564,830 men and women will die of it in 2006 in the United States alone. In 1998-2003, the relative survival rates from cancer in United States by race and sex were: 66.8% for white men; 65.9% for white women; 59.7% for black men; and 53.4% for black women. Based on the rates from 2001-2003, 41.28% of men and women born today will be diagnosed with cancer some time during their life [18].

Cancer treatment varies for each patient, and it takes into account such variables as the type of cancer, the stage of the disease, and the goal of the treatment. Many patients undergo chemotherapy as a part of their treatment. Chemotherapy is a form of treatment involving the use of chemical agents to stop cancer cells from growing. Chemotherapy is considered as a systemic treatment because it affects even those cells that are not in close proximity to the original site [20]. Patients must follow a strict routine of chemotherapy treatment, including frequent visits to the hospital for a simple status check. If the routine checks are not consistently followed, the treatment may be far less effective. We are proposing a smart system that allows patients to go to the healthcare center less often and perform their status checkups more regularly. Through such a smart system patients would have active participation in their treatments via continuous monitoring. The system will also help keep down the tremendous costs of long term medical care.

Other universities and institutions are working on similar pervasive health care projects [6-10,15,16]. However, instead of focusing on patient monitoring (and specifically cancer monitoring) as Wellness Monitor does, they are targeted mainly toward providing assistance to the elderly. The Center for Future Health [14] has implemented a home complete with infrared and other biosensors. The Center for Aging Services Technologies (CAST) [16] has several projects, one of which is a smart home that tracks the activities of its inhabitants, another is a sensor-based bed that can track sleeping patterns and weight. We also describe multiple projects similar to WM, such as TERVA, IST VIVAGO ®, and WWM in the related works section [6-8]. Also, Holzinger et al [23, 24, 25] have addressed the importance of mobile devices for healthcare. There has been good progress in manufacturing new bio-analysis sensors used for monitoring diseases. Sicel Technologies of Raleigh, NC has developed a prototype of biosensor that is wearable and generates data about a tumor's response to chemotherapy or radiation treatment [19]. Related works section has more details of its prospects. A nano-sensor, developed by scientists at the Center for Molecular Imaging Research at Harvard University and Massachusetts General Hospital, can detect whether a drug is working or not by looking at individual cells [21]. The MBIC (Molecular Biosensors and Imaging Center) group in CMU has been collaborating with UC Berkeley and Princeton on the project to develop optical biosensors in living cells. They are exploring the areas with probes on cardiac function, vivo tumors and lymphatic movement [22].

With all the sensor devices in mind we are implementing a smart software system, Wellness Monitor, which will be divided into the following functionalities: sensing, communication, data interpretation, and event management. The goal is to develop a system that will be able to analyze and transmit the data from small wearable sensors on the cancer patients, with an emphasis on reliability, privacy, and ease of use.

A cancer patient may have sensors on several parts of her body depending on the type and purpose of treatment. These sensors will pass the data to a small, wearable mote running TinyOS. The information collected by the mote will then be sent to a handheld device such as PDA or cell phone for analysis. WM features role based access control to insure the privacy of the cancer patient's collected data. The collected data will be readily available for analysis and evaluation by the patient's medical personnel. In this paper, we present an overview of WM, its design and architecture, and a partial implementation. The paper is organized as follows: We provide the features of WM in Section 2; 3 will focus on the functionalities of WM and 4 will describe the design part of the system. Current state of the art is stated in Section 5. Section 6 presents a real world application of our model. A user evaluation of our proposed WM system is presented in Section 7, followed by conclusions.

2 Features of WM

2.1 Security and Reliability Versus Energy Efficiency

Because handhelds and motes have low computing powers, a complex encryption scheme is not feasible. Also, limited battery power allows for only so much system security and reliability. Our system has to trade-off with both the aspects.

2.2 Authentication

In order to avoid false data, all biosensors must be authenticated before data can be treated as reliable. If the mote is collecting data from an unauthenticated sensor, the incorrect data may lead to a misdiagnosis or other health risks for the patient.

2.3 Role Based Information Access

In order to ensure privacy of the patient's data, WM will be including role based information access, whereby different users will have different levels of access to the data. For example, patients may not need to see every detail of the sensors, only major ones that directly affect them. Doctors will be able to access all the data they need to correctly treat the patient. In this way the privacy of the data can be ensured.

2.4 Priority Based Information Representation

WM will be able to dynamically analyze data, detect critical readings, and alert a doctor if necessary. This will allow proactive measures to be taken without having a patient being constantly monitored at a hospital.

2.5 Data Interpretation

We are going with the saying that "a picture is worth a thousand words." Proper data representation is very important to rapid analysis. Easily interpretable data motivates and encourages the user to lead a healthier lifestyle.

2.6 User Friendly Interface and Minimal Interaction

Many of the patients may have a reluctance to use new technology, particularly because this field is generally occupied by older adults. Their fears are primarily due to poor user interfaces on hand held devices such as cell phones. Representations of the data, suggestions, and entire user interface must be extremely simple, self explanatory and easy to use. Such a requirement would make it more likely that a user would disregard the manual and potentially use the system incorrectly.

The data should be presented in a way that requires minimal interaction on the user's part while viewing data. This means that important data must be filtered out so the user is not overwhelmed by a wealth of information.

2.7 Ceaseless Connectivity

The entire system must provide uninterrupted connectivity between the hand held device of the user and the wearable sensor device attached to the body and responsible for collecting data. Also, a doctor or medical representative should have ready access to the patient's data in case of an emergency.

3 Functionalities of WM

3.1 Activation of Functions

WM will be able to respond to the biosensor devices readily and accurately. Although in this paper, WM has been depicted as a monitoring tool for the cancer patients, it is also useable in multiple situations requiring periodic monitoring such as diabetes, obesity, irregular heartbeat, hypertension, and anything that can be considered for a periodic check-up with equal efficiency. The options are customizable and the patient can choose which sensors she wants to use. The application is designed in a way that the mote doesn't show any performance degradation with additional sensors.

3.2 Scheduled Reports and Emergency Alarms

WM can schedule reports on a daily, weekly, or monthly basis depending the problem. For example, the cancer patient may want to view the status of chemotherapy treatment two or three times a week to avoid an emergency situation. Fixed thresholds can be set up to check for an emergency situation (according to gathered data) and generate an alarm. The doctor can also be immediately notified of the situation and can take measures before it is too late. The support groups are identified to apply a policy based alert system conforming to the varying needs.

3.3 Notification and Advising Services

Through WM reminders can scheduled. This functionality is helpful when the patient needs to take a medicine twice a day and forgets. A voice message can be incorporated to notify the patient of the tasks he needs to perform. Also, the patient can be advised by professionals through a specialized messaging system.

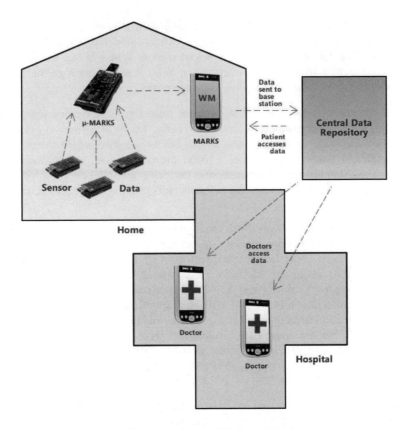

Fig. 1. Overview of Wellness Monitor

3.4 Maximum Protection of Data

The transfer of information from the sensors to the mote and from the mote to the handheld device must be secured. Data encryption should be energy and memory efficient without degrading performance.

All of the data which is collected is formatted and stored in a central data repository located in the handheld device for short term storage and offloaded to a medical center server or the user's personal computer for long term storage and analysis. User levels are provided to prevent unauthorized access to the data. Doctors may request specific data that others cannot view. The doctor has to be authenticated to use the database and request specific information. The units run in a distributed manner.

4 Our Approach

The basis of WM is a network of sensors communicating the patient's status. Many research projects are developing or have developed sensors to test the effectiveness of cancer drugs after chemotherapy or radiotherapy treatments. There are sensors for both intrusive and non-intrusive detection of malignant cancer cells. Though it is beyond the scope of this article to delve into the various subtleties that make a sensor unique, we will be detailing the generic types of sensors WM will use and how they will be used. Figure 1 shows a brief overview of WM. Let's go through the various parts of WM implementation and explore their integration:

Collection of Data through Sensing: As evident from figure 1, various sensors are connected to multiple motes located around the body. The WM System is extensible to allow many different kinds of motes, and so far we are working on integrating a blood oximeter, a temperature sensor, a blood pressure sensor, and various other biosensors. For chemotherapy patients, there is currently research being conducted for a simple biosensor that can measure the patient's cells response to treatment, gauging its effectiveness.

Accumulating Authenticated Data: Each mote gathers data from its sensor and relays the data to the collection mote. Before the collection mote receives data from a sensor node, the node must be authenticated through LPN (Learning Parity with Noise). A smaller version of MARKS (Middleware Adaptability for Resource Discovery, Knowledge Usability and Self-healing) named μ-MARKS [13] will be running on the collection mote which will be responsible for communication and periodic data collection from the sensor nodes. The middleware MARKS has already been developed to perform these core services and provide transparency over the ad hoc protocol. There are five major modules in μ-MARKS implementation. The LPN authenticator is responsible for the authentication mechanism between the sensors and the mote.

Setting up Automated Alerts Based on Roles: After authentication, the data collector module receives data from the various sensor nodes. All the data packets are then sent to the PDA and base station through a data sender module. Data sent to the mote is then inspected through the Threshold Value Comparator, which generates a special 'threshold data packet' if the collected data value crosses a predefined threshold value. Thus, an alert notification is sent early on in the chain, and the patient and doctors can be alerted as soon as possible. From the mote, collected data will be sent to the patient's handheld device. All of the major data processing and display occurs in Wellness Monitor running on the handheld device. The first and most important task of WM is to interpret the wealth of data received from the sensors by way of the collection mote. For this reason the context processor is the core part of the WM architecture. It performs role based and meaningful information representation, which means that the user only sees information which is relevant to his daily life, whereas the doctor has access to all the data. The context processor is also responsible for problem based information representation, which generates a configurable warning to the user and sends an alert to the doctor when an emergency threshold level for a sensor is passed. In addition, all data sent to the handheld device is then relayed to the base station, which will be discussed in further detail later on.

Visual Representation of Data: The second task for WM is visual representation. The user should be able to look at a short term history of her vital signs in an easy to read, graphics-based report. A sample mockup of a graph is shown in Figure 2 (under

Prototype Implementation) on the left. The patient will be able to easily interpret data presented in a visual manner and alter her lifestyle accordingly. If the user wants to access data older than that stored on the handheld device, the device connects to the base station and queries for the older data.

Professional Advice through Messaging: The patient will also be able to receive advice from professionals through WM. For instance, she can receive the day's nutrition outline and an exercise routine, both of which would encourage her to lead a healthy lifestyle and remind her if she forgets. The messaging/notification module also allows the user to reply back to the advisors for clarification or to report changes in her schedule that should be taken into account. The notification module is shown in figure 2 on the right.

Scheduler: Since treatment of any sort usually requires a patient to take medicine periodically, a schedular program is also built into WM. The schedular allows the user to enter in tasks such as 'Take medicine' or 'Call the doctor', and be reminded of them when the specified time comes. A mockup of the scheduler program is shown in Figure 3.

Enhancing Motivation and providing Psychological Aid: WM also investigates ways to increase motivation and improve acceptance of the overall system. A survey in the evaluation section demonstrates the requirements of the different target age groups, particularly considering the usefulness of the design. It attempts to identify the support group to ensure active participation in the WM based healthcare service. The support group includes the doctors, nurse, other caregivers and close relatives and friends of the patient. The identification of the support group is necessary for assigning roles during the critical hours and routine checkup besides self-awareness of the patient himself. The biggest benefit of such a scheme is that it provides the patient a feeling of being connected. Through identification of this support group WM offers a policy based alert system which allows the alerts to be customized to the specific group or person. For example, during an emergency situation, such as a sudden decrease in white blood cell count, the caregiver in charge along with the family members will be in the direct automated alert list. Again, additional alerts need to be set up in away to avoid the danger of failure to attend the call in time.

Staying Updated - Message Board: As an additional feature, WM will incorporate a message board with feedback from doctors, nutrition experts and workout advisors specific to the patient. The message board will also give hope to patients by displaying the state of the art achievements and processes used to overcome cancer..

Storing Data for Record Keeping and Long-term Analysis: The version of WM running in base station will mainly handle a large database which will store all the collected information for a longer period of time and periodically send a summary report to medical personnel. For long-term analysis, data can be aggregated and analyzed in a variety of ways to show historical trends. For example, medical personnel can be aware of the frequency of the drug response deterioration by examining the patient's vitals over a period of time and can proactively determine the next frequency of the therapy for the patient or even switch to an alternate therapy.

Preserving Privacy – the Need of Time: As the information representation will be based on the role of the user, the first step is to choose a role. LPN authentication will be responsible for all types of authentication mechanism and serve the required characteristics of authentication. 'Privacy of Data and Presentation Functions' will be provided by MARKS as one of its core services. It is noteworthy to mention the issuance of the patient privacy protections as part of the Health Insurance Portability and Accountability Act of 1996 (HIPAA) by the Federal Health and Human Services (HSS). The importance of the privacy preservation led to statutory enforcement. The latest methods of electronic transaction require new safeguards to protect the security and confidentiality of health information and require compliance with privacy rules stated with respect to Health plans, health care clearinghouses, and health care providers [25].

5 Related Work

A long term monitoring system known as Terva [8] has been implemented to collect critical health data such as blood pressure, temperature, sleeping conditions, and weight. The problem with Terva is that although it is self-contained, it is housed in a casing about the size of a suitcase, which seriously dampers mobility. As a result, Terva is only practical inside the home. IST VIVAGO® is a system used to remotely monitor activity and generate alarms based on received data [6]. In contrast with Terva, the user only has to wear a wrist unit, which communicates wirelessly with a base station that manages the alarms and remote data access. The wrist unit can generate an alarm automatically (when the user is in danger), or can also be activated manually. Another system, Wireless Wellness Monitor, is built specifically to manage obesity [7]. The system has measuring devices, mobile terminals (handheld devices), and a base station home server with a database. It uses Bluetooth and Jini network technology and everything is connected through the internet. The MobiHealth project [15] is similar to WM as it monitors a person's health data using small medical sensors which transmit the data via a powerful and inexpensive wireless system. A combination of these sensors creates a Body Area Network (BAN), and the project utilizes cell phone networks to transmit a signal on the fly from anywhere the network reaches. Improving the connectivity of home-bound patients is the goal of the Citizen Health System [16]. The project consists of a modular system of contact centers with the purpose of providing better health care services to patients. Other researchers have depicted several required characteristics of wearable health care systems, along with a design, implementation, and issues of such a system. Their implementation, however, is too expensive and utilizes special, proprietary hardware. Our focus is on combining and improving these implementations, to produce a service that is flexible, inexpensive, and deployable on existing systems.

Sicel Technologies (Raleigh, NC) is working on a sensor-based monitoring technology that could be used to assess the effectiveness of cancer treatments. A prototype has been developed that is wearable and can fit in biopsy needle. That can be implanted with the tumor easily. Combined with a miniature transmitter and receiver, the sensor would generate data about the tumor's response to chemotherapy or radiation treatment. The system will eliminate the need for oncologists' guesswork to plan ongoing therapy. Initially, the sensors will be used to monitor radiation levels and temperature within tumors. Radioactive tags attached to cancer drugs would allow the sensor to measure uptake of the drug by the tumor [19]. A nano sensor, developed by scientists at the Center for Molecular Imaging Research at Harvard University and Massachusetts General Hospital, can directly signal whether a drug is working or not by looking at individual cells [20]. Ching Tung, associate professor of radiology at Harvard Medical School, who developed the nano sensor, says his group should also be able to see the sensors inside the human body using MRI. A patient having been undergone cancer treatment for a few days, he could be given the nano sensor and an MRI scan to compare the status of the tumor cells [21].

The researchers at Medical University Hospital in Graz [23] are working towards developing an integrated mobile solution for incorporating touch based questionnaire responses by skin cancer patients (especially the handicapped). It is necessary for both the clinical information system and for a scientific database intended for research in skin cancer. While designing the system the influence of a technical environment, physical surrounding and social and organizational contexts have been considered. The work mostly focuses on HCI, usability engineering issues providing an insight into technological possibility in the clinical field. With an aim to explore the factors affecting an improvement to the quality of life of elderly people [22] the researchers are working with the psychological and sociological aspects of user interface. The design guideline in the work [22] will raise awareness among the developers and the designers to understand the importance of the cognitive impairment of the elderly people while using the mobile applications.

6 An Illustrative Example

Mr. Rahman has been suffering from leukemia and is undergoing chemotherapy treatment. He is scheduled to take some medicine three times a day. He is a new user of WM and chooses his role as a normal user/patient. He wears ergonomic biosensors as part of his daily outfit and the sensors keep him updated of his vital conditions.

Based on the data received from the sensors Mr. Rahman gets a report containing graphs and curves. Using a voice message, WM also reminds him to take his medication on time. In addition, WM generates an instant alarm in the evening mentioning that "Your white blood cell count has gone up, cells are not responding to the therapy." and it has crossed a specified threshold for upper limit. Based on this warning, Mr. Rahman changes his plan for going out and rests instead. He makes an appointment with the doctor and takes regular treatment. He is motivated now and is very mindful of taking medicine due to regular reminders and feedback of personal information. Also, he has more time to himself because WM's sensors send his vital data to the doctor, eliminating the need for frequent visits for regular checkups.

7   Evaluation

7.1   Prototype Implementation

To evaluate the effectiveness of our WM we are developing a   prototype. We are
using Dell Axim X50v PDAs, C# .NET Compact framework, the ZigBee wireless
specification, Moteiv's Tmote Sky as a sensor board and bio sensors for the
implementation.

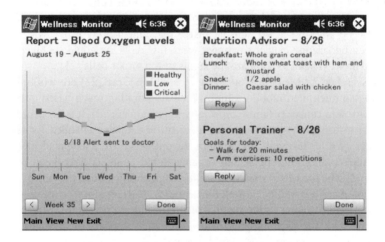

Fig. 2. WM Report and Notification modules

Fig. 3. The Scheduler Module

The previous two mockups show the prototype Wellness Monitor running on the
handheld device. Figure 2 on the left shows a graph of daily blood oxygen levels for a
week, in an easy to read format using colors to show critical levels. The right side is a

sample of the messaging/notification service, which allows the user to receive daily information from a professional, such as a nutrition guide or exercise plan.

Figure 3 details the scheduler module, and a sample form used to create a new task. Both are described in further detail in section 4. Not pictured is a middleware service known as MARKS [21] which we have already developed to provide core services such as connectivity and context processing.

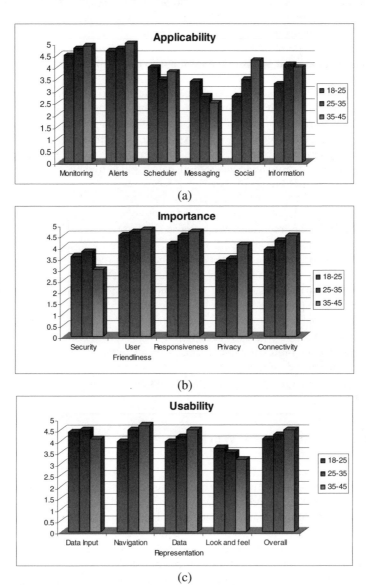

(a)

(b)

(c)

Fig. 4. Survey data collected on the Wellness Monitor concept

7.2 Cognitive Walkthrough

The user experience and opinion of the WM Application has been examined by means of cognitive walkthrough among people from various age groups. The survey involved 24 people of three different age groups with a questionnaire regarding the features of the application. It consisted of questions about the applicability of the offered services, the apparent usability of the prototype, and the overall importance of certain concepts in relation to WM. The survey results were very helpful because they gave us a better understanding of the user experience and helped in prioritizing the features from users' point of view. Figure 4 depicts the results of the survey. Here we briefly highlight the main points of the results with categorized listing.

Applicability: The continuous monitoring and automated alert system to different support groups and stake holders has been in the highest priority list when rating the applicability features. The scheduler part has not been much of a concern though which may be because it requires longer experience and a more accustomed user. The messaging part has been welcomed by the younger age group, while the older group preferred more social motivation and to remain connected to their support group.

Importance: This category primarily covers data confidentiality and privacy issues along with the user friendliness and responsiveness of the application. A user friendly interface was most the important, and users seemed to be less worried about security, especially in the higher age group.

The lack of concerns about security could be due to the fact that impacts are not readily understood, although the older age group did seem to be concerned about the privacy of their own information. All age groups alike wanted to have uninterrupted connectivity and responsiveness while using the system.

Usability: The usability category reveals that it the prototype requires enhancement in navigation and data representation, along with an updated visual style. The user emphasis on the meaningful and convenient data representation and the overall user experience reflects a good rate from the elders considering the utility of the application.

8 Discussion and Future Work

This paper has presented a software solution called Wellness Monitor intended to facilitate the monitoring of a patient's health, particularly a cancer patient when she is not at a medical center, and provides her with meaningful feedback on her status. WM also provides other useful features to the user, including a scheduler and notification system. The software is designed to be open and flexible to allow for many different types of sensors to monitor countless conditions. The scheduler and messaging modules of WM have already been implemented. Our efforts in developing the sensor network is aided by various sensors such as a blood oximeter, Tmotes [26], PDAs, and a short range, energy efficient, wireless protocol called Zigbee. We are also focusing on the resolution of other research challenges including privacy, reliability, and more user interface issues to offer better user experience. To achieve this, we intend to perform more realistic evaluation of the application which involves a real patient and caregivers to reflect the requirements of the real world users of the system.

In the field of pervasive health care and sensor networks the following issues still need to be sorted out for such a system as WM to be practical in a patient's everyday life. First, we must find the most effective security techniques given the mote and handheld devices low memory and power consumption patterns. Second the collected data must be represented in an informative matter with minimal storage and user interaction. Third, the data must be filtered to provide only relevant information to the user so that she is not overwhelmed. Ubiquity is the future of health care, and as more of these issues are addressed in systems such as WM, it will become more pervasive everywhere, not only for cancer patients but also other day to day health monitoring.

References

1. Weiser, M.: Some Computer Science Problems in Ubiquitous Computing. Communications of the ACM 36(7), 75–84 (1993)
2. Kindberg, T., Fox, A.: System Software for Ubiquitous Computing. In: IEEE Pervasive Computing, pp. 70–81. IEEE Computer Society Press, Los Alamitos (2002)
3. Ahamed, S.I., Haque, M.M., Stamm, K.: Wellness Assistant: A Virtual Wellness Assistant using Pervasive Computing. In: SAC 2007, Seoul, Korea, pp. 782–787 (2007)
4. Sharmin, M., Ahmed, S., Ahamed, S.I.: An Adaptive Lightweight Trust Reliant Secure Resource Discovery for Pervasive Computing Environments. In: Proceedings of the Fourth Annual IEEE International Conference Percom, Pisa, Italy, pp. 258–263. IEEE Computer Society Press, Los Alamitos (2006)
5. Ahmed, S., Sharmin, M., Ahamed, S.I.: GETS (Generic, Efficient, Transparent, and Secured) Self-healing Service for Pervasive Computing Applications. International Journal of Network Security 4(3), 271–281 (2007)
6. Korhonen, S.I., Lötjönen, J., Sola, M.M.: IST Vivago—an intelligent social and remote wellness monitoring system for the elderly. In: ITAB. Proceedings of 4th Annu. IEEE EMBS Special Topic Conf. Information Technology Applications in Biomedicine, April 24–26, pp. 362–365 (2003)
7. Parkka, J., Van Gils, M., Tuomisto, T., Lappalainen, R., Korhonen, I.: A wireless wellness monitor for personal weight management, Information Technology Applications in Biomedicine. In: Proceedings IEEE EMBS International Conference, pp. 83–88. IEEE Computer Society Press, Los Alamitos (2000)
8. Korhonen, I., Lappalainen, R., Tuomisto, T., Koobi, T., Pentikainen, V., Tuomisto, M., Turjanmaa, V.: TERVA: wellness monitoring system, Engineering in Medicine and Biology Society. In: Proceedings of the 20th Annual International Conference of the IEEE, vol. 4, pp. 1988–1991 (1998)
9. http://www.centerforfuturehealth.org
10. http://www.cc.gatech.edu/fce/ahri/
11. Hopper, N., Blum, M.: Secure Human Identification Protocols. In: Boyd, C. (ed.) ASIACRYPT 2001. LNCS, vol. 2248, pp. 52–66. Springer, Heidelberg (2001)
12. Weis, S.A.: Security parallels between people and pervasive devices. In: Pervasive Computing and Communications Workshops, 2005. Third IEEE International Conference, pp. 105–109. IEEE Computer Society Press, Los Alamitos (2005)

13. Sharmin, M., Ahmed, S., Ahamed, S.I.: MARKS (Middleware Adaptability for Resource Discovery, Knowledge Usability and Self-healing) in Pervasive Computing Environments. In: ITNG. Proceedings of the Third International Conference on Information Technology: New Generations, Las Vegas, Nevada, USA, pp. 306–313 (2006)
14. Cognitive Walkthrough Strategy, http://www.pages.drexel.edu/ zwz22/CognWalk.htm
15. van Halteren, A., Bults, R., Wac, K., Dokovsky, N., Koprinkov, G., Widya, I., Konstantas, D., Jones, V., Herzog, R.: Wireless body area networks for healthcare: the MobiHealth project. Stud. HealthTechnol. Inform. 108, 181–193 (2004)
16. Maglaveras, N.: Contact centers, pervasive computing and telemedicine: a quality health care triangle. Stud. Health Technol. Inform. 108, 149–154 (2004)
17. Yao, J., Schmitz, R., Warren, S.: A wearable point-of-care system for home use that incorporates plug-and-play and wireless standard. IEEE Transactions of Information Technology in Biomedicine 9(3), 363–371 (2005)
18. National Cancer Institute: SEER (Surveillance, Epidemiology and End Results) Statistics of Cancer from all sites, http:/seer.cancer.gov/statfacts/html/all.html
19. http://www.devicelink.com/mddi/archive/00/05/011.html
20. http://www.Chemotherapy.com/treating_with_chemo/treating_with_chemo.jsp
21. http://www.technologyreview.com/read_article.aspx?id=16467&ch=nanotech
22. Cernegie Mellon University: Molecular Biosensors and Imaging Center: http://www.mbic.cmu.edu/home.html
23. Holzinger, A., Errath, M.: Mobile Computer Web-Application Design in Medicine: Research Based Guidelines. Springer Universal Access in Information Society International Journal 1, 31–41 (2007)
24. Holzinger, A., Searle, G., Nischelwitzer, A.: On some Aspects of Improving Mobile Applications for the Elderly. In: Coping with Diversity in Universal Access, Research and Development Methods in Universal Access. LNCS, vol. 4554, pp. 923–932. Springer, Heidelberg (2007)
25. Kleinberger, T., Becker, M., Ras, E., Holzinger, A., Müller, P.: Ambient Intelligence in Assisted Living: Enable Elderly People to Handle Future Interfaces. LNCS, vol. 4555, pp. 103–112. Springer, Heidelberg (2007)
26. HIPAA, http://www.hhs.gov/ocr/hipaa
27. Tmotes: http://www.moteiv.com

Usability of Mobile Computing in Emergency Response Systems – Lessons Learned and Future Directions

Gerhard Leitner, David Ahlström, and Martin Hitz

Klagenfurt University, 9020 Klagenfurt, Austria
Institute of Informatics Systems
{gerd,david,martin}@isys.uni-klu.ac.at

Abstract. Mobile information systems show high potential in supporting emergency physicians in their work at an emergency scene. Particularly, information received by the hospital's emergency room well before the patients' arrival allows the emergency room staff to optimally prepare for adequate treatment and may thus help in saving lives. However, utmost care must be taken with respect to the usability of mobile data recording and transmission systems since the context of use of such devices is extremely delicate: Physicians must by no means be impeded by data processing tasks in their primary mission to care for the victims. Otherwise, the employment of such high tech systems may turn out to be counter productive and to even risk the patients' lives. Thus, we present the usability engineering measures taken within an Austrian project aiming to replace paper-based Emergency Patient Care Report Forms by mobile electronic devices. We try to identify some lessons learned, with respect to both, the engineering process and the product itself.

Keywords: Human–Computer Interaction, Usability Engineering, Medical Informatics, Emergency Response, Mobile Devices.

1 Introduction

Integrating computerized systems into different areas of application has its potential benefits, e.g. the enhanced quantity and quality of patient data at hand in health care. However, the benefits of digital data to support better treatments of patients also have a potential drawback – bad usability. Bad usability prevents people from using devices, and in case of suboptimal implementation may also cause errors. In contrast to office, web, and entertainment applications, usability problems in safety and health critical contexts could have a severe impact on health, life and environment. Several examples are documented, e.g., the accident in the power plant of Three Mile Island [1], the Cali aircraft accident [2], but also in the medical sector, e.g., the Therac-25 accident series [3], which are at least partly related to bad usability. Other examples of the potential impact of usability in medical contexts are illustrated by J. Nielsen in his Alertbox article entitled "How to kill patients with bad design" [4].

In this paper, we describe the major parts of a usability engineering workflow (evaluations performed and methods applied) carried out within a project which is

A. Holzinger (Ed.): USAB 2007, LNCS 4799, pp. 241–254, 2007.

focused on the optimization of data acquisition, data storage, and workflow in medical emergency cases.

After a short description of the project, an overview of related work is given, followed by the presentation of relevant results of the usability engineering measures. Finally, within the conclusion section the lessons learned are discussed on the basis of which future work is proposed.

2 Background

In Austria, as in many other countries, documentation of the workflow and critical events in emergency cases is still paper based. For example, Grasser et al. [5] collected 47 different report forms used in 23 different European countries. Several shortcomings of paper based documentation are discussed in the literature, cf. e.g [6, 7]. The major challenge of the project CANIS (short for Carinthian Notarzt Information System) was to establish a framework supporting the interconnection of different technical systems for manipulating and storing patient data, sketched in Figure 1. The goal was to interconnect the systems of (for example) the Emergency Call Center (ESCC) and the Hospital Information Center (HIS) directly to the device (CANIS Mobile) the emergency physician (EP) uses on the accident site.

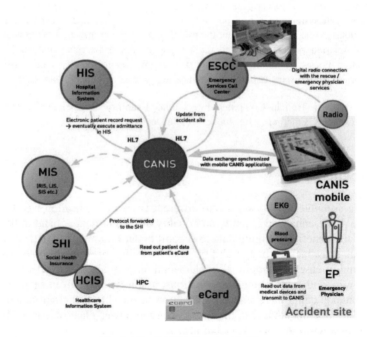

Fig. 1. The CANIS framework [5]

Besides the technical challenge it has been of high importance to optimally design the user interface the emergency physician has at hand. A careful selection of hardware, the evaluation of alternative interaction methods and an optimal design of the software have been the challenges for the usability team and constitute the focus of this paper.

2 Related Work

Integration of computer based devices in medicine has a long history. Decades ago, the discussion regarding usability gained importance, especially in relation to negative examples, such as the aforementioned Therac-25 accident series [3], where bad usability led to mistreatment of cancer patients. The necessity of usability in medical contexts is not only illustrated by the outcome of several field evaluations [8, 9] but has presently also been acknowledged for the development of medical devices, e.g. [10, 11].

In the context of emergency response, the paper of Holzman [6] is noticeable. He discusses the theoretical criteria a system supporting emergency response should fulfil. One of the most important statements in relation to our work is referring to the quality of systems supporting emergency response, which could make "...the difference between life and death..." for the patient. A general problem in emergency response documentation is seen in the low quantity and quality of acquired data. Holzman cites studies that showed that emergency sheets are typically filled out on a level of only 40%. Various reasons could be responsible for this. Hammon et al. [7] observed that nurses documented events not when they occurred, but noted a kind of summary at the end of their shifts. Grasser et al. [5] evaluated 47 report forms and found big differences in their layouts and levels of detail of the information to be filled in. Chittaro et al. [13] assume one of the reasons of bad documentation in the design of the paper sheets. The paper protocol they used for their evaluations seemed to be designed not on the basis of task orientation and logical workflow but: "... to get the most out of the space available on an A4-sized sheet". All these examples lead to the assumption that paper documentation is suboptimal and that a computerized system could enhance the quality of documentation, if the weaknesses of paper sheets are taken into consideration.

Kyng et al. [14] present an approach based on the integration of systems of the different groups involved in emergency response, e.g., firefighters, ambulances, and hospitals. The HCI issues discussed by Kyng et al. are of specific interest. The authors define "challenges" which are relevant in emergency response on different levels. For example, Challenge 7 reads as follows: "Suitability and immediate usability determines what equipment and systems are actually used." In general, the related literature can be summarized by the fact that usability issues in emergency response are dependent on multidimensional factors. Besides the aspects discussed above, also the selection of interaction methods has to be carefully evaluated [6]. For instance, Shneiderman [15] discusses the limitation of speech input related to the increased cognitive load related to speech communication and the changes in prosody in stressful situation, which makes it difficult for the systems to recognize keywords.

Finally, mobility is a huge challenge by itself and the discussion of all mobility related usability aspects would not fit in the limited space of this paper. However, some important things are to be considered, e.g. navigation, font size and shape, interactive elements – cf e.g. [16, 17].

Besides the related scientific literature, we also examined systems on the market which are comparable to our specification, especially the Digitalys 112 RD [18].

3 Method

The project CANIS was subdivided in several phases. The platform developed in Phase one was based on a Tablet PC. A beta version of the system has been evaluated by a cognitive walkthrough and a usability test [19, 20].

We had two different prototype versions for the evaluation – a Compaq Tablet PC and a rugged Panasonic Tablet PC. The operating system was Windows XP Tablet PC Edition, running a prototype version of the software implemented in .NET technology.

The evaluation of the system consisted of two different steps. The first step has been a cognitive walkthrough [19, 20] performed by members of the usability team in order to identify potential pitfalls in the system and to refine the tasks which were defined beforehand by the physicians in the project team.

After the usability walkthrough, minor corrections to the software were performed. Furthermore, due to several hardware related interaction problems encountered with the Compaq Tablet PC during the walkthrough, it was decided that the subsequent usability test should be performed on the other hardware, the Panasonic Tablet PC. After the walkthrough the user test has been designed. The original plan has been to mount the prototype Tablet PC in a car comparable in size to an ambulance car, e.g. a van, and carry out the usability test with emergency physicians who are in stand-by duty. We planned to visit them at the control point and to carry out the test on-site, in order to test the system in a maximal realistic situation.

However, this plan has been abolished for various reasons. Firstly, we didn't want to risk the test being interrupted when an emergency call arrives, resulting in incomplete data. Secondly, we had to consider the ethical aspect that instead of relaxing between emergency cases, the physician would have to participate in our test. This could have influenced his or her performance in a real emergency case. Other reasons were technical ones. For a realistic test it would have been necessary to establish a working wireless infrastructure in the car, which would have tied up too many resources of the development team necessary for other tasks in the project. On the other hand, publications such as [21] show that usability studies in the field and in lab environments may yield comparable results and it seems therefore – in some circumstances – acceptable to get by with lab studies only.

Based on these considerations, the test has been carried out in a usability lab, however we tried to establish an environment within which we were able to simulate a semi-realistic situation.

The first difference to a standard lab test has been that the participants were not sitting on a desk and operating the Tablet PC on the table but were seated in a position on a chair which was similar to sitting in a car. Moreover, the subjects just had a tray for the devices, but mainly had to hold the Tablet PC in their hands or place it on their lap. The situation simulated was designed to be comparable to the situation in an emergency vehicle where the emergency physician or a member of the ambulance staff does parts of the paperwork on the way to or from the emergency scene, as can be observed in real emergency cases.

The usability test was performed with eight subjects (four female, four male; average age 30.8 years), three of which where physicians experienced in emergency cases. The other five persons were people with different medical backgrounds, but also with experiences in emergency response, e.g. professional ambulance staff members and ambulance volunteers, emergency nurses and other emergency room staff. At the beginning of the test session, the subjects were asked to answer some general questions regarding their work in the context of emergency and their computer skills, the latter because we wanted to consider potential effects of different levels of computer literacy on the performance with the prototypes.

Fig. 2. The Usability *Observation Cap*, and the custom made observation software

The devices for observing the scene and recording data were also non-standard lab equipment, but a system that has been developed by ourselves to support usability observations especially in mobile contexts and making it possible for the subjects to freely move around. Interactions on the user Tablet PC were tracked via a VNC-based custom made software, the interactions with fingers, stylus or keyboard were tracked by the *Observation Cap* [22], a baseball cap equipped with a WLAN-camera. The equipment worn by the subject and the software tracking the different data sources are shown in Figure 2.

The test also included the simulation of a realistic emergency case. After briefing the subjects regarding the nature and goals of usability tests and asking the questions regarding their relation to emergency response and their computer skills, they were asked to sit down in the simulated environment and a fictional emergency case was read to them.

The simulated case was a traffic accident of a male motorcyclist, causing major injuries on one leg, resulting in problems with blood circulation, consciousness, and respiration. The scenario was defined by physicians involved in the project to ensure that the simulated case would be realistic and did not contain wrong assumptions.

Subsequently, the subjects had to perform a series of tasks with the system which were based on the fictional emergency case. The subjects had to sequentially fill out form fields containing data related to the different steps of an emergency case, i.e., arriving on the scene – evaluation of the status of the injured patient – preliminary diagnosis – treatment – hand over information to the hospital. Although a realistic situation has been simulated, the subjects were asked to focus on the handling of the device rather than concentrating on a medical correct diagnosis and therapy. It was more important to identify potential problems of the interface which may influence task performance in real situations, than getting correct data. Therefore, the subjects were asked to state their opinion on different aspects of the system whenever they wanted. In contrast to a real accident, time and accuracy was not important, although metrics such as task completion time, number of errors, and kinds of errors have been recorded.

To test the suitability of different interaction methods, the subjects were asked to use finger input at the beginning of the test, and in the middle of the test they were asked to switch to stylus interaction. After performing the tasks, the subjects were asked to summarize their personal opinion on the system.

After the core usability test another evaluation was carried out. At that time, there were already two prototypes of project Phase 2 available, i.e., a PDA with speech input functionality and a digital pen. It was of interest to ask the participants how they found the alternative devices in comparison to the Tablet PC which they had used beforehand. This part of the test was not considered as a "real" usability test, because the maturity of the two devices has not been comparable to the development status of the Tablet PC based application. However, some exemplary tasks could be performed in order to show the devices in operation.

The digital pen was used in combination with a digital form similar to a conventional paper based emergency response form and capable of storing and transferring digitalized data. The second device was a QTEK Smartphone with Windows Mobile and a custom made form based software developed by Hafner, cf e.g. [23], combined with a speech recognition engine which made it possible to fill in the emergency response form with spoken keywords. After performing the exemplary tasks, the subjects were asked to compare the alternative devices with the Tablet PC, to state their preferences and reasons for their decision, the possible drawbacks they could identify, etc.

4   Results

In this section, we present the most relevant results from the usability evaluations carried out – with respect to both, the engineering process itself and the product.

4.1   General Results

A general outcome which has been expected by the team carrying out the usability walkthrough was confirmed by the test results. The system is too much GUI-like and not optimally adapted to the special context of use. This shortcoming could be observed with several aspects of the interaction with the system. As mentioned, all subjects were asked to use finger input at the beginning of the test and in the middle of the test they were asked to switch to stylus interaction. It could be observed, that finger input did not work properly. Even the subjects with very small fingers (especially women) had problems to hit the right widgets and some functions which were based on drag&drop interactions could not be performed with fingers. The majority of subjects said that they preferred stylus input; however, they found it less practical than finger input with respect to the possibility of loss of a stylus.

In general, the subjects found the Tablet PC system useful and adequate to use a computerized system in emergency response. But all of them also stated that such a system would have to be very stable, robust and fault-tolerant to be applied in this context. One of the outcomes of the simulated situation was that the Tablet PC was too heavy. This was stated by the majority of the participants (including both female and male subjects).

Other findings were some typical usability problems related to the design of GUIs. Some information has not been grouped according to the workflow of the tasks, but has rather been positioned on the grounds of available screen space. For example, a comment field was positioned in a specific tab, but it was also meant to contain comments of the other tabs relevant in the current context. When the subjects selected the category "other..." in a group of symptoms (e.g. respiratory) the comments field appeared but also the tab was switched, because the comments field was implemented in the tab containing heart symptoms. The subjects couldn´t recognize this context switch and thought their inputs were lost when they closed the comments field and the entered contents were not visible anymore.

Other typical usability problems such as inconsistencies in layouts, input mechanisms and navigation also occurred. One of them is discussed in detail and illustrated in the following, corresponding figures. When opening a comment field, this was empty and text could be filled in, as shown in Figure 3.

Fig. 3. Entering text the first time

When the field was opened again to add comments, the old comments were marked. When the user tried to write in this mode, all the text has been deleted and overwritten by the new text, as shown in the two parts of Figure 4.

Another usability problem is worth discussing. The prototype included different possibilities to select a date. The widget shown in Figure 5 below seems to be designed considering the requirement to provide alternative input mechanisms for different kinds of input devices and user preferences, as defined in platform guidelines [24]. However, the different alternatives (plus/minus-buttons, text input

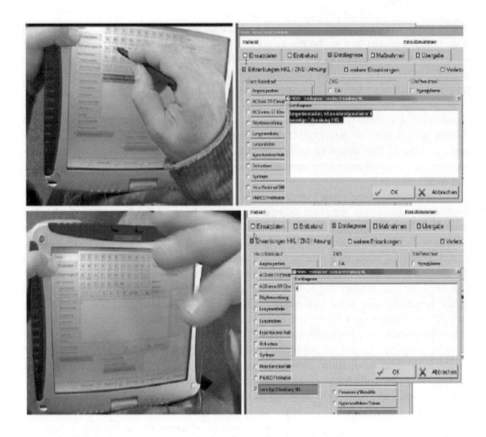

Fig. 4. When new text is entered, existing text is marked and deleted undeliberately

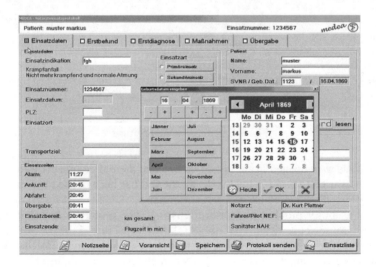

Fig. 5. Date input widget providing different input mechanisms

fields, months buttons, calendar widget, today button) and combinations to fill in a date of birth led to cognitive overload of the subjects and resulted in a slow and error-prone completion of the tasks.

4.2 Specific Results

Besides the standard widgets discussed, the prototype includes some specific widgets which are of special interest with respect to usability. Figure 6 below shows a sketch of the human body which had to be used to mark different kinds of injuries on the corresponding body part. The subjects had severe problems to mark and unmark the corresponding parts because the affordance of the widgets was suboptimal. The task could be performed either by marking a symbol on the tool bar on the left side of the figure and then clicking in the corresponding body part or by dragging and dropping a symbol in the corresponding part of the body.

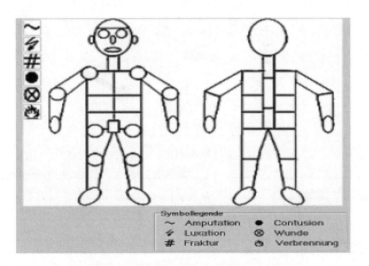

Fig. 6. Body Sketch widget to mark injuries

Many subjects tried to mark injuries with elements of the legend (below the body sketch) because these seemed to look more like functional widgets. Most participants tried to use the widgets as in painting applications, i.e., they tried to select a symbol in the taskbar and then to mark the part, none of them thought, that drag&drop would be possible. In summary, an intuitive and fast usage of this widget and its interactions seems not to be suitable to an emergency situation where stress level is high.

Another specific interaction feature of the system which also was error-prone in the test is shown in the middle of Figure 7. The layout is structured in two parts, the upper part symbolizes a timeline of the emergency case, where important points of time, e.g. arrival time on the emergency site and measured values such as heart rate and blood pressure are shown. The data can be entered in different ways, either by clicking a button below the white area and filling in a form field in a modal dialog, or by clicking directly in the white area, after which a similar modal dialog appears. Besides the difficulty of understanding the different possibilities of input, changing inputs that have been made previously or deleting values has been very clumsy, because of the difficulty to select an icon. This was caused by the size of the icon as well as the fact that it was necessary to change to an editing mode by pressing the stylus down for a few seconds (the different features available in this mode are shown by the arrow in Figure 7). When the subject did not hit an existing icon or did not press the stylus long enough (which would probably occur even more often in reality due to the physician's hurry), the modal dialog to initiate a new entry appeared.

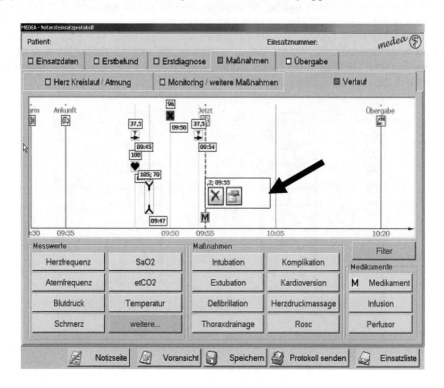

Fig. 7. Emergency case timeline widget

4.3   Results Regarding the Different Prototypes

Compared the different prototypes, the subjects preferred the digital pen, because the handling was quite similar to a conventional paper/pen combination with the additional benefit of producing digital data. However, this benefit was theoretical at the time of the test, because only structured data (e.g. number codes, check boxes) could be transferred, handwritings were not processed by OCR and therefore the subjects ratings were based on (as of yet) unrealistic assumptions.

The speech recognition had been working satisfactorily in the development phase, but within the test too many problems occurred, thus the recognition rate has been very low. Another problem which was disturbing in the test but also increased the

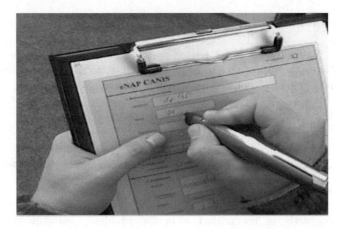

Fig. 8. Digital pen text input from the perspective of the *Observation Cap*

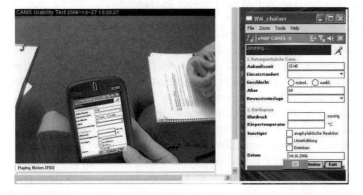

Fig. 9. Screenshots from the perspective of the *Observation Cap*, showing the PDA in the subjects´ hand. On the right the screen of the tracking software is shown, where the entered details could be observed by the test supervisor.

level of reality was the following: In standard operation, the subjects could see whether the system had recognized their input, because recognized values were filled in the corresponding fields.

Within the test, the system sometimes didn´t switch pages, therefore the subjects could not visually observe if the values they spoke were correctly entered. This made the task more difficult, however, also more similar to the situation the system is designed for – just speaking values without getting visual but just acoustic feedback (e.g. "beep") that the keywords are recognized.

The problems described negatively influenced the subjects rating of the speech input functionality. Thus, they stated that they could easily imagine the benefits of hands free entering data, provided it works on a satisfactory level.

5 Discussion and Future Activities

Some of the findings were not expected, but the majority of the results was not surprising. Within the following paragraphs we will try to extract lessons we learned which hopefully help us and others in preventing such errors in the future.

Lesson 1: Apply usability engineering as early as possible
Because of the difficult technical challenge of the project, a situation occurred which can be observed in many other IT-projects. The priority of user interface design and usability engineering has decreased and these issures were shifted to a later stage in the project. Another reason for the delay has been that the originally considered Tablet PC hardware was not available in Austria and very late in the order process the import was banned. This led to the situation that problems which could have been identified very early in the project were carried along and then could not be completely corrected because of time constraints. However, some of the GUI related problems discussed, e.g., the text input problem or the design of the date input widget, could have even been identified on different hardware.

Lesson 2: Usability is not just considering guidelines
Software developers, not being usability experts, are often in a difficult situation. As discussed, usability engineering activities had to be shifted and the hardware platform was not confirmed. Therefore the programmers had to rely on information they had at hand at that point in time. We observe this phenomenon in different contexts – usability laymen think that the consideration of standards and guidelines alone will lead to good usability. Some of these general guidelines (e.g., concerning color and font usage) and platform specific guidelines (e.g., how a window is designed, alignment of widgets and their labels) are nowadays integrated into development platforms such as Visual Studio, which has been used in the current project. These features seem to persuade the developer of the adequate consideration of usability aspects and that it does not seem to be necessary to cooperate with real users. This can be observed in CANIS, because the interface elements are, on first sight, well aligned and grouped together. The problems discussed show that other aspects than layout features have to be considered.

Lesson 3: Don´t forget the context of use
Literature shows that context is a critical factor. But context isn´t only the environment where a system is used, but also what input mechanisms are used, what kind of users are intended to use the system, especially in contexts where time pressure and stress is high. Our results show that aspects such as the weight of the device, input modalities and feedback mechanisms are of high importance. Some of the problems identified in the test did not lead to problems in the development phase, because the situation was not comparable to a real situation. The findings also support our assumption, that the simulation in the lab has been the right choice for the evaluation, however it is clear that in future other approaches are also to be considered.

Lesson 4: Avoid too many alternatives and modes.
The degree of freedom for the users of the system is high, especially for some widgets. However, more alternatives often lead to higher stress and lower performance [25] and modelessness is often claimed as a requirement, cf. e.g. Raskin [26]. Especially in stressful contexts alternatives and different modes overwhelm the user and it would be better to provide only one single method but using it consistently in the whole system.

Based on the findings of the evaluation we come to the conclusion that in future a combination of the platforms and input devices would be the optimal solution. We try to sketch a scenario for illustration: When an emergency call is received, the team jumps into the ambulance vehicle where a Tablet PC is mounted. They take a clipboard and fill in the basic information of the emergency case with digital pen on digital paper. The data is transmitted to the Tablet PC which is connected to the receiving hospital. On the emergency site, the physician wears a headset which is connected to an integrated PDA and enters information regarding the status of the patients with speech input. The data is also transferred to the Tablet PC mounted in the ambulance car. After finishing the patient treatment on the way back to the hospital, the data is completed and – if necessary – corrected on the Tablet PC by using finger input.

References

1. Meshkati, N.: Human factors in large-scale technological systems' accidents: Three Mile Island, Bhopal, Chernobyl. Organization Environment 5, 133–154 (1991)
2. Gerdsmeier, T., Ladkin, P., Loer, K.: Analysing the Cali Accident with a WB-Graph. Human Error and Systems Development Workshop. Glasgow, UK. (last access: 2007-09-07), http://www.rvs.uni-bielefeld.de/publications/Reports/caliWB.html
3. Levenson, N., Turner, C.S.: An Investigation of the Therac-25 Accidents. IEEE Computer 26(7), 18–41 (1993)
4. Nielsen,J.: How to kill patients with bad design. Online at: http://www.useit.com/alertbox/20050411.html (last access: 2007-09-07)
5. Grasser, S., Thierry, J., Hafner, C.: Quality Enhancement in emergency Medicine through wireless wearable computing. Online at: http://www.medetel.lu/download/2007/parallel_sessions/presentation/0419/CANIS.pdf (last access: 2007-09-07)
6. Holzman, T.G.: Computer-Human Interface Solutions for Emergency Medical Care. ACM Interactions 6(3), 13–24 (1999)
7. Hammond, J., Johnson, H.M., Varas, R., Ward, C.G.: A Qualitative Comparison of Paper Fiowsheets vs. A Computer-Based Clinical Inform. System. Chest 99, 155–157 (1991)
8. Schächinger, U., Stieglitz, S.P., Kretschmer, R., Nerlich, M.: Telemedizin und Telematik in der Notfallmedizin, Notfall & Rettungsmedizin 2(8), 468–477.
9. Zhang, J., Johnson, T.R., Patel, V.L., Paige, D.L., Kubose, T.: Using usability heuristics to evaluate patient safety of medical devices. J. of Biomed. Informatics 36(1/2), 23–30 (2003)
10. Graham, M.J., Kubose, T.M., Jordan, D., Zhang, J., Johnson, T.R., Patel, V.L.: Heuristic evaluation of infusion pumps: implications for patient safety in Intensive Care Units. Int. J. Med. Inform. 73(11-12), 771–779 (2004)

11. Hölcher, U., Laurig, W., Müller-Arnecke, H.W.:Prinziplösung zur ergonomischen Ge-
    staltung von Medizingeräten – Projekt F 1902. Online at: http://www.baua.de/
    nn_11598/sid_C4B270C1EE4A37474D3486C0EDF7B13A/nsc_true/de/Publikationen/
    Fachbeitraege/F1902,xv=vt.pdf (last access: 2007-09-07)
12. U.S Food Drug Administration (FDA).: Do it by Design. An Introduction to Human
    Factors in Medical Devices. Online at last access: 2007-09-07, http://www.fda.gov/cdrh/
    humfac/doit.html
13. Chittaro, L., Zuliani, F., Carchietti, E.: Mobile Devices in Emergency Medical Services:
    User Evaluation of a PDA-based Interface for Ambulance Run Reporting. In: Proceedings
    of Mobile Response 2007: International Workshop on Mobile Information Technology for
    Emergency Response, pp. 20–29. Springer, Berlin (2007)
14. Kyng, M., Nielsen, E.T., Kristensen, M.: Challenges in designing interactive systems for
    emergency response. In: Proceedings of DIS 2006, pp. 301–310 (2006)
15. Shneiderman, B.: The limits of speech recognition. Commun. ACM 43(9), 63–65 (2000)
16. Holzinger, A., Errath, M.: Designing web-applications for mobile computers: Experiences
    with applications to medicine. In: Stary, C., Stephanidis, C. (eds.) User-Centered
    Interaction Paradigms for Universal Access in the Information Society. LNCS, vol. 3196,
    pp. 262–267. Springer, Heidelberg (2004)
17. Gorlenko, L., Merrick, R.: No wires attached: Usability challenges in the connected mobile
    world. IBM Syst. J. 42(4), 639–651 (2003)
18. Flake, F.: Das Dokumentationssystem der Zukunft. Digitale Einsatzdatenerfassung mit
    NIDA, Rettungsdienst 29, 14–18 (2006)
19. Nielsen, J.: Usability Engineering. Morgan Kaufmann Publ. San Francisco (1993)
20. Holzinger, A.: Usability Engineering for Software Developers. Communications of the
    ACM 48(1), 71–74 (2005)
21. Kjeldskov, J., Skov, M.B., Als, B.S., Høegh, R.T.: Is It Worth the Hassle? Exploring the
    Added Value of Evaluating the Usability of Context-Aware Mobile Systems in the Field.
    In: Brewster, S., Dunlop, M.D. (eds.) MobileHCI 2004. LNCS, vol. 3160, pp. 61–73.
    Springer, Heidelberg (2004)
22. Leitner, G., Hitz, M.: The usability observation cap. In: Noldus, L.P.J.J., Grieco, F.,
    Loijens, L.W.S, Zimmerman, P.H. (eds.) Proceedings of the 5th International Conference
    on Methods and Techniques in Behavioral Research, Noldus, Wageningen, pp. 8–13
    (2005)
23. Hafner, C., Hitz, M., Leitner, G.: Human factors of a speech-based emergency response
    information system. In: Proceedings of WWCS, pp. 37–42 (2007)
24. Sun Microsystems, Inc and Javasoft: Java Look & Feel Design Guidelines, 1st edn.
    Addison-Wesley Longman Publishing Co., Inc, Redwood City (1999)
25. Schwartz, B.: The paradox of choice: Why more is less, Ecco, New York (2004)
26. Raskin, J.: The Humane Interface: New Directions for Designing Interactive Systems.
    ACM Press/Addison-Wesley Publishing Co, New York (2000)

Some Usability Issues of Augmented and Mixed Reality for e-Health Applications in the Medical Domain

Reinhold Behringer[1], Johannes Christian[2], Andreas Holzinger[3], and Steve Wilkinson[4]

[1,2,4] Leeds Metropolitan University, Leeds, LS6 3QS,
United Kingdom
Innovation North – Faculty for Information and Technology
r.behringer@leedsmet.ac.uk, johannes.christian@gmail.com,
s.wilkinson@leedsmet.ac.uk
[3] Medical University Graz, 8036 Graz, Austria
Institute for Medical Informatics, Statistics & Documentation (IMI)
Research Unit HCI4MED
andreas.holzinger@meduni-graz.at

Abstract. Augmented and Mixed Reality technology provides to the medical field the possibility for seamless visualization of text-based physiological data and various graphical 3D data onto the patient's body. This allows improvements in diagnosis and treatment of patients. For the patient, this technology offers benefits and further potential in therapy, rehabilitation and diagnosis, and explanation. Applications across the whole range of functions that affect the health sector from the physician, the medical student, to the patients are possible. However, the quality of the work of medical professionals is considerably influenced by both usefulness and usability of technology. Consequently, key issues in developing such applications are the tracking methodology, the display technology and most of all ensuring good usability. There have been several research groups who extended the state of the art in employing these technologies in the medical domain. However, only a few are addressing issues of Human-Computer Interaction, Interaction Design, and Usability Engineering. This paper provides a brief overview over the history and the most recent developments in this domain with a special focus on issues of user-centered development.

Keywords: Human–Computer Interaction, Augmented Reality, Mixed Reality, Visualization, User-Centered Development, e-Health, Medicine.

1 Introduction

The concept of Mixed and Augmented Reality provides the fusion of digital data with the human perception of the environment: using computer-based rendition techniques (graphics, audio, and other senses) the computing system renders data so that the resulting rendition appears to the user as being a part of the perceived environment. Most applications of this paradigm have been developed for the visual sense, using

A. Holzinger (Ed.): USAB 2007, LNCS 4799, pp. 255–266, 2007.

computer graphics technology to render 3D objects and presenting them registered to the real world. The difference between Augmented Reality (AR) and Mixed Reality (MR) is in the degree of virtualization: in AR the real world is the main operating framework, with a limited number of virtual objects added into the display, hereby solely augmenting the real-world view of the user. In MR, the emphasis of the virtual environment is much larger, which can manifest itself either in a larger number of virtual objects or in having the virtual environment (VE) as the dominant interaction framework.

Both paradigms have interesting applications in the medical domain: AR can be used to visualize complex 3D medical data sets (e.g. form ultrasound or NMR scans) in a direct projection on the patient, hereby allowing the physician to examine simultaneously the data and the patient. This can be taken a step further in tools for surgery, where the imagery can support the procedures to be undertaken. MR technology on the other hand can be used for off-line simulation and training, with the option to use virtual patients to simulate complex surgical procedures.

R&D has been carried out for enabling the use of AR and MR in combination with information and communication technology (ICT) in the medical area since the early 1990s. This paper gives an overview over the development and the current state of the art, highlighting specific examples of such e-health applications.

2 Basic Concepts of AR and MR

At the core of AR and MR technology is the seamless fusion of the perception of the real world with the computer-generated data. In order to achieve this, it is important that the system hardware supports this fusion. For the visual sense, this can be achieved with a head-worn display which is placed into the field of view of the user, who hereby sees both simultaneously the real world and the display. For such a system to be called "AR" or "MR", it is essential that the information shown in the display is registered with the real world, which means that when the user changes the viewing direction, the content of the display also changes, in accordance with what the user sees. This is necessary to create the impression that the display content is in fact connected to the real world. Other types of displays are hand-held see-through displays, which can be used to overlay information onto the background. A different approach is to use projective displays, which project the information directly into the environment. This approach relies on walls or environmental structures which are relatively close to the projector and have reflective surfaces. Uneven complex surfaces cause problems in the projection approach, as the projection has to be corrected for the user's position and depends on the shape of the surface.

For achieving a seamless mix of the perception of the real world together with the computer-generated graphical information, AR and MR systems need to track the user's viewing direction and position. This can be achieved with a variety of technologies: active IR illumination in connection with IR tracking cameras can be used to track the user's head pose; computer vision techniques can be applied to track the user's position and orientation either from the captured images of the user's environment, captured by a head-worn camera, or from cameras in the environment, pointing at the user. For medical applications it is not crucial that the tracking is done ubiquitously: since medical procedures are generally performed in a designated area or room, it is acceptable to employ technologies which rely on certain infrastructure available at the location, such as markers, projectors, reflectors or similar.

3 Medical Applications of AR and MR

The paradigm of AR and MR is very well suited for the medical domain, as the fusion of 3D medical scan data with the view of the patient will have benefits in diagnosis (through on-patient visualization) and performance support in general. Specifically, AR can be used for support of surgery, while MR and Virtual Reality (VR) are very suitable for simulation without the actual patient. Any of these technologies can be used for training of physicians and medical students, as they improve the situational and spatial awareness of the practitioner. Further on the patient also can be supported by a variety of applications through this technology.

3.1 Visualization of Ultrasound

The research group of Andrei State at University of North Carolina (UNC) has developed medical visualization applications using AR since 1992. Their first project dealt with passive obstetrics examinations [1]. Individual ultrasound slices from a pre-natal fetus were overlaid onto the video of the body of the pregnant woman – both live image streams were then combined using chroma-keying. Since the ultrasound imagery was only 2-dimensional in nature, the overlay resulted only in a correct registration from one particular viewing direction. An improvement was introduced with the volumetric rendition of the fetus. The original real-time ultrasound rendition proved to produce blurry images [2], but the introduction of off-line rendition allowed to render the volumetric data at a better quality at higher resolution [3]. As a further development, UNC used AR for ultrasound-guided needle biopsies for both training and actual performance [4]. Based on the experiences and practice of this technology, the group at UNC has also contributed to development of technology components for such systems, e.g. improvements of the video-see-through head-worn display [5] and tracker accuracy improvements. They have moved their system into actual performance of minimally invasive surgery (e.g. laparoscopic surgery [6]) and are working towards improvements of their system for being used in everyday practice.

3.2 Visualization Using CT Scans

Stockmans [7] demonstrated how CT scans can be used within a 3D visualization package to determine how very complex bone deformities within arms may be corrected as effectively and as accurately as possible. The technique involved determining how wrist joint mobility could be improved by correcting and re-sighting the forearm bone. The software used a virtual skin mesh and an accurate representation of joint movement to view various strategies for the operation. After the operation had been completed and recovery had been taken place it was possible to match the final result with the previous virtual visualization to determine the success of the operation. The software proved to be a valuable tool in decision making, preoperative testing and surgical planning.

3.3 Simulation for Training

MR technology does not need the human patient to be present, as it deals more with off-line simulation and employs more virtual reality (VR) technology. This makes it very suitable for training. One example of such a simulation for training is the birth simulator developed by Sielhorst et al. at the TU München [18]: a complete scale model of a woman's womb is designed as a torso, and a set of birth pliers is providing haptic feedback to emulate the forces when extracting the baby. With the head-worn display, the person to be trained can have an X-ray view into the body torso to get a better learning experience.

Another application of MR technology is training paramedics for disaster situations: Nestler et al. are developing a simulation of "virtual patients" [31], to train paramedics for large-scale disasters. In this simulation, the paramedics learn to deal with the various possible injuries of disaster victims and to apply first aid procedures. These virtual victims are simulated on a large multi-touch table-top in which one patient at a time can be displayed in full size. In this application, tracking precision is not very relevant, but solely the fact that procedures are applied correctly.

A different application of MR in the medical area could be the information of patients, as demonstrated in the VR-only system developed by Wilkinson et al. [9]: this system is designed to educate patients about the upcoming surgery procedures. In a study, this system has been used in a game and demonstrated to be able to reduce children patients' fear of an upcoming hand surgery, as with this system they were able to become more informed about the procedures. This concept could be expanded to use MR or AR technology, to show the procedure directly on the patient's hand.

3.4 Surgery and AR

Surgery is a very complex application for AR: life and death of the patient depend on precise registration of the data for the surgeon to perform the operation properly. The technical requirements for this application have been investigated by Frederick [12] from CMU. A more recent review of surgery and AR has been published by Shuhaiber [10] in 2004. His conclusions pointed out that these systems are still not yet practical in clinical applications, but hold a promising future by providing a better spatial orientation through anatomic landmarks and allowing a more radical operative therapy.

The German government funded the project MEDical Augmented Reality for Patients (MEDARPA) [13], which has the goal of supporting minimally invasive surgery. In this system, AR and VR technologies are used to improve the navigational capabilities of the physician. The system was being evaluated at 3 different hospitals in Germany (Frankfurt, Offenbach, Nürnberg) as a support for placing needles for biopsies (bronchoscopy) and interstitial brachytherapy (to irradiate malignant tumors). In addition, the system also was intended to be used with a surgical robot, to allow a more precise alignment of the incision points.

For computer-aided medical treatment planning, Reitinger et al. [20] present a set of AR-based measurement tools for medical applications. Their Virtual Liver Surgery Planning system is designed to assist surgeons and radiologists in making informed decisions regarding the surgical treatment of liver cancer. By providing quantitative assessment in measuring e.g. distance, volume, angle and adding automation algorithms, it improves the way of performing 3D measurements.

3.5 Therapy/Rehabilitation

Riva et al. [21] expect the emergence of "immersive virtual telepresence (IVT)" and see a strengthening of 3rd generation IVT systems including biosensors, mobile communication and mixed reality. This kind of IVT environments are seen to play a broader role in neuro-psychology, clinical psychology and health care education.

While VR therapies (like e.g. provide in the project NeuroVR [24])and exposure in vivo have proven to be effective in treatment of different psychological disorders such as phobia to small animals (like spiders or cockroaches), claustrophobia or acrophobia, AR offers a different way to increase the feeling of presence and reality judgment. Juan, M.C. et al. [23] have developed an AR system for the treatment of phobia to spiders and cockroaches.

Concerning mental practice for post-stroke rehabilitation Gaggioli et al. [25] present a way of applying augmented reality technology to teach motor skills. An augmented reality workbench (called "VR-Mirror") helps post-stroke hemiplegic patients to evoke motor images and assist this way rehabilitation combining mental and physical practice. An interesting project for healthy living basing on computerized persuasion, or captology, combined with AR technology is the "Persuasive Mirror" [30]. In an augmented mirror, people get help to reach their personal goals like leading a healthier lifestyle by regular exercise or quitting smoking.

3.6 Education / Edutainment

Nischelwitzer et al. [22] have designed an interactive AR application for children, called "My inside the body Book (MIBB)". In a physical book, the human alimentary system is described by interfacing reality and virtuality. In order to be able to interact with the content, buttons were mounted at the bottom of the book. The possibility to examine organs of the human body from different perspectives showed the positive potential of AR for the area of learning. From this result we also envision that for example a medical encyclopedia could be designed for medical students, for example to provide 3D information on anatomy.

4 Technical Issues

4.1 Registration and Tracking

For registration of the data overlay, seamless tracking is essential, to keep the view of the data registered with the patient's body. For medical applications, precision is especially important in order to match the data (e.g. CT or NMR scans [11]) to the patient's body precisely for enabling a correct interpretation and diagnosis. One difficulty is that the human tissue is not rigid and often does not have distinctive features. For camera-based tracking, one can attach markers to the body, to provide fiducial markers as visual anchor points.

The calibration effort should be very small, as the physician needs to focus on the patient and not on the technology. Tracking approaches need to be resistant to possible occlusion, as the physician may need to manipulate instruments [17].

The MEDarpa system [13] employs different tracking methods: an optical tracker is used for tracking the display and the physician's head. The instruments themselves are being tracked by magnetic tracking systems, to avoid problems with occlusion through the display.

4.2 Display Technology

Today the AR/MR community and the end-user have the possibility to choose from a variety of display technologies which best suits their application demands. Different optics can be used as information-forming systems on the optical path between the observer's eyes and the physical object to be augmented. They can be categorized into three main classes: head-attached (such as retinal displays, head-worn displays and head-mounted projectors), hand-held and spatial displays. [26]

Head-worn displays (HWD) provide the most direct method of AR visualization, as the data are directly placed into the view and allow hands-free activity without any other display hardware between the physician's head and the patient. However, they were in the past often cumbersome to wear and inhibited the free view and free motion of the surgeon. Latest research and design tries to bring HWDs (especially the optical see-through type) into a social acceptable size like e.g. integrated into sun- or safety-glasses and has already resulted in some available prototypes. This creates the expectation that in the near future, light weight and low-cost solutions will be available [28][29]

Video-see-through (VST) displays have a lower resolution than optical see-through (OST), but are easier to calibrate. The overlay can be generated by software in the computing system, allowing a precise alignment, whereas an OST system has additional degrees of freedom due to the possible motion of the display itself vs. the head.

The group of Wolfgang Birkfellner from TU Vienna has developed the Varioscope [19], a system which allows visualization of CT data through a head-worn display, with the goal of pre-operative planning. In the design of this display, care was taken to produce a light-weight and small device which would be suitable for clinical use.

Alternatives to HWDs are hand-held or tablet displays, which could be mounted on a boom for a hand-free process. In general, these displays are semi-transparent to allow optical see-through of the patient's body without the need for a separate camera.

The MEDarpa system [13] did not use head-worn displays, but instead developed a novel display type which is basically a half-transparent display. This display can be freely positioned over the patient, providing a "virtual window" into the patient by overlaying CT and MRT data from earlier measurements into the view [14]. This requires that both the display and the physician's head are being tracked.

Most of these examples show that the prediction and model of an e-Assistant [27] is step-by-step becoming true and will be available soon with ongoing miniaturization process of devices, improvements in technology, and becoming available to the users.

4.3 Interaction

Similar as for visualization, the variety of interaction possibilities is large (unless the display does not have an integrated interaction device). Different interfaces like traditional desktop interfaces (e.g. keyboard, mouse, joystick, speech recognition), VR I/O devices (e.g. data glove, 3D mouse, graphics tablet etc.), tangible user interfaces (TUI), physical elements and interfaces under research (like brain computer interface (BCI)) can be used. Further research also concentrates on medicine and application-specific user interfaces.

5 User-Centered Development Issues of AR for Medicine

The quality of the work of physicians is considerably influenced by the usability of their technologies [32]. And it is very interesting that in the past many computer science pioneers have been tempted to ask "what can the computer do?" instead of asking of "what can people do?" [33]. It is obvious that medical procedures can greatly benefit from improved visualization, navigation, interaction and from enhanced decision support [34]. By the application of User Centered Development Processes (UCDP) it is possible to study the workflows of the end users and to involve them into development from the very beginning. However, a serious threat is that medical professionals have extremely little time to spend on such development processes. One possibility is to make them aware about the potential benefits and to demonstrate how this technology solves specific problems for their daily routine [35].

It is essential that the developers of AR applications for the medical domain must understand not only the technology they support and the limitations of the techniques they use, but also the medical professionals, their workflows and their environments. Consequently, Usability Engineering Methods (UEM), which includes the end users in the development from the beginning, are becoming more and more important [36]. However, although approaches for the application of User Centered Development have been around for most of the time [37], in practice today, a gap still exists between these theories and practice. Either software engineers concentrate on software development, or usability experts concentrate on design, they rarely work together [38].

Basically, standard UEMs can be applied for AR environments. However, when looking for differences, we found several adapted methods, although mostly for *evaluating* the usability of AR applications – not User-Centered Development. For example, Hix developed checklists of criteria for assessing VE design and adapted methods for evaluating usability of standard user interfaces for Virtual Reality applications [43], while Kalawsky [45] created a questionnaire audit of VE features derived from a usability checklist for GUIs. Several experimental studies have assessed the effectiveness of different interactive devices (e.g. [46]) and haptic interaction in collaborative virtual environments. A principal difference in assessing user interaction with VEs in comparison with traditional user interfaces is the sense of presence created by immersion in virtual worlds. Assessment of presence has primarily focused on measuring the effects of VE technologies on a person's sense of presence via questionnaires. Slater et al. [47] concluded that presence itself was not associated with task performance. However, measures of presence have concentrated on user perception of VE technology and have not considered the impact of errors. Errors can disrupt the sense of illusion, and hence impair perception of the presence.

Freeman et al. [44] defined presence as "the observer's subjective sensation of 'being there' in a remote environment", while Lombard and Ditton [48] investigated presence and immersion, and defined presence as "the perceptual illusion of non-mediation [which] involves continuous ('real-time') responses of the human sensory, cognitive and affective processing systems". The effects of being immersed have been investigated in a variety of VE system configurations. More recent work has examined presence in collaborative VEs from the viewpoints of self and perceptions of others' presence. Although considerable previous research has been carried out in fully immersive VEs and into how users interact with objects in such VEs, such work has concentrated on head-mounted display (HMD) -based systems. Few studies have investigated fully immersive CAVE systems or Interactive WorkBenches (IWBs), apart from the work of Blach et al. [49] who concluded that realistic perception and interaction support were important for effective interaction in CAVE environments.

While evaluation methods and techniques have acknowledged the importance of presence [43], they have not included techniques for measuring it, apart from simple ratings of the "naturalness" of an environment. One of the weaknesses of questionnaire-based assessment is that people's ratings may be biased by their experience.

Summarizing, the user experience within AR is essential, any problems in interaction can reduce the illusion of presence thereby reduce the advantages of AR [39].

Several researchers made serious concerns about the use of AR (e.g. [40]: The already present information overload can result or be accompanied by a sensory overload; end users, especially in the medical domain might find themselves overwhelmed by the wealth of experience they find in the AR; and most of all: Either too little experience with technology per se, or lack of acceptance. Social effects (beliefs, attitudes and feelings) should not be neglected.

Some interesting questions for further research could be:

- Which interaction metaphors are most appropriate in AR for the medical domain?
- Which influences do AR applications have on performance of the end users in the medical domain?
- What are the effects of adaptation that people might have to make cognitively to believe in and cope within an AR environment?
- How will multimodal interaction through a number of input and output channels enhance or detract from the reality/Virtuality experience?

In comparison to standard HCI work, in AR we must deal with physical interactions, social interactions and cognitive interactions. It is definitely more difficult to isolate variables to investigate them in controlled trials, and there are less constraints on performance, which lead to a lower predictability of behaviors to set up field observations or experiments [39]. According to Dünser et al. [41], for a successful development of AR systems all research domains involved have to be considered and integrated properly with a user-centered design focus. As AR and MR technology is going to be used more and more outside of laboratory settings and becoming integrated in everyday life, the systems need to:

- be more accessible,
- be usable for the everyday end-user,
- follow the notion of pervasive and ubiquitous computing,
- implement the basic ideas of social software,
- be designed for use by users who lack a deep knowledge of IT systems [42].

6 Summary and Outlook

Augmented and Mixed Reality technology has a promising potential to be used in medical applications, as it provides seamless integration of data visualization with the patient's body. This allows improved methods for medical diagnosis and treatment. The technological hurdles of displays and registration are still not completely solved for realistic clinical application in a regular medical environment, but progress in various projects is encouraging. With the special focus also on the user-centered development issues and research on AR and MR design guidelines, the future success for various e-Health applications in the medical domain has promising perspectives. Of specific interest are developments of smaller and lighter head-worn displays, with a larger field of view and higher level of detail. More attention needs to be paid to the actual usability of such systems, avoiding a sensory overload and making the visualisation experience more controllable. Such systems could benefit from an integration of other related HCI technologies in the domain of ubiquitous computing, using speech and gesture recognition for a multimodal interaction with the medical information system.

Acknowledgments. This work was supported by the European Commission with the Marie Curie International Re-Integration grant IRG-042051.

References

1. Bajura, M., Fuchs, H., Ohbuchi, R.: Merging Virtual Objects with the Real World: Seeing Ultrasound Imagery within the Patient. In: Proceedings of SIGGRAPH 1992. Computer Graphics, vol. 26, pp. 203–210 (1992)
2. State, A., McAllister, J., Neumann, U., Chen, H., Cullip, T., Chen, D.T., Fuchs, H.: Interactive Volume Visualization on a Heterogeneous Message-Passing Multicomputer. In: Proceedings of the 1995 ACM Symposium on Interactive 3D Graphics (Monterey, CA, April 9-12, 1995), Special issue of Computer Graphics, ACM SIGGRAPH, New York, pp. 69–74 (1995) Also UNC-CH Dept. of Computer Science technical report TR94-040, 1994
3. State, A., Chen, D.T., Tector, C., Brandt, A., Chen, H., Ohbuchi, R., Bajura, M., Fuchs, H.: Case Study: Observing a Volume-Rendered Fetus within a Pregnant Patient. In: Proceedings of IEEE Visualization 1994, pp. 364–368. IEEE Computer Society Press, Los Alamitos (1994)

4. State, A., Livingston, M.A., Hirota, G., Garrett, W.F., Whitton, M.C., Fuchs, H., Pisano, E.D.: Technologies for Augmented-Reality Systems: realizing Ultrasound-Guided Needle Biopsies. In: ACM SIGGRAPH. Proceedings of SIGGRAPH 1996 (New Orleans, LA, August 4-9, 1996). In Computer Graphics Proceedings, Annual Conference Series, pp. 439–446 (1996)
5. State, A., Keller, K., Fuchs, H.: Simulation-Based Design and Rapid Prototyping for a Pallax-Free, Orthoscopic Video See-Through Head-Mounted Display. In: Proc. ISMAR 2005, pp. 28–31 (2005)
6. Fuchs, H., Livingston, M.A., Raskar, R., Colucci, D., Keller, K., State, A., Crawford, J.R., Rademacher, P., Drake, S.H., Anthony, A., Meyer, M.D.: Augmented Reality Visualization for Laparoscopic Surgery. In: Wells, W.M., Colchester, A.C.F., Delp, S.L. (eds.) MICCAI 1998. LNCS, vol. 1496, pp. 11–13. Springer, Heidelberg (1998)
7. Stockmans, F.: Preoperative 3D virtual planning and surgery for the treatment of severe madelung's deformity. In: Proc of Societé Française Chirurgie de la main, XLI Congres, vol. CP3006, pp. 314–317 (2005)
8. State, A., Keller, K., Rosenthal, M., Yang, H., Ackerman, J., Fuchs, H.: Stereo Imagery from the UNC Augmented Reality System for Breast Biopsy Guidance. In: Proc. MMVR 2003 (2003)
9. Southern, S.J., Shamsian, N., Wilkinson, S.: Real hand© - A 3-dimensional interactive web-based hand model - what is the role in patient education? In: XLIe Congrès national de la Société Française de Chirurgie de la Main, Paris (France), (15-17December, 2005)
10. Shuhaiber, J.H.: Augmented Reality in Surgery. ARCH SURG 139, 170–174 (2004)
11. Grimson, W.E.L., Lozano-Perez, T., Wells III, W.M., Ettinger, G.J., White, S.J., Kikinis, R.: An Automatic Registration Method for Frameless Stereotaxy, Image Guided Surgery, and Enhanced Reality Visualization. In: Computer Vision and Pattern Recognition Conference, Seattle (June 1994)
12. Morgan, F.: Developing a New Medical Augmented Reality System. Tech. report CMU-RI-TR-96-19, Robotics Institute, Carnegie Mellon University, (May 1996)
13. http://www.medarpa.de/
14. Schwald, B., Seibert, H., Weller, T.: A Flexible Tracking Concept Applied to Medical Scenarios Using an AR Window. In: ISMAR 2002 (2002)
15. Schnaider, M., Schwald, B.: Augmented Reality in Medicine – A view to the patient's inside. Computer Graphic topics, Issue 1, INI-GraphicsNet Foundation, Darmstadt (2004)
16. International Workshop on Medical Imaging and Augmented Reality (MIAR), http://www.miar.info
17. Fischer, J., Bartz, D., Straßer, W.: Occlusion Handling for Medical Augmented Reality. In: VRST. Proceedings of the ACM symposium on Virtual reality software and technology, Hong Kong (2004)
18. Sielhorst, T., Obst, T., Burgkart, R., Riener, R., Navab, N.: An Augmented Reality Delivery Simulator for Medical Training. In: International Workshop on Augmented Environments for Medical Imaging - MICCAI Satellite Workshop (2004)
19. Birkfellner, W.M., Figl, K., Huber, F., Watzinger, F., Wanschitz, R., Hanel, A., Wagner, D., Rafolt, R., Ewers, H., Bergmann, H.: The Varioscope AR – A head-mounted operating microscope for Augmented Reality. In: Proc. of the 3rd International Conference on Medical Image Computing and Computer-Assisted Intervention, pp. 869–877 (2000)
20. Reitinger, B., Werlberger, P., Bornik, A., Beichel, R., Schmalstieg, D.: Spatial Measurements for Mecial Augmented Reality. In: ISMAR 2005. Proc. of the 4th IEE and ACM International Symposium on Mixed and Augmented Reality, pp. 208–209 (October 2005)

21. Riva, G., Morganti, F., Villamira, M.: Immersive Virtual Telepresence: Virtual Reality meets eHealth. In: Cybertherapy-Internet and Virtual Reality as Assessment and Rehabilitation Tools for Clinical Psychology and Neuroscience, IOS Press, Amsterdam (2006)
22. Nischelwitzer, A., Lenz, F.J., Searle, G., Holzinger, A.: Some Aspects of the Development of Low-Cost Augmented Reality Learning Environments as Examples for Future Interfaces in Technology Enhanced Learning. In: Universal Access to Applications and Services. LNCS, vol. 4556, pp. 728–737. Springer, New York
23. Juan, M.C., Alcaniz, M., Monserrat, C., Botella, C., Banos, R.M., Guerrero, B.: Using Augmented Reality to Treat Phobias. IEEE Comput. Graph. Appl. 25, 31–37 (2005)
24. http://www.neurovr.org
25. Gaggioli, A., Morganti, F., Meneghini, A., Alcaniz, M., Lozano, J.A., Montesa, J., Martínez, J.M., Walker, R., Lorusso, I., Riva, G.: Abstracts from CyberTherapy 2005 The Virtual Mirror: Mental Practice with Augmented Reality for Post-Stroke Rehabilitation. CyberPsychology & Behavior 8(4) (2005)
26. Bimber, O., Raskar, R.: Spatial Augmented Reality: Merging Real and Virtual Worlds. A. K. Peters, Ltd (2005)
27. Maurer, H., Oliver, R.: The future of PCs and implications on society. Journal of Universal Computer Science 9(4), 300–308 (2003)
28. http://www.lumus-optical.com
29. http://www.trivisio.com
30. del Valle, A., Opalach, A.: The Persuasive Mirror: Computerized Persuasion for Healthy Living. In: Proceedings of Human Computer Interaction International, HCI International, Las Vegas, US (July 2005)
31. Nestler, S., Dollinger, A., Echtler, F., Huber, M., Klinker, G.: Design and Development of Virtual Patients. In: 4th Workshop VR/AR, Weimar (15 July, 2007)
32. Elkin, P.L., Sorensen, B., De Palo, D., Poland, G., Bailey, K.R., Wood, D.L., LaRusso, N.F.: Optimization of a research web environment for academig internal medicine faculty. Journal of the American Medical Informatics Association 9(5), 472–478 (2002)
33. Hesse, B.W., Shneiderman, B.: eHealth research from the user's perspective. American Journal of Preventive Medicine 32(5), 97–103 (2007)
34. Rhodes, M.L.: Computer Graphics and Medicine: A Complex Partnership IEEE. Computer Graphics and Applications 17(1), 22–28 (1997)
35. Holzinger, A., Geierhofer, R., Ackerl, S., Searle, G.: CARDIAC@VIEW: The User Centered Development of a new Medical Image Viewer. In: Zara, J.S.J. (ed.) Central European Multimedia and Virtual Reality Conference (available in Eurographics Library), pp. 63–68, Czech Technical University (2005)
36. Holzinger, A.: Usability Engineering for Software Developers. Communications of the ACM 48(1), 71–74 (2005)
37. Gould, J.D., Lewis, C.: Designing for usability: key principles and what designers think. Communications of the ACM 28(3), 300–331 (1985)
38. Seffah, A., Metzker, E.: The obstacles and myths of usability and software engineering. Communications of the ACM 47(12), 71–76 (2004)
39. Sutcliffe, A., Gault, B., Shin, J.E.: Presence, memory and interaction in virtual environments. International Journal of Human-Computer Studies 62(3), 307–332 (2005)
40. Wilson, J.R., D'Cruz, M.: Virtual and interactive environments for work of the future. International Journal of Human-Computer Studies 64(3), 158–169 (2006)

41. Dünser, A., Grassert, R., Seichter, H., Billinghurst, M.: Applying HCI principles to AR systems design. In: MRUI 2007. 2nd International Workshop at the IEEE Virtual Reality 2007 Conference, Charlotte, North Carolina, USA (March 11, 2007)
42. Billinghurst, M.: Designing for the masses. INTERFACE HITLabNZ, Issue 14 (July2007)
43. Gabbard, D.H., Swan, J.E.: User-centered design and evaluation of virtual environments. IEEE Computer Graphics and Applications 19(6), 51–59 (1999)
44. Freeman, S.E.A., Pearson, D.E., Ijsselsteijn, W.A.: Effects of sensory information and prior experience on direct subjective ratings of presence. Presence Teleoperators and Virtual Environments 8, 1–13 (1999)
45. Kalawsky,: VRUSE A computerised diagnostic tool for usability evaluation of virtual/synthetic environment systems. Applied Ergonomics 30(1), 11–25 (1999)
46. Bowman, D., Hodges, L.: An Evaluation of Techniques for Grabbing and Manipulating Remote Objects in Immersive Virtual Environment. In: Proceedings of the ACM Symposium on Interactive 3D Graphics, ACM Press, New York (1997)
47. Slater, M., Linakis, V., Usoh, M., Kooper, R.: Immersion, Presence and Performance in Virtual Environments: An Experiment Using Tri-Dimensional Chess. Available online at www.cs.ucl.ac.uk/ staff/m.slater/Papers/Chess/index.html (1996)
48. Lombard, M., Ditton, T.: At the heart of it all: the concept of presence. Journal of Computer Mediated Communication 3(2), http://www.ascusc.org/jcmc/vol3/issue2/Lombard.html
49. Blach, R., Simon, A., Riedel.: Experiences with user interactions in a CAVE-like projection environment. In: Proceedings Seventh International Conference on Human-Computer Interaction (1997)

Designing Pervasive Brain-Computer Interfaces

Nithya Sambasivan and Melody Moore Jackson

College of Computing, Georgia Institute of Technology,
Atlanta, GA 30332, United States
{nithya, melody}@cc.gatech.edu

Abstract. The following paper reports on a prototype Brain-computer Interface designed for pervasive control by paralyzed users. Our study indicates that current control and communication devices for users with severe physical disabilities restrict control and independence, offer little articulation and communication capabilities. Integrating multiple devices and services, our application is based on the functional Near-Infrared Imaging technology. Based on the overarching Value-sensitive design framework, our solution is informed by the usage patterns of technology, living habits and daily activities of the disabled users. By designing the context-aware pervasive control solution, we create a venue for communication, environmental control, recreation, assistance and expression among physically disabled patients. The evaluations results of the prototype are also discussed.

Keywords: User-centered Design, Value-sensitive Design, Brain-computer Interfaces, Pervasive Computing.

1 Introduction

Over 2 million people throughout the world are affected by neural diseases such as Amyotrophic Lateral Sclerosis, stroke, multiple sclerosis, brain or spinal cord injury, cerebral palsy, and other diseases impairing the neural pathways that control muscles or impair the muscles themselves. These diseases do not harm the person cognitively, but cause severe paralysis and loss of speech. Brain-Computer Interfaces (BCI) sense and process neural signals through external electrodes around the motor cortex area in the brain. They help in operating control and communication systems, independent of muscles or nerves. Over the past 15 years, much progress has been made in creating assistive technologies. BCIs have been studied in laboratories for nearly two decades but are just now beginning to be reliable and effective enough to use in real-world scenarios. Previous research has investigated the use of BCIs as communication devices and in arts and gaming [3] [4] [6]. Unfortunately, little research has been carried out in understanding the users of BCIs and the conditions and settings of deployment to create a universal control device. The design is further compounded by the increased social isolation, restricted mobility and the ability to carry out fewer Activities of Daily Living. There is also significant impact on the four basic pleasures of life - physiological, psychological, social and ideological pleasures [2]. Smart devices and inter-connected software services have started to proliferate around us.

A. Holzinger (Ed.): USAB 2007, LNCS 4799, pp. 267–272, 2007.

Mark Weiser's original vision of invisible, everywhere computing embedded in the environment has become a reality [5]. Integration of these devices and services with BCIs opens out an entirely new world to the paralyzed users. Context-aware interactions such as controlling temperature automatically, notifications of birthdays and the ability to send gifts and recreation options when the user is bored, offer a plethora of possibilities unrestricted by the disability of the user. The goal of this research is two-fold: first, how do we design better Brain-computer systems taking into account the values, habits and everyday activities of the user? Second, we explore the value-sensitive design framework in an unexplored yet relevant context of BCI design. In order to answer these questions, we conducted a sociological study on the target user group and their families. Based on the findings, we prototyped a user-centered context-aware pervasive application offering control of a wide variety of services.

2 Methodology

We recruited twelve participants with varying levels of motor skill impairment, ranging in age from twenty-one to sixty, divided evenly by gender. The data were triangulated with family members as well as disabled patients. Interviews and questionnaires were the primary methods of data collection. Telephone and in-person interviews were conducted when possible, but Instant Messaging and E-mail were used when the informant was unable to speak. The underlying methodology behind this study is Value-sensitive Design, which is a tripartite methodology involving conceptual, empirical and technical investigations. The central construct of this methodology is that technology either supports or undermines values. These values are centered on human well-being, dignity, justice and human rights. It not only has a moral standpoint, but also combines usability, understanding biases and predilections of the user and the designer [1].

Empirical Investigation. In this stage, we tried to understand the social context of technology and interactions between the system and people. Semi-structured interviews were conducted to gain insight into their lifestyle, everyday activities and recreational activities, with grand-tour questions such as "describe a typical day". Specifically, we uncovered what values users want embedded in computing applications. Domestic technology's value-laden meaning is important because of the home's multiple connections with a range of discourses such as work, leisure, religion, age, ethnicity, identity, and sex. We analyzed the qualitative data using grounded theory, which revealed salient challenges that the locked-in community faces when using current technology and the context of these challenges.

Conceptual Investigation. During the second stage of the project, we conducted an analysis of the constructs and issues. We uncovered the values that have standing from the data we gathered during the empirical phase. Historical awareness of technical and humanistic research helped us consciously choose which themes bear repeating and which to resist.

Technical Investigation. In this stage we tried to understand the state of the current technology and the usage patterns among users. We analyzed how the current technology fosters or inhibits the values from the previous stage. We used questionnaires structured on Activities of Daily Living to figure out current technology and the control of TV, Radio, emergency procedures, personal cleansing, and environmental control.

3 Key Findings

Our investigations throw light on the limitations of the current technology. From our analysis, we consider the following to be key in influencing the quality of life:

Restrictive control: Since most BCIs have been studied in laboratory settings, there are very few practical applications for the user. Although BCIs have advanced immensely in the technology, they are still limited in offering control. *Reduced independence:* In turn, this increases dependence on the care-giver or family member, which is further complicated due to limited/lack of mobility. *Lack of articulation:* There are very few venues designed especially for those with motor difficulties, leading to frustration due to lack of an outlet. *Reduced recreation:* Current technology does not offer much recreation designed especially for paralyzed users, such as controlling TV channels, radio stations and playing instruments. *Lack of customization:* Since not much work has gone into the design of the applications, there are very few customization options, such as changing the voice or accent of the text-to-speech converter. In essence, we find the following values to be crucial for design of BCIs:

Family-centricity: From our data, we find that family is prime in the users' lives. Since the activities are usually confined to inside the home and family members help in day-to-day activities, there is a strong attachment with the family members.
Community: Our users were extremely involved in sustaining communities, i.e. groups of people bound by a common interest. They were constantly in touch with each other, had kind words to offer and would help each other out.
Altruism: There is a heightened sense of altruism in spreading the word about the neural diseases and minimizing the fear in newly-diagnosed patients.
Intellectual Involvement: There is a dire need for intellectual activities in the current system, such as reading or painting exclusively for this user group.
Autonomy: The most affected value is the autonomy of the user.

4 Prototype

The implications from our user research led to the prototype pervasive control interface. We started this phase with different methods of navigation, visualization and selection. After prioritizing our requirements, we came up with a solution that combines ease of interaction, simplicity and cohesion of various services/functions.

The prototype design caters to a wide range of users in terms of age, familiarity of technology and operates on neural signals. It capitalizes on the screen space for displaying large icons on a familiar television-like screen, offering a more personable and less intimidating experience. Our design was based on the following criteria: *learnability* (the system should be easy to learn), *flexibility* (how much tolerance for errors), *ease of use*, *responsiveness* (the system should offer feedback), *observability* (the system should allow the user to recognize current progress), *recoverability* (when the user is in an undesired state, the system should allow them to revert to a previous stage) and *consistency* (consistent appearance and functions). The features were categorized into four key themes: communication, recreation, articulation and environmental control, to fill in the gap uncovered in the investigations. The visualization is based on the radio dial metaphor. Since most people are used to this method of increasing and decreasing quantities, we felt that the slow movement of the dial can alert the user to the next option, as opposed to a linear visualization. The *Selection Window* auto-navigates through the icons. In addition, a dashboard is also displayed, so that the user can have access to the most-frequently accessed functions, such as alarms, time and notes.

Technology: The prototype is based on the functional Near-Infrared Imaging technology, which measures the oxygen level in the blood (Hemo-dynamic response function) which in turn depends on neuronal activity of the brain [2]. Electrodes are placed externally over the Broca's area of the brain, as shown in Figure 1 Pre-determined and pre-trained sets of thoughts are used to control the movement of the cursor on the screen. By performing a mental task, such as sub-vocal (silent) counting to activate the language area, the blood oxygenation is increased. This in turn causes an increase in the waveform. The height of the wave is thresholded. When the slope changes from positive to negative or vice versa, a state change is detected.

 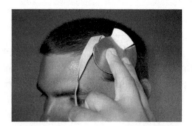

Fig. 1. Functional Near-Infrared Imaging technology being used by a subject in the left figure. An electrode held against forehead in the right figure.

States of the system. The home page provides a high-level access to the four main categories. The selection window cycles through various icons in a clock-wise direction. By performing a mental task of counting numbers, the appropriate icon is selected. The movement of the selection window is slow, so that the user can reach the threshold in order to make a selection. The movement and direction of the window are depicted in Figure 2. Context-awareness is a key feature of the system, allowing automatic and intelligent functions depending on the state of the environment, user and system. The system is embedded with a range of sensors - thermal, proximity, light sensors and a microphone. When a person approaches the user, it automatically brings up the speech application. When ambient light decreases, it pops up an option to turn on the lights. The system notifies the user of birthdays and special occasions, with options to gift. Context-aware interactions such as controlling temperature automatically, notifications of birthdays and the ability to send gifts and recreation options when the user is bored, offer a plethora of possibilities unrestricted by the disability of the user. On selection of an icon, the associated page is opened. Icons are arranged in order of priority. The environment page displays options to call emergency, call for care-giver or family assistance, adjust temperature, and create alerts. The entertainment page displays options to watch Television (using a TV Tuner or online television), listen to radio, listen to online music (create playlists and stay updated with latest music), play computer games or watch sports, read news and shop online. The Communication page displays options to send e-mail, instant messages, SMS and page a care-giver, friend or family member. The Expression page allows the user to write blogs/notes, create notes, send gifts to someone, read e-books and articles.

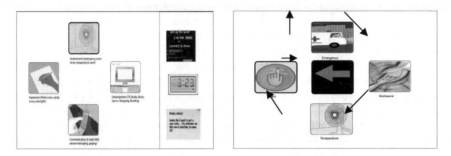

Fig. 2. The left figure displays the home page (the yellow selection window is currently at the first icon), the right figure displays the environmental control page

A drawback with using BCIs is that they do not offer direct manipulation. Therefore, there is a tendency for the user to commit mistakes and incorrectly select options. To correct this, the selection window cycles through various icons and then over the Back button. Thus, the user can revert to a previous stage easily without disrupting the flow.

5 Evaluations

In order to assess the effectiveness of our interface, we utilized the Heuristic Evaluation [7] with the twelve subjects. The following heuristics were considered to be prime metrics of the effectiveness of our system:

Visibility of system status – The system should always keep the user informed of the status, through appropriate feedback within reasonable time.
Consistency and standards – Not having to wonder what different words, situations, or actions mean. Does the system follow conventions?
User control and freedom – Does the system offer control when mistakes are made and options are incorrectly selected?
Flexibility and efficiency of use – Is the system flexible and efficient?
Recognition rather than recall – Does the user have to remember information each time?

We asked our twelve users to evaluate the heuristics on a scale of zero to four, zero for no problem and four for a severe problem. The Heuristic Evaluation provided excellent feedback. Because each of the heuristics we selected were associated with our usability criteria and other non-functional requirements, the usability bugs reported by our heuristic evaluators were, for the most part, very indicative of our design failures (and successes) with respect to our evaluation criteria. It also gave us insight into how well our system design met our usability requirements (some more than others, naturally): *learnability, flexibility, ease of use, responsiveness, observability, recoverability* and *consistency*.

Our results have been positive overall, with strong appreciation for the novel visualization and intuitiveness of the application. The concept of pervasive control of devices and services was very well-received. One of our participants commented, "*I suppose devices making both verbal and written communication easier and much faster are paramount. To be able to speak or write fast enough to participate in a normal conversation would be shear joy. My wildly entertaining wit and humor could once again shine through! Thanks!*" Another participant commented, "*"Excellent central device for controlling all things around me. I feel so much in control. I am sure this will benefit the other PALS"*. The use of icons and the dashboard idea were also appreciated. However, we received constructive feedback on providing more information on how to use, which we will incorporate into our future versions.

One of the issues with using functional Near-Infrared Imaging technology is determining whether the user wishes to control the application. Termed as the "Midas touch" problem, users cannot easily turn off their thoughts and can activate the language area by ambient conversation. One solution to this problem is to train the application with the user's neural signals and threshold it higher than external stimuli or deterrents. Research efforts on techniques for obtaining clean and artifact-free signals are being undertaken by our lab, by means of noise reducing and more sophisticated filters.

Our prototype Brain-computer Interface application for pervasive control of various services and devices was well-received. The use of Value-sensitive Design in the intersection of humanistic and technical disciplines provides a fresh perspective on the design of Brain-computer Interfaces. Although our application is informed by and designed for the domestic context, it could be easily extended to the hospital scenario.

References

1. Friedman, B.: Value Sensitive Design. In: Encyclopedia of human-computer interaction, pp. 769–774. Berkshire Publishing Group, Great Barrington (2004)
2. Jackson, M.M.: Pervasive Direct Brain Interfaces. IEEE Pervasive Computing 5,4 (2006)
3. Nijholt, A., Tan, D.: Playing with your brain: brain-computer interfaces and games. In: Conference on Advances in Computer Entertainment Technology (2007)
4. Orth, M.: Interface to architecture: integrating technology into the environment in the Brain Opera, Conference on Designing interactive Systems: Processes, Practices, Methods, and Techniques (1997)
5. Weiser, M.: The Computer for the 21st Century - Scientific American Special Issue on Communications, Computers, and Networks (September 1991)
6. Wolpaw, R.J., Birbaumer, N., Mcfarland, D.J., Pfurtscheller, G., Vaughan, T.M.: Brain–computer interfaces for communication and control. Clin. Neurophysiol 113, 767–791 (2002)
7. http://www.useit.com/papers/heuristic/heuristic_evaluation.html

The Impact of Structuring the Interface as a Decision Tree in a Treatment Decision Support Tool

Neil Carrigan[1,*], Peter H. Gardner[2], Mark Conner[2], and John Maule[3]

[1] Department of Computer Science, University of Bath, Bath, BA2 7AY, UK
n.carrigan@bath.ac.uk
[2] Institute of Psychological Sciences, University of Leeds, Leeds, LS2 9JT, UK
p.h.gardner@leeds.ac.uk,
m.t.conner@leeds.ac.uk
[3] Leeds University Business School, University of Leeds, LS2 9JT, UK
jm@lubs.leeds.ac.uk

Abstract. This study examined whether interfaces in computer-based decision aids can be designed to reduce the mental effort required by people to make difficult decisions about their healthcare and allow them to make decisions that correspond with their personal values. Participants (N=180) considered a treatment scenario for a heart condition and were asked to advise a hypothetical patient whether to have an operation or not. Attributes for decision alternatives were presented via computer in one of two formats; alternative-tree or attribute-tree. Participants engaged in significantly more compensatory decision strategies (i.e., comparing attributes of each option) using an interface congruent with their natural tendency to process such information (i.e., alternative-tree condition). There was also greater correlation (p<.05) between participants' decision and personal values in the alternative-tree. Patients who are ill and making decisions about treatment often find such choices stressful. Being able to reduce some of the mental burden in such circumstances adds to the importance of interface designers taking account of the benefits derived from structuring information for the patient.

Keywords: Patient decision making, Interface design, Decision-making process, computer based decision support.

1 Introduction

In recent years it has become well established that patients be more involved in decisions about their healthcare. According to Ubel and Loewenstein [1], this is because of a moral concern that informed patients are best placed to decide which of a number of competing healthcare options most closely correspond to their personal values regarding treatment. Studies have shown that involving patients in treatment decisions improves psychological well-being [2], and sometimes physical symptoms [3]. One way in which health professionals have attempted to increase patient involvement is to support their decision-making with a decision aid.

---

* Corresponding author.

A. Holzinger (Ed.): USAB 2007, LNCS 4799, pp. 273–288, 2007.
© Springer-Verlag Berlin Heidelberg 2007

A number of such interventions encourage patients and their families to consider information regarding potential options and then evaluate these options according to their personal values. The present research demonstrates that the way in which people interact with and process health information is affected by the structure of the information presented on the computer interface. By presenting information about treatment options in a manner that reduces the mental effort required to make difficult decisions, people are more likely to make decision in line with their personal values. The finding is especially important given that in many instances people making decisions about their healthcare treatment are stressed and anxious and less likely to process information in a systematic or rational manner [4,5]. Designing technology to reduce the mental effort to make decisions in such settings allows them to make more reasoned choices than might otherwise be the case.

The main way in which previous decision support/aids have been evaluated is through outcome measures such as patient anxiety about treatment before and after the intervention [6-9]; knowledge about treatment options [7, 10, 11]; and patient satisfaction [12, 9]. However, it has been argued that only by examining the decision making processes adopted by a patient while using a decision aid, will the researcher understand why the decision aid led to beneficial outcomes for the decision maker [13-15]. Similarly, the present study focuses not just on outcomes but also decision-making process in order to understand how a decision aid interface can be designed to enhance patient decisions about treatment.

Previous research has demonstrated the benefits of supporting patients' decisions using computer-based applications. Llewellyn-Thomas et al. [16] suggest that the reasons why the interactive format of computer-based decision aids may enhance decision making is that the vivid nature of the presentation on a computer screen attracts participants' attention to a greater extent than the same information in a written format. They also suggest that menus and sub-menus in the computer program allow patients to form an overall outline of the problem they are facing and that because they can control the pace and sequencing of information acquisition they can use their own strategies for processing complex information to develop a uniquely organised schema or cognitive map of the decision problem. Similarly, Street et al. [17] suggest that an interactive computer program involves patients more directly so that they actively explore the information in accordance with their preferences for information and their learning style. While these studies demonstrated the benefits to patients of providing treatment decision information via a computer, they were only able to infer the reasons for success as they measured decision outcomes rather than the process of decision-making.

Payne, Bettman and Johnson [18], argue that for a decision aid to lead to more 'accurate' decisions it should reduce the cognitive effort required to engage in cognitively effortful processing [18, 19]. This is achieved by structuring the information presented in the interface so that it allows people to more easily engage in systematic information processing that ordinarily would require a great deal of mental effort. In doing so, people are less susceptible to cognitive biases or employing simplifying strategies (heuristics). However, the 'accuracy' of a decision is rarely the most important consideration when a patient is comparing two equally effective treatments. Amount of pain or quality of life are likely to be important factors to a patient when making their choice.

Given the inadequacies of normative or 'rational choice' models of decision-making (e.g. Subjective Expected Utility (SEU) theory) in describing how people *actually* make decisions [20-24], a *'reasoned-choice'* model can be seen to encompass a more 'naturalistic' approach (e.g. [25]). This model proposes decision quality should be based on the process by which it was arrived at, not just the choice itself (i.e., the outcome) [26-28]. According to this approach the effectiveness of a decision aid is dependent upon whether or not i) the decision is based on relevant information about the available decision options (i.e., an informed decision), ii) the likelihood of the various options occurring and their attractiveness to the patient are evaluated accurately (i.e., is congruent with their values), and iii) there is a 'trade-off' between these factors (i.e., involves compensatory decision processes) [27, 29, 30]. The research presented here utilizes this framework to evaluate the effectiveness of the decision support provided to participants.

Previous research has shown that the format in which information is presented influences the type of decision-making strategy employed [31]. In the present study a hypothetical scenario about a patient facing heart surgery was used to present information about the effects of surgery on various aspect of the patient's lifestyle. This information was presented on a computer in the form of a decision tree that was manipulated across two conditions. Structuring the interface in the form of a decision tree was seen as a way of reducing the cognitive effort required to engage in cognitively effortful decision strategies [31]. This should manifest itself in an increase in the amount of compensatory decision processing as measured by process tracing methodology. However, by manipulating the format of this tree structure it is possible to test hypotheses about particular aspects of the interface format that influences information processing rather than simply the provision of relevant information about the decision.

In the first condition a tree structure was designed that places the attributes of each alternative in close proximity to each other (see Figure 1) in order to make it easier for participants to compare attributes within an alternative, before considering the next alternative in the same manner (i.e., alternative-based processing). By reducing the cognitive effort required to engage in this alternative-based processing, it was hypothesised that participants would be more likely to adopt this type of processing in line with previous work by Jarvenpaa [39]. In a second condition a tree structure was developed to elicit the opposite type of processing (i.e., attribute-based processing) by placing corresponding attributes from both alternatives next to each other, making it easier for the decision maker to compare within attributes across conditions (see Figure 2). Both attribute- and alternative-based processing are compatible with compensatory decision strategies [32], hence the format itself does not predict which format will result in more compensatory decision-making. However, when values of the attributes are presented as numbers, attribute processing is likely to dominate, whereas when values are presented verbally, alternative-based processing is more likely [31]. In this study, information was presented verbally in both conditions; hence participants presented with the attribute-tree should have a conflict between the tendency to process word values in an alternative-based manner and a tree structure congruent with attribute processing. Participants in the alternative-tree condition used an interface congruent with their tendency to process verbal information in an alternative-based way and hence should require less cognitive effort to adopt this strategy. As a result, participants in the alternative-based tree condition should be more likely to engage in compensatory-based search strategies than those in the attribute-tree condition.

Hence, the present study evaluated not only the decision outcomes but also the decision processes that participants engaged in to arrive at these outcomes. In terms of outcome measures, this study applies O'Connor et al.'s [25] criteria that a good decision is defined as congruence between values and choice, and uses the approach adopted by Holmes-Rovner et al. [15] (that is, the correspondence between participants' decision and the choice derived from SEU) to measure this. The level of satisfaction with the decision is a further outcome measure used within O'Connor et al.'s framework to measure decision quality.

In an attempt to understand how the level of personal involvement with a decision influences outcomes of the decision making process, previous studies have manipulated the objective level of involvement with the decision by altering the decision context faced by a decision maker. A study by Billings and Scherer [33], showed that participants in the high involvement condition searched more of the available information, whereas low involvement participants employed more variable search strategies, suggesting the use of non-compensatory decision strategies. A similar finding was obtained by Maheswaran and Myers-Levy [34], who found that participants in a high involvement condition generated more message-related thoughts (i.e., thoughts with explicit reference to information presented in the message) and recalled more statements from a message about levels of cholesterol.

In the present study, it was hypothesised that participants who were not 'involved' with the material would be less likely to engage in more cognitively effortful decision strategies. However, given that the alternative-tree should make these difficult strategies easier to engage in, it was hypothesised that participants who were less involved should none-the-less be more likely to engage in compensatory processing in this condition than their counterparts in the attribute-tree condition.

In summary, the study hypotheses were that participants would engage in more compensatory decision strategies in the alternative-tree condition resulting in a more 'reasoned' decision (as measured by increased correlation between the decision and expectancy-value ratings (EV) and the use of more compensatory decision processes). Furthermore, the type of interface would moderate the effect of personal involvement. No specific hypotheses were formulated about level of satisfaction with the decision, but it was hoped that both conditions would be similar with regards to this measure.

2 Method

2.1 Participants

One hundred and eighty participants (70 male/110 female) took part. Mean age was 20.70 (SD=2.44) with an age range of 18-36 yrs old. Participants were recruited by sending e-mails to all the academic departments at the Universities of Manchester and Sheffield in the UK. Most participants were from the UK (165), with 4 from Northern Europe, 5 from Southern Europe, 2 from North America and 4 from Asia.

Consent was assumed once they clicked on the link to continue with the study at the bottom of the introductory page.

2.2 Design

The study consisted of two independent groups with the method of presentation (alternative-tree or attribute-tree) as a between-subjects factor. The information in the alternative-tree consisted of ten aspects of lifestyle that might be affected by participants' decision to have an operation or not (see Figure 1). These lifestyle considerations were derived from interviews with participants who had undergone major surgery as well as information drawn from the Internet on the impact of heart surgery on lifestyle. No numerical probabilities were presented to participants (e.g., when participants clicked on the 'Diet' box of the tree they received the information, "If you have the operation it is likely you can continue eating normal family meals").

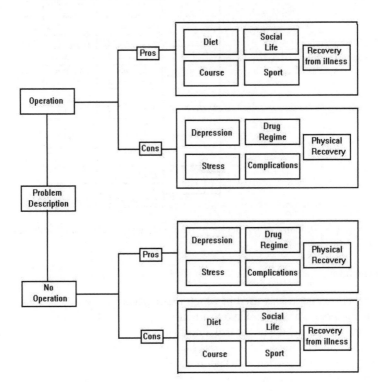

Fig. 1. The outline structure of the decision tree presented to participants in the alternative-tree condition. Participants could click on any of the boxes to receive the information on that attribute.

Exactly the same information was presented in the attribute-tree condition. However, the attributes from each alternative were presented alongside one another to encourage attribute-based processing of the information (see Figure 2).

To control for the potential influence that their decisions might have on subsequent EV ratings, two further conditions were added where the order of making a decision and then EV ratings was reversed. Hence the study initially had a 2x2 design with tree and order of ratings as between-subjects factors.

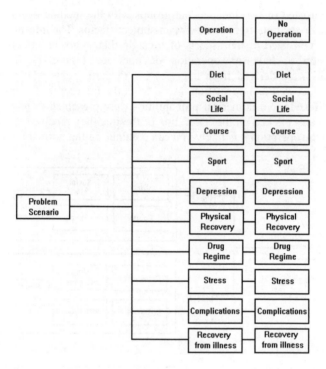

Fig. 2. The outline structure of the decision tree presented to participants in the attribute-tree condition

2.3 Materials

A hypothetical scenario was used in which participants were asked to imagine advising a patient about whether to have heart surgery or not. The material informed participants until recently the patient had a normal active lifestyle. However, a small deficiency in a valve of the patient's heart had been discovered that might be rectified by surgery. Ultimately participants had to decide whether to opt for the surgery or not, but prior to that decision they were asked to consider how each option may change certain aspects of the patient's life, e.g., diet, social life, leisure activities. Each box in the tree in Figure 1 represents one of the aspects that participants could click on to receive more information.

In the "Operation" arm of the alternative-tree there were five aspects of lifestyle that were positively valenced in favour of an operation (Pros), and five aspects of lifestyle that were valenced against an operation (Cons). Each of the ten aspects was replicated in the "No operation" arm of the tree but their valence reversed. Hence neither option had dominance.

The study was accessed by participants through a web browser on any computer linked to the Internet. In order to track participants movements through the website a Common Gateway Interface (CGI) computer program was used written in Perl. The remaining data regarding the decision, their ratings of expectancy-value, involvement and personal details (age, occupation, country of origin) were submitted at the end of the study to the participants unique data file held on the host server.

2.4 Measures

Decision Measures. The decision about whether to have the operation or not was rated on a 9-point rating scale from 1 "I will definitely not have the operation" to 9 "I will definitely have the operation". The "satisfaction with decision" scale was a three item 7-point Likert-type scale (0= 'Not at all' to 6= 'Extremely') taken from Michie et al. [7]. Involvement with the decision was measured using a eight item 7-point scale asking participants to rate from 1= 'agree' to 7= 'disagree' items such as, "I found the material presented in this study engaging". Four of the eight items were reversed.

Expectancy-Value Measures. Participants were asked to rate on a 7-point scale each of the aspects of their lifestyle that were presented in the decision tree. Ten items considered how likely they thought having the operation would change a particular aspect of lifestyle (e.g., "Having the operation will mean I become depressed", 1=Unlikely to 7=Likely) and a further ten items considered the "No-Operation" option (e.g., "Not having the operation will mean I become depressed"). Finally ten items asked for evaluation of each outcome (i.e., "Being depressed will be... ", 1=Bad to 7=Good).

Process Tracing Measures. The total number of 'pages' visited was measured, as well as the depth of search and the reacquisition rate. Depth of search was calculated as the proportion of available information searched; reacquisition rate as the total number of pages inspected minus the total number of first inspections, divided by the total number of pages inspected. A high number of unique page visits along with a high depth of search and reacquisition rate would indicate the use of more compensatory decision strategies. The Payne Index [18] measures the degree to which participants search information primarily by alternative or by attribute. The Index is calculated by subtracting the number of search movements to the same attribute across alternatives from the number of search movements within the same alternative then dividing by the sum of the two counts. The measure has a range of +1 to –1, with positive values indicating alternative-based processing and negative values indicating attribute-based processing.

Variability of search and variability of time spent searching, measures the degree to which the amount of information and time spent searching information per alternative is consistent or variable. It is calculated as the standard deviation of the proportion of attributes examined per alternative, based on first inspections. If a decision maker acquires the same information for all alternatives, the processing is termed consistent and is assumed to reflect a compensatory strategy. If however, the decision maker acquires a varying amount of information for each alternative, the processing is termed variable and reflects a non-compensatory strategy, as the decision maker appears to eliminate alternatives on the basis of only a partial amount of the information.

Westenberg and Koele's [36] Compensation Index (C) is calculated by using the variability of search measure (V) and the depth of search (D), as both are indicative of compensatory strategies: C=D x (1-2V). The index ranges from 0 to 1, with higher values indicating more compensatory strategies.

2.5 Procedure

Participants were randomly allocated to one of the four experimental conditions that opened in a new browser window. They could spend as long as they wanted looking at the information about the treatment until they were ready to click on the button that took them to the decision page. If participants were in the post-EV condition, then they completed their EV ratings prior to making a decision, in the pre-EV condition this procedure was reversed. Participants then completed the satisfaction with decision scale and the level of personal involvement scale. The demographics page asked participants to give the following details; age, sex, occupation, country of origin, and e-mail address.

2.6 Analysis

Each individual likelihood/value rating that related to having the operation was multiplicatively combined and then summed. The same procedure was carried out on likelihood/value ratings related to not having the operation. The mean scores of the summed products for "no operation" were then subtracted from the summed scores for "operation" to give an overall score. This overall score (EV) was a positive value for those who perceived greater utility for having the operation and a negative value for those who perceived greater utility for not having the operation.

To examine the relationship between the decision and EV ratings a Pearson's correlation was calculated between mean EV score and the mean decision rating in each condition. This assesses the degree of correspondence between the reported decision and the perceived relative value of the two decision alternatives. A high positive correlation indicates that the more in favour of a decision alternative a participant is, the greater their perceived value of that alternative. To take account of the potential differences in variance in the two conditions, a moderated regression analysis was used as a further comparison between the conditions [37]. This method tested the difference between unstandardised beta weights from a regression of EV on decision in both conditions using a t-test [38].

3 Results

The mean score for the "satisfaction with decision" scale was above the midpoint of 3, (M=3.91, SD=.70) indicating that participants were satisfied with their decisions. Internal reliability of the satisfaction scale was assessed by Cronbach's alpha, which gave a coefficient of .66. The mean involvement score was 2.76 (SD=.84), with a Cronbach's alpha of .80 suggesting participants were involved with the material.

Across all conditions the means and standard deviations were similar within each scale, suggesting that the scores on these scales were unaffected by whether the tree format was alternative or attribute, or whether participants made their decision prior to making EV ratings or post-EV rating. In all four conditions, the mean decision score was above the midpoint hence participants were generally in favour of having an operation. The same pattern emerged with EV scores, where the mean score in each condition was positive and thus in favour of an operation. The similarity of the mean scores in each of the scales across conditions was confirmed by a 2x2 MANOVA with condition and pre-/post-EV as between-subjects factors. As a result the pre and post-EV conditions were collapsed into the two tree conditions.

The main hypothesis of the study was that differences in interface design would lead to differences in the decision-making strategies participants engaged in. Table 1 presents the mean and standard deviations of the process tracing variables measured in each condition. Participants in the alternative-tree condition viewed more pages and spent more time looking at the information than participants in the attribute-tree condition. In the alternative-tree, participants had a greater depth of search, re-visited more pages, had a higher Compensation Index, and had a lower variability of search and variability of time spent searching alternatives. Taken together, the process tracing data strongly suggests that participants in the alternative-tree condition were using more compensatory search strategies than those in the attribute-tree. A MANOVA with condition as the between-subjects factor showed a significant main effect of condition ($F(8, 171)=15.9$, $p<.05$). The univariate analysis showed that in the alternative-tree condition, participants re-visited significantly more information than they did in the attribute-tree condition ($F(1,178)=5.90$, $p<.05$) and had a significantly different Payne Index score ($F(1,178)= 122$, $p<.05$) compared to the attribute-tree. There was also a significantly greater Compensation Index score in the alternative-tree condition ($F(1,178)=3.81$, $p<.05$) and lower variability of search ($F(1,178)=10.2$, $p<.05$) than in the attribute-tree. There were no significant differences in the amount of page visits, time spent looking at information or depth of search.

Further support for the claim that the attribute-tree format is incongruent with the natural tendency to process the verbal information in an alternative-based manner is provided by the distribution of Payne Index scores. The low negative score for the Payne Index in the attribute-tree condition indicates that while participants were mainly following an attribute-based pattern of interaction, they were also using alternative-based interaction strategies. However, as can be seen in Figure 3b, this contrasts sharply with participants in the alternative-tree condition who were almost exclusively using alternative-based interaction strategies as is shown in Figure 3a.

Another of the study's hypotheses was that because participants in the alternative-tree were more likely to be processing information in a compensatory manner they would show greater correlation between decision and expectancy-value ratings. The correlation between decision and EV in the alternative-tree condition was $r=.45$ ($N=86$, $p<.05$) and in the attribute-tree condition $r=.18$ ($N=94$, ns). The correlation in the alternative-tree condition was also significantly greater than in the attribute-tree ($z=2.05$, $p<.05$), and remained significant when the unstandardised beta-weights from a regression of EV on decision were compared ($t(176)=1.70$, $p<.05$, one-tailed).

Table 1. Mean (SD) scores for process tracing measures in *alternative* and *attribute-tree* conditions

Process Tracing Measure	Alternative Tree		Attribute Tree		
Total page visits	17.74	(6.09)	16.95	(6.80)	
Total time taken(Secs)	136.30	(83.04)	133.83	(86.77)	
Depth of search	74.14	(24.51)	72.24	(28.45)	
Re-acquisition rate	.03	(.05)	.02	(.04)	†
Payne Index	.81	(.46)	-.15	(.67)	†
Compensation Index	.69	(.30)	.59	(.40)	†
Variability of search	.05	(.12)	.13	(.18)	†
Variability of time spent searching alternatives	.11	(.14)	.13	(.14)	

†*p*-value <.05 Alternative Tree vs. Attribute Tree.

Table 2. Mean (SD) process tracing measures for high and low involvement by condition

	Alternative Tree				Attribute Tree			
	High inv.		Low inv.		High inv.		Low inv.	
Total page visits	19.48	(5.49)	17.15	(6.47)	18.56	(5.49)	14.55	(7.12)
Total time taken(Secs.)	161.35	(79.86)	129.23	(77.20)	140.31	(99.41)	112.38	(79.71)
Depth of search	83.02	(22.88)	71.15	(25.51)	78.37	(22.69)	61.96	(29.64)
Reacquisition rate	.02	(.04)	.02	(.05)	.03	(.04)	.01	(.03)
Payne index	.85	(.38)	.76	(.54)	-.09	(.68)	-.20	(.66)
Compensation index	.73	(.36)	.66	(.30)	.72	(.29)	.45	(.38)
Variability of search	.09	(.17)	.05	(.11)	.06	(.13)	.16	(.20)
Variability of time spent searching	.10	(.14)	.10	(.11)	.12	(.15)	.16	(.15)

The median split score used to categorise participants as either high or low in involvement was 2.56. From Table 2 it is clear that participants who were high in involvement in the attribute-tree condition demonstrated more compensatory strategies as they had the highest number of page visits, spent more time looking at information, showed a greater depth of search, had the highest Compensation Index and had the lowest variability in time spent searching alternatives. Participants who were high in involvement in the attribute-tree condition also demonstrated more compensatory decision strategies than those low in involvement in this condition

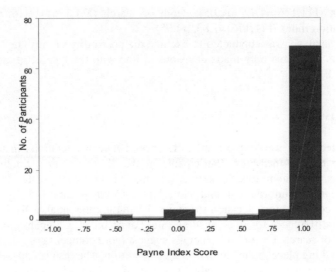

Fig. 3a Distribution of Payne Index scores for participants in the alternative-tree condition

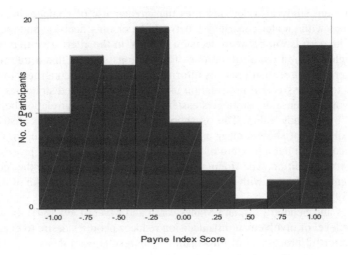

Fig. 3b Distribution of Payne Index scores for participants in the attribute-tree condition

To examine whether the process tracing data for the high involvement participants was significantly different from the other groups a 2x2 MANOVA was conducted with condition and level of involvement as between-subjects factors. It showed a significant main effect of condition ($F(8,169)=15.5$, $p<.01$) and a significant main effect for involvement ($F(8,169)=2.61$, $p<.05$) but no significant interaction ($F(8,169)=1.11$, $p>.05$).

Univariate analysis showed that there were significant differences between participants high and low in involvement for total page visits ($F(1,176)=9.35$, $p<.05$), total time taken (F(1,176)=4.72, p<.05), depth of search (F(1,176)=11.15, p<.05), and Compensation Index (F(1,176)=7.83, p<.05).

Further analysis was conducted to exclude the possibility of "myside" bias, where participants may have only made decisions in line with the type of information they had viewed.

4 Discussion

As predicted there were significant differences between conditions in terms of the processing tracing measures. Participants in the alternative-tree condition viewed more pages, spent more time looking at information, showed a greater depth and less variable search of information, and had a higher Compensation Index; all consistent with more compensatory decision strategies. From the inferential analysis, there were significant differences for re-acquisition rate, Payne Index, Compensation Index and variability of search. Overall the findings suggest more compensatory decision strategies were taking place in the alternative-tree condition. The results add weight to conclusions drawn from previous studies that the structuring of information consistent with cognitively demanding strategies, is more likely to lead to the adoption of such strategies [14, 18, 31, 19].

The present study also demonstrates that greater use of compensatory strategies corresponds with greater consistency between a person's decision and their personal values. The correlation between decision and EV in the alternative-tree was significantly higher than in the attribute-tree. The findings suggest that a tree format such as the alternative-tree, that presents information in a manner congruent to the natural tendency to process verbal information using alternative-based strategies, makes the adoption of compensatory strategies easier; hence a greater correlation between decision and expectancy value. This is supported by the Payne Index distributions in each condition that show a clear use of alternative-based processing in the alternative-tree condition but a mixture of alternative- and attribute-based processing in the attribute-tree condition. The finding supports previous work where the 'congruence' of a presentation format with a decision strategy led to greater uptake of that strategy [31, 32, 39].

Billings and Scherer [33], and Maheswaran and Myers-Levy [34], showed that altering the level of involvement in a decision reduces people's desire to engage in cognitively effortful processing of information. The results from this study show a main effect of involvement in the process tracing data suggesting that participants high in involvement processed information in a more systematic way, in line with Billings and Scherer [33]. However, this effect was not moderated by the alternative-tree condition.

A limitation of the present work is that participants were considering a hypothetical scenario. In facing such a decision in real life, anxiety is likely to affect the way people process information. A more anxious decision maker may process information in a less systematic way than those taking part in the alternative-tree condition here. However, the fact that the alternative-tree condition improved participants' ability to make decisions in line with their values to a greater extent than the attribute-tree condition suggests a benefit of presenting decision information in the alternative-tree format. Future research should consider people actually facing such decisions in order to examine the potential impact of the tree format on their decision strategies. This would also increase the generalisability of the findings, however, given the similarity of the findings with other populations (e.g., [15, 40], there is no reason to suspect that the findings here are not generalisable.

The empirical evidence presented in this study show that structuring information in a computer based alternative-tree format improves the correspondence between a person's decision and their values. A "reasoned choice" approach [27, 29], which considers the process of making a decision and not just the outcome of the process, was adopted to evaluate the interfaces in each condition. Taken together, the findings show the benefit of structuring the interface in a way that supports the *process* of making a decision. In line with Payne, Bettman and Johnson [18], the alternative-tree format presents information in a way that reduces the cognitive effort of engaging in demanding decision strategies and hence leads to the uptake of these strategies. In the present climate of engaging patients more fully in decisions about their treatment, these findings highlight the need for healthcare professionals to take account of way in which they format the information they give patients. Furthermore, it should be presented in a format congruent to the way a patient is likely to try and process that information. With more and more healthcare information being provided online (for example NHS Direct in the UK), it becomes crucial, for the patients' sake that those providing such information are informed about how best to present it.

References

1. Ubel, P.A., Lowenstein, G.: The role of decision analysis in informed consent: Choosing between intuition and systematicity. Social Science and Medicine 44, 647–656 (1997)
2. Fallowfield, L.J., Hall, A., Maguire, P., Baum, M., A'Hern, R.P.: Psychological effects of being offered choice of surgery for breast cancer. British Medical Journal 309, 448 (1994)
3. Molenaar, S., Sprangers, M.A.G., Rutgers, E.J.T., Luiten, E.J., Mulder, J., Bossuy, P.M.M., et al.: Decision support for patients with early stage breast cancer: Effects of an interactive breast cancer CDROM on treatment decision, satisfaction, and quality of life. Journal of Clinical Oncology 19, 1676–1687 (2001)
4. Darke, S.: Effects of anxiety on inferential reasoning task performance. Journal of Personality and Social Psychology 55, 499–505 (1988)
5. Bekker, H., Legare, F., Stacey, D., O'Connor, A.M., Lemyre, L.: Evaluating the effectiveness of decision aids: is anxiety a suitable measure (unpublished Work)
6. Margalith, I., Shapiro, A.: Anxiety and patient participation in clinical decision-making: The case of patients with ureteral calculi. Social Science and Medicine 45, 419–427 (1997)
7. Michie, S., Smith, D., McClennan, A., Marteau, T.M.: Patient decision-making: An evaluation of two different methods of presenting information about a screening test. British Journal of Health Psychology 2, 317–326 (1997)
8. Murray, E., Davis, H., See Tai, S., Coulter, A., Gray, A., Haines, A.: Randomised controlled trial of an interactive multimedia decision aid on hormone replacement therapy in primary care. British Medical Journal 323, 1–5 (2001)
9. Thornton, J.G., Hewison, J., Lilford, R.J., Vail, A.: A randomised trial of three methods of giving information about prenatal testing. British Medical Journal 311, 1127–1130 (1995)

10. Rostom, A., O'Connor, A.M., Tugwell, P., Wells, G.A.: A randomized trial of a computerized versus an audio-booklet decision aid for women considering post-menopausal hormone replacement therapy. Patient Education and Counselling 46, 67–74 (2002)
11. Schapira, M.M., Meade, C., Nattinger, A.B.: Enhanced decision-making: The use of a videotape decision-aid for patients with prostate cancer. Patient Education and Counselling 30, 119–127 (1997)
12. Mitchell, S.L., Tetroe, J., O'Connor, A.M.: A decision aid for long-term tube feeding in cognitively impaired older persons. The Journal of the American Geriatrics Society 49, 313–316 (2001)
13. Broadstock, M., Michie, S.: Processes of patient decision-making: Theoretical and methodological issues. Psychology and Health 15, 191–204 (2000)
14. Carrigan, N.A., Gardner, P.H., Conner, M., Maule, J.: The impact of structuring information in a patient decision aid. Psychology and Health 19, 457–477 (2004)
15. Holmes-Rovner, M., Kroll, J., Rovner, D.R., Schmitt, N., Rothert, M., Padonu, G., et al.: Patient decision support intervention: Increased consistency with decision analytic models. Medical Care. 37, 270–284 (1999)
16. Llewellyn-Thomas, H.A., Thiel, E.C., Sem, F.W.C., Wormke, D.E.: Presenting clinical trial information - A comparison of methods. Patient Education and Counselling 25, 97–107 (1995)
17. Street, R.L., Voigt, B., Geyer, C., Manning, T., Swanson, G.P.: Increasing patient involvement in choosing treatment for early breast-cancer. Cancer 76, 2275–2285 (1995)
18. Payne, J.W., Bettman, J.R., Johnson, E.J.: The adaptive decision maker. Cambridge University Press, Cambridge (1993)
19. Todd, P., Benbasat, I.: Inducing compensatory Information processing through Decision aids that facilitate effort Reduction: An experimental Assessment. Journal of Behavioral Decision Making 13, 91–106 (2000)
20. Klayman, J.: Children's decision strategies and their adaptation to task characteristics. Organizational Behavior and Human Decision Processes 35, 179–201 (1985)
21. Redelmeir, D.A., Shafir, E.: Medical decision making in situations that offer multiple alternatives. Journal of the American Medical Association 273, 302–305 (1995)
22. Simon, H.A.: Rational choice and the structure of the environment. Psychological Review 63, 129–138 (1956)
23. Tversky, A., Kahneman, D.: Judgement under uncertainty: Heuristics and biases. Science 185, 1124–1131 (1974)
24. Thorngate, W.: Efficient decision heuristics. Behavioral Science 25, 219–225 (1980)
25. O'Connor, A.M., Tugwell, P., Wells, G.A., Elmslie, T., Jolly, E., Hollingworth, G., et al.: A decision aid for women considering hormone therapy after menopause: decision support framework and evaluation. Patient Education and Counselling 33, 267–279 (1998)
26. Baron, J.: Thinking and Deciding, 3rd edn. Cambridge University Press, Cambridge (2001)
27. Frisch, D., Clemen, R.T.: Beyond expected utility: Rethinking behavioral decision research. Psychological Bulletin 116, 46–54 (1994)
28. Zey, M.: Rational choice theory and organizational theory: a critique. Sage Publications, USA (1998)
29. Bekker, H.: Understanding why decision aids work: linking process with outcome. Patient Education and Counselling 50, 323–329 (2003)
30. Janis, I.L., Mann, L.: Decision making: a psychological analysis of conflict, choice and commitment. The Free Press, New York (1977)

31. Stone, D.N., Schkade, D.A.: Numeric and Linguistic Information Representation in Multiattribute Choice. Organizational Behavior and Human Decision Processes 49, 42–59 (1991)
32. Sundstroem, G.A.: Information search and decision making: The effects of information displays. In: Montgomery, H., Svenson, O. (eds.) Process and structure in human decision making, John Wiley & Sons, New York (1989)
33. Billings, R.S., Scherer, L.L.: The effects of response mode and importance on decision-making strategies. Organizational Behavior and Human Decision Processes 41, 1–19 (1988)
34. Maheswaran, D., Meyers-Levy, J.: The influence of message framing and issue involvement. Journal of Marketing Research 27, 361–367 (1990)
35. Flesch, R.: A new readability yardstick. Journal of Applied Psychology 32, 221–233 (1948)
36. Westenberg, M.R.M., Koele, P.: Multi-attribute evaluation processes: Methodological and conceptual issues. Acta Psychologica 87, 65–84 (1994)
37. Baron, R.M., Kenny, D.A.: The moderator-mediator variable distinction in social psychological research: Conceptual, strategic, and statistical considerations. Journal of Personality and Social Psychology 51, 1173–1182 (1986)
38. Edwards, A.L.: An introduction to linear regression and correlation. Freeman and Company, New York (1976)
39. Jarvenpaa, S.L.: The effect of task demands and graphical format on information processing strategies. Management Science 35, 285–303 (1989)
40. O'Connor, A.M., Rostom, A., Fiset, V., Tetroe, J., Entwistle, V., Llewellyn-Thomas, H.A., et al.: Decision aids for patients facing health treatment or screening decisions: systematic review. British Medical Journal 319, 731–734 (1999)

Appendix

Scenario Description Given to Participants

You are asked to imagine that you are a doctor who has recently had a patient referred to you with a heart ailment. The disease is sufficiently serious to force them to make quite radical changes to their way of life if they decide to undergo any treatment. There is, however, the possibility of having an operation which, if successful, could relieve the condition and allow a reasonably normal lifestyle. However the success of the operation cannot be assured and it might exacerbate the condition. There is even a small chance that there could be difficulties during the operation which could result in death.

If the patient does not have the operation they will be given medication that will relieve some of the symptoms and may possibly control the condition, they will have to take it for the rest of their life. The patient has asked you to advise them on which option they should take.

When making the decision about which option they should choose you should assume the patient has the following personal details and characteristics. Up to now the patient has been generally healthy. The patient is a student and has been performing well in class.

The patient enjoys the student social life, like going to pubs and clubs etc., and has a wide circle of friends. The patient's partner is also a student and they have been seeing each other for some time. Both of them enjoy socialising and eating out when they can afford it, especially at Indian and Italian restaurants. Sport is important to the patient, enjoying outdoor activities and going to the gym when they have time.

Satisfaction with Decision Scale

Participants were asked to respond to the following items on a 7-point Likert-type scale (0= 'Not at all' to 6= 'Extremely')

- To what extent was your decision a good one
- How satisfied are you with the way you made the decision
- How sure are you that the decision was the right one

Involvement with Decision Scale

Participants were asked to respond to the following items on a 7-point Likert-type scale (1= 'Strongly agree' to 7= 'Strongly disagree').

- I found the scenario presented in the study believable
- I found it hard to imagine myself in this scenario
- I thought the material presented was engaging
- I did not think hard about the decision I made
- Faced with a similar decision in real life I would make the same decision
- I did not take the study seriously
- Faced with a similar decision in real life I would think about the same considerations as those presented here
- I found the material in this study interesting

Dynamic Simulation of Medical Diagnosis: Learning in the Medical Decision Making and Learning Environment MEDIC

Cleotilde Gonzalez[1] and Colleen Vrbin[2]

[1] Dynamic Decision Making Laboratory, Carnegie Mellon University
Pittsburgh, Pennsylvania, United States
conzalez@andrew.cmu.edu
[2] Center for Pathology Quality and Healthcare Research, University of Pittsburgh Medical
Center, Pittsburgh, Pennsylvania, United States
vrbincm@upmc.edu

Abstract. MEDIC is a dynamic decision making simulation incorporating time constraints, multiple and delayed feedback and repeated decisions. This tool was developed to study cognition and dynamic decision making in medical diagnosis. MEDIC allows one to study several crucial facets of complex medical decision making while also being well controlled for experimental purposes. Using MEDIC, there is a correct diagnosis for the patient, which provides both outcome and process measures of good performance. MEDIC also allows us to calculate cue diagnosticity and probability functions over the set of hypotheses that participants are explicitly considering, based on assumptions of local (bounded) rationality. MEDIC has served in a series of studies aimed at understanding learning in dynamic and real-time medical diagnostic situations. In this paper, we outline the tool and highlight results from these preliminary studies which set out to measure learning.

Keywords: Decision Making, Simulation, Medical Diagnosis.

1 Introduction

We accomplish long term goals by making multiple decisions over time. Dynamic decision making (DDM) is about making decisions in an environment that is changing while the decision maker is collecting information about it [1]. Decision makers in dynamic environments make multiple decisions that are intended to reach some goal, and to keep the system under control within a performance range.

Consider a medical dynamic decision making problem. A patient presents symptoms that indicate possible high blood sugar. Tests indicate high blood sugar and low insulin (i.e., hyperglycemia). The physician's goal is to stabilize the patient's health (keep the blood sugar within an acceptable range). The patient can be diagnosed with diabetes (type 1) as the symptoms (cues) develop over time. Once a diagnosis is made a treatment is given, for example, to take insulin. Insulin often takes a moderate amount of time to have an effect in the body.

A. Holzinger (Ed.): USAB 2007, LNCS 4799, pp. 289–302, 2007.

If the amount of insulin is not well calibrated to the state of the body as it is changing, it is possible that the patient would have too much insulin in the body, low blood sugar, and suffer a hypoglycemic crisis. At that point, the solution needs to come quick, to take some sugar by mouth or drink some orange juice, which often have a fast effect on the body. The ideal situation here is to use feedback about the patient's state to keep the system in balance and under control by adding insulin or sugar without over or undershooting. However, we often know that the perception of feedback about the patient's state is inaccurate and the control of the system is often challenging.

Work on the psychology of decision making suggests people have difficulty managing dynamic systems with multiple feedback processes, time delays, nonlinearities, and accumulations [4]. Researchers have found that decision makers remain sub-optimal even with repeated trials, unlimited time, and performance incentives [3, 5, 6, 7]. We believe that more research is necessary to understand the learning process by which individuals improve their decisions after repeated choices in dynamic tasks. Learning is the process that modifies a system to improve, more or less irreversibly, its subsequent performance of the same task or of tasks drawn from the same population [8]. Learning, among other processes (individual differences in cognitive capacity, biases in general reasoning strategies, complexity of dynamic systems), can help explain much of the variance in human performance on dynamic decision tasks. Research in DDM indicates that although individuals may follow very diverse strategies they tend to evolve towards better control policies after an extended number of practice trials [9, 10].

Our research aims at determining how decision makers learn in DDM tasks. In particular, this paper describes a dynamic simulation in a medical context, called MEDIC and presents three behavioral studies to describe how individuals learn using this simulation.

2 MEDIC: A Dynamic Medical Decision Making and Learning Environment

Decision scientists have typically focused on simple and static laboratory tasks. For instance, in the typical laboratory experiment, participants are asked to make likelihood judgments or select among a small number of usually experimenter provided alternatives. Moreover, participants are often provided with one or a sequence of independent choices, where one choice does not influence the next one. Rarely, decision making research has used tools that are representative of the dynamics of the decision making conditions we experience in the real world. Dynamic tools and methods are needed in order to study the dynamics of human behavior. One area of study that can lead to significant improvements in medicine, specifically medical training, is the development of virtual reality simulators for complex medical procedures [11].

We have developed many tools (microworlds, learning environments, management simulators, etc.) for studying DDM [12]. MEDIC is one of those learning environments that was created to study DDM in simulated medical diagnosis.

The development of MEDIC was inspired by Kleinmuntz's [13] paradigm to test the performance of heuristics in a complex dynamic setting. MEDIC includes all the characteristics of a dynamic task described in Kleinmuntz [13]. Using MEDIC we can manipulate the statistical structure of the DDM task in a manner commensurate with Kleinmuntz's analysis: task complexity (i.e., number of diseases and symptoms), disease base rates, time pressure, test diagnosticity, treatment effectiveness, and treatment risk. In addition, we also designed MEDIC to incorporate several dynamic factors not considered by Kleinmuntz, including feedback delays in tests and treatments, dynamic diagnostic cues (cues that have diagnostic patterns unfolding over time), and temporally sensitive symptoms (symptoms that appear at various stages in the progression of the disease). Thus, MEDIC simulates realistic components to a medical diagnosis task, involving multiple feedback loops and possible delays, from the presentation of symptoms to the modification of the patient's health through treatment.

MEDIC has five phases: 1) presentation of symptoms, 2) generation of a diagnosis, 3) test of a diagnosis 4) submission of a treatment and 5) analysis of outcome feedback. The goal in the task is to diagnose and cure patients who are suffering from one disease. The patient is drawn from a population of patients with a configuration of symptoms-diseases and diseases-treatments associations. These configurations are chosen randomly for each patient.

The main measure of performance in MEDIC is the health meter. The health meter is defined as a percent of health, in a scale from 0 to 100. The value of zero indicates death while a value of 100 indicates full recovery. The user of MEDIC can check the status of the patient's health by looking at a graph which continuously monitors the patient's health. MEDIC presents a sequence of patients with an initial health that can vary, but was kept around 50 for the studies reported here. Each patient can be fully cured if the right treatment is applied for the correct disease according to the probabilities defined in the symptoms-diseases and diseases-treatments matrices. MEDIC is a real-time DDM task in the sense that the state of the system deteriorates unless an action is taken. In this case, the patient's health decreases steadily with the passing of time in the simulation.

Initially, a new patient is shown with his or her various contextual data (age, gender and a picture of the patient). Then, participants are presented with probability tables indicating the symptom-disease associations for the patient. According to the probabilities of association of a symptom to a disease, participants decide to conduct tests that would determine the actual presence or absence of a symptom. However, conducting tests for symptoms takes time. Each test is currently configured to return results in 30 simulation minutes. When the test completes, the result of either "present" or "absent" will display, and only then can a second test be issued. Participants are then asked to make their hypothesis of the probability of the different diseases and then they are asked to decide on the most appropriate treatment for the hypothesized disease. This is done by selecting the treatment according to the disease-treatment probability matrix. At the end of each trial (i.e, patient), feedback is provided. Feedback indicates the accuracy of each of the actions taken, such as the real disease the patient suffers from and the belief assigned to that disease by the participant; the correct treatment according to the effectiveness of that treatment for the hypothesized disease. A score structure is defined for participants to associate their actions to their effects. This score is a main source of learning and it is explored in the initial studies as we will explain below.

Although MEDIC allows one to study several crucial facets of complex medical decision making that are often lost in the laboratory, this simulation is also well controlled for experimental purposes. Using MEDIC, we know the correct diagnosis of the patient, which gives us the ability to derive both outcome and process measures of good performance. Although we cannot develop models that prescribe an optimal set of actions, as is the case in many dynamically complex tasks, we can derive Bayesian models that provide benchmarks of "good" behavior. Also, at any point in time (current information state) we can calculate the probability distribution over the diagnostic hypotheses. Moreover, we can assume local (bounded) rationality and calculate cue diagnosticity and probability functions over the set of diagnostic hypotheses the participants are explicitly considering. The studies investigate the learning process in MEDIC from data collected from university students.

3   Three Learning Studies in MEDIC

3.1   Study 1: Learning the Probability Associations

The objective of the first study was to determine whether participants could learn the probabilistic associations of symptoms and diseases in order to make a diagnosis and provide effective treatment for patients in a simulated dynamic medical decision making task.

Methods. The first study was conducted with six graduate and undergraduate students in a research university. They all came to a laboratory where they were trained in MEDIC, and they were asked to diagnose and treat patients for 1 ½ hours.

Participants were presented with a sequence of patients suffering from one of four diseases. The patient and disease associations were selected randomly from one of the four diseases according to the base rates (0.25). Each of the patients in the sequence had a symptom-disease matrix indicating the probabilistic associations between the symptoms and diseases as shown in Table 1. This matrix is seen by participants in the top part of the screen in the MEDIC simulation and it was the same for all the patients in the sequence.

Table 1. Probability matrix with disease-symptom associations used in Study 1

Disease 1	Disease 2	Disease 3	Disease 4	
0.25	0.25	0.25	0.25	Base Rates
0.5	0.5	0.5	0.5	Symptom 1
0.9	0.1	0.5	0.5	Symptom 2
0.9	0.9	0.1	0.1	Symptom 3
0.5	0.5	0.9	0.1	Symptom 4

In this table, diseases 1, 2 3, and 4 presented above each have different associations with the four symptoms. It has been found that participants tend to use positive-testing strategies, suffering from *confirmation bias* [14,15] and *pseudodiagnostic selection* [16]. Thus, an expected behavior is the tendency to issue tests (for symptoms) that have a high likelihood of confirming a hypothesis. However, participants are not restricted in the number or order of tests they can issue. They are allowed to run up to four tests to identify the presence or absence of up to four symptoms.

Participants issue tests (which take time to execute while the patient's health decreases) to determine which symptoms are present. After receiving the test results, the participant adjusts a "belief meter" to reflect his/her assessment of the probability of the disease being present (associations based on the symptom-disease matrix), on a scale of 0 (not present), to 1 (certainly present).

After completing the belief meter for all four diseases, the participant can either conduct more tests or administer treatment. Once a participant has adjusted the belief meters to indicate the likelihood of each disease and is finished testing, effective treatment must be administered according to disease-treatment probability associations defined in another matrix. In this study, the disease-treatment probabilities were fully diagnostic, indicating that one and only one treatment could be effective for each of the possible diseases.

Scoring methods. Human behavior in MEDIC was measured by calculating a score of behavior. We defined three methods of calculating a score.

Scoring Method 1. In the first method, the score included components that rewarded efficient decisions. A cumulative score was presented to each participant based on the following factors: the participant begins with 2000 points for the first patient; if the participant administers an effective treatment, the patient's health is multiplied by 2 (for an ineffective treatment, no points are added); if the patient dies, the participant loses 100 points; each test issued costs 25 points; and points from the belief meter were based on the normalized value assigned on the belief meter compared to the actual value for the correct disease. The equation for scoring method 1 is below.

$$\textbf{Score1} = 2000 + (2 * \text{end health if effective treatment OR } 0 \text{ if ineffective treatment OR } -100 \text{ if patient dies}) - (25 * \text{number of tests performed}) - 50 - (50 * (1 - (\text{actual probability of disease} - \text{participants normalized, believed probability of that disease})^2). \tag{1}$$

This score replaced the value of 2000 in the equation for the second trial, and so on. The actual cumulative score was plotted and compared with (1) the maximum possible score, which assumed the patient's health did not deteriorate, no tests were performed and the normalized believed probability exactly matched the actual probability for the correct disease; (2) the maximum possible score adjusted for two tests, which are necessary to run to provide an accurate diagnostic assessment; (3) a neutral score that represented the score per trial if the participant made no decisions, meaning that the patient died, no tests were run, and the belief meter was not adjusted; and (4) the minimum possible score, which assumed the patient died, all four tests were run, and the normalized, believed probability for correct disease was 0.

Scoring Method 2. The second performance measure involved a modified score equation which emphasized task accuracy. This method assigned points for correct diagnosis (+/- 100 points), correct treatment (+/- 100 points), proportion of life preserved, most probabilistic informative testing sequence (i.e. testing for the third symptom, and if present test for the second symptom or if absent test for the fourth symptom) (+/- 100 points), difference between actual probability and believed probability for present disease. The equation for scoring method 2 is below:

Score 2 = 500 +/- 100 +/- 100 + percentage health saved +/- 100 - the absolute
value of the difference between actual probability of present disease and   (2)
guessed probability of present disease .

This score replaced the value of 500 in the equation for the second trial, and so on. The actual cumulative score was plotted and compared with (1) the maximum possible score, which assumed correct diagnosis and treatment, no health deterioration, most probabilistic testing sequence, and accurate probabilistic diagnostic assessment for correct disease on belief scale; and (2) the minimum possible score, which assumed incorrect diagnosis and treatment, patient died, any testing pattern other than the most probabilistic sequence, 0 believed probability for correct disease on belief meter.

Scoring Method 3. The third method to calculate the score focused more on the participants' comprehension of the cues as they relate to the probability matrices. This measure of performance incorporated whether or not the participants correctly identified the most probabilistic options by calculating the frequency with which the participants identified the most probable disease even if that wasn't the disease that was present, as well as the frequency with which effective treatment was provided for the guessed disease, even if the guessed disease wasn't the correct or most probable disease. Testing behaviors were also investigated. The goal of this method was to identify whether the participants understood the probability matrices.

In summary, the different scoring methods above would help us best understand the strategies by which the different individuals attempted to save the life of the series of patients they were presented with, and will allow us to understand their behavior.

Results. Results using Score 1 were plotted against neutral, minimum, maximum and maximum adjusted scores. Examples of two individuals demonstrating best and worst performances in this task over the course of the number of patient trials are shown in Figure 2. These two examples show typical behavior in this task according to Score 1. The bolded line represents each participant's actual cumulative score. In total, 33% of participants performed worse than if they had made no decisions at all in the task, similar to Participant 4 below. The other 77% did not perform much better than the neutral score, similar to Participant 6 (Figure 2).

We then fit the participants' behavior per trial to a simpler score calculation, Score 2, hoping this would help reveal why participants performed so poorly in this task. Results from the Score2 are shown graphically in Figure 3 below for the same two participants that were displayed in Figure 2. Figure 3 displays the participant's recalculated cumulative score (bolded line) as compared to the maximum and minimum cumulative scores.

Using this modified score calculation it appears that the overall performance of the participants is better than using the calculation method in Score 1. The participants, on average, were more accurate than efficient in the task.

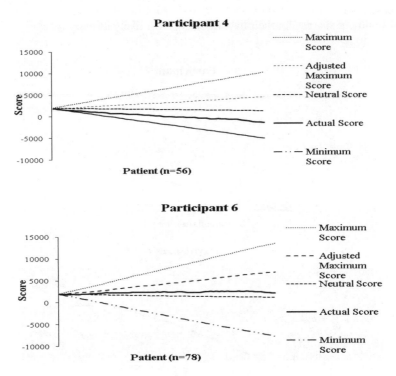

Fig. 2. Best (*participant 6*) and worst (*participant 4*) performances in Study 1 as measured by Score1. The graphs show the maximum score, the adjusted maximum score, the neutral score, the actual score and the minimum score. Participant 6 does not perform much better than the neutral score, and participant 4 performs worse than the neutral score.

A third method for measuring learning was used on the data collected in the first study. This Score 3 helped us identify how participants interpreted the probability matrix. The previous two scoring methods assigned points based on aspects of the task that do not conclusively point to a clear understanding of the probability matrices, such as was the treatment effectiveness and how many tests were run. With scoring method 3, we can identify how well a participant used the cues to identify the most probable disease instead of just adding or subtracting points for accuracy and efficiency.

Table 2 displays the results for Score 3, these are averages across all patients. Since the probabilistic associations between symptoms and diseases have some built in ambiguity (none of the symptoms are 100% associated with any of the diseases), the goal of this method was to determine whether participants could interpret the symptom-disease and treatment-treatment matrices. The results identify a large variability among the six participants in this study. For example, they vary a lot in identifying most probable disease and effective treatment for the chosen disease. Some individuals were very inaccurate in identifying the most probable disease as the real disease. Surprisingly, and despite the fact that the disease-treatment matrix was fully diagnostic, where one and only one treatment was effective for a particular disease, individuals were also ineffective at selecting the treatment with the highest probability of success for their hypothesized, most likely disease.

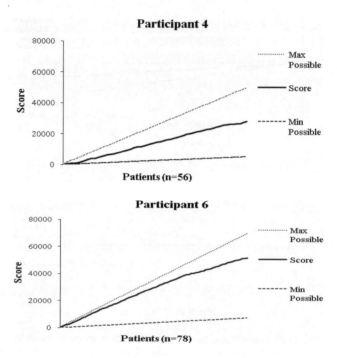

Fig. 3. Performance in Study 1 for two participants measured by Score 2. The graphs show the maximum possible score, the actual score, and the minimum possible score.

Table 2. Performance in Study 1 measured by Score 3 – Diagnosis and treatment behavior

	Indicated Highest Probability on Belief Meter for Most Probable Disease (%)	Chose Effective Treatment for the Disease where Highest Probability was Indicated on Belief Meter (%)
Participant 1	79.2	96.2
Participant 2	89.0	75.6
Participant 3	96.6	94.3
Participant 4	75.0	91.1
Participant 5	64.4	72.9
Participant 6	30.8	33.3

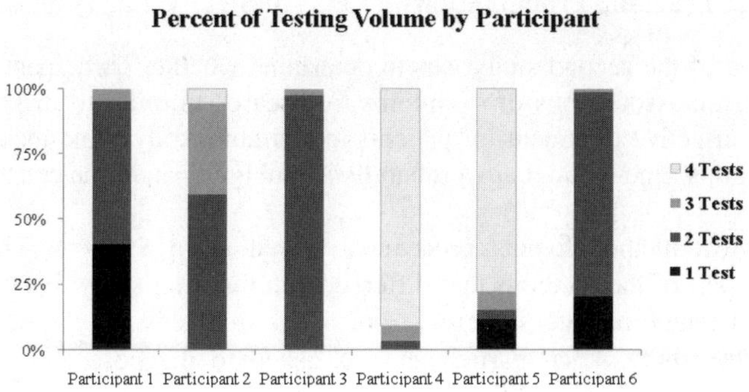

Percent of Testing Volume by Participant

Fig. 4. Performance in Study 1 measured by Score 3

Figure 4 displays the highly variable testing behavior for the six participants. Two of the participants used all 4 tests more than 75% of the trials, whereas two participants used 2 tests for 75% of the trials, and two others used two tests for about half of the trials.

These results are interesting, especially since the test for symptom one does not provide beneficial information, as the symptom is equally associated with all four diseases at 0.5.

Conclusions. We analyzed three different score measures to investigate learning in MEDIC. In the first method, task efficiency was rewarded more heavily. The second method rewarded accuracy.   And the final method rewarded comprehension. Performing highly in one of the score calculation methods does not guarantee high performance in another. The different scores allowed us to investigate human behavior at different levels of detail. Most participants did not demonstrate learning, despite the fact that they were allowed extensive practice and despite the full information they were given (both, the symptom-disease probability matrix was given as well as the disease-treatment matrix). The probability matrices were displayed and made fully available to them and they were provided with complete feedback on their behavior. In fact, some of our participants did worse than if they would have made no decisions in the task.

The first score calculation method identified an overall poor performance by all of the participants, who at best performed slightly better than if they had made no decisions in the task. In other words, the decisions that the participants made were not wholly efficient. However, after recalculating the score using the second method, performance appeared much better. This suggests that the participants were able to optimize accuracy better than efficiency in the task.   The final score calculation method, which demonstrated probabilistic comprehension, showed that not all of the participants selected the most probable diseases and treatments, suggesting that participants had difficulty in interpreting the probability matrices. Each score calculation method highlights different learning strategies for the task.

Finally, there was notable variability in testing patterns between subjects. This variability inspired the manipulations to MEDIC for the second pilot study, described below.

3.2 Study 2: Learning Probability with Less Ambiguity and Time Constraints

The objective of the second study was to determine whether participants could learn the probabilistic associations of symptoms to diseases in order to make a diagnosis and provide effective treatment for patients in a simulated dynamic medical decision making task, this time without any probabilistic ambiguity and time constraints.

Methods. Most methodological procedures stayed as in Study 1. Thus, here we describe the part of the methods that differed from that first study. Nine participants, graduate and undergraduate students, from a research university were recruited to completed this study. Each participant was asked to diagnose and treat multiple patients for one hour in a modified version of MEDIC; it was modified from the first study to remove the time constraints with both testing delay and decreasing patient health. We also modified the ambiguous associations between symptoms and diseases to include probabilities of either 1 or 0, reducing the ambiguity of the symptom-disease association. The relationships used in this study are shown in Table 3. Given that the Disease-Treatment matrix was fully diagnostic, we removed the treatment section. Thus, the task ended with the participant's diagnosis. The feedback provided included a tally of correct diagnoses and total number of patients seen. According to this new association matrix, the best testing strategy included a total of 2 tests. Participants issuing more than 2 tests would be following a suboptimal strategy, and would be demonstrating a complete lack of understanding of probability relationships.

Table 3. Probability matrix with disease-symptom associations used in Study 2

Disease 1	Disease 2	Disease 3	Disease 4	
0.25	0.25	0.25	0.25	Base Rates
0	0	1	1	Symptom 1
1	0	0	1	Symptom 2
1	1	0	0	Symptom 3
0	1	1	0	Symptom 4

Performance was measured using a score presented to the participant. This score was a tally of correct diagnoses and total number of patients seen.

Results. Table 4 shows each participants performance using the score presented to each participant, which was the total number of correct diagnoses and the total number of trials. Under unambiguous probabilities and no time constraints, all nine participants were able to consistently identify the correct disease.

We then analyzed the testing frequency, as we expected to find a clear and consistent pattern of testing procedures, with a maximum of two tests per participant. Unlike the first study, fewer participants relied on running all four tests, but there was still a considerable amount of variability between subjects who ran 2 versus 3 tests for most of the trials. In this study we also analyzed the range of probabilities individuals indicated for the correct disease when making the diagnosis. Since all ambiguity had been removed from the task, the correct disease was 100% likely to be present. However, 100% was not what participants indicated and surprisingly, ranges varied from .30 to .98. Table 5 displays the results.

Table 4. Performance in Study 2 measured by AVERAGE frequency of correct diagnoses ACROSS TRIALS

	Number of Trials	Correct Diagnosis (%)
Participant 1	244	242 (99.2)
Participant 2	161	161 (100.0)
Participant 3	201	201 (100.0)
Participant 4	211	209 (99.1)
Participant 5	207	206 (99.5)
Participant 6	195	193 (99.0)
Participant 7	170	160 (94.1)
Participant 8	267	259 (97.0)
Participant 9	232	230 (99.1)
Total	1888	1861 (98.6)

Table 5. Pilot Study 2-Measuring learning – Guessed probability for correct disease

	Average Guessed Probability for Correct Disease	Range of Guessed Probability for Correct Disease
Participant 1	0.97	0.91-0.98
Participant 2	0.96	0.82-0.98
Participant 3	0.49	0.30-0.98
Participant 4	0.97	0.97-0.98
Participant 5	0.91	0.80-0.97
Participant 6	0.97	0.88-0.98
Participant 7	0.97	0.97-0.98
Participant 8	0.95	0.55-0.98
Participant 9	0.84	0.69-0.98
Total	0.85	0.30-0.98

Conclusions. Diagnostic accuracy improved from Study1 to Study 2, where the task was over simplified by reducing all the uncertainty in the symptom-disease probabilities and the time constraints. Despite a clear improvement in accuracy with these simplifications, testing patterns were still variable. Individuals were still suboptimal in their testing patterns and in their perception of the probability of the correct disease after testing for symptoms. With these results in mind, modifications were made for a third study described below.

3.3 Study 3: Monetary Incentives

Given that participants are suboptimal learners even in the simplest possible diagnosis task, with unambiguous probabilities and no time constraints, we hypothesized that the only possible explanation left for this performance was motivation. The third study was designed to provide participants with a monetary incentive in the task in order to demonstrate whether participants could learn the probabilistic associations of symptoms to diseases while using the optimal testing strategy and having accurate probability assessments. Participants were assigned to one of two conditions that could earn a bonus. In one condition, a bonus was earned for running two tests, since using the unambiguous probabilities from Study 2 require only two tests to be run in order to make an accurate diagnosis. In the second condition, a bonus was earned by accurately assessing the probabilities of all four diseases each trial.

Methods. This study was identical to the second pilot study, with the exception being the score now represented dollars earned. Nine graduate and undergraduate students were part of this study. The participants were split into 2 conditions. Both conditions incorporated financial incentives of $0.02 per trial to reduce variability while maintaining the level of diagnostic accuracy seen in the second pilot study.

In one condition, which contained 5 participants, a bonus was earned for the ideal testing behavior, which with the unambiguous probability matrix (the same as in Study 2) meant that only two tests were necessary to have complete confidence in a diagnosis.

In the other condition, which contained 4 participants, a bonus was earned for assigning accurate probabilities on the belief scale for all four diseases. With the probability matrix the same as Study 2, the correct disease had a probability of 1, and the other three had a probability of 0.

Each participant was asked to complete 200 trials. Only one of the nine participants was unable to finish, but completed 107 of 200 trials.

Results. Table 6 contains a summary of each participant's performance based on diagnostic accuracy. Although several participants did not perform as well as others, overall the performance was better when individuals earned a bonus for the ideal

Table 6. Performance in Study 3 measured by frequency of correct diagnoses. This table presents the averages across patient trials.

		Number of Trials	Percent of Trials with Correct Diagnosis	Percent of Trials with Bonus Earned
Testing Bonus	Participant 1	200	100%	97%
	Participant 2	200	95%	85%
	Participant 3	200	100%	100%
	Participant 4	200	99%	98%
	Participant 5	200	93%	90%
	Total	1000	97%	94%
Diagnostic Probabilistic Accuracy Bonus	Participant 6	107	70%	0%
	Participant 7	200	100%	93%
	Participant 8	200	100%	95%
	Participant 9	200	57%	0%
	Total	707	83%	53%

testing behavior compared to the participants which earned a bonus for determining the correct probability of the real disease. Surprisingly, 2 out of 4 participants did not earn a bonus for any trial in this second condition.

Both testing behavior and guessed probabilities were analyzed. Providing a financial incentive for optimal testing frequency in this study led to less variability in testing volume. The participants learned that only two tests were necessary, likely by receiving $0.02 cents when the testing strategy was ideal. However, when the financial incentive was not provided for testing strategy, but instead for diagnostic probabilistic accuracy, testing variability was similar to the results seen in Study 2. This suggests that the participants can properly interpret the provided probability matrices, but are willing to run excessive tests in the absence of a financial incentive.

Conclusions. Interestingly, earning a bonus for the ideal testing behavior of two tests greatly reduced the variability of testing behavior between participants earning the same bonus. However, diagnostic probabilistic accuracy did not seem to experience the same decrease in variability when a bonus could be earned.

4 Discussion

MEDIC was developed to study learning in a complex dynamic setting. Many aspects of the simulation can be manipulated, allowing for a variety of experiments. Simulations can be run with or without time constraints, with different levels of feedback, with varying symptoms, tests, diseases and treatments. MEDIC is an important step toward the development of a tool for training and/or reference by medical professionals in decision-making tasks.

The potential applications of MEDIC can be classified into two broad categories: (1) to study cognitive processes underlying physicians' learning of symptoms as they relate to infectious diseases, and, (2) to understand behavior so as to design and implement decision support technology that would assist dynamic decision making under time constraints. Studies that examine cognitive processes focus on understanding the factors that affect hypothesis generation. As described in the studies reported here, decision making using MEDIC can be studied by manipulating the probability matrix that relates symptoms to diseases as well as the types of feedback provided to physicians.

The studies that we ran with MEDIC and that are reported in this paper, demonstrate that even in the simplest possible conditions, with no time constraints and no ambiguity in the symptom-disease probabilities, participants with a high level of education are unable to perform optimally. We also showed that incentives played a key role in their effort and attention they put in to finding the best testing strategy and the determining the appropriate probabilities of the different diseases. The question is then: How can we improve performance in real-world medical diagnosis tasks, where there are immense complexities, time constraints and lack of motivation?

MEDIC allows one to study several crucial facets of complex medical decision-making that are often lost in the laboratory, while also being well controlled for experimental purposes. Using MEDIC, we know the correct diagnosis of the patient, which gives us the ability to derive both outcome and process measures of good performance. Overall, MEDIC provides the necessary paradigm to test the dynamics of hypothesis generation; it also provides data to support the design of medical diagnosis technology that would compensate for deficiencies underlying human cognition under conditions of high workload. We aim at continuing to study the effects of probability uncertainty, time constraints, and feedback on medical diagnosis, and we think MEDIC will support this goal.

Acknowledgments. This research was partially supported by the National Science Foundation (Human and Social Dynamics: Decision, Risk, and Uncertainty, Award number: 0624228) award to Cleotilde Gonzalez.

References

1. Edwards, W.: Dynamic Decision Theory and Probabilistic Information Processing. Human Factors 4, 59–73 (1962)
2. Brehmer, B.: Strategies in Real-Time, Dynamic Decision Making. In: Hogarth, R.M. (ed.) Insights in Decision Making, pp. 262–279. University of Chicago Press, Chicago (1990)
3. Sterman, J.D.: Misperceptions of Feedback in Dynamic Decision Making. Organizational Behavior and Human Decision Processes 43(3), 301–335 (1989)
4. Sterman, J.D.: Business Dynamics: Systems Thinking and Modeling for a Complex World. McGraw-Hill, Boston (2000)
5. Diehl, E., Sterman, J.D.: Effects of Feedback Complexity on Dynamic Decision Making. Organizational Behavior and Human Decision Processes 62(2), 198–215 (1995)
6. Sterman, J.D.: Modeling Managerial Behavior: Misperceptions of Feedback in a Dynamic Decision Making Experiment. Management Science 35(3), 321–339 (1989)
7. Sterman, J.D.: Learning in and about Complex Systems. Systems Dynamics Review 10, 291–330 (1994)
8. Simon, H.A., Langley, P.: The Central Role of Learning in Cognition. In: Simon, H.A. (ed.) Models of Thought, vol. II, pp. 102–184. Yale University Press, New Haven London (1981)
9. Gonzalez, C.: Learning to Maker Decisions in Dynamic Environments: Effects of Time Constraints and Cognitive Abilities. Human Factors 46(3), 449–460 (2004)
10. Kerstholt, J.H., Raaijmakers, J.G.W.: Decision Making in Dynamic Task Environments. In: Ranyard, R., Crozier, W.R., Svenson, O. (eds.) Decision Making: Cognitive Models and Explanations, Routledge, London, pp. 205–217 (1997)
11. Scerbo, M.W.: Medical Virtual Reality Simulators: Have We Missed an Opportunity? Human Factors & Ergonomics Society Bulletin 48(5), 1–3 (2005)
12. Gonzalez, C., Vanyukov, P., Martin, M.K.: The Use of Microworlds to Study Dynamic Decision Making. Computers in Human Behavior 21(2), 273–286 (2005)
13. Kleinmuntz, D.N.: Cognitive Heuristics and Feedback in a Dynamic Decision Environment. Management Science 31(6), 680–702 (1985)
14. Wason, P.C.: Reasoning. In: Foss, B.M. (ed.) New Horizons in Psychology, Penguin, Baltimore, pp. 135–151 (1966)
15. Wason, P.C.: Reasoning about a Rule. Quarterly Journal of Experimental Psychology 20, 273–281 (1968)
16. Doherty, M.E., Mynatt, C.R., Tweney, R.D., Schiavo, M.D.: Pseudodiagnosticity. Acta Psychologica. 43, 111–121 (1979)

SmartTransplantation - Allogeneic Stem Cell Transplantation as a Model for a Medical Expert System

Gerrit Meixner[1], Nancy Thiels[2], and Ulrike Klein[3]

[1] University of Kaiserslautern, Germany
[2] German Research Center for Artificial Intelligence (DFKI), Germany
[3] University of Heidelberg, Germany
meixner@mv.uni-kl.de,
nancy.thiels@dfki.de,
Ulrike.Klein2@med.uni-heidelberg.de

Abstract. Public health care has to make use of the potentials of IT to meet the enormous demands on patient management in the future. Embedding artificial intelligence in medicine may lead to an increase in quality and safety. One possibility in this respect is an expert system. Conditions for an expert system are structured data sources to extract relevant data for the proposed decision. Therefore the demonstrator 'allo-tool' was designed. The concept of introducing a 'Medical decision support system based on the model of Stem Cell Transplantation' was developed afterwards. The objectives of the system are (1) to improve patient safety (2) to support patient autonomy and (3) to optimize the workflow of medical personnel.

Keywords: Human-Computer Interaction in Health Care, Usability of Medical Information Systems, Human Aspects of Future Technologies in Health Care, Cognitive Task Analysis, Usability Engineering, Stem cell transplantation, Decision support system.

1 Introduction

In many areas of human life, computer-based Information Technology (IT) has prevailed and become essential for the coordinated and efficient organization of workflow. This offers numerous advantages for the future, but will also lead to problems for the people confronted with it.

Especially in the field of health care, interaction between human beings and Information Technology is a sensitive subject. Physicians have immense reservations and apprehensions, of being made the slaves of information scientists and of their programmed computer system. Nevertheless medicine has to become more scientific in patient management. Inevitably, people working in health care will have to make use of the potentials of IT in order to meet the enormous demands on patient management.

Within the development of computer-based software for medical personnel, the importance of interpersonal networking between physicians and IT-specialists and with associated occupational groups, had previously been underestimated or ignored.

A. Holzinger (Ed.): USAB 2007, LNCS 4799, pp. 303–314, 2007.

The importance of interdisciplinary advice and discussion, which is a prerequisite for the best possible decision on a treatment strategy, has to be reflected in the application of new information technologies. Embedding artificial intelligence in medicine may lead to an increase in quality and safety in medical treatment and decrease costs significantly. From our point of view there are three major rationales for implementation of IT services into clinical routine.

First of all the software architecture of most German hospital IT systems is deficient. This is displayed by a lack of integration of different software systems leading to redundant data sources in mixed formats. Searching for specific data within a certain subject results in time-consuming queries. Medical expert systems which conflate these different data sources will reduce wasted time of expensive medical staff.

Secondly, although testing medical intervention for efficacy has been done for a long time, Evidence-based medicine (EBM) has become major impact in the medical daily routine in the last years. EBM, also called scientific medicine, is an approach to apply more uniformly the standards of evidence which are gained from the scientific method to certain aspects of medical practice. According to the Centre for Evidence-Based Medicine, *'Evidence-based medicine is the conscientious, explicit and judicious use of current best evidence in making decisions about the care of individual patients'* [1]. Those parts of medical practice that are in principle subject to scientific methods shall be clarified by EBM. In recent years a fulminant explosion of medical data took place. This implicates a complex existence of numerous relevant and irrelevant data [2]. Keeping a complete picture of current expert opinion in mind is therefore only in a small number of faculties feasible. In the future, IT systems are supposed to have the ability to extract the relevant data, to filter irrelevant information, to focus on significant information and to make it available to the user [3].This needs to be part of a medical expert system to make it attractive in the daily routine. Gaining a contemporary update of published, peer reviewed and relevant data will directly lead to an improvement of medical decision and therefore in patient safety [4]. This applies not only for physicians but also for nurses, doctors' assistance and physiotherapists.

Thirdly, while cost effectiveness in health care has had no major pertinence for decades, the financial and socio-political impact in the last twenty years has become most important. Not only the changing age distribution and the increase in pharmaceutical budget have brought up the discussion about responsible distribution of resources. Medical expert systems need to be designed with a clear interest to mind cost effectiveness [5; 6]. This can happen through an easy providing of actual guidelines for evidence based medicine so that the deciding physician conceives support in the best care. In many cases this will show instant results in shorter duration of hospitalization, preventing serious adverse events e.g. through information of drug interactions and omitting interventions which would have been undertaken in order to prevent misdiagnosis. In the long term decision support systems can lead to a reduction of morbidity or mortality which was shown for example in the faculty of hematology [7].

The idea of medical expert systems isn't new [8; 9]. In the past, however, especially medical expert systems have not become popular and utilized. The reason was not that the technology had failed, but that the implementation was inadequate. A prerequisite for success is that we understand 'how medicine thinks', in order to be able to create a 'decision-supporting', and not a 'decision-making' system.

Despite the above mentioned possibilities of improving health care systems through medical expert systems, these systems have to prove their applicability and their cost-benefit relationship through well planned and randomized clinical trials [10]. Beside that these systems have to demonstrate that their computer based recommendations are not misleading in wrong decisions with an increase in morbidity and mortality.

Against the background of the changes mentioned above, the concept of introducing a 'Medical decision support system based on the model of Stem Cell Transplantation', named allo-tool, was developed. The objectives of the system are (1) to improve patient safety through EBM, (2) to support patient autonomy and (3) to optimize the workflow of medical personnel with a favorable result in cost effectiveness. This might lead to a more efficient use of resources without detrimental effects on the relationship between physician and patient or on the physician's autonomy to decide. The allogeneic hematopoietic stem cell transplantation is extremely well suited as a model for this type of system, because of repeating standard procedures, the well defined span of time required and the predictable recovery period as well as recurrent side effects after transplantation.

2   Stem Cell Transplantation and Aim of the Project

Hematopoietic stem cell transplantation (HSCT) or bone marrow transplantation (BMT) is a medical procedure in the field of hematology and oncology. It is most often performed for people with diseases of blood or bone marrow. HSCT remains a risky procedure and has always been reserved for patients with life threatening diseases. Since the availability of stem cell growth factors, most hematopoietic stem cell transplantation procedures have been performed with stem cells collected from the peripheral blood [11; 12]. Most recipients of HSCTs are patients with leukemia or aggressive hematological tumors who would benefit from treatment with high doses of chemotherapy or total body irradiation. Other patients who receive bone marrow transplants include pediatric cases where patients have an inborn defect and were born with defective stem cells [13]. Other conditions that bone marrow transplants are considered for include inherently diseases of the bone marrow.

More recently non-myeloablative or so-called ‚mini transplant' procedures have been developed which do not require such large doses of chemotherapy and radiation [14]. This has allowed HSCT to be conducted in older patients and as a matter of principle without the need for hospitalization. There are two major types of stem cell transplantation maneuvers: Autologous HSCT involves isolation of HSC from a patient, storage of the stem cells in a freezer, high-dose chemotherapy to eradicate the malignant cell population at the cost of also eliminating the patient's bone marrow stem cells, then return of the patient's own stored stem cells to their body. Autologous transplants have the advantage of a lower risk of graft rejection, infection and graft-versus-host disease. Allogeneic HSCT involves two people, one is the healthy donor and one is the recipient. Allogeneic HSC donors must have a tissue type that matches the recipient and, in addition, the recipient requires immunosuppressive medications. Allogeneic transplant donors may be related or unrelated volunteers.

The number of performed HSCT, autologous and allogeneic, is increasing. Due to better anti-infective medication the life-threatening side effects of infectious complications are decreasing, but they still remain as the main risk factor for life threatening side effects [15]. While bacterial infections are the main factor of morbidity and mortality in the first days after transplantation (see Fig. 1), the risk of invasive fungal infections rises in the later time of neutropenia. Often after successful reconstitution of the blood count viral infections occur [16].

Optimizing anti-infective therapy individualized to the specific patient risk profile which is influenced by age, sex, underlying malignancy, tissue type of the donor, previous infectious complications, known allergies and co-morbidities may result in higher patient safety and reducing morbidity and mortality in the setting of HSCT. Randomized trials have to determine, if this leads to cost savings in a longer term.

A further severe complication in allogeneic transplantation is the appearance of graft versus host disease (GVHD). Specific white blood cells, called T cells, can eradicate malignant cells in the recipient. Unfortunately, these T cells recognize the recipient as foreign and employ an immune mechanism to attack recipient tissues in a process known as GVHD. The full therapeutic potential of allogeneic haematopoietic SCT will not be realized until approaches to minimize GVHD, while maintaining the positive contributions of donor T cells, are developed [17].

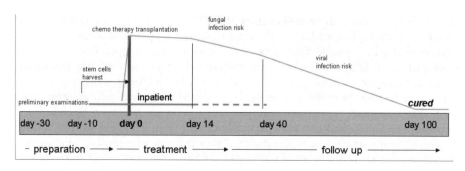

Fig. 1. Risk chart

Therefore one intention of the allo-tool is the improvement of patient safety through an expert system for the decision of optimal chemotherapy, anti-infective therapy and management of graft versus host disease based on evidence based advices. Beside this the support of patient autonomy through modern forms of communications like web-based access to specific patient data and the optimization of the workflow in the complex process of an allogeneic transplantation is the goal of the allo-tool 'smartTransplantation'.

3 Useware Engineering: Process Support for the Systematic Development of User Interfaces

The level of acceptance and efficiency of a modern user interface are not at least determined by the ease of use. System development has been advanced by the development of a comprehensive process [18]. The primary considerations in this evolutionary development process are always the requirements and needs of the user, for whom the user interface is being developed. This is the only guarantee for an efficient use of the system. The process is made up of the following phases: analysis, structuring, general and detailed design, realization, and evaluation (see Fig. 2). The individual phases are not to be viewed in isolation from one another, but rather, as overlapping. The evaluation, as a continuation of the analysis, parallels the whole process.

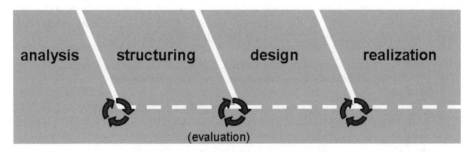

Fig. 2. Useware development process

Within the scope of the allogeneic project, different coordination meetings were conducted and the detailed plans for the follow-on phases were developed prior to the questioning of users. The research and survey data collected through these meetings were used to prepare an interview guideline [19] for the questioning. For a better understanding of the transplantation procedure, a risk chart (see Fig. 1) has been developed explaining the work flow of the stem cell transplantation in detail. Starting with the phases of the transplantation preparation, treatment and follow-up, the required user groups for the analysis could be derived. These were allograft-coordinators, physicians, case-management, ward physicians and nursing staff, of which ten different persons were chosen to be questioned.

After the questioning and characterizing the collected data of the various user groups on the basis of their tasks were compiled. Additional results include an evaluation of existing user interfaces and documentation of a 'wish list' expressed by the users. Problems to be found within the analysis of the clinical situation for the stem cell transplantation were distributed. Redundant data, due to different tools and paper documentation were identified as one major problem as well as standard operating procedures, which are difficult to handle. Finally no automated support for users is available while preparing different documentation.

Further results of the analysis phase are the task models of each user. Those form the basis for developing the use structure within the structuring phase, which will be explained next. Here, we employ an abstract use model, which represents the operations of future system and is, to a large extent, independent of the system hardware.

The modeling language *use*ML [20] is used in structural design. *use*ML defines the use model by use objects and elementary use objects, i.e. 'change', 'release', 'select', 'enter' and 'inform'. With these five elementary use objects it is possible to develop the whole structure for the allo-tool. A part of the use model developed for the allo-tool can be seen in Fig. 3. While developing the use model, it is possible to order different tasks to different user groups. This leads to user group specific perspectives of the use model. It enables developers to filter the tasks on the basis of user groups. This can be seen on the left in Fig. 3. The main screen in Fig. 3 shows the order of different tasks as a part of the use model. The first task is patient list, which is further divided in patient data and history. The list itself consists of different single tasks to choose, show and inform about patients, which are treated with allogeneic stem cell transplantation. Patient data informs about specific data like name, birth date, sex etc. of the chosen patient. Patient history shows details about former diseases, infections etc. Therefore the result of the structural design phase is a platform-independent model, which provides the foundation for the following design phase.

Fig. 3. Part of the use model of the allo-tool

The use concept for the selected hardware platform is prepared in this work step on the basis of the conducted analysis and the use model developed. This includes the tasks. Furthermore, a global navigation concept, a draft of the related structure as well as a proposed layout was developed.

The result is a specific layout of the user interface system, including design of the typical work flow. The layout focuses on standard user tasks and the use structure. In this phase, the layout of the future user interface is developed on the basis of the results from the previous phases. The detailed design results in the specific layout of the user interface, including the design of detailed prototype. This will be presented and explained within the next chapter.

Simultaneous evaluation during all of the formerly mentioned phases enables users to track and assess the development progress at all times on the basis of structures or prototypes. In this way, a timely response to desired changes or modifications is possible. The evaluation included user surveys to determine the validity of the results of structuring and designing.

4 Prototype: Allo-Tool

The intention of the allo-tool is the optimization of the workflow in the complex process of an allogeneic transplantation. This aim will be reached by the data integration of different existing information sources: clinical information system, drug information system and paper patient files. The digitized aggregated data is displayed in a clear structured way, according to the workflow of the allogeneic transplantation, i.e. only relevant data is presented to the user. The user is informed about the medical status and can enter the results directly in the allo-tool, so there is no need to use different tools. Therefore media breaks are avoided.

In future, physicians are supported by a decision support system with integrated knowledge bases (i.e. if a patient has fever of unknown origin, the physician is supported with relevant hints which can lead to the cause and the solution of the problem). The tool shall be able to extract medical information from different sources, to structure the information and to generate automatically discharge letters and further documents, e.g. drug plans. Patients in the follow-up phase will have the possibility to access test results over internet and can communicate directly with their allocated physician via email.

The best starting point for the development of the user interface are the tasks of future users [21]. After the analysis and structuring phase (see Fig. 2) first prototypes, w.r.t. user tasks were developed and tested. According to [24] early testing of prototypes with future users during the user interface development cycle is indispensable.

One of the golden rules of user interface design according to [25] is consistency. As one result, every user group is able to interact in an equal manner with the allo-tool. The structure of the tool consists of four main views: patient, clinical trial, memo and calendar. According to the needed task completeness [22], this structure covers all main use cases, which were derived during the analysis phase.

In the calendar view, users have the possibility to manage appointments, e.g. an appointment with a patient. Physicians and nurses are familiar with the user interface of Microsoft Outlook, so the calendar should be visualized in a similar manner. Already existing appointments of Microsoft Outlook should then be synchronized.

The memo view consists of a large text field, where users have the possibility to take notes. This is very important, because daily life in a hospital environment can be very hectically. Users must have the possibility to take notes very fast and efficient. The memo view is similar to the well known digital yellow post-it note.

In the clinical trial view, users are able to administrate information about current clinical trials. A clinical trial is the application of a scientific method to human health.

For allogeneic transplantations information about conditioning, transplantation, immunosuppression, graft versus host disease prevention and donor lymphocyte transfusions are stated within these clinical trials.

In the patient view, current patients are displayed in a table (the patient list), sorted by appointment. After selecting a patient, users enter a detailed patient screen, which is divided into a summary area at the top and a main work area at the bottom. In the summary area users get consolidated information (see Fig. 4) about the actual status of the transplantation.

According to the workflow of an allogeneic transplantation, the main work area consists of five different chronological parts: patient history, donator search, preliminary examinations, inpatient stay and follow-up treatment. These five parts are clearly separated phases in the time line of an allogeneic transplantation (see Fig. 1). The graphical representation of a time line (see Fig. 4) enables physicians to have a better understanding [26] of the current date according to the complex process of an allogeneic transplantation. The tool supports physicians in keeping deadlines according to a physical examination time schedule.

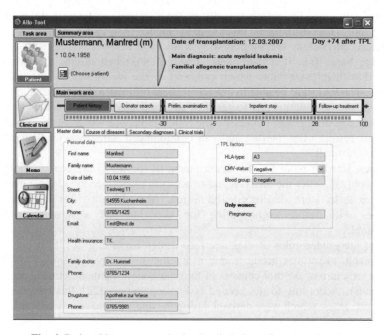

Fig. 4. Patient history screen in the detailed view of a selected patient

The patient's history phase consists of a structured overview of the patient's master data. User do not need to waste time for searching data from the course of disease. Formally, physicians squandered valuable time for searching medical results in different redundant information sources. In the course of diseases, previous examinations and medical results are displayed in a table view. In the allocated clinical trial view, patient are allocated to those, i.e. which patient takes part in which clinical trial and will be medicated and treated according to the advocacies. Secondary diagnoses are an important part of the decision support system and help to minimize adverse effects.

The donator search phase is an important task in the workflow of an allogeneic transplantation. The physician searches for a matching allograft. Once a donator is found allograft coordinators arrange the transportation of the allograft.

The preliminary examination phase contains information about the accomplished examinations according to the physical examination time schedule. The results of external examinations, i.e. examinations in other departments or other clinics, are entered into the clinical information system. The allo-tool shall be able to extract the results of external examinations. A physician evaluates and enhances the results with further information.

In the phase of hospitalization, the patient receives high dose chemotherapy and in selected cases a whole body irradiation. After that the allogeneic stem cell transplantation is performed. During the whole phase, patient data (e.g. vital signs, blood counts or organ functions) are monitored very closely. The documentation of e.g. vital signs is very important for an automatic analysis of the decision support system. Beside others, one important task of the attending physician is to administrate drug plans. Via an implemented interface, physicians are able to use the existing well-known drug information system. They do not need to learn the handling of another system.

In the follow-up phase physicians are reminded to accomplish examinations according to the physical examination time schedule. The time schedule consists of a large number (more than 150) of physical examinations and complex execution logic. Through a digitized time schedule in the prototype, the execution logic is transferred from the physician to the software.

Another identified user group is the patient. Patients in the follow-up phase have the possibility to access test results via internet and can directly communicate with their allocated physician. This will be reached by developing a web-based access for patients. Test results are first evaluated and then released by physicians. Thus patients will have the possibility to view their own data (e.g. blood test results, X-ray photographs). Patients in the follow-up phase have to control e.g. blood pressure values, so they are able to measure blood pressure on their own and email the results over the web-based access directly to their allocated physician. Physicians will have more time for emergencies, and patients don't have to call the hospital via telephone or have to visit their physician on their own.

5 Decision Support

In the future, physicians will be supported by a decision support system with data mining capabilities. Monitoring and interpretation of several parameters such as vital signs or laboratory results will be one main function of the decision support system component. The system thus detects specific clinical situations and pushes unsolicited warnings or reminder and starts the according work flow.

In everyday clinical practice, physicians are often faced with an information overflow rather than a lack of information. Therefore they have to spend valuable time in looking for relevant findings. To reduce this time, a 'Semantic Information

Extraction' is provided. The system looks for findings in the patient's history, which could fit into the context of the current clinical picture. So the physician gets a clearly arranged listing of relevant findings matching the current issue.

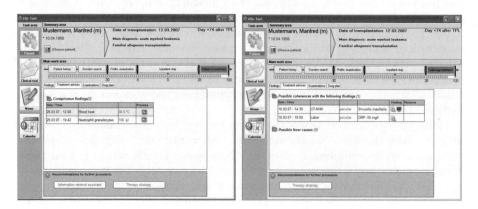

Fig. 5. Conspicuous findings (left) and possible coherences (right)

For a defined amount of clinical pictures, therapy and diagnostic recommendations are provided. These recommendations are on the one hand based on so-called domain knowledge which is patient-independent knowledge on e.g. diseases or processes.

In the follow-up phase physicians are supported by treatment advices (see Fig. 5). The left part of Fig. 5 shows conspicuous findings of a patient. In the example above, a patient in the follow-up phase has fever of unknown origin. The physician is able to view e.g. blood heat process information over the last days or view laboratory results. Additionally she is able to get further information, e.g. there might be possible coherences with other findings (information retrieval assistant). In the right part of Fig. 5, a list with different possible coherences is shown. Findings, e.g. Sinusitis maxillaries may be interesting for further medical decisions. Physicians have the possibility to view original findings e.g. a picture of a computer tomography (CT). Also the system will support physicians by showing possible fever causes. A future version of the decision support system also covers aspects like therapy strategy, i.e. which drugs should the patient take or if the patient should get hospitalized for further examinations.

A critical issue, when it comes to the implementation of decision support systems, is user compliance. Mainly physicians can be seen as relevant users, but also patients in an advanced state of the software. We have to ensure usability of the tool (as described in chapter 4) and ensure that physicians get a benefit from it. The patients are most likely to do everything to improve their therapy outcome because of their severe illness.

One reason why decision support systems often do not prevail in clinical practice is poor work flow integration. Physicians will more likely use a decision support system, if they see a clear benefit from it. This means mainly time-saving as well as convenient access to all relevant information. As the decision support system will seamlessly be embedded as a component into the allo-tool, it can reach an optimal solution to this obstacle.

6 Outlooks

Further steps within this project will be the development of the interfaces from the allo-tool to existing tools in this medical environment, the extraction of data from these tools into the allo-tool as well as the insertion of data from the allo-tool and finally the integration of an expert system to further support the medical staff with their decisions.

7 Conclusions

It is incontestable that people working in health care will have to make use of the potentials of IT in order to meet the enormous demands on patient management in the future. Beside this, the quality of work can be supported by intelligent software which is able to extract, rate and provide relevant data to the user. Furthermore patients require more autonomy of their own health information data. To meet these challenges, the demonstrator of the allo-tool was developed.

Time consuming data search, redundant information and vast numbers of simultaneously used software applications are reduced by displaying all data in one tool. Time schedules, reminder of deadlines and coherent information about study procedures enables medical staff to work efficiently. Taking this as a basis, conditions for a medical decision support system are accomplished. In order to meet the exploding number of scientific perception, decision support systems are needed in the future to maintain the quality of medical decisions.

Through web-based access to selected health information, patients obtain more autonomy and responsibility of their own data. They are able to check and control these and on demand can contact their physician.

Summarizing the potentialities of the planned allo-tool, the goals (1) improvement of patient safety (2) support of patient autonomy and (3) optimization the workflow of medical personnel are illustrated. Studies to evaluate these potentialities are needed to prove the effects of the allo-tool.

Acknowledgements

This work was supported by the Gottlieb Daimler- and Karl Benz-Foundation.

References

1. Glossary of terms in Evidence-Based Medicine. Centre for Evidence-Based Medicine (last access: 2007-08-28), http://www.cebm.net/index.aspx?o=1116
2. Clarke, M., Hopewell, S., Chalmers, I.: Reports of clinical trials should begin and end with up-to-date systematic reviews of other relevant evidence: a status report. J. R. Soc. Med. 100(4), 187–190 (2007)
3. Sneiderman, C.A., Demner-Fushman, D., Fiszman, M., Ide, N.C., Rindflesch, T.C.: Knowledge-Based Methods to Help Clinicians Find Answers in MEDLINE. J. Am. Med. Inform. Assoc. 2007 (2007)
4. Boulware, L.E., Marinopoulos, S., Phillips, K.A., et al.: Systematic review: the value of the periodic health evaluation. Ann. Intern. Med. 146(4), 289–300 (2007)

5. Hope, C., Overhage, J.M., Seger, A., et al.: A tiered approach is more cost effective than traditional pharmacist-based review for classifying computer-detected signals as adverse drug events. J. Biomed. Inform. 36(1-2), 92–98 (2003)
6. Poley, M.J., Edelenbos, K.I., Mosseveld, M., et al.: Cost consequences of implementing an electronic decision support system for ordering laboratory tests in primary care: evidence from a controlled prospective study in the Netherlands. Clin. Chem. 53(2), 213–219 (2007)
7. Bruynesteyn, K., Gant, V., McKenzie, C., et al.: cost-effectiveness analysis of caspofungin vs. liposomal amphotericin B for treatment of suspected fungal infections in the UK. Eur. J. Haematol. 78(6), 532–539 (2007)
8. Shortliffe, E.H., Davis, R., Axline, S.G., Buchanan, B.G., Green, C.C., Cohen, S.N.: Computer-based consultations in clinical therapeutics: explanation and rule acquisition capabilities of the MYCIN system. Comput. Biomed. Res. 8(4), 303–320 (1975)
9. Yu, V.L., Buchanan, B.G., Shortliffe, E.H., et al.: Evaluating the performance of a computer-based consultant. Comput. Programs Biomed. 9(1), 95–102 (1979)
10. Shi, H., Lyons-Weiler, J.: Clinical decision modeling system. BMC Med. Inform. Decis. Mak. 7(1), 23 (2007)
11. Montgomery, M., Cottler-Fox, M.: Mobilization and collection of autologous hematopoietic progenitor/stem cells. Clin. Adv. Hematol. Oncol. 5(2), 127–136 (2007)
12. Korbling, M., Champlin, R.: Peripheral blood progenitor cell transplantation: a replacement for marrow auto- or allografts. Stem Cells 14(2), 185–195 (1996)
13. Satwani, P., Morris, E., Bradley, M.B., Bhatia, M., et al.: Reduced intensity and non-myeloablative allogeneic stem cell transplantation in children and adolescents with malignant and non-malignant diseases. Pediatr Blood Cancer (2007)
14. Del, T.G., Satwani, P., Harrison, L., et al.: A pilot study of reduced intensity conditioning and allogeneic stem cell transplantation from unrelated cord blood and matched family donors in children and adolescent recipients. Bone Marrow Transplant 33(6), 613–622 (2004)
15. Neuburger, S., Maschmeyer, G.: Update on management of infections in cancer and stem cell transplant patients. Ann. Hematol. 85(6), 345–356 (2006)
16. Afessa, B., Peters, S.G.: Major complications following hematopoietic stem cell transplantation. Semin. Respir. Crit. Care Med. 27(3), 297–309 (2006)
17. Shlomchik, W.D.: Graft-versus-host disease. Nat. Rev. Immunol. 7(5), 340–352 (2007)
18. Zühlke, D.: Useware-Engineering für technische Systeme, Berlin (2004)
19. Bödcher, A.: Methodische Nutzungskontext-Analyse als Grundlage eines strukturierten USEWARE-Engineering-Prozesses. Fortschritt-Berichte pak, Band 14, Kaiserslautern: Technische Universität Kaiserslautern (2007)
20. Reuther, A.: useML – Systematische Entwicklung von Maschinenbediensystemen mit XML. Fortschritt-Berichte pak, Band 8, Kaiserslautern: Technische Universität Kaiserslautern (2003)
21. Nielsen, J.: Usability Engineering. Morgan Kaufmann, San Francisco (1994)
22. Dix, A., Finlay, J., Abowd, G., Beale, R.: Human-Computer Interaction, 3rd edn. Pearson, London (2004)
23. Holzinger, A.: Usability Engineering for Software Developers. Communications of the ACM 48(1), 71–74 (2005)
24. Shneiderman, B.: Designing the User Interface, 3rd edn. Addison-Wesley, Reading (1997)
25. Norman, D.: The Design of Everyday Things, Currency (1990)

Framing, Patient Characteristics, and Treatment Selection in Medical Decision-Making

Todd Eric Roswarski[1], Michael D. Murray[2], and Robert W. Proctor[3]

[1] Ivy Tech Community College, Lafayette, Indiana, USA
Office of the Dean, Academic Affairs
troswars@ivytech.edu
[2] University of North Carolina at Chapel Hill, North Carolina, USA
Division of Pharmaceutical Outcomes & Policy
[3] Purdue University, West Lafayette, Indiana, USA
Department of Psychological Sciences

Abstract. The effects of patient characteristics, information, and framing on decision-making were explored using scenarios involving patients with AIDS and lung cancer. Participants were physicians affiliated with a large university medical center and undergraduate psychology students. For the physicians, the roles of experience, workload, fatigue, continuing education, and supervision were examined. In scenario one, physicians showed that the way outcomes were framed affected treatment selection for patients with hemophilia ($p < .0005$), but not for patients who were intravenous drug users ($p = .107$). In scenario two, similar to a previous study, the students showed a significant framing effect ($p = .001$), but the physicians did not ($p = .085$). Patient characteristics and the framing of treatment options can alter decision-making. Experience and additional outcome information also play a role in treatment selection. Computer applications may be a means to eliminate these treatment differences.

Keywords: Decision-Making, Framing, Patient Characteristics, Information, Experience, Decision Support Systems.

1 Introduction

The type and way information is presented to a decision maker generally affects people's decisions. One influence is based on the number of available choices or, in the case of medicine, the number of similar treatment alternatives. Redelmeier and Shafir [1] demonstrated a deferral of choice between medication treatment and/or referral to orthopedics. When physicians were presented with one medication option (Ibuprofen), 53% of the physicians chose to refer to orthopedics without starting any new medication. However, when presented with two options (Ibuprofen or Piroxicam), 72% of the physicians chose to refer without starting any new medication.

In contrast, Roswarski and Murray [2] found in a replication of the previous study [1] that physicians' responses showed a non-significant effect of multiple treatment alternatives ($p = .841$). In the one-medication version, 45.5% of the physicians chose

A. Holzinger (Ed.): USAB 2007, LNCS 4799, pp. 315–322, 2007.

to refer to orthopedics without starting any new medication, with 44.0% choosing this same option in the two-medication version. Further, a significant interaction was found between supervision of medical students and the multiple alternatives effect (p = .012). Physicians who supervised medical students were not influenced by the multiple alternatives, but those who did not supervise students were affected, showing more deferral in the two-medication version. The interaction suggests that supervision of medical students may protect against cognitive bias through the mechanism of explicit and/or implicit learning.

A second influence on decisions is based on the characteristics of the patient (e.g., age, sex, and race). This could lead a physician to different diagnoses or treatments depending on the characteristics of the presenting patient or patient population [3]. A third influence is based on framing (i.e., the way alternatives are presented or structured). Tversky and Kahneman [4] showed reversals of choice based on how outcomes are framed and anchored to a certain value. A situation is first framed and referenced to a certain value. Then, relative to that value, people make decisions based on whether the choice is perceived as a gain or loss. Generally, when situations are framed as a gain, people are risk averse in their decision-making, that is, they will avoid or minimize risk. In contrast, when the situation is framed as a loss, people are risk seeking in their behavior. Levin and Chapman [5] explored decision-making based on framing and characterizations of the groups to receive treatment. Students in their study showed this typical framing effect for a hemophilia group, but not for a homosexual/bisexual/intravenous drug user group.

McNeil, Pauker, Sox, and Tversky [6] explored decision-making based on framing effects for a person with lung cancer. A preference reversal of treatments (surgery or radiation) was found for physicians, patients, and students based upon how the outcomes of those treatments were framed (survival or mortality). McNeil et al. suggested that the mortality data, especially the mortality data associated with surgery were more influential. Further, their study explored life expectancy data, which showed a greater expectancy of 1.4 years of life for surgery versus radiation. The present scenario is important as it includes the cumulative probability and life expectancy data. We propose that the additional data in the form of life expectancy may offset the mortality data.

Thus, the purpose of the present study was to explore physician decision-making using written scenarios and a survey, and to determine whether certain professional characteristics and practices [experience, workload (number of patients seen per day/per week), fatigue (hours worked per day/per week), supervision (medical students, residents), teaching, and continuing education] of physicians would alter their decision-making processes. Specifically, the study examined the effect of patient characteristics on treatment decisions, and the ability to eliminate the framing effect by providing more comprehensive outcome information.

2 Methods

Physician participants were from the Indiana University School of Medicine and student participants were enrolled in Introductory Psychology at Purdue University.

Physicians were given $5.00 *a priori* as a sign of appreciation for their time. Students received credit toward a course requirement.

Scenario one, patterned from Levin and Chapman's [5] Experiment 1 (hereafter referred to as the previous study), explored decision-making based on framing and characterizations of the groups to receive treatment. Scenario two, patterned after McNeil et al.'s [6] lung cancer scenario (hereafter referred to as the original study), explored decision-making based on framing effects (see Appendix). The numerical values used for the cumulative probabilities and life expectancies were identical to the McNeil et al. [6] study. However, the necessity of presenting a currently valid scenario required that certain factors be updated, such as time spent in the hospital, treatment regimen, and possible side effects [7, 8].

The data were analyzed using log-linear procedures. Analogous to the analysis of variance (ANOVA), these procedures test significance of main effects and higher-order interactions based on frequency data [9]. All tests of significance used G^2, which is analogous to the Pearson Chi-Square, and has asymptotically a chi-square distribution. For scenario one, physician responses were recorded for each patient group based on presentation of outcome data (survival versus mortality) and treatment program selected (Program A versus Program B). These variables form the basis of the framing effect. For scenario two, student and physician responses were recorded based on the presentation of the outcome data (survival versus mortality) and the treatment alternative recommended (surgery versus radiation), which form the basis of the framing effect.

3 Results

Of 314 surveys sent to physicians, 192 (61%) were completed and returned. All 145 students completed one version of the lung cancer scenario. For scenario one, similar to the students in the previous study [5], physicians showed a significant framing effect for the hemophilia group ($p = .0005$), but not for the intravenous drug user group ($p = .107$). For the hemophiliacs, only 8.2% of the physicians chose Program B (risky choice) in the survival frame, whereas 54.7% chose this option in the mortality frame. For the intravenous drug users these percentages were 26.3% and 42.9%, respectively. Also, for the intravenous drug users group, a significant interaction of experience and number of patients seen per week with the framing effect was found ($p = .047$). Physicians of 11 years or more showed a significant effect if they saw 61 patients or less per week ($p = .005$), but not if they saw 62 patients or more ($p = .628$), whereas physicians of 10 years or less showed non-significant effects regardless of the number of patients seen (see Table 1).

For scenario two, similar to the students and physicians in the original study [6], the present students showed a significant framing effect ($p = .001$). In the survival frame, 18.1% chose radiation as their recommended treatment, whereas in the mortality frame, 42.5% chose radiation. In contrast, the present physicians showed a non-significant effect ($p = .085$). That is, 17.2% chose radiation as their recommended treatment in the survival frame, with 28.1% choosing radiation in the mortality frame. Experience was shown to be an important variable in the decision-making process. A significant framing effect for physicians of 10 years or less was found ($p = .050$),

Table 1. Participant Responses for Survival and Mortality Frames in Terms of Percentages of Risky Choices (Program B)

| | Percentages of Participants Choosing Risky Choice | | |
	Survival	Mortality	
Hemophilia Patient Group			
Present Physicians (N = 102)	8.2	54.7	p = .0005
Levin & Chapman's Students (N = 47)	21.7	79.2	p < .01
Intravenous Drug User Patient Group			
Present Physicians (N = 87)	26.3	42.9	p = .107
Levin & Chapman's Students (N = 47)	50.0	39.1	*NS
Experience in Years (11 or more)			
Patients Seen per Week			
61 or less (N = 22)	0.0	46.2	p = .005
62 or more (N = 18)	45.5	57.1	p = .628
Experience in Years (10 or less)			
Patients Seen per Week			
61 or less (N = 28)	33.3	31.6	p = .926
62 or more (N = 15)	28.6	50.0	p = .395

*The Chi-squared difference value reported was 0.56.

Table 2. Participant Responses for Survival and Mortality Frames in Terms of Percentage Choosing Radiation over Surgery

| | Percentages of Participants Choosing Radiation | | |
	Survival	Mortality	
Current Study			
Students (N = 145)	18.1	42.5	p = .001
Physicians (N = 176)	17.2	28.1	p = .085
McNeil et al. (1982) Study			
Students (N = 297)	16.8	42.9	*p < .001
Patients (N = 119)	22.0	40.0	*p < .001
Physicians (N = 167)	16.1	50.0	*p < .001
Experience (in years)			
10 or less (N = 100)	16.3	33.3	p = .050
11 or more (N = 76)	18.2	18.8	p = .950

*Individual analyses of each subject population were not reported, however, the main effect of framing was reported at p < .001.

whereas there was no effect for physicians of 11 years or more ($p = .950$; see Table 2). The difference between these two groups was non-significant in the mortality frame ($p = .134$) and in the survival frame ($p = .814$).

4 Discussion

For scenario one, physicians in the present study and students in the previous one [5] exhibited a framing effect for the hemophilia patient group, but neither did for the intravenous drug user group. Thus, the robustness of the framing effect was again demonstrated for the hemophilia group; however, that robustness was overcome by the strength of patient characteristics for the intravenous drug user group. Those physicians with 11 years or more of experience and who see 61 patients or less per week did demonstrate a framing effect with the intravenous drug user patient group. Although these experienced physicians who have a smaller workload were not influenced by patient characteristics, they were nonetheless still influenced by the effects of framing. Thus, overall the treatment selection of physicians was influenced by the effects of framing or patient characteristics.

For scenario two, the most important contribution was the demonstration that a more complete set of outcome information could eliminate the framing effect. Second, this elimination was participant specific, with the students still showing a framing effect. In fact, the framing effect was not even reduced for the present students, who had the additional information, in comparison to the original students [6] who did not. Third, the additional information led to a complete absence of an effect for the more experienced physicians, but not for the less experienced physicians. The framing effect for the less experienced physicians was, however, reduced in comparison to the physicians in the original study [6].

Of the numerous survey variables, only experience and workload seemed to have an impact on the effects of framing and patient characteristics. Thus, this shows the robustness of the effects of patient characteristics and framing. Experience may not protect a decision-maker from the effects of framing when only limited information is provided, but experience does appear to protect that decision-maker when a more complete set of information is provided. Experience may afford this protection by allowing the decision-maker to thoroughly organize, integrate, and analyze information. Also, experience combined with the additional information may allow the decision-maker to check his or her biases [10].

5 Implications for Computer Support

Roswarski and Murray's [2] study stated that encouraging residents and new faculty to engage in active supervision each week may protect them from the cognitive bias associated with multiple alternatives and other biases. For the current scenarios, experience and workload were the only two survey variables that had an effect on treatment, as well as the additional life expectancy data provided for the lung cancer scenario. However, not all physicians will have the opportunity to be supervisors, workload is not easily kept to fewer than 61 patients seen per week, and experience only comes with years of practice; yet treatment decisions are made everyday.

Thus, certain computer applications may assist physicians in eliminating the cognitive biases associated with these treatment decisions. First, a database of cognitive biases and their applications to medicine could be established, along with links to published papers associated with these biases. This would allow medical students and physicians of all experience levels quick and easy access to these materials. Further, since decisions associated with treatment selection involving diseases such as lung cancer often involve the patient, a similar patient site could also be developed. Second, treatment decision scenarios could be developed to allow medical students and residents to examine their treatment choices versus the computer. This would allow them to see how their cognitive biases affect their treatment selection, and then the residents and students would be better equipped to check their own biases in the future. Finally, decision-support systems [11] may be developed to assist physicians by asking for pertinent information, assisting with organizing and analyzing the data and outcome information, and prompting questions that would allow physicians to check for possible biases in their decision-making process.

Roswarski and Murray [2] hypothesized that physicians with more experience (years worked and hours supervising) would be more likely to engage in the type of decision-making described by poliheuristic theory. This theory incorporates elements of both cognitive decision-making (holistic) and expected utility theory (analytical) [12]. According to this theory, the first task of the decision-maker is to eliminate unacceptable alternatives. The second task is to use analytical processing to choose one of the alternatives. Roswarski and Murray [2], based on their results and poliheuristic theory, also postulated that physicians could be instructed on how to approach decisions with multiple alternatives. This would include instructing physicians to eliminate unacceptable alternatives, combine choices into categories, choose among those categories, and then choose within the category [12, 13]. In their scenario, this would result in categories of medication and no medication. If the medication option were chosen, then the physician would select between the two drugs.

Similarly, physicians could and should be taught how to approach decisions involving framing and patient characteristics. However, there are large numbers of potential treatment biases associated with patient characteristics (e.g., women and coronary heart disease) [14, 15], treatment decisions with multiple alternatives (and more than just two alternatives), and treatment decisions involving outcome data that could be framed, as well as the fact that cognitive processes that may result in biases are adaptive in many decisions [16]. Thus, decision support systems based on the elements of poliheuristic theory could be designed to aid physicians in making treatment decisions with multiple alternatives. Further, these systems could provide outcome information to the decision-maker in multiple ways and combinations, and could store and provide information of biases associated with patient populations, such as gender differences in presenting symptoms for certain diseases or conditions. Finally, through the decision-making experience gained through the use of these support systems, physicians' decision-making should be less influenced by cognitive biases during times when the support systems are unavailable or impractical to use due to time constraints.

Acknowledgements. James S. Nairne, James J. Walker, Mark R. Lehto, and Alicia M. Altizer provided helpful comments. Richard Schweickert provided assistance with the data analysis. Herbert E. Cushing III provided his endorsement of the study. Financial support was provided by the Purdue University Department of Psychological Sciences Incentive Research Fund, Human Performance Laboratory, and the School of Liberal Arts Faculty Development Fund, and by R01s AG19105, AG07631, AG021071, & HL69399.

References

1. Redelmeier, D.A., Shafir, E.: Medical Decision Making in Situations that Offer Multiple Alternatives. Journal of the American Medical Association 273, 302–305 (1995)
2. Roswarski, T.E., Murray, M.D.: Supervision of Students May Protect Academic Physicians from Cognitive Bias: A Study of Decision-making and Multiple Treatment Alternatives in Medicine. Medical Decision Making 26, 154–161 (2006)
3. McKinlay, J.B., Potter, D.A., Feldman, H.A.: Non-medical Influences on Medical Decision-making. Social Sciences and Medicine 42, 769–776 (1996)
4. Tversky, A., Kahneman, D.: The Framing of Decisions and the Psychology of Choice. Science 211, 453–458 (1981)
5. Levin, I.P., Chapman, D.P.: Risk Taking, Frame of Reference, and Characterization of Victim Groups in AIDS Treatment Decisions. Journal of Experimental Social Psychology 26, 421–434 (1990)
6. McNeil, B.J., Pauker, S.G., Sox, H.C, Tversky Jr, A.: On the Elicitation of Preferences for Alternative Therapies. New England Journal of Medicine 306, 1259–1262 (1982)
7. National Cancer Institute: Lung Cancer (NIH Publication No. 99-1553). (last access 1999), http://cancernet.nci.nih.gov/wyntk_pubs/lung.htm
8. American Cancer Society: Lung Cancer Resource Center. (last access 2000), http://www.cancer.org
9. Kennedy, J.J.: Analyzing Qualitative Data. In: Log-linear Analysis for Behavioral Research, 2nd edn. Praeger, New York (1992)
10. Smith, P.J., Galdes, D., Fraser, J., et al.: Coping with the Complexities of Multiple-solution Problems: A Case Study. International Journal of Man.-Machine Studies 35, 429–453 (1991)
11. Lehto, M.R., Nah, F.: Decision-Making Models and Decision Support. In: Salvendy, G. (ed.) Handbook of Human Factors and Ergonomics, pp. 191–242. John Wiley, New Jersey (2006)
12. Mintz, A.: How do Leaders Make Decisions? A Poliheuristic Perspective. Journal of Conflict Resolution 48, 3–13 (2004)
13. Dacey, R., Carlson, L.J.: Traditional Decision Analysis and the Poliheuristic Theory of Foreign Policy Decision Making. Journal of Conflict Resolution 48, 38–55 (2004)
14. Frank, E., Taylor, C.B.: Coronary Heart Disease in Women: Influences on Diagnosis and Treatment. Annals of Behavioral Medicine 15, 156–161 (1993)
15. Mark, D.B.: Sex Bias in Cardiovascular Care. Should Women be Treated More like Men? Journal of the American Medical Association 283, 659–661 (2000)
16. Christensen, C., Elstein, A.S., Bernstein, L.M.: Formal Decision Supports in Medical Practice and Education. Teaching and Learning in Medicine 3, 62–70 (1991)

Appendix: Scenario Versions

Scenario One

The United States is expecting the outbreak of a new strain of AIDS, which is expected to kill 6,000 persons. This new strain of AIDS will primarily affect intravenous drug users. There have been two programs developed to combat the disease, and the scientific estimates of their consequences are the following.

Please indicate which program you would choose.
Participants given the positively framed options with intravenous drug user patients had the following options to choose between:

Program A)   If this program is implemented, 2,000 people will be saved.

Program B)   If this program is implemented, there is a one-third probability that all 6,000 people will be saved, and a two-thirds probability that nobody will be saved.

Participants given the negatively framed options were asked to choose between the following options:

Program A)   If this program is implemented, 4,000 people will die.

Program B)   If this program is implemented, there is a one-third probability that nobody will die, and a two-thirds probability that all 6,000 people will die.

For the hemophilia patient group, the participants were given the same options, with the words 'hemophiliacs needing blood transfusions', replacing the words 'intravenous drug users' in the scenario.

Scenario Two

A patient presents with lung cancer. Given their physical condition, and type and stage of lung cancer, your treatment options are either surgery or radiation. With surgery, most patients are in the hospital for one to two weeks, and have some pain around their incisions. Further, some patients experience fatigue, reduced strength, shortness of breath, and fluid buildup in the lungs. They spend about six weeks recuperating at home. After that, they generally feel fine.

With radiation, the patient comes to the hospital about five times a week for six weeks. During the course of the treatment, some patients experience skin tenderness, fatigue, shortness of breath, swallowing difficulties, and loss of appetite. By two weeks post-treatment, these symptoms alleviate and they generally feel fine. Thus, after the initial eight weeks, patients treated with either surgery or radiation therapies feel about the same.

Participants given the positively framed options were provided with the following statistics and had the following options to choose between:

Based upon this patient's profile, of 100 people having surgery, 90 will be alive immediately after the treatment, 68 will be alive after one year, and 34 will be alive after five years. The life expectancy of all patients who undergo surgery is 6.1 years. Of 100 people having radiation therapy, all will be alive immediately after treatment, 77 will be alive after one year, and 22 will be alive after five years. The life expectancy of all patients who undergo radiation therapy is 4.7 years.

Which treatment would you recommend?
A) Surgery
B) Radiation

Participants given the negatively framed options were provided with the following statistics:

Based upon this patient's profile, of 100 people having surgery, 10 will die during treatment, 32 will have died by one year, and 66 will have died by five years. The life expectancy of all patients who undergo surgery is 6.1 years. Of 100 people having radiation therapy, none will die during treatment, 23 will die by one year, and 78 will die by five years. The life expectancy of all patients who undergo radiation therapy is 4.7 years.

The How and Why of Incident Investigation: Implications for Health Information Technology

Marilyn Sue Bogner

Institute for the Study of Human Error, LLC Bethesda, MD
msbogner@verizon.net

Abstract. The potential of health information technology to effectively support the work of health care providers and reduce the likelihood of errors and incidents has not been realized; however, the manner of investigating incidents can provide information to aid in its realization. Implications of negligible findings from extensive research on provider accountability for errors point to the importance of addressing the nature of error incidents. Consideration of the nature of incidents together with lessons learned from industry error research expands the focus of incident investigations to include how and why the event happened. A model to guide incident investigations and examples of the viability of that model to address issues in using health information technology are described. The wisdom of Sherlock Holmes accompanies this sleuthing for methods to enhance information technology to better support care providers in their daily work.

Keywords: Incident Investigation. Health Information Technology, Error, Behavior, Systems, User Support.

1 Introduction

Health Information Technology (HIT) commonly considered as comprised of the Electronic Medical Record (EMR), Computerized Physician Order Entry (CPOE), and drug administration bar coding has been advocated in the United States as the means to enhance if not ensure patient safety.

Despite all the rhetoric that extols the virtues of HIT there is little evidence that HIT or any aspect of it appreciably supports the work of health care providers across institutions. Indeed, the literature advocating HIT is replete with words such as can, should, has the potential. Possibly due to the lack of clearly demonstrated usefulness for providers of varying degrees of computer savvy and more certainly the initial and maintenance costs for the technology, as well as other reasons, the rates for adopting HIT are low. It is estimated that from 17% to 24% of physicians have access to EMR and 4% to 24% of hospitals have adopted CPOE [1].

2 You See, But You Do Not Observe: Sherlock Holmes in "Scandal In Bohemia"

Difficult to read handwriting has long been an issue in medication errors and misunderstanding orders in patients' charts. A means to avoid handwriting issues is

A. Holzinger (Ed.): USAB 2007, LNCS 4799, pp. 323–334, 2007.

provided through the use of technology. Health care specific software often having the same format as the hardcopy transforms a computer into an electronic chart, which often is considered an EMR although ideally that term would refer to a composite of all medical records for a patient. Similarly software was developed for the provider to enter orders including prescriptions into a hand held computer, CPOE, transforming that technology into a vehicle for conveying written orders without subjecting the person receiving them to incomprehensible handwriting. Because the users of EMR and CPOE may have needs that differ from those of the typical computer user, special attention is given to the user interface - be it a desktop, laptop, or mobile computer. Usability is most often defined as the ease of use and acceptability of a system for a particular class of users carrying out specific tasks in a specific environment. Exactly, this ease of use affects the users' performance and their satisfaction. Consequently, it is of great importance that every software practitioner not only be aware of various usability methods, but be able to quickly determine which method is best suited to every situation in a software project [2].

Especially, when design and develop mobile systems for health care, it is essential to obtain empirical insight into the work practice and context in which the proposed mobile system will be used. Consequently, mobile devices are only useful when design and software validation aspects have been taken into account [3, 4]. On the other hand, bar coding for drug administration seems less complex – possibly, what is effective in the supermarket certainly should be effective in the hospital.

Studies of the actual use of CPOE and bar coding suggest that those technologies in their present form may not be the silver bullet they are purported to be. A study in a highly computerized U.S. Veterans Administration hospital found that despite CPOE, bar coded medication delivery, EMR, automated drug-drug interaction and computerized allergy alert, of the 937 admissions in a 20 week period 437 experienced adverse drug events (ADEs). Of those ADEs, 36% were an adverse drug reaction, 33% wrong dose, and 7% inappropriate medication [5]. Errors were identified in 61% of medication orders, 25% in medication monitoring, 13% in drug administration, 1% in dispensing, and 0% in transcription. It was concluded that health care providers "... should not rely on generic CPOE and bar coding". This does not bode well for HIT unless the problems are unique to this study – they aren't.

The impact on error of a commercially sold CPOE software program was studied by comparing patient demographic, clinical, and mortality data before and after implementation of CPOE in a regional, academic, tertiary-care children's hospital. To assess similarity of the pre and post implementation groups, 18 demographic and clinical characteristics of the 1394 children admitted during the 13 months prior to the implementation of CPOE were compared to those of the 548 children admitted during the 5 months post implementation. Statistical analysis found no significant differences between the characteristics of the pre and post implementation groups. A difference in mortality was found, however.

The mortality rate for the 13 months before CPOE was 2.80% whereas the rate was 6.37% for the 5 month study period after implementation. These differences were not attributed solely to errors in specific drug ordering; rather they reflected several unintended consequences from the implementation of CPOE. Delays in treatment occurred that were considered as reflecting the amount of time spent in the physical process of entering orders using CPOE. With CPOE order entry required an average of 10 mouse clicks or 1 to 2 minutes for each order. The same order in written form took only a few seconds. With multiple orders, the total minutes spent using CPOE became non-trivial and delayed time sensitive therapies. Difficulties integrating CPOE into the existing workflow occurred. The interactions among ICU team members were altered and the dynamics of bedside care changed after the implementation of CPOE. From the findings of this study, CPOE seems to be tool rather than a solution for patient safety.

In addressing the viability of bar coding, it was found that multi-dose vials or containers, such as IV solution bottles and bags, insulin bottles, inhalers, and tubes of ointments are difficult to scan as are wristbands that have blood or other substances on them. In addition, bar codes can be absent or difficult to read by the scanner [7]. To increase the viability of bar coding, the study authors advocate the establishment of bar code verification laboratories that would assess bar-code quality and wristband verification. The goal of these laboratories is to enhance the quality and longevity of wristbands to reduce or eliminate workarounds necessary to accommodate problem wristbands and to address problematic medication bar codes.

Introducing new technology into complex medical systems is intended to be beneficial; however, it can present new challenges. We see that technology works well in other applications, or does it? Returning to the use of bar coding at the supermarket – by being attentive at checkout one can observe perturbations in the viability of bar coding – perturbations induced by the shape of the product being scanned. Since the impact of the product shape on bar code scanning is not observed or if observed, not heeded; the common presumption is the technology works well and the issue most likely is with the user or possibly with the interface of the user and the technology.

This has becomes so much a mind-set that it may seem there is nothing else to consider. That belief by focusing on the care provider has constrained the effectiveness of many patient safety endeavors. Valuable lessons can be learned from the investigation of incidents – lessons that can enhance the technology that supports health care professionals.

3   It Is a Capital Mistake to Theorize Before You Have All the Evidence It Biases the Judgment Sherlock Holmes in "Study in Scarlet"

Although the topic under discussion is the investigation of incidents, which are unintended unexpected occurrences, errors also are addressed as a special case of incidents. The nature of incidents and errors provides the evidence base for a model of behavior that affords insight for technology enhancements to effectively and efficiently support the work of health care providers. The nature of error, however, typically is not addressed.

3.1 *Who* Is Responsible for an Incident?

People strive to explain what causes something to happen. This has a survival benefit be it social or physical because the explanation can identify conditions to replicate a pleasant experience or to avoid an unpleasant or dangerous one. When searching for causes of an event a person explores possibilities until he or she comes to a reasonable explanation that can be used to effectively address a similar situation. At that point, the person terminates the search for a cause. This is the manner by which the subjectively pragmatic Stop Rule [8] determines explanations for an occurrence. Typically when a person is associated with the event, the explanation is *that the person caused it.*

Nowhere is this more apparent than in identifying the causes of health care errors and incidents as evidenced by the response to a report on medical error.

The citizens of the United States were outraged by the report by the Institute of Medicine (IOM) of the U.S. National Academy of Sciences that annually 44,000 – 98,000 hospitalized patients die because of errors in their care [9]. That number was considered low because of under-reporting, which added fuel to the outrage. Error in health care is not unique to the U.S.; it was reported that an error occurs in 1 in 10 hospital admissions in U.K. [10].

The IOM report proposed that the issue of medical errors be addressed by research to "...hold providers accountable for their performance, or alternatively provide information that can lead to improved safety" [9, p. 74]. Interestingly, the report recommended that a research program be instituted to determine how to hold providers accountable rather than providing information that can lead to improved safety. The power of the Stop Rule is evident in this recommendation. The report went on to state research should be conducted to meet the goal embedded in the assertion that it is "irresponsible to expect anything less than 50 percent reduction in errors over 5 years" [9, p. 3].

In response to the public consternation, Congress appropriated the unprecedented amount of USD 50 million to fund such research for each of the five years and a vigorous research program commenced under the auspices of the U.S. Agency for Healthcare Research and Quality.

3.2 *What* Was Addressed by the Research

The definitions of error incidents illustrate what was reported and continues to be reported. Error has been defined by Bosk [8] as technical errors reflecting skill failures, judgmental errors involving the selection of an incorrect strategy of treatment, and normative errors that occur when the larger social values embedded within medicine as a profession are violated. A definition of errors in terms of intentions, actions, and consequences by Reason [9] includes slips, mistakes, lapses, latent errors, resident pathogens, and active errors. Another definition of error is provided by Gibson & Singh [10] in terms of the aspect of health care in which the event occurs, such as errors of missed diagnosis, mistakes during treatment, medication mistakes, inadequate postoperative care, and mistaken identity.

Error was defined in the IOM report [9, p. 23] as " ... the failure of a planned action to be completed as intended (i.e., error of execution) or the use of a wrong plan to achieve an aim (i.e., error of planning)". It is worth noting that in all these definitions of error, the care provider is the implicit cause. Thus, it is not surprising that errors were reported in the research in terms of the person who caused the event or who did what in keeping with the emphasis on provider accountability. At the end of the five years, a meeting was convened to determine the outcome of five years and USD 250 million of research focused on provider accountability – the bulk of that research essentially addressed error in terms of who did what. Efforts to attain the 50% reduction in error not only did not meet that goal, but the impact of those efforts on error was negligible.

It might be said that the major contribution of the efforts that considered error as caused solely by the heath care provider is the perpetuation of the blame culture. Because the causes of errors were presumed hence found to be the care providers, the seemingly only possible conclusion was that those individuals alone are to blame for errors. The approach of provider accountability exists today and the conclusion that the health care providers are to blame for errors persists. Notable progress in reducing the incidence of errors continues to be elusive.

The wisdom of Holmes quoted at the beginning of this section, that it is a mistake to theorize before having evidence because it biases judgment speaks directly to the implications of what is considered error. A hard look at the previously reported definitions of error finds that they are only descriptions of errors, they describe the evidence that an error occurred, but indicate nothing about the nature of error, that is, what occurs that is manifest in such descriptions. Because of that they provide no information as to what might be changed to prevent recurrence of the incident except to train or chastise the perpetrator of the error, the care provider.

Describing an error as a skill failure or a mistake provides no information about the nature of the process by which that skill failure or mistake occurred. This is another instance of the Stop Rule – those descriptions satisfy the need to explain an adverse outcome, but not to understand its dynamics. Such understanding is necessary to identify actionable items to address and prevent the recurrence of the event. Insight into the nature of error comes when an error incident is considered as what it actually is, an act by a person – an act which is a behavior. Although this statement is simplistic it has profound implications for incident investigations and for technology that assists health care workflow.

The implications are profound because the discipline that has formally studied human behavior for over two centuries, psychology, has found that behavior, B, is a function, f, of the person, P, interacting with the environment, E [14]. This interaction can be represented as equation (1).

$$B = f \{P \times E\} \tag{1}$$

Representing the nature of behavior as an equation (1) drives home the necessity of considering more than the person who is associated with the event. Indeed, as the equation illustrates addressing only the person is incomplete and misleading. The environment must be considered. It is to be noted that although the definition of behavior comes from psychology, all science including the physical sciences considers the focus of their study as it functions within an environmental context. The lack of findings from the 5 years of research on provider accountability can be interpreted as demonstrating the fallacy of an approach that does not consider context.

The context for health care is arguably more complex than any other in which human behavior is addressed. This can be disquieting "The scale of the problem and the complexity of the system will require sustained effort if solutions are to be found"

[15]. The complexity of context however, should not, and indeed cannot, be ignored when investigating incidents.

A further fact is that information retrieval is an important part of the daily work of medical professionals and the body of knowledge in medicine is growing enormously, consequently clinicians must rely on various sources of medical information and increasingly they are faced with the problem of too much rather than too little information and the problem of information overload is rapidly approaching.

4   How Often Have I Said to You That When You Have Eliminated the Impossible, Whatever Remains, *However Improbable*, Must Be the Truth Sherlock Holmes in "The Sign of Four"

Complexity of health care need not be off-putting because the basic unit to be investigated is the person who was associated with the event in relation to factors in the context. [16]. The distal factors in an event influence the person so he or she acts to be the proximal cause of the incident. Context although complex, is less onerous when it is considered only as that affecting the person whose behavior is being addressed. The context to be addressed thus is a microcosm representative of all of health care. Nonetheless, it is a Gordian knot albeit a relatively small one to be unraveled. The unraveling is accomplished through findings from error research by industries other than health care such as manufacturing, nuclear power, and aviation [17, 18, 19].

Those findings identify categories of interacting factors that impact the performance of a person – categories that are analogous across those industries. Because such consistency suggests that human performance is affected by comparable factors across tasks, the categories of factors are considered appropriate for addressing health care incidents [20].

In terms appropriate for health care, the categories of characteristics with examples of characteristics for each are:

- *Patient* – age, gender, weight, co-morbidity, unreported meds, surgical site, name because of same name mix-ups;
- *Means of Providing Care* – medications, hand operated medical devices, therapeutic radiation devices, diagnostic devices, exercise equipment, therapy devices;
- *Provider* – experience, stamina, anthropometric characteristics including height and hand size, human constraints such as handedness and human cognitive limitations;
- *Ambient Conditions* – illumination, temperature, noise, altitude, sound, air quality;
- *Physical Environment* – room size, clutter, placement of devices and equipment, arrangement of supplies;
- *Social Environment* – other care providers, family and friends of the patient, professional culture;
- *Organization* – workload, reports, policies, organizational culture; and

- *Legal-Regulatory-Reimbursement-National Culture Factors* – litigation threat, regulatory concerns, fiscal constraints, societal culture.

Those categories of interacting factors comprise the environmental context of systems that influence the person performing the task – systems that are hierarchal with perturbations in one system affecting others. For example, changes in reimbursement affect how the organization operates, which affects the other systems and the care provider. These eight systems of factors are illustrated as Figure 1.

Factors in the eight systems impinge on everyone involved in providing health care; however, all factors in all categories do not always impinge on a person. The factors change as the person performs different tasks in different locations caring for different patients using various means of providing care. Factors that influence a person at a specific time are considered to be in that person's life space [21].

If a factor is not in a person's life space at the time of an incident, then it does not functionally exist for the person hence it does not affect the error behavior. When investigating an incident it is necessary to identify the factors in those systems that affected the behavior of the care provider to whom the incident is attributed – the factors in his or her life space – at the time of the incident.

Thus, this approach to investigating an incident seeks to understand the behavior that lead to the accident by addressing the person interacting with factors in his or her environment [22, 23] represented by the systems in Figure 1.

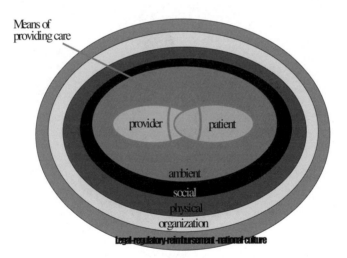

Fig. 1. The context of systems of factors that influence a person performing a task.

To assist in conceptualizing factors in the life space of a care provider, those factors in the concentric circles representing the systems of factors in a person's environment that affect him or her at each specific point in time [23] are represented as the concentric layers of leaves of an artichoke. The person affected by those factors, the care provider, is the center or focus of those factors, the heart of the artichoke. This systems model of behavior, the Artichoke model, is illustrated as Figure 2.

4.1 *How* and *Why* and the Artichoke

In investigating an incident it is necessary to determine not only how it occurred, but also why it occurred so changes can be made to prevent its recurrence.

To determine *how* an incident occurred, the care provider involved in the incident should consider each of the systems of factors and note the factors that very likely affected him or her contributing to the actions that culminated in the incident or error. Those factors indicate *how* the incident or error occurred.

To determine *why* the incident occurred, a Five Whys analysis of each of the incident inducing factors is performed. That analysis consists of writing down one of the factors involved in the incident and simply asking why that factor contributed to the incident. The response is recorded and becomes the target for another inquiry of Why. This process continues five times, hence the name Five Whys analysis, or until at least one actionable item is identified for each defining factor. It is to be noted that when investigating an incident involving more than one individual, the life space Artichoke of each individual is considered separately. The how and why are determined by each person only for his or her self. Factors identified by that process by each person are compared for any common error provoking factors that are central to the incident. The perception of error inducing factors by more than one person cannot be considered in one life space. The Artichoke has only one heart. The perception of another person, however, can be a factor in the social environment of a person's life space.

Fig. 2. The Artichoke context of systems factors that affect the health care provider and induce behavior that can lead to an error or incident

To effectively reduce error, the list of actionable items identified by the Why process for all incident defining factors is referred to a designated person who is empowered to see to it that each item is addressed so that a recurrence of the incident might be prevented.

Although an incident investigation typically is closed at this point if not earlier at the point of identifying how the incident occurred, the Artichoke behavior systems approach to investigating an incident is not closed until changes to prevent recurrence of the incident are implemented and evaluated. Simple examples illustrate the value of this approach.

Bar Coding: Ointment misadministration. Consider the incident of using an ointment that wasn't ordered for that patient to dress a wound.

Although no long-term adverse effects resulted from this, the wound required more time to heal than projected using the appropriate ointment.

To determine how the wrong ointment was used on a patient when all medications are bar coded and scanned for accurate delivery, the care provider involved in the incident responded to the systems of factors of the Artichoke to identify factors involved in the incident. First the care provider noted that the bar code scan was accomplished as directed although he had some difficulty scanning the bar coded label that was adhered directly to the tube. This identifies the entities involved in this incident.

To determine why the incident occurred, the provider asks himself *"Why does the label being on the tube contribute to the incident?"* He answers that the label was wrinkled apparently because the part of the tube it was on was uneven as if it had been squeezed when held. Why wasn't the tube in an individual box with the label on the less easily disturbed box surface? The reply was that the tubes of that ointment were purchased in bulk. Why were they purchased in bulk? Because that has a lower cost than buying tubes in individual boxes.

This is an actionable item – criteria for purchasing medications. To address that issue, pharmacy, purchasing, and the care provider involved in the incident can discuss if the bulk purchase cost savings are sufficient to offset the risk of inappropriate bar code scans or if there might be way to ensure a solid flat area on a tube for the bar coded label removing the difficulty in scanning.

CPOE: Wrong drug ordered. In addressing an incident of ordering of the wrong drug using CPOE, rather than chastise the provider for the act, the Artichoke model is used to guide the quest for *how* the incident occurred and *Why*. The care provider involved in the incident responds to the systems of the Artichoke that are relevant to the incident to describe how it occurred. In considering ambient conditions, the care provider indicated that it was quite noisy with several very loud startling crashes at the time and location where she ordered a drug using her CPOE device. This describes how the incident occurred. Given that hospitals can be very noisy places with loud crashes not uncommon and CPOE is used to avoid errors in drug ordering, the care provider sought to delineate why the incident occurred by directing the Five Whys technique to the factors that described how the incident occurred.

The care provider asked "why was there such noise?" The answer was that an area of the hospital was being renovated. This is not an actionable item because the construction is necessary to preserve the infrastructure of the hospital and sudden loud noises can occur during construction. Looking to herself, the provider, she asked why she was startled by the crashes.

The reply was she was concentrating intently on ordering the drug because of the distracting level of the ambient noise and didn't anticipate the extreme deviation in the ambient sound of the crashes. Why might this contribute to the incident? The noise of the crash startled the care provider causing her hand to jerk.

Why is that relevant? The jerking of the care provider's hand lead to her inadvertently indicating a drug that was adjacent on the list to the drug she intended to order. Why did this happen? The CPOE software presents the drug names quite close together in alphabetical order. Why is that relevant?

An adjacent drug inadvertently may be ordered. This might occur not only when the care provider is startled by a loud noise and jerks her hand, but also when a provider's attention is distracted by being asked a question. This line of reasoning has identified an actionable item – the CPOE software. The design of the CPOE software should be revised to eliminate the condition that extreme precision must be used in designating a drug or another drug will be inadvertently ordered. The precision may not be a problem under ordinary conditions when the ambient noise although possibly loud is of a consistent intensity and when there are no interruptions. When a loud crash or other marked distraction occurs, that level of precision becomes a hazard, an incident waiting to happen. To prevent recurrence of the incident, the CPOE software should be designed so it is extremely difficult to order the wrong drug. That software should be tested by an active care provider in a variety of environmental conditions – simulated ones if necessary.

4.2 Lessons from the Artichoke

The determination of how and why incidents occur using the Artichoke as a guide leads to the identification of factors that contributed to if not provoked the incident. Those factors are targets for change to reduce the likelihood of recurrence of the incident. In the CPOE incident, the target for change was the design of the software. An additional advantage of this approach is that information about why an incident occurred such as identifying software that invites error by requiring undue precision in use – precision that is difficult to next to impossible to execute in the patient care environment – can be applied to other technologies to avoid errors throughout the facility. Because the same devices are used in other facilities, the issues and solutions for CPOE and other technologies can be shared with colleagues in other locales thus multiplying the positive impact of the findings from the incident investigation.

The value of a behavior systems approach is underscored when it is considered with respect to the activities from investigating an incident in terms of provider reliability. The latter investigation addresses only the care provider involved in the incident, so remedial activities are directed to the specific care provider. Because of that focus, software issues that affect all providers using CPOE and other software-driven devices are not addressed so those issues continue to contribute to if not induce incidents.

5 You Know My Method It Is Founded Upon the Observance of Trifles Sherlock Holmes in "The Boscombe Valley Mystery"

Although only bar coding and CPOE examples were discussed, EMR issues can be analyzed similarly by using the Artichoke and the Five Whys. The factors that characterize the actual problem, how the incident occurred, can be determined and those factors that contributed to why it occurred can be identified and addressed. The impact of those activities can be evaluated and incorporated into the technology involved in the incident so it is refined to effectively support the endeavors of the health care provider. The nature of error as behavior, of the person interacting with factors in the environment, provides the conceptual clarification that is a prerequisite for efficient experimentation [15]. To effectively reduce error and incidents in health care, it is mandatory that investigations be conducted as efficient experimentation to determine how an incident occurred and identify the factors involved in why it occurred as well as who and what were involved in it. It is emphasized that to truly address error, factors in the context of health care that contribute to hazards as well as near misses should be identified and subjected to the same analytical process as errors. By determining how and why a hazard exists or near miss occurs as well as what is involved and the types of care providers whose behavior could be affected, errors and possible adverse consequences can be prevented. No special activities are necessary to addressing hazards and near misses – health care providers are aware of such threats to patient safety and compensate for them daily. Factors that contribute to hazards and near misses are in the life spaces of health care providers.

Activities to prevent incidents should not be limited to incidents that have occurred. Conscientious and continual investigations should be conducted for near misses and hazards or accidents waiting to happen that are associated with HIT as well as devices and procedures. As with incidents that have occurred, these investigations should determine how and why there are HIT related near misses and hazards, the identified contributing factors addressed and the resulting changes incorporated into the technology. Such investigations of near misses and hazards together with investigations of incidents that have occurred can create a program of continuous quality improvement for HIT.

6 Conclusion

It is time, indeed past time, to acknowledge errors for what they are; the results of behavior and vigorously identify and address those factors that contribute to such behavior in hazards and near misses as well as errors with adverse outcomes. By identifying such factors through continuous vigilance as well as incident investigation and effectively addressing them, patient safety can be enhanced. Such enhancement can occur by designing technology including HIT so its use can be optimal in a variety of patient care settings. The enhancement should not stop with the design of HIT; the context of care should be designed to be free of known error inducing factors that could affect even optimally functioning HIT.

References

1. Robert Wood Johnson Foundation: Health Information Technology in the United States: The Information Base for Progress. Executive Summary. Robert Wood Johnson Foundation, Washington D.C (2006)
2. Holzinger, A., Geierhofer, R., Errath, M.: Semantic Information in Medical Information Systems - from Data and Information to Knowledge: Facing Information Overload. In: Proc. of I-MEDIA 2007 and I-SEMANTICS 2007, pp. 323–330 (2007)

3. Holzinger, A.: Usability Engineering for Software Developers. Communications of the ACM 48(1), 71–74 (2005)
4. Holzinger, A., Errath, M.: Mobile computer Web-application design in medicine: some research based guidelines. Universal Access in the Information Society International Journal 6(1), 31–41 (2007)
5. Nebeker, J., Hoffman, J., Weir, C., et al.: High Rates of Adverse Drug Events in a Highly Computerized Hospital. Archives of Internal Medicine 165, 111–116 (2005)
6. Han, Y., Carcillo, J., Venkataraman, S., Clark, R., et al.: Unexpected Increased Mortality After Implementation of a Commercially Sold Computerized Physician Order Entry System. Pediatrics 116, 1506–1511 (2005)
7. Mills, P., Neily, J., Mims, E., Burkhardt, M., Bagian, J.: Improving The Bar-Coded Medication Administration System at the Department of Veterans Affairs. American Journal of Health-System Pharmacists 63 (August 2006)
8. Rasmussen, J.: Human Error and the Problem of Causality in Analysis of Accidents. Philosophical Transactions of the Royal Society of London 337, 449–462 (1990)
9. Kohn, L., Corrigan, J., Donaldson, M. (eds.): To Err Is Human: Building a Safer Health System. National Academy Press, Washington (1999)
10. Expert Group on Learning from Adverse Events in the NHS (National Health Service): An Organization with a Memory. (last access, 2007 -08-31), http://www.dh.gov.uk/PublicationsAndStatistics/Publications/
11. Bosk, C.: Forgive and Remember: Managing Medical Failures. University of Chicago Press, Chicago (1979)
12. Reason, J.: Human Error. Cambridge University Press, New York (1991)
13. Gibson, R., Singh, J.: Wall of Silence: The Untold Story of the Medical Mistakes That Kill and Injure Millions of Americans. Lifeline Press, Washington, D.C (2003)
14. Lewin, K.: Principles of Topological Psychology. McGraw-Hill, New York (1966)
15. Buckle, P., et al.: Issues in Health Care, International Ergonomics Meeting (2006)
16. Heider, F.: The Psychology of Interpersonal Relations. Wiley & Sons, New York (1958)
17. Rasmussen, J.: Human Errors: A Taxonomy for Describing Human Malfunction in Industrial Installations. Journal of Occupational Accidents 4, 311–333 (1982)
18. Senders, J., Moray, N.: Human Error: Cause, Prediction, and Reduction. Lawrence Erlbaum Associates, Inc, Mahwah (1991)
19. Moray, N.: Error Reduction as a Systems Problem. In: Bogner, M.S. (ed.) Human Error in Medicine, pp. 67–92. Lawrence Erlbaum Associates, Inc, Mahwah (1994)
20. Bogner, M.: A Systems Approach To Medical Error. In: Vincent, C., DeMol, B. (eds.) Safety in Medicine, pp. 83–100. Pergamon Press, Amsterdam (2000)
21. Lewin, K.: Behavior and Development as a Function of the Total Situation. In: Cartwright, D. (ed.) Field Theory In Social Science, pp. 238–303. Harper & Row, New York (1964)
22. Bogner, M.S.: Stretching The Search for the Why of Error: The Systems Approach. Journal of Clinical Engineering 27, 110–115 (2002)
23. Bogner, M.S.: Misadventures in Health Care:Inside Stories. In: Bogner, M.S. (ed.), pp. 41–58. Lawrence Erlbaum Associates, Inc, Mahwah (2004)

Combining Virtual Reality and Functional Magnetic Resonance Imaging (fMRI): Problems and Solutions

Lydia Beck[1], Marc Wolter[2], Nan Mungard[1], Torsten Kuhlen[2], and Walter Sturm[1]

[1] University Hospital, RWTH Aachen University
lbeck@ukaachen.de,
nan.mungard@rwth-aachen.de,
sturm@neuropsych.rwth-aachen.de
[2] Virtual Reality Group, RWTH Aachen University
wolter@rz.rwth-aachen.de,
kuhlen@rz.rwth-aachen.de

Abstract. Combining Virtual Reality (VR) and functional magnetic resonance imaging (fMRI) offers great possibilities to researchers. Brain activation in VR can be studied and more realistic stimuli can be used in fMRI studies. Unfortunately, no standard solution exists for the combination of both methods. As part of an interdisciplinary project addressing the diagnostics and treatment of neglect, we created a neuroscientific VR-fMRI experiment. Our experiences are reported in this paper. The description of problems and our solution are intended as substantial facilitation and help for other researchers interested in creating VR-fMRI experiments.

Keywords: Virtual Reality, fMRI, New Research Methodologies, Human–Computer Interaction.

1 Introduction

Attentional impairments are among the most frequent symptoms following brain damages. One special case is unilateral visual spatial neglect. Patients with neglect typically ignore the left side of their visual field after right hemisphere damage which leads to a reduced range of visual exploration. Research results indicate that there is a dissociation of neglect symptoms in near space (within arms' reach) and far space (beyond arms' reach), i.e., there are patients showing neglect symptoms in near or in far space exclusively [1,2]. Case studies with these neglect patients show that using tools (e.g., a stick) to operate beyond arms' reach results in similar performances as within arms' reach [2,3]. If the extend of arms' reach influences spatial processing, neglect patients who suffer from neglect symptoms in near or far space exclusively could use their respective unimpaired spatial dimension compensatorily.

The idea of our research project was to treat neglect patients by manipulating their reach, resulting in an extended range of vision. To realize this form of therapy, we wanted to apply Virtual Reality (VR) to create an interactive task in which objects could be moved in space by a virtually elongated arm.

A. Holzinger (Ed.): USAB 2007, LNCS 4799, pp. 335–348, 2007.

For a validation of our approach, we decided to examine the cerebral networks underlying spatial processing in VR for healthy subjects using functional magnetic resonance imaging (fMRI). Thus, in the context of our project an integration of VR and fMRI was necessary. To cope with the significant technical challenges inherent to the combination of both methods, we established a cooperation between neuroscientists and VR experts.

The combination of VR and fMRI is of great general interest for VR as well as for fMRI researchers. On the one hand, fMRI represents a method for VR researchers to evaluate VR's capability to simulate reality and to validate the therapeutic and diagnostic value of VR. On the other hand, fMRI researchers could use VR to create more realistic stimuli for fMRI experiments. Combining VR and fMRI is not a trivial task, as common fMRI software tools such as Presentation[1] do not offer VR support and MRI scanners prohibit head movement. Previous research was focussed on demonstrating *that* it is possible to combine both methods. But these studies do not sufficiently address *how* researchers can combine these two technologies efficiently. Also, tailor made hardware was employed that is not available to other researchers. Finally, only few of the previous studies have used the VR-fMRI combination as an actual *tool* for neuroscientific research.

The intent of our paper is to document how to combine VR and fMRI. We discuss technological and methodological constraints inherent to both methods, how we realized our experiments, and what experiences we gathered. We aspire a holistic view which considers both the technical and the neuroscientific perspective. This should help other researchers interested in combining both methods.

2 Background

2.1 Functional Magnetic Resonance Imaging (fMRI)

In fMRI, neuronal activity is visualized indirectly. When neurons are active and need energy, surrounding blood vessels widen to provide more oxygenated blood. FMRI utilizes this phenomenon by creating a strong magnetic field which can detect differences in the oxygenation of blood. These differences can be detected because hemoglobin, the blood pigment, has different magnetic characteristics depending on the oxygenation of blood. In fMRI images, more oxygenated blood looks brighter because it has a higher contrast value.

FMRI is an indirect method because changes in the oxygenation of blood can only be assessed relatively. Therefore, images are usually created under two experimental conditions. The activation condition differs from the control condition only in demanding one more cognitive process of interest. The resulting images are then contrasted by statistical methods. The result is a contrast that only contains those brain areas that are relevant for the process of interest. An example contrast is shown in Figure 1.

One complete brain scan lasts about three seconds. During this so-called time of repetition (TR), about 30 different slices of the brain are scanned consecutively. To get a clear image of activations in all of the different brain areas, and because

---

[1] http://www.neurobs.com/

measured effects in fMRI are relatively small, many repetitions (around 40) of the presented stimuli in different scan times are required. For a generalization of the obtained images, data of several persons has to be assessed, merged and analyzed.

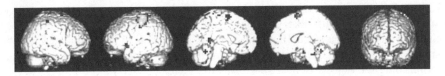

Fig. 1. Brain activations during a reaction time task

2.2 Virtual Reality

Virtual Reality (VR) has found its way into psychological and neuroscientific research in the last decade. VR applications are used as research methods in neuropsychology, clinical psychology, motor rehabilitation, and in various other applied disciplines. Compared to two-dimensional representation VR offers many advantages. It delivers a high ecological validity, offers stimulus control, and facilitates the implementation and conduction of experiments [4]. More and more areas of application are discovered especially in therapy, rehabilitation, and basic research. Typical characteristics of VR systems are stereoscopic projection, multimodality and interactivity. Especially head tracking for a user-centered projection allows for immersive environments. Both interactivity, that is a fast response time to user action, as well as a high framerate lead to a feeling of immersion in the participant of a VR experiment.

3 Analysis

3.1 Project Description

The aim of our project was to create realistic scenarios for diagnostics and therapy of neglect patients. Because differences in the spatial processing of near and far space were hypothesized these scenarios were to be presented in different distances. For an evaluation of the diagnostic value of the scenarios and to be able to evaluate possible reorganization effects, the underlying cerebral networks for spatial processing in near versus far space had to be examined. Therefore, a combination of the methods of VR and fMRI was chosen.

The development and programming of two VR-fMRI paradigms was planned. The first paradigm was intended to be an everyday life-like VR diagnostics tool for neglect patients. One of the currently used neglect diagnostics tool is the line bisection test which requires subjects to mark horizontal lines in the middle.

Neglect patients normally show a bias to the right because they ignore parts of the left visual field.

Instead of the abstract line bisection task, we wanted to create a more ecologically valid tool. Therefore, everyday life-like objects should be created. These objects were to be judged regarding their spatial position, i.e., if they were positioned in the center or shifted to the left or right. All objects were to be presented in near space as well as in far space. To assess the underlying cerebral network of spatial processing in near versus far space in this VR tool, fMRI was to be applied.

In the second paradigm, an interactive therapy tool for neglect patients was planned. By means of a virtual hand avatar which moves according to real hand movement, everyday life-like objects should actively be moved to the perceived center. In this task, the same 3D objects were to be used as in the first paradigm. We planned to create three different conditions of the task. In the first condition, a virtual arm reaching the objects in near space was to be created (see Figure 2, left side). In the second condition, a virtual arm reaching an object in far space by the means of a tool (e.g., a stick or laser beam) was to be depicted. In the third condition, a virtual elongated arm reaching objects in far space was to be presented (see Figure 2, right side). To clarify whether spatial manipulation in far space with a virtual elongated arm involves cerebral processes similar to those found in near space with normal arm's length, we wanted to apply fMRI. In both paradigms, fMRI studies were to be carried out with healthy subjects first.

For the realization and presentation of VR scenarios in an MRI scanner, no standard software solution existed. A cooperation was therefore established between the University Hospital and the Virtual Reality Group at the Center for Computing and Communication (CCC), i.e., between neuroscientists and computer scientists. The purpose of the cooperation was to create a presentation system for VR scenarios in an MRI scanner and to realize the paradigms described above in VR.

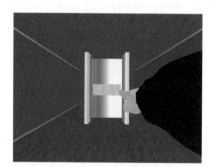 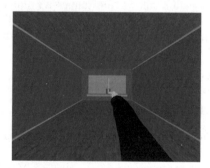

Fig. 2. Two example images from the experiment. Left: The virtual arm controlled by the user's data glove manipulates a near vertical object. Right: For manipulating far objects, the arm was elongated up to three meters.

3.2 Constraints

The realization of our project was restricted by several constraints of fMRI and VR as methods and the given project context. These constraints substantially impact the development of VR scenarios and their presentation in an MRI scanner.

Functional Magnetic Resonance Imaging. Due to the strong magnetic fields created by an MRI scanner, only specialized (MRI-ready) hardware can be used in the scanner room. The hardware must not contain ferromagnetic materials and is substantially more expensive. Movement is greatly restricted in the scanners due to the narrowness of the tube. More importantly, any head movement of more than a few millimeters leads to unsuitable data. Therefore, head-tracking in an MRI scanner is impossible. Although motion tracking is theoretically possible, e.g., by means of optical trackers, the choices for a tracking system are very limited. Addtionally, any tracking system would have to be installed permanently which would interfere with other research groups that require other hardware setups. Finally, MRI scanners create loud noises during operational state. This impairs the presentation of auditory stimuli.

Statistics. Standard fMRI experiments require a minimum of 12 valid data sets per group to carry out a random effects analysis which allows for the generalization of results. As mentioned before, enough repetitions (around 40) of the stimuli of interest are to be presented. Control conditions have to be created to rule out irrelevant brain processes. When comparing different conditions, the contrasts should be masked, i.e., in the resulting contrast, only those activations are to be visualized which were initially included in the contrast of interest. Otherwise, deactivations of the subtracted comparison contrast will show as an increase in activation. Finally, usually only male and right-handed subjects are tested. This is because women's hormone status varies and it is hypothesized that sex hormones influence the connectivity of brain hemispheres. Furthermore, left-handed subjects usually have a different lateralization of brain hemispheres compared to right-handed subjects.

Limited Resources. The acquisition costs of the technical equipment are high. An MRI scanner including the necessary software costs about 2-3 million Euros. The MRI-ready Head Mounted Display (HMD) we used cost about 65,000 Euros. The power consumption of MRI scanners is very high. In operating state, the consumption is about 40-100 kWh, in standby about 10 kWh. To maximize the benefit of an MRI scanner it is shared by many different researchers. Access to the scanner is normally allocated evenly among them. In our case, we had access to the MRI scanner for one or two sessions of 60 minutes each per week. Frequent changes of the system or the paradigm are discouraged due to the associated costs, the resulting delays and the invalidation of the previously collected data. The limited access to the scanner also restricts testing on the target system. For this purpose dedicated simulation PCs exist. In general, the high costs of fMRI research implicate that fMRI research projects are very dependent of funding and cannot be carried out in the context of routine research.

Context University Hospital. The staff at our University Hospital has only very limited VR knowledge. Furthermore, there is only limited hardware available. Especially, cutting edge hardware is not available. Moreover, due to the high aquisition costs, hardware cannot be renewed every few years, but has to be used over a long period. Because the MRI scanner is used by different research teams, customizations of the hardware cannot be installed permanently. Instead, changes to the configuration have to be undone after each session. The constraints we found in our University Hospital can be transferred to similar diagnostics and intervention settings.

4 Solution

The solution consists of two parts: the VR presentation system and the development of the required VR scenes. The development of the VR presentation system is a main task of the Virtual Reality Group, while the development of the required VR-scenes is an interdisciplinary task involving both neuroscientists and VR experts.

4.1 VR Presentation System

The VR presentation system involves the integration of the fMRI system and external hardware such as HMD and data glove. Part of the integration is the synchronization of the virtual environment with the fMRI brain scan. This is necessary to correlate brain activity data and stimulus presentation. Synchronization is achieved with an active trigger signal sent by the MRI scanner before each scan starts.

Technological Environment. The VR presentation system had to be integrated with the following systems. The University Hospital uses two MRI scanners: a 1.5 Tesla Philips Gyroscan Intera and 1.5 Tesla Achieva MRI system. For the presentation of images, an MRI-compatible HMD (VisuaStim XGA, Resonance Technology Inc., Los Angeles) with a field-of-view of 30 degrees and a maximum resolution of 1024×768 was used. Corrective lenses were available for myopic subjects. For simple decision interaction (e.g., left, middle, right - answer) a fiber optic-connected button user-input device with three buttons was available. For direct interaction with the virtual environment (e.g., manipulation of objects), an MRI-compatible 5DT data glove with 16 sensors (5DT Inc., Irvine, CA) was employed. A PC (Athlon XP, 2000 MHz, 2GB) with a Matrox Millenium P750 graphics card functioned as host system. An identical PC was available at the University Hospital for simulation purposes. The basic technological environment is shown in Figure 3.

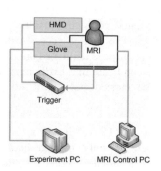

Fig. 3. Experiment setup. Left: Schematic view of setup. Experiment PC controls all peripheral devices (HMD, glove), the trigger signal is its only connection to the MRI, which is controlled by an own PC. Right: Image of subject lying in scanner.

Requirements. The constraints described in the analysis result in several requirements that a VR presentation system needs to observe. We defined the five main requirements to be performance, usability, flexibility, portability and testability.

Performance. To maintain immersion VR applications depend on high frame rates and on fast response times to user interaction (< 100 ms). While older hardware suffices for the presentation of images, videos or simple scenes as found in most common fMRI experiments, the interaction with complex scenes desired in VR requires up to date hardware. Current graphics cards support modern computer graphics techniques and thereby improve the ecological validity of the experiment. For example, additional depth cues with shadows can be integrated, different lighting techniques can be applied, or geometrical complex environments can be presented. The same is valid for processing power. While the presentation of 2D scenes or video does not require high computing performance, interactive Virtual Reality applications benefit from fast and concurrent execution.

Usability. The medical staff should be able to operate the VR presentation system independently and create scenarios on their own. While configuration of such a system needs technical expertise, the design, testing, and execution of VR-based experiments by non-technicians directly shortens development time.

Flexibility. A VR presentation system for fMRI experiments should be flexibly applicable to different neuroscientific questions. This is important, as the setup cost in time and money for new experiments decreases significantly with increasing reuse of the system.

Portability. Input and output devices of a VR presentation system can change over time, e.g., a new HMD can be bought or a different MRI scanner used for a given experiment. Thus, portability is a key requirement for a VR presentation system. In addition, portability allows the execution of a designed experiment at different sites and therefore the exchange of studies between researchers or medical staff.

Testability. To save valuable MRI scanner time (see Section 3.2), components and paradigms should be testable individually outside the scanner. Independent tests of the experiments including all special components (HMD, data glove, trigger signal) should be possible.

ReactorMan. In our experiments we extended the existing ReactorMan software [5] to support the needed functionality for fMRI experiments. The basic concept of ReactorMan is to provide a software tool enabling medical researchers and neuroscientists to design VR-based experiments. ReactorMan bases on open software packages: OpenGL is used for graphics, OpenSG is used for scene graph management and rendering, and OpenAL is applied for auditive stimuli. For Virtual Reality functionality, ReactorMan is based on the VR toolkit ViSTA. Animated virtual humanoids are based on the h-anim standard as provided by the VRZula module of ViSTA [6].

The integration of open software modules allows for an easy deployment process and a good system portability. As portability between display systems and operating systems is a basic principle of ViSTA, ReactorMan also runs on different platforms and was already applied with different VR display systems (e.g., HMDs, PowerWalls, a Workbench, a CAVE-like environment), tracking hardware (e.g., Flock of Birds,

Polhemus, A.R.T. and Qualisys optical tracking) and special VR hardware (e.g., different data gloves, spacemice).

ReactorMan uses the concept of separating content and functionality. Functionality is provided by the ReactorMan core software, which handles all time-critical and complex tasks. The content is specified by the experiment designer using Lua, an easy to learn scripting language. Lua is used to enrich the host environment at very dedicated points. This is a contrast to most APIs that try to stub their C/C++ environment completely and only with minor modifications. This has direct impact on performance and does not support comprehension for novice users.

To further increase the design convenience, the experiment designer can make use of a special ReactorMan language on top of Lua (see Figure 4a). This language consists of a restricted vocabulary commonly used by psychological experimenters, e.g. session, blocks, trial, or stimuli.

```
-- Reactorman example

testscene = NewScene ()
  AddGeometry ("board.wrl")
  object = AddGeometry ("testtube.wrl")
  AddAvatar ("hand.wrl", „glove")

NewTrial ()
  DoScene (testscene , 3000 ms, middle)
  RegisterEvent ("glove", „move_object")
```

(a) Lua script   (b) Resulting image

Fig. 4. Example of the ReactorMan scripting language. A simple trial with two geometries and a moving virtual arm is created. Events of the data glove are processed by a Lua function named "move_object".

Most basic experiments can be built using this restricted language. At any time, the full Lua language can be applied to implement more sophisticated sequences. A more detailed description can be found in [5]. We decided to use the ReactorMan tool as VR presentation system, as it fulfills most of the requirements. Additional input devices (the MRI synchronization signal and the dataglove) had to be integrated, as well as special data output methods for easier use with the SPM statistical analysis software package. For a more realistic hand avatar, a skinning method based on vertex blending was added to the already existing avatar functionality. For an easier testing, the software included a software simulation of the fMRI trigger signal.

4.2 VR Scenario Development

ReactorMan provides a framework for the definition of VR scenes. For the definition of the scenes and creation of 3D models, the neuroscientists had to be instructed by VR experts. In the context of our cooperation, a process evolved which allocated specific tasks and responsibilities to each cooperation partner.

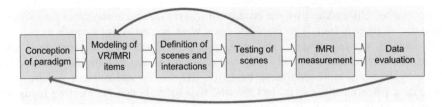

Fig. 5. Activities during experiment development

In the development of VR-based fMRI studies, several distinct activities can be identified (see Figure 5). The initial conception of the paradigm is done by neuroscientists. They define the relevant items, the sequencing of items and the possible interactions with the items. However, VR experts need to give early feedback concerning the technical feasibilty of the concepts. Given the specification of the paradigm, the neuroscientists design geometric 3D models of the items and implement the actual VR scenes by placing the 3D models in a control flow and defining the possible interaction techniques. Both tasks are supported by VR experts.

The implemented paradigm has to be tested extensively by VR experts as well as neuroscientists. Testing involves the technical part as well as the conceptual part. On the technical side, software performance, correct control flow, and latencies are evaluated by VR experts. Neuroscientists evaluate the validity of the implemented paradigm. Testing should be possible on a system similar to the measurement system. Many iterations of testing of the experiment are necessary before an fMRI measurement is possible. A prerequisite for the actual execution of the paradigm as fMRI experiment is the approval of the paradigm by all cooperation partners.

In the next step, subjects perform the paradigm offline on a PC as pretest under the instruction of neuroscientists. The pretest allows to filter unsuitable subjects and serves as a final test of the paradigm. Also, the behavioral reaction time data gained in a pretest is more valid than reaction time data gained in the MRI scanner because of the prevailing noise and limited space.

After the pretest, the actual paradigm is performed by the subjects in the MRI scanner under supervision of neuroscientists. In the last step, the combined behavioral and brain activity data are analyzed by neuroscientists.

5   Evaluation

We achieved our goal to realize the planned paradigms. The VR scenes were defined without greater difficulties and the paradigms were executed as fMRI studies. Suitable data from twelve healthy men for each of the paradigms was collected and is currently being analyzed.

5.1   Requirements

For the requirements identified as relevant for our solution we can make the following statements.

Performance. Difficulties evolved because of insufficient performance of the existing hardware at project start. The graphics cards were not powerful enough to process high-detailed, moving models. The processing power was unsatisfactory, such that fluent user-interaction, especially with the data glove, was impossible. Because all fMRI experiments until then were based on 2D scenes and did not require high computing performance, our project was the first to challenge the existing hardware of the University Hospital. Therefore, the host PC connected to the MR as well as the simulation PC had to be renewed (Athlon 64, 3800+, 4GB) and new graphics cards (NVIDIA GeForce 7600 GT) were employed.

Usability. The VR scenes were executed easily, so that neuroscientists were able to carry out the experiment independently. Because a simple script language was used for the definition of scenes, neuroscientists were able to modify the created scenes after a short introduction. For modeling of 3D objects, graphic design skills are still needed. But the development and design of the paradigm can be performed by neuroscientists independently. The complexity of ReactorMan software is comparable to Presentation which is commonly used by neuroscientists for the definition of 2D fMRI paradigms. Regarding the installation and deinstallation of the host system, at the beginning technicians were needed, but in the end neuroscientists were able to install hardware and peripheral devices on their own.

Flexibility. The developed solution is flexible enough to allow other researchers to realize different VR-fMRI experiments. At the moment, the VR presentation system is used by another neuroscientist research group at the University Hospital to investigate spatial attention in 3D space.

Portability. ReactorMan runs on different platforms and can be applied with different VR display systems, different tracking hardware and special VR hardware. The portability of the ReactorMan software counteracts the compatibility weakness of other software systems mentioned by Rizzo et al. [7] (see Section 6). Concerning the project, the created VR scenes were executed in two different MRI systems, with two different types of graphics cards and PCs. For evaluation purposes the VR scences were also tested on an active stereo Holobench.

Testability. The defined VR scenes were tested several times and necessary adjustments were made before actually executing the scenes in the MRI scanner. Still, the simulation of the target system was not completely possible so that a few adjustments had to be made during fMRI time. However, for future projects these problems should be minimal due to the experiences gathered in this project.

5.2 Problems

In the course of the project, difficulties evolved that we did not anticipate. These related to technical and conceptual issues.

One problem arose with the synchronization signal of the scanner. While it was sent correctly, the signal itself was quite short (several milliseconds), shorter than the time needed to render a new frame. If the detection of this signal was done frame-synchronous, the signal was easily missed. Our solution was to stretch the signal over a longer period of time. Other possible solutions are asynchronous signal checking or the usage of interrupts. These problems occured with both types of scanners we used.

Another problem concerned the calibration of the data glove. Normally, data gloves have to be calibrated to the user in order to operate correctly. As these gloves typically are one-size-fits-all, they are made out of stretchable material. This induces cross-sensor errors depending on the measured joint and material as discussed by Kahlesz et al. [8]. We experienced the cross-coupling of the 5DT data glove MRI as worse than the measured CyberGlove by Kahlesz et al. Differences in hand size had a severe effect on the sensors output. For small persons the glove was not close-fitting and they had to be eliminated from the sample after the pretest. A possible solution could be to use a set of differently sized gloves, but the associated costs are high and the availability of MRI-ready equipment is limited. The pretest also showed that fitting for a single subject changed during the experiment. To compensate, we recalibrated the data glove several times during a running experiment. The data glove also produced massive amounts of data as each single finger movement which exceeded $3°$ was recorded requiring sophisticated data analysis.

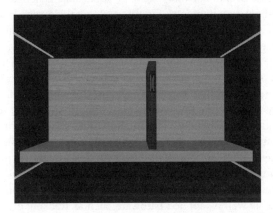

Fig. 6. De-central position of the book stimulus. Judging of offset from center is easy in 3D.

Apart from technical problems VR also posed a substantial conceptual restriction. In our initial concept, we wanted to present everyday objects such as a book or a CD case in different positions on a shelf. However, if the reference object is located at the center of the screen, the position of rectangular objects can be judged by determining which side is visible (see Figure 6). In general, the visible side face of an object is a strong indication of its position in comparison to a reference object. As a solution, we decided to use round objects with no clearly delimited side faces, e.g. a test tube.

6 Related Work

In most studies employing VR and fMRI simultaneously, the focus lies on the general feasibility. For example, Hoffman et al. conducted several studies using a VR environment to treat burn patients during woundcare by distraction [9, 10]. In [9] the authors combined both methods and assessed the immersion by subjective ratings. In another study fMRI was used as a validation tool and results showed that pain-related brain activity was reduced during VR distraction condition [10]. Other research topics assessed by the combination of VR and fMRI include driving behavior [11], memory-guided navigation [12] and smoking craving [13]. There are further studies where fMRI is applied as a validation tool before and after a VR therapy for evaluation of effects, but the methods are not combined simultanously [14,15]. All mentioned studies had a sample size lower than twelve and only few repetitions of conditions, i.e., the results cannot be generalized. The technical realization including constraints and requirements involved in the combination of methods are not discussed.

Mraz et al. [16] conducted two combined fMRI-VR studies, a spatial navigation task in a virtual 3D city and a finger tapping task. They identify research questions which can be answered by the combination of both methods. For instance, whether brain activity mirrors what is produced by actions in the real world or how brain activity relates to conventional tools such as paper and pencil tests. Moreover, the authors specify three requirements. Firstly, peripheral devices must be capable of operating at high magnetic fields. Secondly, VR experiments during fMRI must be optimized so that brain activity can be determined effectively. Finally, efficient interplay between researchers in multiple disciplines is required. The authors identify several contributions and requirements from a neuroscientist's perspective, but lack an interdisciplinary perspective, especially technical details are missing. Rizzo et al. [7] systematically assess strengths, weaknesses, opportunities, and threats to the field of VR rehabilitation and therapy. The authors state the following weaknesses of current VR rehabilitation and therapy systems: cost and complexity, immature engineering process, platform compatibility, and side effects. Typical VR solutions are one-off solutions for a specific problem, which confines re-usability and increases software development costs for new projects. Platform compatibility concerns operating systems as well as applied hardware such as trackers or input. The number of available VR hardware is restricted and existent systems should be useable for different tasks. In addition, common users are only familiar with a specific operating system. The authors remind that "rehabilitation therapists and professionals are often not programmers" [7] and that the system's frontend should be adapted to that fact. Our specific constraints found for the combination of VR and fMRI coincide with several of the above mentioned weaknesses. Baumann et al. [17] built a Virtual Reality system for neurobehavioral and functional MRI studies which provides a predefined virtual world with interconnected environments such as an apartment and a restaurant. Riva et al. [18] recently introduced the NeuroVR toolkit, a Virtual Reality platform that allows non-expert users to adapt the content of a predefined virtual environment to meet the specific needs of a clinical or experimental setting. Using an adapted modeling program, researchers can generate virtual environments which are then presented using the NeuroVR player. While created for easy usability, both frameworks lack flexibility and only Baumann et al. support fMRI. As an open-source project, NeuroVR may be enhanced with regard to certain experimental questions, but this demands expertise in programming and computer graphics.

7 Conclusion

The contribution of our paper is a detailed analysis of constraints, requirements and difficulties associated with the combination of VR and fMRI. We identified performance, usability, flexibility, portability and testability of the VR software as main requirements for use with fMRI. Moreover, we presented the solution we developed for our project involving technical as well as interdisciplinary aspects. Previous studies primarily focused on the feasibility of combining VR and fMRI. In some of the VR-fMRI studies, information about methods were missing or the studies were methodologically flawed. Some constraints and requirements were reported, but not from an interdisciplinary perspective. No technical or process solution was provided. Our results indicate that near space processing in VR was different from real world. The activations indicate that objects in near space were not processed spatially. One possible explanation is that the level of immersion was not good enough, either due to the resolution of the HMD or the lack of context clues, e.g., reference objects or shadows. Future research is needed to evaluate the effect of higher levels of immersion on spatial processing. The development of the ReactorMan software continues. It is currently employed by another research group at the University Hospital and new features such as context clues are implemented. In the future we intend to share ReactorMan as open source software tool.

Acknowledgments. The project was supported by a grant from the Interdisciplinary Center for Clinical Research 'BIOMAT.' within the faculty of Medicine at the RWTH Aachen University (TV N 59).

References

1. Morris, R., Mickel, S., Brooks, M., Swavely, S., Heilman, K.: Recovery from neglect. Journal of Clinical and Experimental Neuropsychology 7, 609 (1985)
2. Berti, A., Frassinetti, F.: When far becomes near: Remapping of space by tool use. Journal of Cognitive Neuroscience 12, 415–420 (2000)
3. Ackroyd, K., Riddoch, M., Humphreys, G., Nightingale, S., Townsend, S.: Widening the sphere of influence: using a tool to extend extrapersonal visual space in a patient with severe neglect. Neurocase 8, 1–12 (2002)
4. Loomis, J., Blaskovich, J., Beall, A.: Immersive Virtual Environment Technology as a Basic Research Tool in Psychology. Behavior Research Methods, Instruments, & Computers 31(4), 557–564 (1999)
5. Wolter, M., Armbruester, C., Valvoda, J.T., Kuhlen, T.: High ecological validity and accurate stimulus control in vr-based psychological experiments. In: EGVE 2007. Proceedings of Eurographics Symposium on Virtual Environments/Immersive Projection Technology Workshop, pp. 25–32 (2007)
6. Valvoda, J.T., Kuhlen, T., Bischof, C.: Interactive Virtual Humanoids for Virtual Environments. In: Short Paper Proceedings of the Eurographics Symposium on Virtual Environments, pp. 9–12 (2006)
7. Rizzo, A., Kim, G.J.: A SWOT Analysis of the Field of Virtual Reality Rehabilitation and Therapy. Presence - Teleoperators and Virtual Environment 14(2), 119–146 (2005)

8. Kahlesz, F., Zachmann, G., Klein, R.: 'Visual-Fidelity' Dataglove Calibration. In: CGI 2004. Proceedings of the Computer Graphics International, pp. 403–410 (2004)
9. Hoffman, H., Richards, T., Coda, B., Richards, A., Sharar, S.: The illusion of presence in immersive virtual reality during an fMRI brain scan. CyberPsychology & Behavior 6(2), 127–131 (2003)
10. Hoffman, H., Richards, T., Coda, B., Bills, A., Blough, D., Richards, A., Sharar, S.: Modulation of thermal pain-related brain activity with virtual reality: evidence from fMRI. Neuroreport 15(8), 1245–1248 (2004)
11. Carvalho, K., Pearlson, G., Astur, R., Calhoun, V.: Simulated driving and brain imaging: combining behavior, brain activity, & virtual reality. CNS Spectrums 11(1), 52–62 (2006)
12. Pine, D., Grun, J., Maguire, E., Burgess, N., Zarahn, E., Koda, V., Fyer, A., Szeszko, P., Bilder, R.: Neurodevelopmental aspects of spatial navigation: a virtual reality fMRI study. NeuroImage 15(2), 396–406 (2002)
13. Lee, J., Lim, Y., Wiederhold, B., Graham, S.: A functional magnetic resonance imaging (fMRI) study of cue-induced smoking craving in virtual environments. Applied Psychophysiology and Biofeedback 30(3), 195–204 (2005)
14. You, S., Jang, S., Kim, Y., Kwon, Y., Barrow, I., Hallett, M.: Cortical reorganization induced by virtual reality therapy in a child with hemiparetic cerebral palsy. Developmental Medicine & Child Neurology 47(9), 628–635 (2005)
15. You, S., Jang, S., Kim, Y., Hallett, M., Ahn, S., Kwon, Y., Kim, J., Lee, M.: Virtual reality-induced cortical reorganization and associated locomotor recovery in chronic stroke: an experimenter-blind randomized study. Stroke 36(6), 1166–1171 (2005)
16. Mraz, R., Hong, J., Quintin, G., Staines, W., McIlroy, W., Zakzanis, K., Graham, S.: A platform for combining virtual reality experiments with functional magnetic resonance imaging. Cyberpsychology & Behavior 6(4), 359–368 (2003)
17. Baumann, S., Neff, C., Fetzick, S., Stangl, G., Basler, L., Vereneck, R., Schneider, W.: A virtual reality system for neurobehavioral and functional MRI studies. CyberPsychology & Behavior 6(3), 259–266 (2003)
18. Riva, G., Gaggioli, A., Villani, D., Preziosa, A., Morganti, F., Corsi, R., Faletti, G., Vezzadini, L.: NeuroVR: An open-source virtual reality tool for research and therapy. Medicine Meets Virtual Reality 15, 394–399 (2007)

Cognitive Task Analysis for Prospective Usability Evaluation in Computer-Assisted Surgery

Armin Janß, Wolfgang Lauer, and Klaus Radermacher

Chair of Medical Engineering, Helmholtz-Institute for Biomedical Engineering,
RWTH Aachen University, Pauwelsstr. 20, 52074 Aachen, Germany
{Janss, Lauer, Radermacher}@hia.rwth-aachen.de

Abstract. Within the framework of the INNORISK (Innovative Risk Analysis Methods for Medical Devices) project, a twofold strategy is pursued for prospective usability assessment of Computer-Assisted Surgery (CAS) systems in the context of a risk management process. In one approach ConcurTaskTrees are applied to accomplish a hierarchical task analysis including temporal relations. In the other approach, based on the Cognitive Task Analysis method CPM-GOMS (Cognitive Perceptual Motor – Goals Operators Methods Selection Rules), a new technique for detecting potential contradictions and conflicts in the use of concurrent cognitive resources is generated. Within this model-based approach, extrinsic and intrinsic performance shaping factors are comprised, taking into account the specific context of modern surgical work systems. Additionally, a computer assisted usability analysis tool including the above-mentioned methods is developed to provide support for small and medium-sized enterprises in early stages of the design development process of risk sensitive Human-Machine-Interfaces in medical systems.

Keywords: Cognitive Task Analysis, ConcurTaskTree, Human Error, CAS-System, Cognitive Workload, Model-based User Interface Design.

1 Introduction

Rapidly evolving technological progress and automation in the domain of Computer-Assisted Surgery implicate basic alteration in Human-Computer-Interaction [1][2]. Moreover, increasing functionality and complexity of technical equipment within a clinical context cause latent human errors. Harvard Medical Practice Studies confirm the important role of human factors in safety aspects of Human-Machine-Interaction in clinical environment. Following their statement, 39.7% of avoidable mistakes in operating rooms arise from "carelessness" [3]. 72% of preventable failures concerning the usage of technical systems in orthopedic interventions are due to user error [4]. Thus, human-oriented interface design plays an important role in the introduction of new technology in medical applications.

In order to increase reliability in medical systems it is necessary to perform usage-oriented risk management and to implement methodical usability assessment. Because of high time efforts and infrastructural costs caused by interaction-centered usability

A. Holzinger (Ed.): USAB 2007, LNCS 4799, pp. 349–356, 2007.

tests - especially in areas where highly-qualified specialists are required - manufacturers of medical devices and especially small and medium-sized enterprises (SME), are interested in formal-analytical and systematic prospective usability investigation methods. Moreover, failures in product design have to be detected and removed in an early developmental stage because post-development corrections can only be handled with higher time and cost effort [5][6]. By applying prospective usability analysis in early development stages it is possible to reduce development time up to 40% [7]. According to former studies, expenses for improvement on a finished product can be up to ten times higher compared to those for measures taken at an early developmental stage [8].

2 State of the Art

Currently, risk analysis for medical products (requirements in EN ISO 14971) on the basis of FMEA (Failure Mode and Effect Analysis) is a standard procedure for manufacturers; however FMEA (product- and process FMEA) only considers the technical product (components) and the production process, leaving out usability problems and therefore human factors in accordant clinical context. Within the framework of risk management for medical devices, ever since the introduction of the European standard IEC 60601-1-6 for usability analysis, a human-centered approach is implemented to fulfill safety requirements and assure ergonomic optimization as well as minimization of usage hazards.

There are diverse procedures to evaluate usability in interactive systems, emphasizing to a greater or lesser extent the user or system perspective. A method overview according to Whitefield, Wilson & Dowell [9] is given in Table 1.

Table 1. Diverse approaches for usability evaluation according to Whitefield, Wilson & Dowell [9]

		User	
		Model	Real
System	Model	Formal-analytical approach	User-centered approach
System	Real	Product-centered approach	Interaction-centered approach

The interaction-centered methods are frequently used; however a precondition for applying these techniques is an interactive prototype and qualified testing subjects. The disadvantages for accomplishing these experimental tests are excessive time and infrastructural costs. In addition, potential design failures cannot be detected in an early developmental stage. In contrast to this, formal-analytical methods offer the possibility to evaluate the working system by means of modeling. There are various methods like Cognitive Architectures (e.g. ACT-R, SOAR and EPIC), Cognitive Task Analysis (e.g. GOMS) and Task Analysis (e.g. Hierarchical Task Analysis and ConcurTaskTree) which allow prospective Human-Machine-Interface evaluation. A lifecycle-centered view shows classification of applied usability examination methods in different development stages [10] (Figure 1). The Criteria-based (product-centered) approach and the experimental (user- and interaction-oriented) evaluation are applied later in development process than formal examination methods.

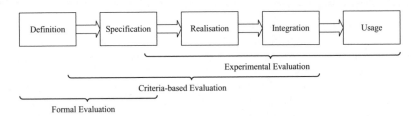

Fig. 1. Classification of usability evaluation methods for different development stages

3 Bottlenecks

Earlier studies in the field of reliability of planning and navigation systems in Computer-Assisted Surgery confirm that experimental usability assessment is extremely time-consuming and produces a large amount of infrastructural costs [11]. The question comes up if experimental analyses are practical in the framework of a comprehensive risk analysis, comprising identification of potential technical and human errors. The conclusion is that recognizing endangerments and risks as well as deriving design rules early in the development process, by providing appropriate tools, should be investigated and improved to minimize the number of iteration loops in interaction-based investigation.

A further disadvantage in interaction-centered usability testing is the fact that logically not all operations of a modern Human-Machine-System with a high amount of applications can be evaluated, especially in a complex CAS system. Here, a systematic and methodological acquisition of all task steps/types, and subsequently potential human errors in a formal-analytical approach, can be expedient and useful for medical device manufacturers.

Another handicap concerning experimental usability methods is the fact that evaluation results are often acquired too late for a sufficient integration and application in the current development process. However, due to a high validation grade, before bringing a new product to the market, there still remains the need for testing prototypes (requirement for approval according to IEC 60601-1-6). These approaches should be minimized for the above-named reasons.

Derived from earlier studies concerning interaction-centered usability investigation [12], the question is posed whether valid statements can be made solely on the basis of observational research, video analysis and questionnaire or if the focus has to be put also on the current state of cognitive resources workload.

An early usability evaluation can be a significant advantage with regard to saving time and resources. Therefore, the above-mentioned facts emphasize the need for further examination and enhancement of formal-analytical usability investigations in interactive medical systems as a helpful support in addition to classic approaches like user- and guideline-based assessment.

4 Approaches

Within the framework of the INNORISK project, a software-assisted tool for prospective usability evaluation of medical products (particularly with respect to the technical equipment of CAS systems) is developed to provide feasible and practical usability-testing methods within the entire risk management process. Here, in particular the special context of Computer-Assisted Surgery systems shall be considered.

Cognitive Engineering is becoming ever more a part of medical equipment design, allowing the improvement of patient safety measures [13]. Consideration of cognitive information processing in Human-Machine-Interaction regarding surgical work systems, particularly within the scope of multifunctional applications, is a special requirement for clinical usability assessment. The surgical team (surgeon, anesthesiologist, nurses and, if necessary, further personnel) has to work in an environment where multidimensional information transfer is often the basis for efficient communication and coordination [14]. Various information sources (e.g. physiological and medical imaging data, verbal briefings and alarm signals) have to be perceived by all participants and the surgeon has to operate with the CAS system via complex input devices (e.g. based on speech recognition, touch screen, remote control and tracked surgical tools). Taking also into account the life-critical and stressful situation, alongside long intra-operative times during most surgical interventions, the working conditions in CAS can easily lead to mental overload [15] and therefore provide a basis for unintentional "clumsy automation" in a safety-critical area [16]. Thus, Cognitive Task Analysis can constitute a helpful tool in the usability assessment of complex working systems [17].

The below described formal method for prospective usability examination is divided into two consecutive parts, beginning with an overview of performed task steps (system- and user-related) necessary to reach a specific goal, followed by the approach of ConcurTaskTrees (CTT) [18] where a diagrammatic notation for the specification of task models is generated. The usage of task analysis has recently been recognized as an important contribution to support user interface design of envisioned systems [19]. According to Bomsdorf & Szwillus ConcurTaskTrees are quasi-standard for the notation of task modeling [20].

The graphical syntax (tree-like structure) and the formal specification facilitate the utilization for designers and developers, even for systems with a high amount of complex applications. Sibling tasks on the same level of decomposition can be linked in the structure of ConcurTaskTrees. This differs from previously developed task models, where operators (tasks) only act on parent-child relationships. Within this model-based approach, hierarchical and temporal relations as well as classification of task types are implemented. A CTT includes four task types: abstraction, application, interaction and user tasks. An abstraction task is a high-level task that can be further divided into application, interaction or user tasks. An application describes performance completely done by the system whereas an interaction task is any task during which the user is interacting with the system (initiated by the user). Moreover, a user task refers to a cognitive or a physical (excluding interaction with the system) task performed by the user (e.g. when a user decides between two options). The concerning temporal relations such as approval, contemporaneousness, blocking, discontinuity and optional accomplishment are also mapped to the model.

Applying ConcurTaskTrees allows the designer to concentrate on the main tasks by using an individually adjustable abstraction level. Thus, by means of creating a clear understanding of all task steps (system and user tasks), CTT prepares the way for subsequent cognitive workload analyses.

The second part of presented prospective usability assessment is based on the results of the CTT analysis which are restricted to high-level perceptual and motor activities of the user. Breaking down the previously detected high-level decompositions into low-level operators makes the acquisition of involved cognitive processing steps more feasible, albeit success is dependable on accuracy of prior examinations. In the suggested cognitive-oriented method, based on the psychological cognition model of Rasmussen [21], different levels of regulation for human information processing are distinguished. Rasmussen differentiates skill-, rule- and knowledge-based levels of operator behaviour. In terms of Cognitive Engineering Process according to the GOMS model (**G**oals **O**perators **M**ethods **S**election Rules), a Cognitive Task Analysis based on the Model Human Processor (MHP) [22] is accomplished. MHP (inspired by computer architecture) is based on psychological studies on human information processing and integrates three parallel working processors (perceptual, motor-driven and cognitive). Furthermore four kinds of storage are included (visual and auditory buffer, working and long-term memory). Characteristic processor cycle times are derived from various studies and are based on several specific rules (e.g. Fitt's Law, Power Law of Practice and Hick's Law).

Within a Cognitive Task Analysis different task steps are matched with appropriate cognitive processing levels. In the course of these analyses potential contradictions and conflicts in concurrent cognitive resource utilisation can be detected and several risks and hazards in Human-Machine-Interaction can be identified and assessed.

In contrast to conventional GOMS techniques, a parallel modeling of operators as in CPM-GOMS (**C**ritical **P**ath **M**ethod or **C**ognitive **P**erceptual **M**otor) [23] is implemented to receive more detailed statements and to model information-flow constraints. In project Ernestine CPM-GOMS has been validated to real-world task performance by assessing Human-Machine-Interfaces of telephone companies' workstations [24]. Within this methodology not only cognitive workload is examined but also perceptual and motor-driven activities. Here, three integrated processors work independently but in each case only sequent processing with primitive operators according to MHP specifications is possible. Especially the parallelism between the processors/operators can be helpful in order to model multimodal Human-Machine-Interaction occurring in the context of Computer-Assisted Surgery. In CPM-GOMS dependencies between the users' perceptual, cognitive and motor-driven activities are mapped out in a schedule chart, where the critical path represents the minimum required execution time. John and Gray have provided templates for cognitive, motor-driven and perceptual activities alongside their dependencies under various conditions [23].

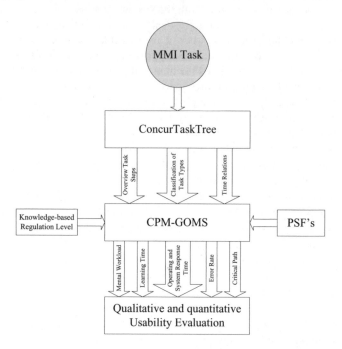

Fig. 2. Method for prospective usability assessment

Additionally, in contrast to the conventional CPM-method, performances shaping factors (PSF's) as well as knowledge-, rule- and skill-based behaviour are distinguished in the PERT-Chart (**P**rogram **E**valuation **R**eview **T**echnique). Naturally, time performance predictions can only be made for primitive skill-based operators without special contextual PSF's, but nonetheless there is a great benefit. In addition, a process bar is integrated into the CPM-Chart in order to transcribe human and system activities (obtained in CTT analysis) to facilitate decomposition of high-level tasks and to create additional help for the user. The inclusion of these new features into CPM-GOMS is an alternative providing more detailed qualitative evaluation criteria and an extensive overview of the complete working system. In combination with existing quantitative statements on learning and operating time, as well as on failure rates and efficiency, the usability assessor faces a powerful extension of CPM-GOMS, supplemented with foregoing CTT analysis (overview in Figure 2). Implementation and evaluation of the above-mentioned methodologies within the framework of an application-oriented risk analysis software tool is subject of future research, since adequate methods for prospective usability examination need quickly and easily applicable computer-based modeling tools in order to enable quantitative measurable criteria [25].

5 Discussion

The overall motivation for cognitive modeling in Human-Computer-Interaction is to provide engineering models of human performance for early optimization of interface design. Cognitive Task Analysis can provide useful information (e.g. survey of task steps, potential hazards, information-flow constraints) in an early developmental stage. Taking into account the high cognitive workload for operating room personnel during CAS interventions, the aforementioned two-fold strategy is intended to be a useful approach for modeling cognitive information processing in Human-Machine-Interaction.

However, despite high expenses and several methodological disadvantages, interaction-based evaluation is necessary and essential during the validation process of products used in risk-sensitive areas like Computer-Assisted Surgery. Performance Shaping Factors such as social aspects, individual professional qualifications and fatigue states are difficult to represent in a model-based approach but can only be measured with experimental techniques. A combination of formal-analytical and interaction-based usability investigation should optimize risk and hazard detection of Human-Machine-Interfaces. In conclusion, the developed Cognitive Task Analysis provides comprehensive information for subsequent user-based approaches and therefore may lead to extensive effort minimization in experimental tests by reducing iterative loops.

The aim of the INNORISK project is to support SME's of medical products by creating systematic methodologies and smart software tools which allow an accurate and user-guided prospective usability evaluation within the framework of the risk management process. As part of ongoing research, the suggested formal method will be validated in conjunction with medical industrial partners.

Acknowledgments. The AiF/FQS project INNORISK is funded by the German Ministry for Economy and Technology (BMWi).

References

1. Cook, R.I, Woods, D.D.: Adapting to the new technology in the operating room. Human Factors 38, 593–613 (1996)
2. Sarter, N.B., Woods, D.D., Billings, C.E.: Automation Surprises. In: Salvendy, G. (ed.) Handbook of Human Factors and Ergonomics, 2nd edn., pp. 1926–1943. Wiley, Chichester, New York (1997)
3. Brennan, T.A., Leape, L.L., Laird, N.M., et al.: Incidence of adverse events and negligence in hospitalized patients: Results of the Harvard Medical Practice Study-I. N. Engl. J. Med. 324, 370–376 (1991)
4. Rau, G., Radermacher, K., Thull, B., Pichler, C.v.: Aspects of an Ergonomic System Design of a Medical Work system. In: Taylor, R., Lavallée, S., Burdea, G., Moesges, R. (eds.) Computer Integrated Surgery, pp. 203–221. MIT-Press, Cambridge (1996)
5. Holzinger, A.: Usability Engineering Methods for Software Developers. In: Communications of the ACM, vol. 48(1), pp. 71–74. ACM Press, New York (2005)
6. Nielsen, J.: The Mud-Throwing Theory of Usability(2000) http://www.useit.com/alertbox/20000402.html

7. Mayhew, D.J.: The Usability Engineering Life Cycle. Morgan Kaufmann Publishers, San Francisco (1999)
8. Freise, A.: Der Nutzen gut bedienbarer Produkte. Siemens – Pictures of the Future, 2:65. Siemens AG, München (2003)
9. Whitefield, A., Wilson, F., Dowell, J.: A framework for human factors evaluation. Behaviour and Information Technology 10(1), 65–79 (1991)
10. Kraiss, K.-F.: Modellierung von Mensch-Maschine Systemen. In: Willumeit, H.-P., Kolrep, H. (eds.) Hrsg.: ZMMS-Spektrum, Band 1, Verlässlichkeit von Mensch-Maschine-Systemen, S, pp. 15–35. Pro Universitate Verlag, Berlin (1995)
11. Zimolong, A., Radermacher, K., Stockheim, M., Zimolong, B., Rau, G.: Reliability Analysis and Design in Computer-Assisted Surgery. In: Stephanides, C., et al. (eds.) Universal Access in HCI, pp. 524–528. Lawrence Erlbaum Ass, Mahwah (2003)
12. Radermacher, K., Zimolong, A., Stockheim, M., Rau, G.: Analysing reliability of surgical planning and navigation systems. In: Lemke, H.U., Vannier, M.W., et al. (eds.) International Congress Series 1268, CARS, pp. 824–829 (2004)
13. Woods, D.D.: Behind Human Error: Human Factors Research to Improve Patient Safety. National Summit on Medical Errors and Patient Safety Research, Quality Interagency Coordination Task Force and Agency for Healthcare Research and Quality (2000)
14. Berguer, R.: The application of ergonomics in the work environment of general surgeons. Rev Environ Health 12, 99–106 (1997)
15. Woods, D.D, Cook, R.I, Billings, C.E: The impact of technology on physician cognition and performance. J. Clin. Monit. 11, 5–8 (1995)
16. Wiener, E.L.: Human factors of advanced technology ("glass cockpit") transport aircraft. (NASA Contractor Report No. 177528). Moffett Field, CA: NASA-Ames Research Center (1989)
17. Rasmussen, J.: A Framework for Cognitive Task Analysis in Systems Design. In: Hollnagel, E., Mancini, G., Woods, D.D. (eds.) NATO AS1 Series on Intelligent Decision Support in Process Environments, vol. 21, Springer, Heidelberg (1986)
18. Paternò, F., Mancini, C., Meniconi, S.: ConcurTaskTrees: A Diagrammatic Notation for Specifying Task Models. In: Proc. of IFIP Int. Conf. on Human-Computer Interaction Interact 1997, Sydney, July 1997, pp. 362–369. Chapman & Hall, London (1997)
19. Diaper, D., Stanton, N.: The Handbook of Task Analysis for Human-Computer Interaction. Lawrence Erlbaum Associates, Mahwah, London (2004)
20. Bomsdorf, B., Szwillus, G.: Tool support for task-based user interface design. In: 'Proceedings of CHI 1999, Extended Abstracts', pp. 169–170. Pittsburgh PA (1999a)
21. Rasmussen, J.: Skills, Rules, Knowledge: Signals, Signs, and Symbols and other Distinctions in Human Performance Models. IEEE Transactions on Systems, Man and Cybernetics SMC-3, 257–267 (1983)
22. Card, S.K., Moran, T.P., Newell, A.: The psychology of Human-Computer Interaction. Lawrence Erlbaum Associates, Hillsdale, New Jersey (1983)
23. John, B.E., Gray, W.D.: CPM-GOMS: An Analysis Method for Tasks with Parallel Activities. In: Conference companion on Human factors in computing systems, pp. 393–394. ACM Press, New York, NY, USA (1995)
24. Gray, W.D., John, B.E., Atwood, M.E.: The precis of Project Ernestine or an overview of a validation of GOMS. In: Proceedings of CHI, Monterey, California, May 3- May 7, 1992, pp. 307–312. ACM, New York (1992)
25. Schweickert, R., Fisher, D.L., Proctor, R.W.: Steps toward building mathematical and computer models from cognitive task networks. Human Factors 45, 77–103 (2003)

Serious Games Can Support Psychotherapy
of Children and Adolescents

Veronika Brezinka[1] and Ludger Hovestadt[2]

[1] Department of Child and Adolescent Psychiatry, Zürich University, Switzerland
[2] Computer Aided Architectural Design, ETH Zürich, Switzerland
veronika.brezinka@ppkj.uzh.ch

Abstract. Computers and video games are a normal part of life for millions of children. However, due to the association between intensive gaming and aggressive behavior, school failure, and overweight, video games have gained negative publicity. While most reports centre upon their potential negative consequences, little research has been carried out with regard to the innovative potentials of video games. 'Treasure Hunt', the first psychotherapeutic computer game based on principles of behavior modification, makes use of children's fascination for video games in order to support psychotherapy. This interactive adventure game for eight to twelve year old children is not meant to substitute the therapist, but to offer attractive electronic homework assignments and rehearse basic psycho-educational concepts that have been learnt during therapy sessions. While psychotherapeutic computer games may prove to be a useful tool in the treatment of children and adolescents, unrealistic expectations with regard to such games should be discussed.

Keywords: psychotherapy, childhood disorders, computer-based treatment, cognitive behavior therapy, serious games.

1 Introduction

Computers and internet are a normal part of life for millions of children. Every year, more than 30 million children use the internet, more than any other age group [1]. Daily video-gaming is reported for toddlers [2], school children [3] and adolescents [4]. With regard to computers, the generation of adults – parents and teachers – has been labeled 'digital immigrants', whereas children and adolescents are considered to be 'native speakers' [5].

However, in the scientific community commercial computer games for children have gained mainly negative publicity due to the reported association between intensive gaming and aggressive behavior, game addiction, school failure and overweight [6-8].

In fact, most reports on the effects of video games appear to centre upon their potential negative consequences [9], with a substantial part of research focusing exclusively on the use of violent games [10-12].

On the other hand, surprisingly little research has been carried out with regard to the innovative potentials of computer games [13]. Yet, these potentials exist.

A. Holzinger (Ed.): USAB 2007, LNCS 4799, pp. 357–364, 2007.

Computer games improve spatial performance in children, adolescents and adults [14]. Action-video-game playing with so called first person shooters has been reported to alter a range of visual skills and to enhance visuospatial attention [15]. Studies also show that commercial computer games can be used innovatively at school [16]. Civil engineering students enjoy game-based learning and profit at least as much as from traditional learning methods [17]. In the medical sector, computer games have been therapeutically successful with physically handicapped children and children with chronic disease like asthma [18], diabetes [19] and cancer [20, 21]. The newly developed game Dybuster has yielded significant improvements in writing skills of dyslexic children [22].

As to psychotherapy, no child therapist can ignore how fascinated children are by computers and video games. Yet, very few initiatives make use of this fascination to enhance psychotherapeutic treatment, whereas for adults, computer- or internet-based treatments become quite common [23-26]. These programs often consist of a combination of face-to-face contact with the therapist and a computer-supported self-help program.

2 How Can Video Games Support Psychotherapy?

In contrast to adults, developing a computer-based treatment for children does not simply mean putting a treatment manual on the screen. In order to motivate children for psychotherapy, the challenge lies in designing serious games that incorporate therapeutic goals as well as various skills training. Serious games are defined as 'entertaining games with non-entertainment goals' [5]. Serious games have been developed for both adults and children with regard to issues as aids prevention, conflict management, prevention of child predating, nutritional education, diabetes management, relaxation or immunology (see http://www.socialimpactgames.com for an overview of serious games).

By including therapeutic concepts into a game, children can be offered attractive electronic homework assignments that enable them to repeat and rehearse basic psychoeducational concepts they have learnt during therapy sessions. What is more, using a psychotherapeutic game that matches the theoretical orientation of the therapist can help him / her to structure therapy sessions and to explain important theoretical concepts in various and child-friendly (=user-friendly) ways.

Cognitive behavior therapy is one of the best-researched and empirically supported treatment methods for adults and children. It's theoretical framework is based on the assumption that emotions and behavior are largely a product of cognitions; thus psychological and behavior problems can be reduced by altering cognitive processes.

Cognitive behavior therapy offers several intervention programs for children from which learning goals for serious games can be derived. For example social problem solving, a standard therapeutic intervention for young children [27, 28] could be incorporated into a game to support psychotherapy with children as young as five years. Such a game might even be used for the prevention of behavior problems in young children. Another example for the integration of learning goals into a serious game are strategies of healthy thinking such as outlined in cognitive-behavioral treatment programs like 'Coping Cat' [29], 'Friends for Children' [30] or 'Think good – feel good' [31]. As will be shown further down, these strategies form the basis of Treasure Hunt, the first computer game developed to support cognitive-behavioral treatment of children with various disorders [32]. Incorporating elements of anger management programs [33, 34] into video games could be an even greater challenge for the development of serious games. As most of the children treated for anger and aggression problems are boys, and boys are reported to show considerably more fascination for computers than girls [14], creating serious games that include anger management strategies might support treatment of this notoriously difficult and non compliant group. This would also hold for serious games that integrate Dodge's theory of social-cognitive biases of aggressive children. If such games could help aggressive children to reduce hostile attributional biases and to ameliorate cognitive processing of potentially threatening situations [35, 36], treatment of a chronic and difficult group of clients might become easier. Last but not least, psychotherapy with migrant children could be made more sustainable through serious games translated into foreign languages, as these games would give children the opportunity to repeat the psychoeducational elements of therapy sessions at home and in their own language (and eventually play them with parents and siblings).

3 Treasure Hunt – The First Computer Game to Support Cognitive-Behavioral Treatment of Children

3.1 Theoretical Background

Treasure Hunt is a serious game that is being developed by the Department of Child and Adolescent Psychiatry of Zürich University [32]. It is a psychotherapeutic computer game based on principles of cognitive behavior modification and is designed for 8 to 12 year old children who are in cognitive-behavioral treatment for various disorders. Each of the six levels of the game corresponds to a certain step in cognitive-behavioral treatment. The maximum amount of time needed to solve all tasks of a level is about twenty minutes.

Treasure Hunt is not meant to substitute the therapist, but to support therapy by offering attractive electronic homework assignments and rehearsing basic psychoeducational parts of treatment. The game integrates mainstream cognitive behavior therapy for children as described in 'Coping Cat' [29], 'Friends for Children' [30], 'Think good – feel good' [31] and 'Keeping your cool' [34].

3.2 Story

Treasure Hunt takes place aboard an old ship inhabited by Captain Jones, Felix the ship's cat and Polly the ship's parrot. The metaphor of an old ship is expected to be attractive for both boys and girls. Captain Jones, an experienced sailor - but not a pirate - leads the child through the game, whereas the (female) parrot embodies the help-menu. Captain Jones has found an old treasure map in the hull of his ship. However, to solve its mystery, he (the adult expert) needs the help of a child. Thus, while the child in treatment is guided through the game by an adult, he/she is an expert him/herself by helping to solve the mystery of the treasure map. Tasks take place in different parts of the ship – on deck, in the galley, in the dining room of Captain Jones and in the shipmates' bunks. Each task corresponds to a certain step in cognitive-behavioral treatment, implying a linear structure of the game. For each completed task, the child receives a sea star. The old treasure map has a dark spot in the shape of a sea star at six important places. The child and Captain Jones will only be able to read what is written there when they place the missing sea star on the map. After having solved all the tasks, the last mission consists of a recapitulation of the previous exercises. Once the child has solved this last problem, he/she will find out where the treasure is buried. Before joining Captain Jones on the final search for the treasure, the child receives a sailor's certificate that summarizes what he/she has learnt through the game and that is signed by Captain Jones and the therapist. One of the most interesting parts of the game is dedicated to the hunting of unhelpful (automatic) thoughts by means of an ego-shooter. The child has to catch a flying fish and to read the unhelpful thought written on it and replace it by a helpful one (see Fig. 1).

Fig. 1. In the fifth level of Treasure Hunt, the principle of an ego-shooter is used to teach the child to hunt unhelpful (automatic) thoughts which appear as flying fish

3.3 Software and Tools

Treasure Hunt is a 2.5 D Flash adventure game programmed with C++ and XML. Flash was used to guarantee platform independence, as only a Flash compatible internet browser is needed and no program has to be installed. This facilitates giving homework to children independent from their computer hardware and operating system at home. User interaction will be recorded in XML files to help therapists analyze children's choices and / or progress. The game follows a linear model.

Design and music were specially developed for the game in order to maximize immersion of the player / child and thus enhance motivation. While the voice of a man (Captain Jones) and a woman (the parrot) lead the child through the game, the tasks are spoken by children's voices in order to maximize immersion into the game.

Images were rendered (3D Studio max) with diverse plugins and finished by hand in photoshop. Sound effects are realized with Logic Audio in mp3 format; music was registered beforehand in .aff format and then integrated in mp3 format.

3.4 Evaluation

Playability tests with an experimental version showed that children appreciate the game and its diverse tasks. Several therapists in our department have used pilot versions of the game; they all reported positive reactions of the children in treatment. Originally, Treasure Hunt was developed to offer attractive homework assignments in between therapy sessions. However, the pilot showed that therapists like to use the game as reinforcement during therapy sessions – 'if you work well, we will play Treasure Hunt for the last ten minutes'. Moreover, the game seems to help young or less experienced therapists to structure therapy sessions and to explain important cognitive-behavioral concepts like the influence of thoughts on our feelings or the distinction between helpful and unhelpful thoughts. In Treasure Hunt, basic concepts of cognitive behavior therapy are being explained playfully and in a metaphor that is attractive for children. These basic concepts are important for the treatment of internalizing and externalizing disorders, so that the game should be able to support treatment of a broad array of disorders. However, as the professional version of the game is not finished yet, conclusions about its effectiveness are premature.

A word of caution should also be issued. A psychotherapeutic computer game will never be able to cure or ameliorate childhood disorders on its own. Moreover, Treasure Hunt is not designed as a self-help instrument. Only a behavior therapist can make optimal use of the game during treatment, as the underlying concepts are self-explanatory merely in a superficial way. Various exercises for further therapy sessions can be derived from Treasure Hunt, such as 'help us to design a next level' or 'draw flying fish with more unhelpful thoughts'. Using a computer game in psychotherapy sessions does not mean that classic therapeutic methods like writing, drawing or role-playing lose their significance in the treatment of children and adolescents.

4 Discussion and Conclusion

Psychotherapy of children and adolescents is an area in which innovative use of computers in the form of psychotherapeutic video games may enhance child compliance and offer new ways of treatment. Treasure Hunt, the first psychotherapeutic computer game based on principles of behavior modification, has yielded positive reactions of children and therapists in a small pilot. However, as the professional version of the game is not finished yet, final conclusions about the usability and / or effectiveness of the game are premature.

Still, Treasure Hunt is the first serious game designed to support cognitive behavioral treatment of children between the age of eight and twelve years. As has been outlined above, the innovative potentials of serious games for psychotherapy are numerous. They may enhance child compliance, offer attractive homework assignments, structure therapy sessions and support treatment of migrant children who could play the games in their own language and share their content with parents and siblings. Yet, there is still a long way to go and considerable resistance to overcome. Not all game-designers are positive about the concept of serious games - some think that because children are required to play the game in psychotherapy, it might lose its attractiveness. On the other hand, many academics and health professionals are not used to view computer games as something different from 'pure fun' or 'only a game' and doubt that a computer game can teach useful skills. Moreover, there is fear that if psychotherapeutic games are successful, computers might replace therapists in the long run; this, however, is irrational, as these games show their maximum potentials only under guidance of a therapist.

Last but not least, unrealistic expectations with regard to a psychotherapeutic game should be discussed and cleared up. In September 2006, we presented the prototype of another game that has never been carried out but was described in the media; we keep receiving demands of parents where they can buy the game, often paired with long descriptions of the psychological problems of their child [37]. As stated above, no psychotherapeutic game will be able to alleviate childhood problems on its own. Psychotherapeutic games are a tool, but no magic.

Even so, these games may prove to be a useful tool in the treatment of children and adolescents. Serious games incorporating therapeutic knowledge and strategies have the potential to support psychotherapy of children and adolescents with various disorders. Undoubtedly, development of more serious games that can support psychotherapy is only a question of time. Over several years a whole array of psychotherapeutic computer games, ideally labeled with a quality seal for therapists, will support psychotherapy of children and adolescents.

Acknowledgements. We thank the Department of Child and Adolescent Psychiatry (Head: Prof. Dr. Dr. H.C. Steinhausen) of Zürich University and the Technology Transfer Fund UNITECTRA of Zürich University for financial support. Special thanks are due to Christoph Wartmann (programming), Denis Chait (design) and Wolfgang Güdden (music).

References

1. Bremer, J.: The internet and children: advantages and disadvantages. Child & Adolescent Psychiatric Clinics of North America 14, 405–428 (2005)
2. Jordan, A.B., Woodard, E.H: Electronic childhood: the availability and use of household media by 2- to 3-year-olds. Zero-To-Three 22, 4–9 (2001)
3. Livingstone, S., Bovill, M.: Children and their changing media environment: A European comparative study. Lawrence Erlbaum Associates, Mahwah (2001)
4. Annenberg Public Policy Center: Media in the home: the fifth annual survey of parents and children. Annenberg Public Policy Center, Philadelphia (2000)
5. Prensky, M.: Digital game-based learning. McGraw Hill, New York (2001)

6. Browne, K., Hamilton-Giachritsis, C.: The influence of violent media on children and adolescents: a public-health approach. Lancet 365, 702–710 (2005)
7. Huesmann, L., Moise-Titus, J., Podolski, C., et al.: Longitudinal relations between children's exposure to TV violence and their aggressive and violent behavior in young adulthood: 1977-1992. Developmental Psychology 39, 201–221 (2003)
8. Slater, M., Henry, K., Swaim, R., et al.: Violent media content and aggressiveness in adolescents: A downward spiral model. Communication Research 30, 713–736 (2003)
9. Anderson, C., Funk, J., Griffiths, M.: Contemporary issues in adolescent video game playing: brief overview and introduction to the special issue. Journal of Adolescence 27, 1–3 (2004)
10. Carnagey, N., Anderson, C.: The effects of reward and punishment in violent video games on aggressive affect, cognition, and behavior. Psychological Science 16, 882–889 (2005)
11. Funk, J.: Children's exposure to violent video games and desensitization to violence. Child & Adolescent Psychiatric Clinics of North America 14, 387–404 (2005)
12. Gentile, D., Lynch, P., Linder, J., et al.: The effects of violent video game habits on adolescent hostility, aggressive behaviors, and school performance. Journal of Adolescence 27, 5–22 (2004)
13. Griffiths, M.: The therapeutic use of videogames in childhood and adolescence. Clinical Child Psychology and Psychiatry 8, 547–554 (2003)
14. Subrahmanyam, K., Greenfield, P., Kraut, R., et al.: The impact of computer use on children's and adolescents' development. Applied Developmental Psychology 22, 7–30 (2001)
15. Green, C.S., Bavelier, D.: Effect of Action Video Games on the Spatial Distribution of Visuospatial Attention. Journal of Experimental Psychology: Human Perception & Performance 32, 1465–1478 (2006)
16. Jayakanthan, R.: Application of computer games in the field of education. The Electronic Library 20, 98–102 (2002)
17. Ebner, M., Holzinger, A.: Successful implementation of user-centered game based learning in higher education - an example from civil engineering. Computers & Education 49, 873–890 (2007)
18. Lieberman, D.: Management of chronic pediatric diseases with interactive health games: Theory and research findings. Journal of Ambulatory Care Management 24, 26–38 (2001)
19. Brown, S., Lieberman, D., Gemeny, B., et al.: Educational video game for juvenile diabetes: Results of a controlled trial. Medical Informatics 22, 77–89 (1997)
20. Redd, W., Jacobsen, P., Die Tril, J., et al.: Cognitive-attentional distraction in the control of conditioned nausea in pediatric cancer patients receiving chemotherapy. Journal of Consulting & Clinical Psychology 55, 391–395 (1987)
21. Vasterling, J., Jenkins, R., Tope, D., et al.: Cognitive distraction and relaxation training for the control of side effects due to cancer chemotherapy. Journal of Behavioral Medicine 16, 65–80 (1993)
22. Gross, M., Voegeli, C.: The display of words using visual and auditory recoding. Computers & Graphics (submitted)
23. Celio, A., Winzelberg, A., Wilfley, D., et al.: Reducing Risk Factors for Eating Disorders: Comparison of an Internet- and a Classroom-Delivered Psychoeducational Program. Journal of Consulting & Clinical Psychology 68, 650–657 (2000)
24. Lange, A., Rietdijk, D., Hudcovicova, M., et al.: Interapy: A Controlled Randomized Trial of the Standardized Treatment of Posttraumatic Stress Through the Internet. Journal of Consulting & Clinical Psychology 71, 901–909 (2003)

25. Marks, I., Mataix-Cols, D., Kenwright, M., et al.: Pragmatic evaluation of computer-aided self-help for anxiety and depression. British Journal of Psychiatry 183, 57–65 (2003)
26. Proudfoot, J., Swain, S., Widmer, S., et al.: The development and beta-test of a computer-therapy program for anxiety and depression: hurdles and lessons. Computers in Human Behavior 19, 277–289 (2003)
27. Shure, M., Spivack, G.: Interpersonal problem-solving in young children: A cognitive approach to prevention. American Journal of Community Psychology 10, 341–356 (1982)
28. Webster-Stratton, C., Reid, M.: Treating conduct problems and strengthening social and emotional competence in young children (ages 4-8 years): The Dina Dinosaur treatment program. Journal of Emotional and Behavioral Disorders 11, 130–143 (2003)
29. Kendall, P.: Coping Cat Workbook. Temple University, Philadelphia (1990)
30. Barrett, P., Lowry-Webster, H., Turner, C.: Friends for Children Workbook. Australian Academic Press, Bowen Hills (2000)
31. Stallard, P.: Think good - feel good. A cognitive behaviour therapy workbook for children and young people. John Wiley & Sons, Chichester (2003)
32. Brezinka, V.: Treasure Hunt - a psychotherapeutic game to support cognitive-behavioural treatment of children. Verhaltenstherapie vol.17, 191–194 (2007)
33. Lochman, J., Lenhart, L.: Anger coping intervention for aggressive children: Conceptual models and outcome effects. Clinical Psychology Review 13, 785–805 (1993)
34. Nelson, W., Finch, A.: 'Keeping Your Cool': Cognitive-behavioral therapy for aggressive children: Therapist manual. Workbook Publishing, Ardmore (1996)
35. Dodge, K., Price, J., Bacharowski, J., et al.: Hostile Attributional Biases in Severely Aggressive Adolescents. Journal of Abnormal Psychology 99, 385–392 (1990)
36. Dodge, K.A.: Translational science in action: Hostile attributional style and the development of aggressive behavior problems. Development and Psychopathology 18, 791–814 (2006)
37. Brezinka, V.: Das Zauberschloss - zur Medienrezeption eines verhaltenstherapeutischen Computerspiels. In: Brezinka, V., Goetz, U., Suter, B. (eds.) Serious Game Design für die Psychotherapie, pp. 73–79, edition cyberfiction, Zürich (2007)

Development and Application
of Facial Expression Training System

Kyoko Ito[1,2], Hiroyuki Kurose[2], Ai Takami[2], and Shogo Nishida[2]

Osaka University, 1-3 Machikaneyama-cho Toyonaka Osaka 5608531, Japan
[1] Center for the Study of Communication-Design
[2] Graduate School of Engineering Science
ito@cscd.osaka-u.ac.jp

Abstract. The human's facial expression plays an important role as media that visually transmit feelings and the intention. In this study, the purpose is to support the effective process for facial expression training to achieve the target expression using computer. And, an interface for users to select a target facial expression and a whole development of an effective expression training system is proposed, as a first step toward an effective facial expression training system.

Keywords: Interfaces, Communication channels, Training, Systems design, Experiment.

1 Introduction

Nonverbal information, such as that contained in facial expressions, gestures, and tone of voice, plays an important role in human communications [1]. Facial expressions, especially, are a very important media for visually transmitting feelings and intentions [2] [3]. At least one study has shown that more than half of all communication perceptions are transmitted through visual information [4].

However, the person transmitting a facial expression cannot directly see his or her own expression. Therefore, it is important to understand the expression being transmitted and identifying the target facial expression in order to ideally express it. Facial muscles play a critical role in human facial expressions.

Facial expression training has recently garnered attention as a method of improving facial expressions [5] [6] [7] [8]. In facial expression training, exercises are performed that target a specific part of the face and those facial muscles. The muscles used in facial expressions are strengthened by training and when the facial expression is softened, the ideal facial expression can be expressed. Facial expression training has effective applications not only in daily communications, but also within the realm of business skills and rehabilitation. Facial expression training can take multiple forms, one of which is a seminar style experience with a trainer. Another method uses self training books or information on the Internet to serve as a guide. Some seminars are very expensive, and time and space are restricted. Alternatively, in a self training venue, it is difficult to clearly see your target facial expression when alone, and to compare your ideal facial expression with the present one.

A. Holzinger (Ed.): USAB 2007, LNCS 4799, pp. 365–372, 2007.

The aim of this study is the proposal of an effective expression training system using the computer to achieve the target facial expression. As an initial step, an interface to select the target facial expression is proposed. Next, the expression training system, including the target expression selection interface, is developed.

A previous study [9], one that used a virtual mirror, developed a facial expression training system that created a support system for facial expression training by utilizing a computer. The virtual mirror study was a facial expression training system displaying a facial expression by emphasizing the person's features with a virtual mirror. This study, however, is different from the virtual mirror study because this study instead selects the target facial expression of an actual face.

This study will be considered to enhance toward the application to the medical treatment field in the future. Specifically, the targets would be the applications for rehabilitation after the treatment in the orthodontics, an abnormal diagnosis according to an expression change in the undergoing training, and so on.

2 A Facial Expression Training System That Achieves the Target Facial Expression

2.1 Support Process

In this study, the following steps are taken within facial expression training to achieve the target facial expression.

(1) The target facial expression is identified.
(2) The current facial expression is expressed in an attempt to achieve the target facial expression.
(3) The current facial expression is compared with the target facial expression.
(4) The muscles used for facial expression are trained.

In this study, the above enumerated processes are supported by a computer. This study also aims to develop a facial expression training system that can achieve target facial expressions. A computer is utilized to support each process of this method as follows:

(1) For support in making the target facial expression;
(2) For support in recognizing the current expression;
(3) For support in comparing the current facial expression with the target facial expression; and
(4) For support in understanding the facial expression muscles that require training to achieve the target facial expression.

A primary objective of this study is the first item, above, to support making the target facial expression, and a user interface is proposed to achieve this goal. A proposal for the user interface is described below.

2.2 Support Process

In this study, the following parameters are observed in making the target facial expression:

(a) Your own face is used.
(b) Your target facial expression must correspond to the movement of a real human face, the expression which is achievable.

The first item above, (a), that the target facial expression be made by using a real face is important because each human face is different from the next one. Your facial expression corresponds to your facial features.

The second item above, (b), necessitates making the target expression one that can actually be anatomically expressed using your real face. It is also important to be able to naturally make a satisfactory target facial expression.

Using these parameters, a user interface to select the target facial expression is considered. In addition, the following two stage approach is considered so that the user may select a satisfactory target facial expression:

- First stage: Rough expression selection
- Second stage: Detailed expression selection

During the first stage, an interface (as illustrated in Figure 1) is proposed. Such an interface protects the user from having to worry about details and allows the user to intuitively select the target expression.

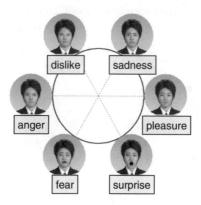

Fig. 1. Interface for selecting the target facial expression

As shown in Figure 1, the interface consists of six facial images, each displaying a different potential facial expression, all arranged around a big circle in the center [10]. These six images are designed to correspond with six basic facial expressions: pleasure, surprise, displeasure, anger, fear, and sadness.

Additionally, it is possible to both mix expressions from two different images and choose an expression strength level by selecting a point on the circle at the center. In the second stage, for the user's satisfaction, another user interface is employed that enables a user to be able to control further details. In this stage, an action unit (AU), derived from Ekman et al's [11] previous research about facial expression features is used. The elements of each feature as measured in AU can be thus established in detail. These features are comprised of eyebrows, eyes, cheeks, mouth, and mandibles. Action units of 3, 5, 2, 10, and 5 are used in each feature. In total, 25 kinds of AU are used.

3 Design and Development of the Facial Expression Training System

A facial expression training system including the above mentioned processes and the interface for selecting the target facial expression has thus been developed.

First, a personal computer camera is used to capture the current facial expression of the user. Visual C++6.0 is used as the software for the development of this system. In order to fit the user's face with a wire frame model, FaceFit [12] is used. The facial expression training system developed [13] [14] has been named "iFace." The following procedures are utilized in iFace:

(1) User registration;
(2) Selection of the target facial expression;
(3) Expression of the current facial expression; and
(4) Comparison of the current facial expression and the target facial expression.

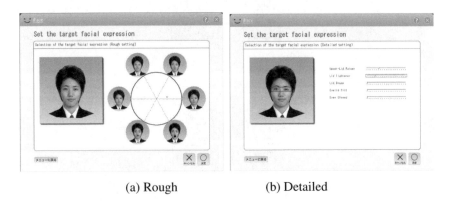

(a) Rough   (b) Detailed

Fig. 2. Rough setting screen

Figure 2 shows the screens for selecting the target facial expression with a rough setting (a) and a detailed setting (b) respectively.

4 Evaluation Experiment of "iFace"

4.1 Purposes

In order to examine the effectiveness of the target facial expression selection interface and the facial expression training system, an evaluation experiment will be conducted. This evaluation experiment specifically will aim to examine the following points:

- The effectiveness of the target facial expression selection interface for a facial expression training system; and
- The potential for a facial expression training system.

4.2 Methods

A Experimental Procedure

(1) The target facial expression is selected by the user by employing the facial expression training system.
(2) A user attempts to enact the self-selected target facial expression, while the user's current facial expression is digitally captured.
(3) The user's current facial expression is then compared with the target facial expression, and the results are presented.

B Experiment Participants

- ● 12 females (medical professionals)

Mean age: 28.4
Age range: 24-34
Profession: dentist
Experience: orthodontics

C Methods of Analysis

The results of a questionnaire administered both before and after the experiment are used. In addition, data obtained during use of the facial expression training system are used.

4.3 Results

- Effectiveness of the target facial expression selection interface

Table 1 shows the results of the questionnaire regarding the effectiveness of the target facial expression selection interface. Answers to the questionnaire are ranked on a seven point scale of +3 to -3. Questionnaire results reflected an average satisfaction rating scores of -0.3 for the first target expression attempt (#1) and 1.7 for the second target expression attempt (#2). The questionnaire results also show that the

Table 1. Questionnaire results regarding the effectiveness of the interface

#	Questionnaire item	Avg.
1	Did you satisfactorily achieve the target expression? (Asked regarding the first target expression)	-0.3
2	Did you satisfactorily achieve the target expression? (Asked regarding the second target expression)	1.7
3	Was the facial expression selection interface both intuitive and comprehensible?	2.2
4	Did you experience the synthesized facial expression in a way that felt natural?	-0.3

satisfaction rating on the smile selection is lower than that of the elective choice expression. In addition, the time required for selecting the target facial expression is shown in Figure 3 with separate amounts for a rough setting and a detailed setting at the smile selection. In a comparison between a rough and a detailed setting, it was demonstrated that there were many people who spent more time in a detailed setting.

- Facial expression training system potential

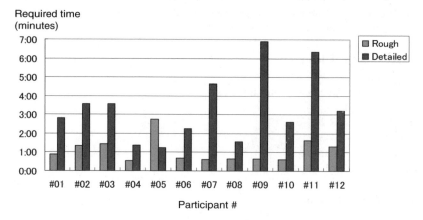

Fig. 3. The time required for rough and detailed selections of target facial expressions by each user

Table 2. Questionnaire results regarding facial expression training system potential

#	Questionnaire item	Avg.
5	Was the training method for deciding the target facial expression both concrete and comprehensible?	2.6
6	Was it easy to express your actual expression by comparing the current facial expression with the target facial expression?	2.5
7	Would you want to use the facial expression training system daily?	1.8

Table 2 shows the questionnaire results for facial expression training system potential. The questionnaire answers are ranked on a seven point scale of +3 to -3. The results demonstrate that the facial expression training system was positively evaluated.

The following comments were obtained from comments written on the questionnaire:

- I think that they can use it [a facial expression training system] to undergo rehabilitation for a paralytic face.

- My motivation for facial expression training increased when I made the target facial expression.

Each user can select two target facial expressions. One target is the user's ideal smile, and the other is an elective choice by the user. The user's current facial expression is expressed three times in attempting to achieve each target expression. Therefore, a total of six expressions of the current facial expression are attempted.

Further experiments will be conducted and analyzed; In addition, the important point will be considered to apply to the medial fields.

5   Conclusion

In this study, a target facial expression selection interface for a facial expression training system and a facial expression training system were both designed and developed. Twelve female dentists used this facial expression training system, and evaluations and opinions about the facial expression training system were obtained from these participants.

In the future, we will attempt to improve both the target facial expression selection interface and the comparison of a current and a target facial expression. Successful development of an effective facial expression training system can then lead to actual and varied usage in the medical domain.

Acknowledgement

We are grateful to Prof. Takada and Prof. Yagi, who major in orthodontics and dentofacial orthopedics, division of oral developmental Biology, graduate school of dentistry in Osaka University for their important contributions to the development and application.

References

1. Kurokawa, T.: Nonverbal interface, Ohmsha, Ltd., Tokyo (in Japanese) (1994)
2. Yoshikawa, S.: Facial expression as a media in body and computer, Kyoritsu Shuppan Co., Ltd., Tokyo (in Japanese), pp. 376–388 (2001)
3. Uchida, T.: Function of facial expression, Bungeisha, Co., Ltd., Tokyo (in Japanese) (2006)

4. Mehrabian, A.: Silent messages, Implicit Communication of Emotions and Attitudes, 2nd edn. Wadsworth Pub. Co, Tokyo (1981)
5. Inudou, F.: Facening Official Site, http://www.facening.com/ (in Japanese) (last access: 2007-07-01)
6. Inudou, F.: Facening, Seishun Publishing Co., Ltd., Tokyo (in Japanese) (1997)
7. COBS ONLINE Business good face by facial muscles training, http://cobs.jp/skillup/face/index.html (in Japanese) (last access: 2007-07-01)
8. Practice of Facial Expression, http://www.nikkeibp.co.jp/style/biz/associe/expression/ (in Japanese) (last access: 2007-07-01)
9. Miwa, S., Katayori, H., Inokuchi, M.: Virtual mirror: Proposal of Facial expression training system by face image data processing. In: Miwa, S. (ed.) Proc. 43th conference of the institute of system, control and information engineers, pp. 343–344 (1999)
10. Schlosberg, H.: The description of facial expression in terms of two dimensions. Journal of Experimental Psychology 44 (1952)
11. Ekman, P., Friesen, W.V.: The Facial Action Coding System. Consulting Psychologists Press (1978)
12. Galatea Project. http://hil.t.u-tokyo.ac.jp/galatea/index-jp.html (in Japanese) (last access: 2007-07-01)
13. Parke, F. I.: Techniques for facial animation, New Trends in Animation and Visualization, pp.229–241(1991)
14. Waters, K.A: Muscle Model for Animating Three dimensional Facial Expression. Computer Graphics, SIGGRAPH 1987 2(4), 17–24 (1987)

Usability of an Evidence-Based Practice Website on a Pediatric Neuroscience Unit

Susan McGee, Nancy Daraiseh, and Myra M. Huth

Cincinnati Children's Hospital, 3333 Burnet Ave, Cincinnati, OH, USA 45229 ML 11016
susan.mcgee@cchmc.org,
nancy.daraiseh@cchmc.org,
myra.huth@cchmc.org

Abstract. Evidence-based practice (EBP) is an established method for improving clinical practice and has been shown to improve cost-effectiveness of patient care. Despite the evidence that EBP promotes positive outcomes, nurses have been slow to incorporate this process into practice. One major barrier to nurses implementing evidence in their daily work is the lack of time to search the literature. The objective of this study is to design, implement, and evaluate an inpatient unit-specific website which allows nurses and other direct care providers to easily access literature on specific nursing and pediatric neuroscience care issues. The goal is to minimize time dedicated to the literature search as a barrier to implementing EBP.

Keywords: Evidence-based practice, usability, nurses, healthcare providers.

1 Introduction

Evidence-based practice (EBP) is an established method for improving clinical practice and has been shown to improve cost-effectiveness of patient care, but nurses have been slow to incorporate this process into practice [1]. One major barrier to using EBP is lack of time to search the literature [2]. Evidence indicates that evaluation techniques vary in reliability [3] and that multiple evaluation methods may provide better data to guide redesign [4]. The objective of this study is to design, implement, and evaluate a unit-specific website which allows nurses and other direct care providers to easily perform EBP literature searches on specific nursing and pediatric neuroscience care issues. The goal is to minimize time as a barrier to implementing EBP.

2 Methods

This website was designed specifically for the healthcare providers (HCPs) on a neuroscience unit of a pediatric hospital. A four phase process provided the structure to create, implement, and evaluate this site. In Phase I, brainstorming sessions (n=30) elicited the HCPs' needs. In Phase II the site was built based on feedback from Phase I. In Phase III, the website was activated and the number of hits tallied monthly for a

A. Holzinger (Ed.): USAB 2007, LNCS 4799, pp. 373–374, 2007.

five month period. Phase IV includes a satisfaction survey of the site and a time study. The time study compared the speed of obtaining evidence through the study website compared to other internet websites. The subjects are randomized to two groups: one using the study website and the other group using other internet options.

3 Data Analysis

The monthly hit tallies and multiple choice survey items were analyzed by frequency distributions and measures of central tendency. Type 3 Tests of Fixed Effects was used to check the time difference between the two groups.

4 Results

The brainstorming sessions, attended by a total of 30 HCPs, provided specific ideas which were included in the website design. The hit tally averaged 204.5 per month (range137-306) over 5 months. The survey was completed by 51 subjects who were predominantly female (93.9%) with a mean age of 31 years (SD=8.56). Most subjects had less than 6 years experience at the study hospital (68.6 % ≤ 1-5 years, 23.5% ≤ 6-10 years, and 7.8 % > 10 years) and described themselves as competent internet users (76.5 % competent, 19.6 % novice, and 3.9 % expert). A majority (74.51%) indicated that they understood how to access the website and approximately half (49.02 %) disagreed that using the website takes too much time. Three questions on relevance to practice noted that subjects found the information helpful in planning patient care (68.63 %), in communicating with patients, families and other HCPs (60.78%), and was relevant to their patients' care (84.31 %). We found that use of the EBP website significantly reduced the time required to search for evidence. Internet skill was not significantly related to time. Future directions include implementing EBP websites throughout the medical center and measuring the impact on patient outcomes.

References

1. Fineout-Overholt, E., Levin, R.F., Melnyk, B.M.: Strategies for Advancing Evidence-Based Practice in Clinical Settings. J. N. Y. State Nurses Assoc. 35, 28–32 (2004)
2. Pravikoff, D.S., Tanner, A.B., Pierce, S.T.: Readiness of U.S. Nurses for Evidence-Based Practice. Am. J. Nurs. 105, 40–52 (2005)
3. Molich, R., Ede, M.R., Kaasgaard, K.: Comparative Usability Evaluation. Behavior & amp: Information Technology 23, 65–74 (2004)
4. Yang, C.Y., Woodcock, A., Scrivener, S.A.: The Application of Standard HCI Evaluation Methods to Web Site Evaluation. Contemporary Ergonomics, 515–519 (2004)

Cognitive Load Research and Semantic Apprehension of Graphical Linguistics

Michael Workman

College of Business, Florida Institute of Technology
Melbourne, FL, USA
workmanm@fit.edu

Abstract. In knowledge-work, there are increasing amounts of complex information rendered by information technology, which has led to the common term, *information overload*. Information visualization is one area where empirically tested semantic theory has not yet caught up with that of the underlying information storage and retrieval theory, contributing to information overload. In spite of a vast body of cognitive theory, much of the human factors research on information visualization has overlooked it. Specifically, information displays have facilitated the data gathering (ontological) aspects of human problem-solving and decision-making, but have exacerbated the meaning-making (epistemological) aspects of those activities by presenting information in linear rather than in graphical (holistic) forms. Drawing from extant empirical research, we present a thesis suggesting that cognitive load may be reduced when holistic information is imbued with transformational grammar to help alleviate the information overload problem, along with a methodological approach for investigation.

Keywords: Human–Computer Interaction, Graphical Linguistics, Cognitive Load, Medical Informatics, Decision Support Systems.

1 Introduction

By gathering data from disparate sources and making those data available to human consumers, the implementations of technologies ranging from data integration middleware and business process management software to data warehouses and data mining technology, have facilitated the ontological aspects of human problem-solving and decision-making, however, have exacerbated the epistemological aspects of those activities [1], [2]. This is because in knowledge-work, people must "make sense" from the increasing amount of complex information rendered by these information technologies, which has led to the common term, *information overload;* which can be devastating during critical events where situational awareness and decision-making must occur accurately under stress conditions in which the understanding of potentially thousands of time-sensitive variables is required [1]. Under these conditions, people must consider the relationships among the many variables (integrated tasks) as well as the values or states of the individual variables (focused tasks) in order to make timely decisions and take appropriate actions [3], [4], [5].

A. Holzinger (Ed.): USAB 2007, LNCS 4799, pp. 375–388, 2007.

There are no shortages of examples of such critical situations, including emergency room and intensive care physicians who must evaluate the complex relationships among indicators of illnesses when they review signs and symptoms, laboratory information, and results of specialized diagnostic studies or cases in order to diagnose acute patient conditions and decide on treatments (as was noted when physicians removed the wrong kidney from a patient –see [6]); transportation control rooms such as railroads where trains are dispatched and switched among myriads of tracks according to situational variables such as train and crew operability and container contents and schedules (as was noted when two BNSF railroad trains collided –see [7]); power grids where tens-of-thousands of different kinds of generators must be monitored for electrical power output, temperature, vibration and heat (as was noted during the New York and Northeast blackout –see [8]); and disaster monitoring and recovery centers (as was noted during Hurricane Katrina "post-disaster disaster" –see [9]). Studies [10], [11] have shown that visual information can be structured to reduce information overload and enhance understanding of data and improve retention and recall of the information. Consequently, to try to address the information overload problem, non-textual forms of information representations have evolved from the conventional line graphs and pie charts to the more sophisticated three and four-dimensional models, heat maps, fisheyes, fractal displays, spanning trees, "face" displays [12], [13], and the star-like Kiviat displays [14], [15], [16]. More recently, "dashboards" have emerged in a variety of business applications using colored shapes (glyphs) and metaphors such as gauges and stoplights. A dashboard is a visual display containing important information needed to achieve objectives, consolidated and arranged on a single screen so the information can be monitored at a glance [17].

While clearly these "visual" forms of information representations have mitigated the information overload problem in many ways –in particular most have the ability to show comparative proportion and gradation of relatively few data elements –they tend to either lack the ability to show high-density data efficiently because they must rely on the *linear* aspects of the data to convey the epistemological characteristics [18], for example, a timeline graph relies on the association of X and Y axes and data fluctuations of singular variables over time, or they lack the relational properties of transformational grammar [19] that enable objective interpretations of the data [20], [2]. To illustrate, a stoplight metaphor is lit "red" to indicate a negative condition, but if the metric refers to the concept "inventory," then either the over-stocked or under-stocked condition may transition the state of the metaphor to "red" leaving ambiguity about the meaning of the condition without additional cues or further data gathering. Using a gauge metaphor instead of a stoplight adds additional semantic context, but takes up additional screen real estate and reduces the display data density.

Hence these conventional display methods can be very rich for low density data or satisfy situations where there is sufficient time to drilldown into the details, but they do not convey sufficient context in high-density to reduce information overload [9], [21], [22]. Thus in practice while tremendous importance has been placed on visual displays with respect to physical layout and the encapsulation of what might be termed "world semantics," they have neglected the relational and contextual information of the data in high-density displays that might be termed "display semantics" [4], [5], [21], [23].

There are many anecdotes and expert opinions about effective information displays (c.f. [24], [25], [26]), and a very extensive body of empirically tested human factors research into the design of conventional "world semantics" displays (e.g. [27], [28], [29] –just to name a few). However, what is still lacking is a theoretical grounding to the empirical research that can assist practitioners with the development of more effective display design and technology to fit a particular set of time-sensitive problem-solving tasks [30]. Despite the very large and mature stream of cognitive and neuroscience theory literature on visual perception and attention (c.f. [31], [32]), memory (c.f. [18], [33] [34]), and linguistics (c.f., [35], [36]), this is one aspect within the area of information visualization and human factors research where empirically tested semantic theory has not yet caught up with that of the underlying information storage and retrieval theory (c.f. [36], [38]). For example, underlying storage and retrieval research (c.f. [39]) has been utilizing semantic and cognitive theory to drive the current implementations of ontology markup using the resource description framework (RDF) and Web ontology language (OWL) for over a decade.

This disparity between the semantically rich underlying description logics and the representation of the information models in visual displays begs for theory-driven research into display semantics. This is especially relevant to situational awareness and decision-making from high-density time-sensitive data as evidenced by the fact that information overload in these settings continues to be a significant problem in spite of having all the latest information display technology and best-design practices [1], [40], [41].

1.1 An Unanswered Research Question

An important distinction has been made between design for "data availability" and design for "information extraction" relative to the epistemological nature of information model representation [49], [50]. Information display designs that consider data availability alone often leave the decision-maker with the burden of collecting and identifying relevant data, maintaining these data in working memory, integrating these data, analyzing them, and arrive at a decision. These mental processes tax available cognitive resources (the tax is called cognitive load, [18], [42]).

Data availability versus data extraction is an important distinction because performance on a task is inversely related to cognitive load required to carry out that task [43]. When cognitive load increases there are deteriorations in performance as observed in lower response times and increased errors because performance crucially depends on the relationship between cognitive resources and cognitive load in a task [44]. The deteriorations often appear as a gradual decline in task performance rather than a calamitous breakdown [46], but the decline is measurable [47]. Given these premises, the research question of interest that remains unanswered is that, drawing from cognitive processing theory, how do information displays designed for "data availability" and those designed for "information extraction" affect cognitive load and performance in a decision-making task from time-sensitive high-density data typically found in high-density displays?

1.2 Theoretical Underpinning to the Research Question

Underpinning the study of the research question is a theory of implicit and explicit cognition [18], [48]. Implicit cognition results from automatic cognitive processes, which are effortless, unconscious, and involuntary [33]. It is rarely the case however, for all three of these features to hold simultaneously (c.f. [49] for review), but it should be pointed out that *ballisticity* [51], a feature of a cognitive process to run to completion once started without the need of conscious monitoring, is common to all implicit processes [18].

Explicit cognition results from intentional processing that are effortful and conscious [52]. Conscious monitoring in this context refers to the intentional setting of the goals of processing and intentional evaluation of its outputs [18]. Thus, according to this conceptualization of cognition, a process is implicit if it (due to genetic "wiring" or due to routinization by practice) has acquired the ability to run without conscious monitoring, whereas intentional cognition requires conscious monitoring and relies on short-term working memory [47].

Taking this into account, Baddeley and Hitch [48] proposed a model of working memory comprised of a number of semi-independent memory subsystems that function implicitly, which are coordinated centrally by a limited capacity "executive" that functions explicitly. Their model suggests that there are separate stores for verbal and visual information; for example, a "visuospatial sketch pad" (VSSP) is responsible for temporary storage of visual-spatial information, with the central executive being responsible for coordinating and controlling this, and other peripheral subsystems [53].

The Baddeley and Hitch [48] model highlights the effects of explicit cognitive processing of information encoded serially. Human cognition works in this fashion essentially as a linear scanning system [54]. For instance, in an auditory channel, people use an "articulatory loop" to rehearse and elaborate on information they hear to form cognitive schema. In a visual channel, people make brief scans across the series of symbols and then fixate momentarily (saccades) while they encode the information into cognitive schema [55]. These encoding processes consume working memory resources, and the effect on performance is a product of the available working memory resources [53]. As information complexity increases, there is greater serialization of information increasing cognitive load, which drains cognitive resources, and task performance deteriorates [44].

Next, Anderson's [56] model of human cognitive architecture asserts that only the information to which one attends and processes through adequate elaborative rehearsal is spread to the long-term memory. Long-term memory can store schemata and subsequently retrieve them with varying degrees of automaticity [57], [58]. The capacity of long-term memory is, in theory, virtually unbounded but people are not directly cognizant of their long-term memories until they retrieve the schema into their working memory, which is greatly limited –with seven concepts (plus or minus two) being the upper bound [54], [59].

Since durable information is stored in the form of organized schemata in long-term memory, rendering information effectively to people can free up working memory resources and hence allow the limited capacity of explicit ("attentional") cognition to address anomalies or attend to the more novel features in the information conveyed, and as these schemata allow for enriched encoding and more efficient information transfer and retrieval from the long term memory, they allow cognitive processes to operate that otherwise would overburden working memory [54], [60].

2 Linearity, Holism, and Transformational Grammar

As suggested, information *complexity* contributes to *information overload* [50], [61], and information complexity is partially the result of how information is presented [3], [44]. That is, a primary aspect of information complexity is *linearity*; which is to say that concepts must be relationally connected in order for the concepts to "make sense" [5], [57], [62], [63]. For example, the word "tear" makes sense only when relationally connected to other concepts such that *you have a tear in your eye* versus *you have a tear in your shirt*. Linearity is also a characteristic of data point relationships over time such as a line graph showing call center call volumes, or a medical patient's heart rate, or bar charts showing financial gains and losses over a given period.

To add to the epistemological capacity of graphical displays, additional information must be supplied in one or more displays such as found in typical financial statements or medical laboratory reports with the concomitant textual information. As the number of variables increase (e.g. number of lines on a graph) and as linearity increases (e.g. number of contextual components of a given variable), information complexity grows exponentially [54]. As the amount and complexity of information increases and are presented in these conventional forms, people have greater difficulty extracting the relevant and important information from the available data because of the information overload [54], [61].

However, Langer [64] proposed that some forms of information representation are more *holistically* apprehended than others. This is consistent with prevailing neurocognitive theory and research [65], [66], [67]. Holism involves the extent that all of the properties of a concept or idea are perceived or apprehended simultaneously, rather than as constituent parts [22]. This proposition is further supported by the fact that when examining artwork, people take in the image more holistically than when reading prose, and when looking at an image on a computer screen, people do not *read* the picture pixel by pixel [5], [21]. It is important to realize, however, that when research subjects are asked to describe the meaning of such artwork or pictures, they are generally incapable of describing them in objective terms [68]. They are left with vague impressions and subjective interpretations of the information because there are no commonly accepted rules or vocabulary to describe what the artist intended to convey and what observer apprehended.

Studies of computer visualization technologies (e.g. [22], [69]) suggest at least anecdotally that interpreting glyphs may reduce information overload compared to traditional computer-rendered information representations because they present information more holistically [54]. Glyphs are signs that have an associated meaning, but not syntax. Examples of glyphs may be a red circle containing a white line through the center that means, "Do not enter" or computer icons such as spyglasses, clipboards, or file folders, or computer symbols such as a yellow triangle containing a white exclamation point to indicate "caution."

Hence, glyphs share several features with simultaneous representational forms described by Langer [64]. In particular, they can assert symbolic meaning to which people have associated cognitive schema. And because they are discrete forms, they are visually apprehended more or less at once as suggested in studies of comprehension and visualization [70] and examinations of eye saccades in scanning prose versus glyphs [21], [22], [71]. Nevertheless, these studies also indicate that people have difficulties relative to objective interpretation, since in many if not most instances the glyphs lack a well-defined, universal, or standard grammar – i.e. they cannot be juxtaposed in relational or subject-predicate form.

Thus vocabulary rules used to convey information determines the objective measures by which people draw conclusions and make inferences about an intended meaning from complex information. The basis for these rules is Chomsky's [19] transformation grammar in which he pointed out that sentences exist on at least two levels, a deep structure and a surface structure. The deep structure is essentially that of meaning (or intended meaning) encoded with the surface structure, which is that of syntax. To formulate a conception of meaning, the relationships among the words or concepts must be known.

Chomsky [19] divided transformational grammar into sets of rules for encoding syntax into a cognitive schema (top down) and generativity for assembling syntax to represent an encoded cognitive schema (bottom up), where S represents a phrase structure grammar comprised of a noun phrase (NP) and a verb phrase (VP) such that S → NP +VP for top down processes, and for bottom up processes, where D represents a deep structure (semantic component) and N represents a noun phrase and V represents a verb phrase, such that NP → D + N and VP→ V + NP. Transformational grammar is therefore the system of rules that transforms a sentence from one structural level to another [71], [72].

2.1 Graphical Linguistics and Cognitive Load

Given that semantic enrichment is the product of transformational grammar, we propose the notion of "graphical linguistics," which possesses the holistic properties of glyphs and the subject-predicate relational properties of transformational grammar. We assert that graphical linguistics should reduce cognitive load when apprehending complex, high-density, time-sensitive information compared to the same information represented in either linear form or as glyphs.

Graphical linguistics requires some illustration. A graphical linguistic should be able to express declarative and procedural domain general knowledge [74], [75] that does not require any special expertise other than a trial in which the grammatical rules are learned. Nouns may be expressed by their static placement on a display. The grammar structure of graphical linguistics is such that S → NP +VP may be asserted in subject-predicate form as shape and color. For example, the color green may represent "meets expectations," the color blue may represent "below expectations," and the color orange may represent "above expectations."

Depending on the context, above or below expectations may denote positive or negative; for example, in a medical context if low-density lipoprotein or "bad" cholesterol is above expectations then the interpretation is negative, whereas if high-density lipoprotein ("good" cholesterol) is above expectations then the interpretation is positive, and in a business context if the noun represents "inventory," then either the above or below expectations may be negative. The predicates are modified using color-coded shapes embedded in the symbols. If we were to express conditions of a power generator, for instance, nouns for the expression may include: heat, vibration, cost of energy, power reserve, and power output (See Figure 1).

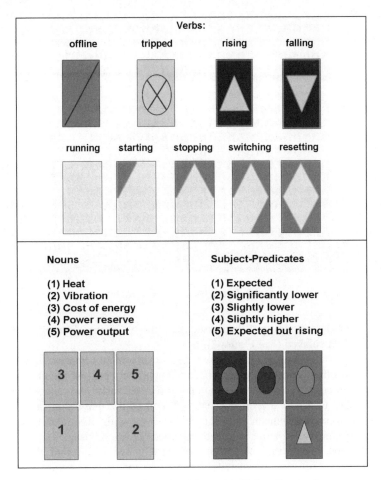

Fig. 1. Graphical Linguistic Syntax Applied to Generators

The states in which nouns might exist include: much lower than the expected level, lower than the expected, expected, higher than expected, much higher than expected, lower than expected and falling, lower than expected and rising, higher than expected and rising, higher than expected and falling, offline (maintenance), and tripped.

The verbs may consist of running, starting, stopping, switching, and transitioning. Thus a single symbol can be used to express subject-predicate relationships; for example a symbol can assert that heat is severely higher than expected, and rising,

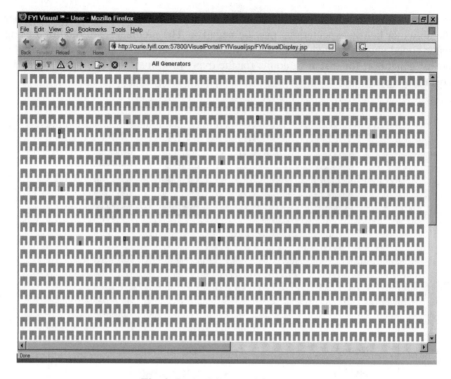

Fig. 2. Dense Display of Generators

and the generator is stopping. When assembled into a sentence, complex information can be densely displayed (See Figure 2).

The more simultaneous (holistic) the forms of information representation, the less cognitive effort that is utilized to perceive and cognitively process the information [5], which frees up cognitive resources to attend to focused tasks and reduces the amount of time it takes to assess a critical event and reduces the number errors [52]. As an example, Bourke and Duncan [61] found that even with dissimilar tasks, there is cognitive interference when performed together. Using a visual search task, they investigated the underlying causes of the interference and found that when the complexity of the information increased, there was increasing cognitive interference even among dissimilar concurrent tasks. Although the retrieval of words is an implicit automatic process, which may serve as well to cognitively prime other cognitive retrieval [52], the processes of meaning construction and inference are partly intentional integrative processes that tax working memory, and this can be measured by dividing cognitive processes using a dual-task test [44], [48], [72].

3 Methodological Approach to Investigation

Brunken et al. [43] constructed a dual-task test to measure learning under cognitive load with multimedia. This study provided a model for Workman et al's [2] exploratory study into cognitive load in diagnosing patient conditions from extant patient charts compared to an implementation of a graphical linguistic. In classic dual-task experiments, two isolated stimuli are presented with variable stimulus onset asynchronies, and the reaction times are recorded. The general result is that the reaction time to the secondary task is delayed with increased cognitive load. The primary task utilizes cognitive resources needed to initiate or react to the second task [53]. Experiments using dual-tasks in which subjects perform a primary task along with a concurrent secondary task are suitable to understanding cognitive load from processing visual representations of information [61].

An important additional element to note relative to visual information processing and the design of information displays is that people are generally not aware of their eye movements. Many dual-task experiments require concurrent eye movements even though the eye movements are rarely acknowledged as a concurrent task [22]. Yet as Kowler et al [76] pointed out, saccadic and pursuit eye movements invariably require the allocation of working memory resources, which suggests that eye movements should both affect, and be affected by, concurrent tasks –at least at a fine temporal scale. While cognitive load has not been addressed in these dual-tasks involving eye movements (because of the very brief periods involved in individual eye movements) the load is predicted by theory [71]. In summary then, if eye-sweeps are reduced by the display designs for rendering information so that they present information more holistically, this should allow more encoding during saccades and should be less taxing on working memory. Future research into cognitive load with graphical linguistics should utilize a fully randomized design and tasks involving declarative and procedural domain general knowledge [74], [75] which does not require any special expertise (other than a learning trial) in order to generalize the findings. Individual differences such as prior knowledge and cognitive capabilities can be controlled in the research design using pretests as covariates and gain scores to indicate when learning has been achieved. Electronic information could be assembled in textual form, graphical forms as found in typical dashboard configurations, and as graphical linguistics. If these representations of the same information differ in the amount of cognitive load required for information processing in the visual subsystem of working memory, then participants should also exhibit differences in performance when processing a simultaneously presented secondary visual task. Participants working with simultaneously rendered visual information in a dual-task condition should have comparatively more cognitive resources at their disposal for processing a visual secondary task when working graphical linguistics than when participants are working with linear or glyph information.

Lesser [77] conceived of graphical linguistics in the form of knowledge enhanced graphical symbols. Using this conception, a graphical linguistic should be able to express with S → NP + VP declarative and procedural domain general knowledge that does not require any special expertise other than a trial in which the grammatical rules are learned [2]. Drawing from Starren and Johnson's [78] taxonomy of medical information representations, and using the dual-task approach similar to Brunken et al's [43] learning with multimedia, Workman et al [2] conducted an exploratory study to determine physician performance in terms of timeliness and accuracy in diagnosing patient acute care needs from extant patient charts containing textual and graphical information compared to the implementation of the knowledge enhanced graphical symbols implementation of graphical linguistics (GL). Two studies were performed.

The first study assessed the cognitive load from interpreting traditional patient record information compared to GL implemented as a human metaphor, as well as mapped onto a human metaphor. The performance was best when the GL was mapped onto a metaphor compared to any other of the conditions. In this study, a dual task methodology was employed. For the secondary task, a simple, continuous, visual observation was used in which participants were presented with a series of 50 different patient conditions in traditional patient record form and in the two GL forms. The secondary task used a colored letter 'A' displayed in a separate section of the window on the screen of the display. After a random period of 5–10 seconds, the color and the letter were changed (e.g. a black "A" to red "B"). The participants were to press a designated key on the keypad as soon as possible after the letter had changed.

Once the key was pressed, the response time was recorded and the next countdown started. The software automatically recorded the lapse between the appearance of the letter in a new color and the key press. For the analysis of reaction times, the data from the secondary task were first synchronized with the program on the basis of time-stamped log file data. Then the secondary task measures were matched to their corresponding test condition: patient record interpretation primary task versus GL interpretation primary task. The first study indicated that the cognitive load to interpret the GL was less than those presented in the textual, chart and graphic form found in traditional patient charts.

A second exploratory study was conducted to infer the performance effects (as measured in time to diagnose and accuracy of diagnosis) from reducing cognitive load. In this study, three board-certified physicians, trained with GL, participated, and a nurse manager selected the patient materials that would be evaluated. Two of the physicians were compared on their performance, while the third acted as a judge. This study consisted of two stages. In the first stage, participants were told that time was more important than accuracy. For this, both participants evaluated 10 cases each (5 in standard patient chart form and 5 in GL form). For the second stage, participants were told that accuracy was more important than time. For this, both participants evaluated 30 cases each (15 in standard patient chart form and 15 in GL form).

Participant physicians were evaluated by the judge on their assessments of patient information, and then answered a standardized questionnaire regarding diagnosis, physiologic abnormalities and level of illness and treatment plans regarding the patients. The questionnaires were then independently reviewed by the judge for accuracy, who scored the results without knowledge of which questionnaire was filled out by which participant. In both instances, all patient records were picked at random by a nurse manager in the critical care unit and were unknown to the reviewing physicians. No patient examination was allowed; however, the traditional chart and flow sheet physician had access to all patient progress notes, histories and physicals, etc. Under the GL condition, the physician had access to a single display.

The results of this study indicated that when time and accuracy were compared under the condition that more emphasis was placed on time than accuracy, time to diagnosis was shorter for the GL interpretation than that for the traditional patient chart rendering, and accuracy was slightly better (although not statistically better). Hence, there was no statistical difference in accuracy when emphasis was placed on time. For the second test where more emphasis was placed on accuracy rather than time, both time and accuracy of diagnosis was statistically better using GL than that of the traditional patient chart rendering.

4 Conclusion

While this study was exploratory in nature, it raises the provocative question that the ways in which information is presented to physicians may affect the quality of acute care, such as in intensive, critical and emergency care units. When information can be presented in GL form rather than when presented in the conventional textual and chart form, it may be cognitively processed more efficiently when working under the specific condition of working with high-density, time-sensitive information, potentially lowering errors in diagnosis and increasing the responsiveness to patient conditions. This previous work is leading to a new lineage of research into the use of graphical linguistics in other contexts. As the body of literature begins to flesh out, new insights into the construction of display information should emerge. It is certainly promising that by employing the kinds of semantic principles now utilized in underlying storage and retrieval description logics may reduce the information overload problem at the display level as well.

References

1. Killmer, K.A., Koppel, N.B.: So much information, so little time: Evaluating web resources with search engines. Technol. Horizons in Education Journal 30, 21–29 (2002)
2. Workman, M., Lesser, M.F., Kim, J.: An exploratory study of cognitive load in diagnosing patient conditions. International Journal for Quality in Health Care 19, 127–133 (2007)
3. Bennett, K.B., Flach, J.M: Graphical displays: Implications for divided attention, focused attention, and problem solving. Human Factors 34, 513–533 (1992)
4. Chechile, R.A, Eggleston, R.G., Fleischman, R.N., Sasseville, A.M.: Modeling the cognitive content of displays. Human Factors 31, 31–43 (1989)
5. Lohr, L.: Creating visuals for learning and performance: Lessons in visual literacy. Prentice Hall, Upper Saddle River, NJ (2003)
6. Dyer, C.: Doctors go on trial for manslaughter after removing wrong kidney. British Medical Journal 324, 10–11 (2002)
7. National Transportation and Safety Board. Collision of two Burlington Northern Santa Fe freight trains. Washington DC: NTSB Report PB2006-916302 Notation 7793A (2002)
8. CNN. Major power outage hits New York, other large cities, (Retrieved July 07, 2007) http://www.cnn.com/2003/US/08/14/power.outage
9. Bradshaw, L.: Information overload and the Hurricane Katrina post-disaster disaster. Information Enterprises, Fremantle, WA (2006)
10. Larkin, J.H., Simon, H.A.: Why a diagram is (sometimes) worth ten thousands words. Cognitive Science 11, 65–99 (1987)
11. Healey, C., Kellogg, G., Booth, S., Enns, J.T.: High-speed visual estimation using preattentive processing. ACM Trans. on Computer-Human Interaction 14, 107–135 (1996)
12. Chernoff, H.: Using faces to represent points in k dimensional space graphically. Journal of American Statistical Association 68, 361–368 (1973)
13. Kondo, H., Mori, H.: A computer system applying the face method to represent multiphasic tests. Medical Information 12, 217–222 (1987)
14. Morris, M.: Kiviat graphs - conventions and figures of merit. ACM SIGMETRICS Performance Evaluation Review 3, 2–8 (1974)

15. Kolence, K.W., Kiviat, P.J.: Software unit profiles and Kiviat figures. ACM SIGMETRICS Performance Evaluation Review 2, 2–12 (1973)
16. Pola, Pl., Cruccu, G., Dolce, G.: Star-like display of EEG spectral values. Electroencephalography and Clinical Neurophysiology 50, 527–529 (1980)
17. Marcus, A.: Dashboards in your future. Communications of the ACM 13, 48–60 (2006)
18. Posner, M.I.: Chronometric explorations of mind. Erlbaum, Hillsdale, NJ (1978)
19. Chomsky, N.: Human language and other semiotic systems. Semiotica 25, 31–44 (1979)
20. Cooper, G.: Cognitive load theory as an aid for instructional design. Australian Journal of Educational Technology 6, 108–113 (1990)
21. Rehder, B., Hoffman, A.B.: Eye tracking and selective attention in category learning. Cognitive Psychology 51, 1–41 (2005)
22. Komlodi, A., Rheingans, P., Ayachit, U., Goodall, J.R., Joshi, A.: A user-centered look at glyph-based security visualization. IEEE Conference Workshop on Visualization for Computer Security 26, 21–28 (2005)
23. Mayer, R.E.: Multimedia learning. Cambridge University Press, Cambridge (2001)
24. Shneiderman, B.: Designing the user interface: Strategies for effective human-computer interaction. Addison-Wesley Longman Publishing Co, Boston (1992)
25. Tufte, E.R.: The visual display of quantitative information. Graphics Press, Cheshire, CT (1986)
26. Powsner, S.M., Tufte, E.R.: Graphical summary of patient status. The Lancet 344, 386–389 (1994)
27. Bederson, B.B., Shneiderman, B., Wattenberg, M.: Ordered and quantum treemaps: Making effective use of 2D space to display hierarchies. In: Bedderson, B.B., Shneiderman, B. (eds.) The craft of information visualization, pp. 257–278. Morgan Kaufmann Publishers, San Francisco (2002)
28. Bennett, K.B., Payne, M., Calcaterra, J., Nittoli, B.: An empirical comparison of alternative methodologies for the evaluation of configural displays. The Journal of the Human Factors and Ergonomics Society 42, 287–298 (2000)
29. Carswell, C.M., Wickens, C.D.: The proximity compatibility principle: Its psychological foundation and relevance to display design. Human Factors 37, 473–494 (1995)
30. Loft, S., Sanderson, P., Neal, A., Mooij, M.: Modeling and predicting mental workload in en route air traffic control: Critical review and broader implications. The Journal of the Human Factors and Ergonomics Society 49, 376–399 (2007)
31. Johnson, M.H.: The development of visual attention: A cognitive neuroscience perspective. In: Gazzanga, M.S. (ed.) The cognitive neurosciences, pp. 735–750. MIT Press, Cambridge MA (1995)
32. Rafal, R., Robertson, L.: The neurology of visual attention. In: Gazzanga, M.S. (ed.) The cognitive neurosciences, pp. 625–648. MIT Press, Cambridge MA (1995)
33. Schacter, D.L.: Implicit memory: New frontiers for cognitive neuroscience. In: Gazzanga, M.S. (ed.) The cognitive neurosciences, pp. 824–825. MIT Press, Cambridge MA (1995)
34. Tulving, E.: Working memory: An Introduction. In: Gazzanga, M.S. (ed.) The cognitive neurosciences, pp. 751–754. MIT Press, Cambridge MA (1995)
35. Caplan, D.: The cognitive neuroscience of syntactic processing. In: Gazzanga, M.S. (ed.) The cognitive neurosciences, pp. 871–880. MIT Press, Cambridge MA (1995)
36. Garrett, M.: The structure of language processing: Neuropsychological evidence. In: Gazzanga, M.S. (ed.) The cognitive neurosciences, pp. 881–900. MIT Press, Cambridge MA (1995)
37. Gavrilova, T.A., Voinov, A.V.: The cognitive approach to the creation of ontology. Nauchno-Tekhnicheskaya Informatsiya 2, 59–64 (2007)

38. Schroeder, J., Xu, J., Chen, H., Chau, M.: Automated criminal link analysis based on domain knowledge. Journal of the American Society for Information Science and Technology 58, 842–855 (2007)
39. McBride, B.: The resource description framework (RDF) and its vocabulary description language RDFS. In: Staab, S., Studer, R. (eds.) The handbook on ontologies in Information Systems, pp. 223–257. Springer, Heidelberg (2003)
40. Albers, M.J.: Information design considerations for improving situation awareness in complex problem solving. In: ACM SIG Design of Communication, Proc. of the 17th ann. Int. conference on computer documentation, New Orleans, LA, pp. 154–158. ACM Press, New York (1999)
41. Endsley, M.R., Bolte, B., Jones, D.G.: Designing for situation awareness: An approach to user-centered design. Taylor & Francis, NY (2003)
42. Sweller, J.: Cognitive load during problem solving: Effects on learning. Cognitive Science 12, 257–285 (1988)
43. Brunken, R., Steinbacher, S., Plass, J.L., Leutner, D.: Assessment of cognitive load in multimedia learning using dual task methodology. Journal of Experimental Psychology 49, 109–119 (2002)
44. Hazeltine, E., Ruthruff, E., Remington, R.W.: The role of input and output modality parings in dual-task performance: Evidence for content-dependent central interference. Cognitive Psychology 52, 291–345 (2006)
45. Woods, D.D.: The cognitive engineering of problem representations. In: Weir, G.R.S., Alty, J.L. (eds.) Human-computer interaction and complex systems, pp. 169–188. Academic Press, London (1994)
46. Norman, D.A., Bobrow, D.J.: On data-limited and resource-limited processes. Cognitive Psychology 7, 44–64 (1975)
47. Richardson-Klavvehn, A., Gardiner, J.M., Ramponi, C.: Level of processing and the process-dissociation procedure: Elusiveness of null effects on estimates of automatic retrieval. Memory 10, 349–364 (2002)
48. Baddeley, A.D., Hitch, G.J.: Working Memory. In: Bower, G. (ed.) The psychology of learning and motivation: Advances in research and theory, pp. 47–90. Academic Press, New York (1974)
49. Breitmeyer, B.G.: Visual masking: past accomplishments, present status, future developments. Advances in Psychology 3, 9–20 (2007)
50. Holzinger, A., Geierhofer, R., Errath, M.: Semantic information in medical information systems - From data and information to knowledge: Facing information overload. In: Proceedings of I-MEDIA 2007 and I-SEMANTICS 2007, Graz, Austria, pp. 323–330 (2007)
51. Monsell, S., Driver, J.: Control of cognitive processes: Attention and performance XVIII. MIT Press, Cambridge MA (2000)
52. Jacoby, L.L.: A process discrimination framework: Separating automatic from intentional uses of memory. Journal of Memory and Language 30, 531–541 (1991)
53. Barnhardt, T.M.: Number of solutions effects in stem decision: Support for the distinction between identification and production processes in priming. Memory 13, 725–748 (2005)
54. Halford, G.S., Baker, R., McCredden, J.E., Bain, J.D.: How many variables can humans process? Psychological Science 16, 70–76 (2005)
55. Smith, E.E., Jonides, J.: Working memory in humans: Neuropsychological evidence. In: Gazzanga, M.S. (ed.) The cognitive neurosciences, pp. 1009–1020. MIT Press, Cambridge MA (1995)
56. Anderson, J.R.: Cognitive psychology and its implications. Worth Publishers, New York, NY (2000)

57. Reder, L.M., Schunn, C.D.: Metacognition does not imply awareness: Strategy choice is governed by implicit learning and memory. In: Reder, L.M. (ed.) Implicit memory and metacognition, pp. 45–78. Lawrence Erlbaum, Hillsdale, NJ (1996)

58. Sternberg, R.J.: Intelligence, information processing, and analogical reasoning: The componential analysis of human abilities. Erlbaum, Hillsdale, NJ (1977)

59. Cowan, N.: The magical number 4 in short-term memory: A reconsideration of mental storage capacity. Behavioral and Brain Sciences 24, 87–185 (2000)

60. Sweller, J.: Cognitive load during problem solving: Effects on learning. Cognitive Science 12, 257–285 (1988)

61. Bourke, P.A., Duncan, J.: Effect of template complexity on visual search and dual-task performance. Psychological Science 16, 208–213 (2005)

62. Draycott, S.G., Kline, P.: Validation of the AGARD STRES battery of performance tests. Human Factors 38, 347–361 (1996)

63. Wise, J.A., Thomas, J.J, Pennock, K., Lantrip, D., Pottier, M., Schur, A., Crow, V.: Visualizing the non-visual: spatial analysis and interaction with information for text documents. In: Card, S., Mackinlay, J. (eds.) Readings in information visualization: Using vision to think, pp. 442–450. Morgan Kaufmann Publishers Inc., San Francisco (1999)

64. Langer, S.: Philosophy in a new key: A study in the symbolism of reason, rite, and art. Harvard University Press, Cambridge, MA (1957)

65. Bergeron, V.: Anatomical and functional modularity in cognitive science: Shifting the focus. Philosophical Psychology 20, 175–195 (2007)

66. Miller, E.K., Chelazzi, L., Lueschow, A.: Multiple memory systems in the visual cortex. In: Gazzanga, G. (ed.) The cognitive neurosciences, pp. 475–490. MIT Press, Cambridge (1995)

67. Simon, G., Petit, L., Bernard, C., Rebaï, M.: Occipito-temporal N170 ERPs could represent a logographic processing strategy in visual word recognition. Behavioral and Brain Functions 3, 3–21 (2007)

68. Legge, G.E., Gu, Y., Luebker, A.: Efficiency of graphical perception. Perception and Psychophysics 46, 365–374 (1989)

69. Tsang, M., Morris, N., Balakrishnan, R.: Temporal thumbnails: Rapid visualization of time-based viewing data. In: Proceedings of the 15th annual ACM symposium on user interface software and technology, pp. 175–178. ACM Press, New York (2002)

70. Montgomery, D.A.: Human sensitivity to variability information in detection decisions. Human Factors 41, 90–105 (1999)

71. Pollatesk, A., Reichle, E.D., Rayner, K.: Tests of the EZReader model: Exploring the interface between cognition and eye movement control. Cognitive Psychology 52, 1–56 (2006)

72. Shiffrin, R.M., Schneider, W.: Controlled and automatic human information processing: II. Perceptual learning, automatic attending, and a general theory. Psychological Review 84, 127–190 (1977)

73. Bransford, J.D., Franks, J.J.: The abstraction of linguistic ideas. Cognitive Psychology 2, 331–350 (1971)

74. Kozma, R.B.: Learning with media. Review of Educational Research 61, 179–211 (1991)

75. Trafton, G.J., Trickett, S.B.: Note-taking for self-explanation and problem solving. Human-Computer Interaction 16, 1–38 (2001)

76. Kowler, E., Anderson, E., Dosher, B., Blaser, E.: The role of attention in the programming of saccades. Vision Research 35, 1897–1916 (1995)

77. Lesser, M.F.: GIFIC: A graphical interface for information cognition for intensive care. In: Proceedings from the 18th Ann. Symp. on Computers in Applied Medical Care (1994)

78. Starren, J., Johnson, S.B.: An object-oriented taxonomy of medical data presentations. Journal of the American Medical Information Association 7, 1–20 (2000)

An Ontology Approach for Classification of Abnormal White Matter in Patients with Multiple Sclerosis

Bruno Alfano[1], Arturo Brunetti[2], Giuseppe De Pietro[3], and Amalia Esposito[3]

[1] National Research Council (CNR)
Biostructure and Bioimaging Institute (IBB)
via Pansini 5 – 80131 – Napoli, Italy
bruno.alfano@ibb.cnr.it
[2] University "Federico II"
Department of Biomorphological and Functional Sciences
via Pansini 5 – 80131 – Napoli, Italy
arturo.brunetti@unina.it
[3] National Research Council (CNR)
Institute for High-Performance Computing and Networking (ICAR)
via Pietro Castellino 111 – 80131 – Napoli, Italy
{giuseppe.depietro, amalia.esposito}@na.icar.cnr.it

Abstract. Multiple Sclerosis (MS) is an inflammatory autoimmune disease of the Central Nervous System, characterized by development of lesions that cause interference in the communication between brain and the rest of the body. Some techniques using numeric algorithms based on mathematical and probabilistic theories are generally used in order to obtain lesions detection. In this paper we describe an innovative approach for lesions recognition to be applied after segmentation of brain tissues from quantitive evaluation of MR studies. Knowledge about MS lesions is formalized through an ontology and a set of rules: integrating them, automatic inferences can be realized to point out lesions, starting from data about potentially brain abnormal white matter.

Keywords: Multiple Sclerosis, Lesion Detection, Medical Ontology, Rules, Reasoning.

1 Introduction

Multiple sclerosis (MS) is an inflammatory autoimmune disease of the Central Nervous System characterized by damage of myelin (demyelination) and nervous fibers (axons). The characteristic feature of MS pathology is the demyelinated plaque distributed throughout the Central Nervous System. Magnetic Resonance Imaging (MRI) of the brain and the spine can detect the typical MS plaques located within the white matter; recent studies demonstrated that MS involves also the grey matter [1].

MRI shows areas of demyelination as bright lesions on Proton Density and T2-weighted images or FLAIR (fluid attenuated inversion recovery) sequences. Gadolinium contrast is used to demonstrate active plaques on T1-weighted images.

A. Holzinger (Ed.): USAB 2007, LNCS 4799, pp. 389–402, 2007.

Based on its high sensitivity, MRI is routinely used in the clinical workup of MS both for diagnosis and to monitor disease changes over time and response to treatment. Quantitative MRI techniques, including segmentation and volumetric imaging, magnetization transfer imaging (MTI), diffusion tensor imaging (DTI), and proton MR (1H-MR) spectroscopy have greatly improved the possibility to detect subtle changes that cannot be detected with visual assessment.

Recently measurements obtained from MRI studies of the brain have been used to objectively monitor "lesion load" (volume of abnormal white matter) or "active disease" (areas of gadolinium enhancement) and the degree of brain atrophy (an additional important indicator of disease severity). Therefore, Segmentation (tissue classification) procedures have been developed to obtain operator independent assessment of lesion burden and brain volumetry.

We investigated the opportunity of using an innovative method for MS lesions recognition: data derived by segmentation from quantitative evaluations of MR studies is integrated with knowledge formalized through ontology and rules in order to make automatic inferences for pointing out plaques. We realized a knowledge base consisting of an intentional component (TBOX) with all classes and properties of the ontology, a set of rules (RBOX), useful to express the knowledge not included in the TBOX, an Extensional Component (ABOX), containing the instances obtained starting from the results of the brain tissue segmentation, and we used a reasoner to make automatic inferences on them.

This ontology based method allows integrating knowledge belonging to different fields: data about lesions could be enriched with information about sex, age, other possible MS diagnosed cases in family and various clinical manifestations of the patient, which are all information generally used for supporting Multiple Sclerosis diagnosis [2]. This approach is quite different from the most used methods.

The paper is structured as follows: first there is a survey of some techniques used for MS lesions detection and of the use of ontology approach in the medical world, then our approach is presented with the description of realized ontology and rules and the results obtained by the reasoning. The last section contains conclusions and possible future works.

2 Motivations and Related Work

Semantic web technologies are recently been experimented as a useful instrument in the medical field [3], for example to manage information during medical diagnostic process [4]. At present, in the medical field there is an increasing interest for ontology approaches, which are investigated for the following aims:

- to share knowledge in the medical practice;
- to realize decision support methods;
- to research about specific diseases;
- to support knowledge and data integration;
- to construct powerful and interoperable systems for transmitting, sharing and reusing patients' data.

Most of the works concerning medical ontologies use them as shared vocabulary in order to solve the notable problem related to the heterogeneous terminology. FMA

[5], GALEN [6], UMLS [7] and NCI cancer ontology [8] are only some of the most important examples in this direction, others are the so called Open Biomedical Ontologies [9]. Nevertheless ontologies can be used in a more applicative way, as in this work, introducing the concept of rules and reasoning. An ontology, composed of an intentional and an extensional part, can be enriched with a set of rules and a reasoner can inference new knowledge not explicitly expressed.

A promising field for ontologies is medical imaging, particularly for diagnosing diseases, for which usability of the decision support system assume a great importance for the end users: medical professionals. In particular, usability has to be interpreted as learnability, efficiency, memorability, low error rate and satisfaction [10]. If a disease can be pointed out from one or more images, then it has identical image features in any case and there are some specific criteria that give evidence for the disease. So some features can be derived from pathological image entities and from their possible connection. In this sense, ontologies have been used to support inferences on entities of different anatomical levels of granularity, such as for systems used for carcinoma classification [11], or for mammography interpretation [12], and some rules based systems have been developed , for example in order to label computed tomography head images containing intracerebral brain hemorrhage [13]. Some other works were realized to label brain structures by integration of an ontology and rules based approach [14].

Our research can be inserted in this sphere and it is focused on data derived from MR brain images, which become the object of an automatic reasoning. The main goal of our work is to develop an ontology and rule based system to assist radiologists in their decision-making process for an automated measurement of brain lesion load in MS. Information derived by MRI scans is used to manage and treat MS, since it is an useful instrument for studying disease change over time and for monitoring response to treatment.

At present different methods have been realized to make lesion detection. In particular, we referred to a technique integrating a procedure for brain normal tissues segmentation, based on a relaxometric characterization of brain tissues using calculated R1, R2 and proton density maps from spin-echo studies and a procedure using both relaxometric and geometric features of MS lesions for their classification [15]. Some other techniques have been developed with the same purpose. For example in [16] a probabilistic method based on the extraction of region of interest and the application of Gaussian mixture models is described. Another one has been realized through a fuzzy C-means (FCM) algorithm. It combines information derived by a segmentation of an MR image, a mask to discard lesions found outside WM and statistical knowledge for identification of segmented regions through brain atlas [17, 18]. Some other methods for the MS lesion detection are based on probabilistic theories and use numeric procedures.

We propose an innovative approach to obtain MS lesion recognition, realizing an integration of an OWL [19] ontology and a set of SWRL [20] rules to model knowledge about abnormal white matter recognized by brain segmentation and features useful to recognize lesions, starting from voxels of this abnormal tissue.

Considering that the increasing of lesions number provokes a greater damage and a higher risk of disability, having an ontology and rule based system able to verify the distribution and the features of MS lesions can be a useful instrument for helping MS diagnosis or for studying disease advancement.

3 Ontology Modeling

An ontology can be defined as an explicit specification of a shared conceptualization [21]: it describes the concepts in the domain of interest and the relationships that hold between these concepts. A set of rules can enrich ontology allowing deriving knowledge which is not included in it explicitly. A rule has an antecedent (or body) and a consequent (or head): whenever the conditions specified in the antecedent are true, the conditions specified in the consequent must hold. Applying rules can be useful to reason about OWL individuals and to infer new knowledge about them.

In particular, rules allow to set a property of all OWL individuals belonging to a class, or it can be used to classify all individuals of a class characterized by a certain property expressed in the antecedent as belonging to the class indicated in the consequent. We have used rules in this last way.

To explain the concept of rule, we report an example of one that can be integrated with an ontology describing brain anatomical structures [22]:

$$SF(n1) \ and \ SF(n2) \ and \ sulciConnection(s) \ and \ isSFBoundedBy(n1,s) \ and$$
$$isSFBoundedBy(n2,s) \ \rightarrow \ isSFConnectedTo(n1,n2)$$

SF and *sulciConnection* represent respectively the class of the Sulcal folds and of the connections between sulci. *isSFBoundedTo* and *isSFConnectedTo* are binary properties that link two instances of the aforementioned classes: the first expresses that a sulcus is bounded by another one, the second says that a sulcus is connected to another sulcus. The defined rule allows to infer that if n1, n2 and s are instances of the classes *SF*, *SF* and *sulciConnection* respectively and n1 is bounded by s and n2 is bounded by s, then n1 and n2 are connected.

Our ontology based technique needs a preventive tissue segmentation of the MR study that has to be analyzed, in order to obtain clusters of abnormal white matter that will become the object of the reasoning.

3.1 Segmentation

As said before, the proposed ontology approach has to be applied after an automated magnetic resonance segmentation method for identification of normal tissues (grey matter, white matter, cerebrospinal fluid) and individuation of clusters of PAWM (*Potentially abnormal white matter*) voxels, labelled as PL (*Potential Lesion*) [15].

Starting from MR images this procedure for segmentation calculates R1, R2, PD maps and generates QMCI image, characterized by a simultaneous display of MR tissue characteristics, within a single color coded image [23]. Then it creates a preliminary 3D segmentation matrix in which each pixel of the original slices is substituted by the order number of the corresponding tissue, according to their ROI. In the multifeature space MS lesions partially overlap normal tissue distribution, consequently voxels position alone does not permit unequivocal classification of MS plaques, but only allows the definition of a ROI for tissues that can be classified as PAWM. Clusters of PAWM voxels are labelled as PL scanning presegmented 3D matrix, whose elements are marked with numbers corresponding to the various brain tissues after segmentation. All obtained PLs are smaller than 8 ml because a fragmentation of spatial cluster is eventually applied to improve classification of large lesions, where there is a great probability of normal tissue voxels, included in the PAWM, connected to true lesions voxels. Some interesting values for classification are calculated for each PL: WMp, percentage of surrounding white matter; FFD, a shape dimension factor; a distance factor, which represents distance between PL and outline in the R2, PD plane.

3.2 Ontology and Rules Driven Classification

A radiologist takes into account some specific features in order to classify an element of a segmented MR image as a lesion and that these features can be schematized and quantified. This consideration led to the idea to apply an ontology and rules based approach to automate MS lesions detection. These features are the following: a lesion is surrounded by white matter (WM), small MS lesions are roundish (as size increases, the shape becomes more irregular), lesion has a great distance factor, which means that there is a big distance between outline and lesion in the R2, PD plan. In particular, this last characteristic expresses that if the distance between the PL and its border is great in the multiparametric space, then the probability that the PL is different from the surrounded tissue is greater.

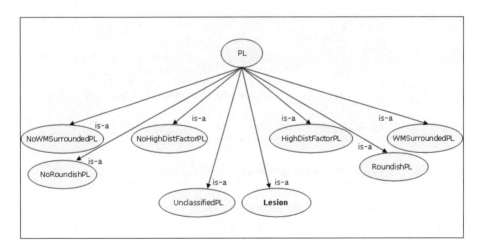

Fig. 1. Taxonomy

We defined necessary and sufficient conditions for the *UnclassifiedPL* and *Lesion* classes: an individual belongs to the *Lesion* class if and only if it belongs to the classes *HighDistFactorPL*, *WMSurroundedPL* and *RoundishPL*, while it belongs to the class *UnClassifiedPL* if and only if it is a *NoHighDistFactorPL* or

$$HighDistanceFactorPL \cap WMSurroundedPL \cap RoundishPL \Leftrightarrow Lesion$$

$$NoHighDistanceFactorPL \cup NoWMSurroundedPL \cup NoRoundishPL \Leftrightarrow UnclassifiedPL$$

Fig. 2. Definition of *Lesion* and *UnClassifiedPL* by set-theoretical formalism

NoWMSurroundedPL or *NoRoundishPL*. In Fig. 3 and in Fig. 3 definition of the classes *UnclassifiedPL* and *Lesion* is expressed using a set-theroretical formalism and OWL language.

```
<owl:Class rdf:ID="Lesion">
 <rdfs:subClassOf>
    <owl:Class rdf:ID="PL"/>
 </rdfs:subClassOf>
 <owl:equivalentClass>
    <owl:Class>
       <owl:intersectionOf rdf:parseType="Collection">
          <owl:Class rdf:ID="HighDistFactorPL"/>
          <owl:Class rdf:ID="WMSurroundedPL"/>
          <owl:Class rdf:ID="RoundishPL"/>
       </owl:intersectionOf>
    </owl:Class>
 </owl:equivalentClass>
</owl:Class>

<owl:Class rdf:ID="UnClassifiedPL">
   <rdfs:subClassOf>
      <owl:Class rdf:ID="PL"/>
   </rdfs:subClassOf>
   <owl:equivalentClass>
      <owl:Class>
       <owl:unionOf rdf:parseType="Collection">
          <owl:Class rdf:about="#NoHighDistFactorPL"/>
          <owl:Class rdf:about="#NoRoundishPL"/>
          <owl:Class rdf:about="#NoWMSurroundedPL"/>
       </owl:unionOf>
      </owl:Class>
   </owl:equivalentClass>
</owl:Class>
```

Fig. 3. Definition of *Lesion* and *UnClassifiedPL* in OWL

In medical field an important question is related to the closed or open world assumption. Semantic Web languages, such as OWL, make the open world assumption: if a question cannot be proved to be true, it may not be assumed false. In this context we need to classify all instances of the class PL as Lesion or UnclassifiedPL, so the reasoner should understand that every instance of the class PL that has been classified as Lesion, is not an UnclassifiedPL.

This is only possible in a closed world. In order to allow this classification by reasoner we made an explicit definition of both of the classes *Lesion* and *UnclassifiedPL*.

We defined some datatype properties to express PLs' features. These properties have PL as domain and Float or String as range: WMp for the percentage of surrounding white matter, FFD for the shape-dimension factor, distance_factor for the homonymic value, Seed, for coordinates of plaque's seed, XMinLim, XMaxLim, YMinLim, YMaxLim, ZMinLim and ZMaxLim for PLs spatial limits, VoxelsNumber for voxels number of the lesion, Barycentre for the homonymic value in the multiparametric space.

First three properties are used in rules as useful element to classify PLs, while the others allow expressing PL dimension and its position in the image.

Table 1. Properties

Property
WMp
FFD
Distance_factor
Seed
XMinLim
XMaxLim
YMinLim
YMaxLim
ZMinLim
ZMaxLim
VoxelsNumber
Barycentre

Classes and properties form TBOX of our ontology, while ABOX will be completed with instances of the classes. When the ABOX is filled with data of PLs of an MRI segmented study, all instances are defined as belonging to the class PL.

After taxonomy and properties, we defined a set of rules (RBOX), using SWRL language. These rules act on the instances and allow to classify them as belonging to a specific subclass of *PL,* according to the numeric values of some of their properties (WMp, FFD and distance_factor). Defined rules are showed in Fig.4, where *valueWMp, valueFFD, valueDF* are experimentally found values. For example, the first one expresses that if an instance of PL is surrounded by a high percentage of white matter (quantified using a threshold called *valueWMp*), then PL is classified as *WMSurroundedPL*. Conversely, for the second rule, if its percentage of surrounding white matter is less than that threshold, then it is classified as *NoWMSurroundedPL*. Similar observations can be made for the other rules: if PL has a small FFD factor, then it is classified as a *Roundish PL*, else it is classified as *NoRoundishPL*.

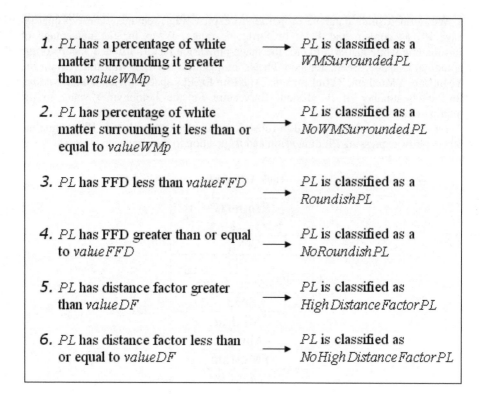

1. *PL* has a percentage of white matter surrounding it greater than *valueWMp* ⟶ *PL* is classified as a *WMSurroundedPL*

2. *PL* has percentage of white matter surrounding it less than or equal to *valueWMp* ⟶ *PL* is classified as a *NoWMSurroundedPL*

3. *PL* has FFD less than *valueFFD* ⟶ *PL* is classified as a *RoundishPL*

4. *PL* has FFD greater than or equal to *valueFFD* ⟶ *PL* is classified as a *NoRoundishPL*

5. *PL* has distance factor greater than *valueDF* ⟶ *PL* is classified as *HighDistanceFactorPL*

6. *PL* has distance factor less than or equal to *valueDF* ⟶ *PL* is classified as *NoHighDistanceFactorPL*

Fig. 4. Rules

4 Test Environment

After ontology modeling and rules development, we tested the realized technique on data obtained from an MR study of a MS diagnosed patient. We used the reasoner KAON2 [24, 25]. It is able to support OWL and SWRL languages and to make inferences integrating knowledge deriving from the knowledge base expressed through ontology and its instances and from rules application. API offered by KAON2 has been used to manage our ontology and rules and to submit SPARQL queries.

We realized a file Lesions.owl containing classes, properties and SWRL rules definition. A file LesionsInstances.owl, built starting from data deduced by results of segmentation was imported into the first in order to able the reasoner to make inferences (Fig. 5). The following information have been provided for each PL: identify number, barycentre coordinates, coordinates of the surrounding white matter barycentre, distance factor, dimension factor, shape factor, dimension-shape factor, percentage of surrounding voxels of white matter, limits, seed, voxels number.

The MR study used for test has been obtained at 1.0 T using data from two spin–echo sequences (TR/TE 640/30; TR/TE 2200/30, 90 msec dual-echo sequence). This

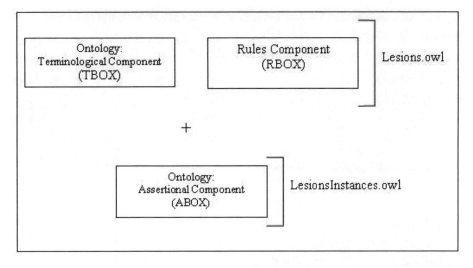

Fig. 5. Knowledge Base

study is characterized by 32 slices, 4 mm thick, covering entire brain. The aforementioned segmentation procedure with PLs individuation has been applied on it, obtaining 4849PLs.

5   Preliminary Results

We tested our ontology and rule based system for lesions recognition on the 4849 PLs of the aforementioned MR study. They were inserted as instances in the ABOX of our knowledge base. We recurred to SPARQL language [26] to formalize queries in order to point out instances of the different classes and their features; in particular, query "SELECT ?x WHERE (?x rdf:type **a**:Lesion)" was used to point out instances of the class Lesion by the reasoning, where **a** stays for the ontology's namespace. 37 instances were given as result. Similar queries were used in order to ask for all instances of the classes *WMSurroundedPL, RoundishPL, HighDistFactorPL* and

Table 2. Results

Class	Instances number
PL	4849
WMSurroundedPL	788
RoundishPL	496
HighDistFactorPL	3530
UnClassifiedPL	4812
Lesion	37

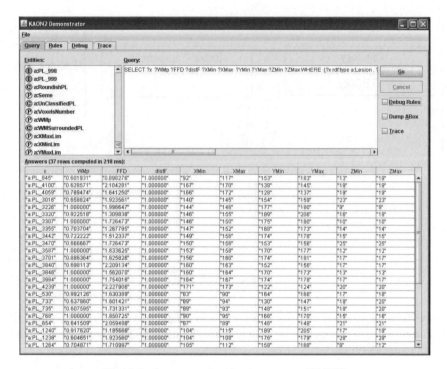

Fig. 6. Lesions recognized by the reasoner KAON2

UnClassifiedPL, obtaining respectively 788 PLs surrounded by a great percentage of white matter, 496 roundish PLs, 3530 PLs with a high distance factor, 4812 PLs to be reclassified (see table 2).

Eventually, after classification, sum of instances number of *UnClassifiedPL* and of Lesion is equal to number of instances of *PL*.

The answer to the more detailed SPARQL query SELECT ?x ?WMp ?FFD ?distF ?Xmin ?Xmax ?YMin ?YMax ?ZMin ?ZMax WHERE {?x rdf:type a:Lesion . ?x a:WMp ?WMp . ?x a:FFD ?FFD . ?x a:distance_factor ?distF . ?x a:XMinLim ?Xmin . ?x a:XMaxLim ?Xmax . ?x a:YMinLim ?YMin . ?x a:YMaxLim ?YMax . ?x a:ZMinLim ?ZMin . ?x a:ZMaxLim ?ZMax} gives not only instances name of the *Lesion* class, but also value of some properties that characterize them: percentage of surrounding white matter, shape dimension factor, distance factor and spatial limits. In particular, spatial limits represent important information for discovering identified lesions in the images.

A screenshot with the results given to this query by reasoner is showed in Fig. 6. Reasoning time was 218 ms.

We compared our method with an approach based on a procedure using relaxometric and geometric features [15]; for sake of simplicity, in the following we will refer to such procedure as "algorithmic procedure". This procedure found 23 lesions from PL starting set, while ontology method gave 37 lesions as result. Lesions found by ontology approach were all the same lesions recognized by the other method plus other 13 lesions, showing a greater sensibility.

To make visual comparisons between images containing lesions found with these two methods, we used a commercial photo-editing program to display slices obtained with QMCI technique, in order to verify reliability of our results. Besides three channels RGB, we used a channel to display all PLs, another one for lesions found by algorithmic procedure, a channel for lesions found by ontology approach. Finally, an additional channel was used for showing lesions found by using both approaches.

Lesions found by the two approaches were selected on slices, and it was asked to an expert to establish if they could be considered as lesions.

The comparison of the results of two methods showed many additional potential lesions found by using the ontology approach; most of such potential lesions were referred to cerebellum area, which is a critical zone where usually there are no lesions. So this area was not considered when the images were examined. We found 5 false-positive (i.e. lesions found by the reasoner but for the expert are not lesions). For two cases we found a lesion which was not discovered with the algorithmic procedure. The other remaining potential lesions were difficult to distinguish into the slices for the human eye, so the expert was not able to classify them.

In figures 7 and 8 some examples of slices with detected lesions are showed. In particular, Fig. 7 shows a lesion that has not be identified by the algorithmic procedure (an arrow indicates it).

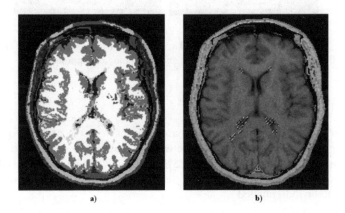

a)   b)

Fig. 7. a) Result of segmentation with all PLs in yellow;   b) QMCI image with selected identified Lesions

Beyond the number of found lesions, caused by greater sensibility of a method than the other, the important thing to be underlined is the great difference between the used approach: by the ontology method, we realized a high level semantic description of the domain of interest, trying to represent (and automatically reproduce) the neuroradiologist reasoning.

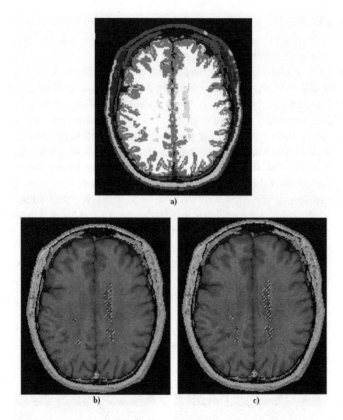

Fig. 8. a) Result of segmentation with all PLs in yellow; b)-c) QMCI images with selected lesions identified by algorithmic procedure (b) and by ontological approach (c)

6   Conclusions and Future Work

In this paper we presented an ontology-based approach to the problem of brain MS lesions recognition. This approach can be used to realize a new technique for lesion detection starting from data of segmented MR images. It can be a simple and effective alternative to methods based on probabilistic or numerical procedures, when MS lesions have to be individuated.

Obviously, this innovative technique, such others realized for the same purposes, cannot replace the expert's knowledge but it could represent a useful help for him, reporting information that cannot be detected with visual assessment. In this way the doctor, who represents the final user, receives a clear benefit. We are investigating the possibility of enriching the set of rules for lesions recognition introducing the concept of their position. At the moment it is not possible saying a priori that a lesion cannot be localized in a specific place of the brain, but it is possible to individuate regions where lesions are more probably. This information could be formalized developing classes for a PL in a specific place of the brain or defining rules useful for reasoning on placement. This could allow to discard PL located in unusual places or to point out the others as sure lesions.

Finally, this work represents a first step for the construction of a more complex system for supporting Multiple Sclerosis diagnosis, considering other elements useful for making inferences, beyond lesions load. For example we could take into account the data registry of the patient (some studies demonstrate a correlation among sex, race, age, presence of other relatives with the MS and the rising of this disease) or some clinical attacks. All these data could concur to establish a greater or a smaller probability to develop MS, supporting the doctors in their decision-making process.

References

1. Pirko, I., Lucchinetti, C.F., Sriram, S., Bakshi, R.: Gray matter involvement in multiple sclerosis. Neurology 68, 634–642 (2007)
2. McDonald, W.I, Compston, A., Edan, G., et al.: Recommended Diagnostic Criteria for Multiple Sclerosis. Annals of Neurology 50, 121–127 (2001)
3. Holzinger, A., Geierhofer, R., Errath, M.: Semantic Information in Medical Information Systems - from Data and Information to Knowledge: Facing Information Overload. In: Proc. of I-MEDIA 2007 and I-SEMANTICS 2007, Graz, Austria, pp. 323–330 (2007)
4. Podgorelec, V., Pavlic, L.: Managing Diagnostic Process Data Using Semantic Web. In: CBMS 2007. Twentieth IEEE International Symposium on Computer-Based Medical Systems, IEEE Computer Society Press, Los Alamitos (2007)
5. Rosse, C., Mejino, J.L.V.: A Reference Ontology for Bioinformatics: The Foundational Model of Anatomy. Journal of Biomedical Informatics 36, 478–500 (2003)
6. Rector, A.L., Rogers, J.E., Zanstra, P.E., Van Der Haring, E.: OpenGALEN: Open Source Medical Terminology and Tools. In: Proc AMIA Symp., p. 982 (2003)
7. Humphreys, B., Lindberg, D.: The UMLS project: making the conceptual connection between users and the information they need. Bulletin of the Medical Library Association 81(2), 170–177 (1993)
8. Golbeck, J., Fragoso, G., Hartel, F., Hendler, J., Parsia, B., Oberthaler, J.: The national cancer institute's thesaurus and ontology. Journal of Web Semantics 1 (2003)
9. Open Biomedical Ontologies, available at http://www.obofoundry.org
10. Holzinger, A.: Usability Engineering for Software Developers. Communications of the ACM 48(1), 71–74 (2005)
11. Kumar, A.Y., Lina Yip, Y., Smith, B., Marwede, D., Novotny, D.: An Ontology for Carcinoma Classification for Clinical Bioinformatics.Medical Informatics Europe (MIE 2005), Geneva, 635–640 (2005)
12. Golbreich, C., Bouet, M.: Classification des compte-rendus mammographiques a partir d'une ontologie radiologique en OWL. In: Extraction et gestion de Connaissances (EGC'2006), RNTI, vol. 1, pp. 199–204 (January 2006)
13. Cosic, D., Loncaric, S.: Rule-Based Labeling of CT Head Image. In: Keravnou, E.T., Baud, R.H., Garbay, C., Wyatt, J.C. (eds.) AIME 1997. LNCS, vol. 1211, Springer, Heidelberg (1997)
14. Mechouche, A., Golbreich, C., Gibaud, B.: Towards an hybrid system for annotating brain MRI images. In: CEUR. Proceedings of the OWLED 2006 Workshop on OWL: Experiences and Directions, Athens, Georgia, USA, November 10-11, 2006, vol. 216, pp. 10–11 (2006)

15. Alfano, B., Brunetti, A., Larobina, M., Quarantelli, M., Tedeschi, E., Ciarmiello, A., Covelli, E., Salvatore, M.: Automated Segmentation and Measurement of Global White Matter Lesion Volume in Patients with Multiple Sclerosis. Journal of Magnetic Resonance Imaging 12, 799–807 (2000)

16. Shahar, A., Greenspan, H.: A Probabilistic Framework for the Detection and Tracking in Time of Multiple Sclerosis Lesions, Macro to Nano, 2004. In: IEEE International Symposium on Biomedical Imaging, pp. 440–443. IEEE Computer Society Press, Los Alamitos (2004)

17. Boudra, A., Dehak, R., Zhu, Y.M., Pachai, C., Bao, Y.G., Grimaud, J.: Automated segmentation of multiple sclerosis lesions in multispectral magnetic resonance imaging using fuzzy clustering. Computers in Biology and Medicine 30(1), 23–40 (2000)

18. Ganna, M., Rombaut, M., Goutte, R., Zhu, Y.M.: Improvement of brain lesions detection using information fusion approach. In: 6th International Conference on Signal Processing (2002)

19. McGuinness, D.L., van Harmelen, F.: OWL Web Ontology Language Overview, W3C Recommendation (February 10, 2004). Latest version is available at http://www.w3c.org/TR/owl-features

20. Horrocks, I., Patel-Schneider, P.F., Boley, H., Tabet, S., Grosof, B., Dean, M.: SWRL: A Semantic Web Rule Language Combining OWL and RuleML, W3C Member (May 21, 2004) (submission).Latest version is available at http://www.w3.org/Submission/SWRL

21. Gruber, T.: Towards Principles for the Design of Ontologies Used for Knowledge Sharing. International Journal of Human Computer Studies, 907–928 (1995)

22. Dameron, O.: Modélisation, représentation et partage de connaissances anatomiques sur le cortex cérébral, Thése de doctorat d'Université, Université de Rennes 1 (2003)

23. Alfano, B., Brunetti, A., Arpaia, M., Ciarmiello, A., Covelli, E.M., Salvatore, M.: Multiparametric Display of Spin-Echo Data from MR Studies of Brain. Journal of Magnetic Resonance Imaging 5(2), 217–225 (1995)

24. KAON2 , http://kaon2.semanticweb.org/

25. Motik, B., Harrocks, I., Rosati, R., Seattler, U.: Can OWL and logic programming live together happily ever after? In: Cruz, I., Decker, S., Allemang, D., Preist, C., Schwabe, D., Mika, P., Uschold, M., Aroyo, L. (eds.) ISWC 2006. LNCS, vol. 4273, pp. 501–514. Springer, Heidelberg (2006)

26. Prud'hommeaux, E., Seaborne, A., SPARQL,: Query Language for RDF, W3C Candidate Recommendation (June 14, 2007). Latest version is available at http://www.w3.org/TR/rdf-sparql-query

The Evaluation of Semantic Tools to Support Physicians in the Extraction of Diagnosis Codes

Regina Geierhofer and Andreas Holzinger

Institute for Medical Informatics, Statistics & Documentation (IMI)
Research Unit HCI4MED
Medical University Graz, A-8036 Graz, Austria
regina.geierhofer@meduni-graz.at,
andreas.holzinger@meduni-graz.at

Abstract. Over the past few years the extraction of medical information from German medical reports by means of semantic approaches and algorithms has been an increasing area of research. Currently, several tools are available that aim to support the physician in different ways. We developed a method to evaluate these tools in their ability to extract information from large amounts of data. We tested two off-the-shelf tools that worked in a background mode. We found that the field of quality management made it necessary that these large amounts of data could be background or batch processed. Additionally, we developed a metric, based on the semantic distance of the ICD codes, in order to improve the comparison of the accuracy of the codes suggested by the tools. The results of our evaluation showed that, at present, the tools are capable of supporting inexperienced physicians, however are still not sophisticated enough to work without human interaction.

Keywords: Human Language Analysis and Natural Language Processing, Evaluation, Semantics.

1 Introduction and Motivation for Research

The coding of diagnoses and procedures in many countries is obligatory in medical practise, because it provides basic information used for the financing and controlling of health care institutions [LKF], [DRG], [1]. In medicine, it is necessary to categorize free text reports for further processing [2], [3], [4], the decision on which code is most accurate is done by the physician. Recently many tools have been developed in order to support physicians and facilitate decision support [5]. Many doctors regard these tools as simply *diagnosis browsers,* which consist of little more than a search engine; few of these tools provide any more functionality. Consequently, these tools are not reliable enough to prepare medicals texts for coding on their own, or reliable enough for the automatic background coding of text.

This is despite the fact that several of these programs are based on network-like structures and are, in principle, capable of analyzing text in a far more sophisticated way that just substring matching [6].

A. Holzinger (Ed.): USAB 2007, LNCS 4799, pp. 403–408, 2007.

Our goal was to develop methods to evaluate such tools with regard their accurateness in the **automatic** analysis of medical texts and their ability at mapping these texts to medical codes. The fact that these texts could be background processed was useful for the following reasons; 1) there were large quantities of reports, 2) it made the performance comparison easier, and 3) because the tools all had different ways in which they interacted with the end user. This interaction can interfere with the accurateness of the final results/codes. Sometimes, users are not able to find a correct code at all, especially true if there are inaccuracies in the knowledgebase or in the filtering/ranking algorithm.

2 Methods and Materials

We used two types of coded reference sets to evaluate two separate off-the-shelf tools. Despite the fact that we would have preferred to have evaluated more tools, only two managed to satisfy our preconditions, specifically: 1) to work with German texts, 2) be capable of background processing, and 3) be available free of charge. However, in order to develop a suitable method of evaluation, it was not necessary to have more than two tools available. Both tools differed in their underlying philosophy in many respects, however, the following were relevant to us:

Different goals: Tool 1 only extracts short phrases from text, sufficient enough to choose a correct diagnosis code, while Tool 2 analyzes the entire text and codes the medical information using semantic axes, analogous to coding in SNOMED [7] .

Different fields: Tool 1 was designed for interactive use as an expert system. It works with a structure based on decision trees; each node is a term or concept. Goal-oriented questions are asked if the extracted information is too limited, according to the rules stored in its structure, in order to make the necessary decisions to reach a leaf (a diagnosis code). These questions form part of the tool's results if used in batch mode. As input it expects short phrases, such a physician's typical diagnosis, or a discharge letter diagnosis.

Tool 2 uses a semantic network. Its focus is not limited to the coding of diagnoses; it provides a knowledge base used by many applications, each with different goals. Goal-oriented questions are not a basic feature of this tool. The tool returns more than one code, both in interactive and in background mode. In principle, medical texts of any length are suitable as input.

Definition of success: In contrast to Tool 2, the first tool has an internal definition of the degree of its success (i.e. it extracts sufficient concepts or terms to reach an unambiguous diagnosis code).

Due to the differences and restrictions of the tools, we chose two types of reference sets: the ICD WHO 2005 descriptions, and sample diagnosis descriptions extracted from various discharge letters.

Scenario 1: ICD descriptions:
The main advantage of using the ICD descriptions is that you have, by definition, an unambiguous code for each description. We used the ICD WHO 2005 German

descriptions that can be downloaded from the German Institute of Medical Documentation and Information (DIMDI). We also allowed alternative descriptions for certain codes (see Table 1).

Table1. Alternative descriptions for ICD-Code D68.0

D68.0 Willebrand-Jürgens-Syndrom	D68.0 Von Willebrand's disease
Angiohämophilie Faktor-VIII-Mangel mit Störung der Gefäßendothelfunktion Vaskuläre Hämophilie	Angiohaemophilia Factor VIII deficiency with vascular defect Vascular haemophilia

If a code's description was not self-documenting, it was replenished according to the rules of the WHO. This task should be easy to compute, due to the fact that these texts already form parts of each tool's respective knowledgebase. Tool 1, however, found that the test data was not suitable; consequently we could not gather any results for scenario 1.

Scenario 2: Diagnosis description:
In this scenario the input consists of several thousand diagnosis descriptions coded by physicians from various medical disciplines. In this scenario, the physician's coding is used for comparison. Their codes, however, cannot be used as a reference value in the same manner as the ICD codes since it is possible that the code chosen by the physician is not the most appropriate code available for the medical text. Therefore, the ICD codes derived from the medical documentation were treated as if they were the results of a third tool.

Scenario 1 and Scenario 2:
In both scenarios we categorized each result as "precise" if the first 4 digits of any of the ICD codes returned by the tools matched the codes provided by us. Discrepancies in the ICD code's 5th digit were not considered, as the WHO itself does not utilize a fifth digit and because the 5th digit differs between the German, Swiss and Austrian editions. The result was considered "imprecise" if and when only the first 3 digits matched the correct classification code for the disease in question. We classified a result as "false" if none of the returned codes (maximum of 10 hits per description) was at least imprecise.

Semantic distance:
Simple string matching provides us a first impression of the quality of the initial results. However, it considers the hierarchical structure of the ICD only; it completely ignores the semantic structure that the ICD provides and the medical closeness of the described disease patterns. This semantic structuring considers the fact that related diseases correspond to codes in various chapters of the ICD.

In the ICD the WHO provides information where these related diseases and codes may be found. We analyzed this structure, converted it to a more suitable form, and developed a metric. Subsequently, we used an adaptation of the a-star-algorithm to calculate their semantic distance [8].

Based on the categorization method mentioned above we refined the evaluation results using the semantic distances. To exemplify the approach, consider the case of a patient who is allergic to her eye shadow.

One possible coding from a medical point of view is

H01.1 Noninfectious dermatoses of eyelid Dermatitis:
allergic another
L23.2 Allergic contact dermatitis due to cosmetics.

The ICD defines H01.1 as being directly related to L23. Without any weight at the edges of the graph, the semantic distance is 2 and is equal to the distance between L23.2 and L20-L30 Dermatitis and Eczema. The hierarchical distance of H01.1 and L23.2 would be 7 and equal to the distance between H01.1 and F20.2 Catatonic schizophrenia.

3 Results

Scenario 1:
We were only able to test some versions of the second tool, due to the reasons mentioned above. The results vary from between 84.2% and 95.27%.

Table 2. Results of the evaluation of Tool 2 as per Scenario 1

			Worst version		Best version	
precise			50928	84.20%	57623	95.27%
	First suggested code		46316		43851	
	Suggested code #2-10		4612		13772	
imprecise			1757	2.90%	850	1.41%
	First suggested code		279		315	
	Suggested code #2-10		1478		535	
false			5114	8.46%	1963	3.25%
	First suggested code		665		497	
	Suggested code 2-10		4449		1466	
		all false	1898		637	
	Suggested code <#10		2551		829	
no code			2683	4.44%	46	0.08%

Scenario 2:
Applying the same evaluation approach and using the physician's coding as a reference set, we got the following results:

Table 3. Results from the Tools from Scenario 1

	Tool 1	Tool 2
Precise	57%	57%
Imprecise	7%	4%
False	21%	27%
no code	15%	12%

It was not possible to compare the results in the granularity above, because the first tool only returns a single code or none at all. To be as fair as possible, we considered only the first hit returned by the second tool. This is, in part, responsible for the noticeable decrease in Tool 1's precise results.

Deficiencies and peculiarities of the tools were, in some cases, responsible for both tools returning different results or returning results that did not match the physician's coding. Sometimes the knowledge base was not complete; in other cases the processing was stopped too early, and in some cases the direction of the interpretation of the text lead to differing results. Text which could not be unambiguously interpreted due to a lack of information was yet another reason. Unambiguousness, is not normally a problem for a physician as they have more information available to them than the tools. The tools can only work on a given phrase, and have no other information available. Because a physician's coding could be incorrect for a particular text, it cannot be used as an absolute reference such as an ICD description or a gold standard. Consequently we built a number of sets of codes that matched either (a) the codes supplied by the other tool or (b) a physician's coding. Additionally, we used our implementation of a semantic distance metric to further refine the evaluation results. The results are presented in table 4.

Table 4. Results for the 3468 diagnosis descriptions mentioned above

	# codes	Median of the semantic distance (tool 1: physician)	median of the semantic distance (tool 2:physician)
tool 1=tool 2 = physician	1587	1,2	0,6
(tool 1 = tool 2) <> physician (3 digit)	152	6,3	6,6
(tool 2 = physician) <> tool 1 (3 digit)	521	6,7	1,3
(tool 1 = physician) <> tool 2 (3 digit)	641	2,7	7,4
tool 1 <> tool 2 <> physician (3 digit) but all return a code	183	6,3	7,3
tool 1 <> tool 2 <> physician (3 digit) one tool returns no code	247	5,76	7,4
(tool 1 = tool 2 = no code) <> physician	137	---	---

In cases where both tools suggested the same ICD code, these codes were more accurate than the physician's code. After manually checking all of the codes which had a short semantic distance between them, we corrected the results. After correction, the tools had a rate of 70.24% and 76.87% respectively (codes which were either *precise* or *imprecise*).

4   Conclusion and Future Work

At present, current tools used for the extraction of diagnoses codes are able to produce practical suggestions for diagnoses, especially if the input is short enough. As an interactive support tool for unskilled or inexperienced (novice) physicians, the benefit to the end user can be reasonable. We think that within a few years the use of such tools could be used to facilitate quality control significantly. At present it makes sense to develop methods to evaluate these tools with a systematic and technological approach. These methods also enable to deal with realistic magnitudes of text. Also of importance is the fact that this evaluation method measures objectively. We discovered that it is not easy to evaluate tools objectively against each other when they work in batch mode, due to their different approaches and, most of all, their various interactive designs. It will also be necessary to focus more on large text passages rather than short phrases, and to refine our way of measuring semantic distances (by using weights for the edges, for example). Adjustments, using other semantic interpretations of the ICD (such as SNOMED or UMLS), are also planned.

References

1. Stausberg, J., Koch, D., Ingenerf, J., Betzler, M.: Comparing paper-based with electronic patient records: Lessons learned during a study on diagnosis and procedure codes. Journal of the American Medical Informatics Association 10(5), 470–477 (2003)
2. Holzinger, A., Geierhofer, R., Errath, M.: Semantische Informationsextraktion in medizinischen Informationssystemen. Informatik Spektrum 30(2), 69–78 (2007)
3. Ruch, P., Baud, R., Geissbuhler, A.: Learning-free text categorization. In: Dojat, M., Keravnou, E.T., Barahona, P. (eds.) AIME 2003. LNCS (LNAI), vol. 2780, pp. 199–208. Springer, Heidelberg (2003)
4. Geierhofer, R., Holzinger, A.: Creating an Annotated Set of Medical Reports to Evaluate Information Retrieval Techniques. In: SEMANTICS 2007, Graz, Austria, September 5-7, 2007, pp. 331–339 (2007)
5. Holzinger, A., Geierhofer, R., Errath, M.: Semantic Information in Medical Information Systems - from Data and Information to Knowledge: Facing Information Overload. In: Proceedings of I-MEDIA 2007 and I-SEMANTICS 2007, pp. 323–330 (2007)
6. Matykiewicz, P., Duch, W., Pestian, J.: Nonambiguous concept mapping in medical domain, In: Artificial Intelligence and Soft Computing. In: Rutkowski, L., Tadeusiewicz, R., Zadeh, L.A., Zurada, J.M. (eds.) ICAISC 2006. LNCS (LNAI), vol. 4029, pp. 941–950. Springer, Heidelberg (2006)
7. Schulz, S., Hanser, S., Hahn, U., Rogers, J.: The semantics of procedures and diseases in SNOMED (R) CT. Methods of Information in Medicine 45(4), 354–358 (2006)
8. Senvar, M., Bener, A.: Matchmaking of semantic web services using semantic-distance information. In: Yakhno, T., Neuhold, E.J. (eds.) ADVIS 2006. LNCS, vol. 4243, pp. 177–186. Springer, Heidelberg (2006)

Ontology Usability Via a Visualization Tool for the Semantic Indexing of Medical Reports (DICOM SR)

Sonia Mhiri[1] and Sylvie Despres[2]

[1] The Computer Science Center of the Paris V University (CRIP5)
75006 Paris, France
[2] The Computer Science Laboratory of the Paris XIII University (LIPN)
93430 Villetaneuse, France
sonia.mhiri@math-info.univ-paris5.fr,
sylvie.despres@lipn.univ-paris13.fr

Abstract. One purpose of our research works is a contribution to a semantic indexing of structured reports in accordance with the DICOM SR[1] standard and we propose to guide this process with an ontology. In this paper, we describe our motivations for building this ontology according to a modularization approach assisted by the reuse of existing ontologies. Moreover, a prototype of a bilingual visualization tool is suggested. It allows specialists during their semantic indexing to load and visualize ontologies or modules from an ontology in a multi-axial way. Currently, six axes are planned: patient context, anatomy, pathology, visual descriptor, technique and recommendation.

Keywords: DICOM SR, Visualization Tool, Semantic indexing, Ontologies.

1 Introduction

For the medical imaging community taking advantage of the DICOM SR standard to improve medical image retrieval systems such as CBIR[2] one become a challenging research issue. A CBIR system refers to the retrieval from image databases using information extracted from the content of images [10]. In this paper, a special emphasis is given to semantic CBIR systems and more particularly to the use of ontologies as a support of indexing [15]. Our initial aim is a contribution to an ontology-based semantic indexing of structured reports in accordance with the DICOM SR standard. DICOM[3] is the only standard that can be used by imaging industry for the exchange and management of multimodalities images (radiology, MRI...). Since 1993, this standard was centered only on the image. In 2000, Structured Reporting (SR) was added to the DICOM standard to provide an efficient mechanism for the management of clinical reports [1][2][3][4]. The main advantage of SR is the ability to link clinical reports with the referenced images for simultaneous

---

[1] Digital Imaging and Communications in Medicine for Structured Reporting.
[2] Content-Based Image Retrieval.
[3] http://medical.nema.org

A. Holzinger (Ed.): USAB 2007, LNCS 4799, pp. 409–414, 2007.

retrieval and display. From the computerized systems perspective, SR has many potential advantages, such as the production of well-organized reports and the ability to communicate results promptly with more speed, reduced costs and fewer errors.

The rest of this paper is organized as follows. In section 2, we discuss our motivations to build an ontology for the semantic indexing of DICOM SR documents according to a modularization approach. In section 3, a first solution is proposed and consists in a prototype of a bilingual visualization tool to help specialists in their semantic indexing of reports. In section 4, a brief overview on related works is given while in the last section, the conclusion of our work and a brief description of our future research in this area are presented.

2   Ontology Usability for the Semantic Indexing of Medical Reports

Ontologies are widely used in semantic CBIR systems especially as a support of indexing or querying. The most referenced definition of the notion of ontology is given by Gruber in [12] as "an explicit specification of a conceptualization". An ontology provides a common, formal and shared knowledge for modeling a domain and is composed of a set of concepts and some specifications about their meanings (properties, relationships...). Thus, potential benefits are considerable such as more unambiguously express indexing, knowledge reuse, etc.

2.1   Specialists Indexing Layers of Reports According to DICOM SR

When we think about developing an ontology for indexing clinical reports related to patient imagery examinations, we need to take into account the diverse viewpoints of the different specialists (the radiologist, the cardiologist, the dermatologist...) in front of its reports with their associated images. Moreover, medical images are very particular because a large number of modalities exist (radiology, MRI, ultrasound...) and inside one modality, the tuning of an imager may lead to significantly varying images [18]. Due to the increasingly various sources, specialists can establish different interpretations of their reports, objective ones or subjective ones. For these reasons, we suggest to represent in Figure 1 these viewpoints according to six abstract layers: the contextual layer, the visual layer, the technical layer, the anatomical layer, the pathological layer and the recommendation layer.

2.2   A modular Ontology as a Support of Semantic Indexing

Building an ontology from scratch with domain experts or from text analysis requires a huge effort of conceptualization and a lot of time. Moreover, we notice an ever-increasing number of online ontologies and libraries available on the web. Search engines such as Swoogle[4] or OntoSearch[5] have also started to appear to facilitate online search. In fact, ontology reuse is nowadays the most promising research area for the knowledge engineering community. In our work, we intend to construct an

---

[4] http://www.swoogle.com
[5] http://www.ontosearch.com

ontology to assist specialists in indexing their medical reports with respect to the definition of the six layers (Fig. 1). This ontology is very large and typically requires collaboration among multiple individuals or groups with expertise in specific areas, with each participant contributing only a part of the ontology. That's why in our approach, instead of a single and centralized ontology, we would like to build our ontology according to a modularization approach [13] [14]. Because, no single ontology can meet the needs of all specialists under every conceivable scenario, the ontology that meets the needs of a specialist or a group of specialists needs to be assembled or unified from several independently developed ontology modules.

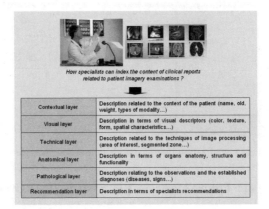

Fig. 1. A representation of specialists indexing layers

3   A Visualization Tool for the Semantic Indexing of Medical Reports

The major goal of this paper is to demonstrate a specific ontology visualization tool designed for specialists to assist the semantic indexing of reports according to DICOM SR. Its main characteristics and its future incorporation inside a medical CBIR system are mentioned.

3.1   Tool Characteristics

We have developed a bilingual tool (English/French) which loads and visualizes ontologies or a modular ontology in a multi-axial way. As shown in Figure 2 and in respect of our definition of specialists indexing layers (Fig. 1), six axes are created. After charging a DICOM SR file and before the indexing process, two scenarios are possible. The first one allows specialists to load for each axis an ever-increasing number of ontologies. The second one offers the possibility to load modules extracted from a specific ontology. Off-line extraction activity will be conducted by tools. Ontologies or ontology modules are showed according to an arborescence view. For each ontology element, a textual description is given (Fig. 2). Currently, OWL ontologies are load and only concepts are showed.

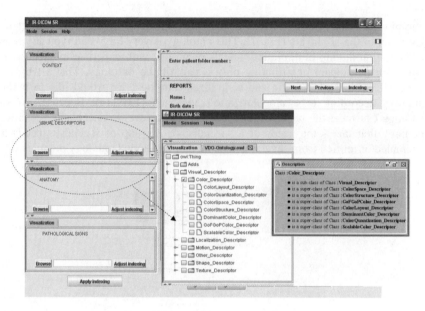

Fig. 2. The ontology visualization interface with an arborescence view of an ontology

3.2 A Component of a Future Medical CBIR System (IR-DICOM SR)

To improve medical diagnostic quality, an overview of our future medical CBIR system called IR-DICOM SR (Indexing and Retrieval) is illustrated in Figure 3. Three modules will be sequentially designed: the administration module (connection to the source of imagery, recovery starting from the source, storage into a database), the semantic indexing module (the multi-axial ontology visualization tool, the visualization of indexing results) and the retrieval module (query formulation, similarity measures, visualization of results). DICOM SR files will be firstly recovered from specific imaging devices and specifically stored in a database. Then, specialists with the help of semantic indexing tools will index with the help of the

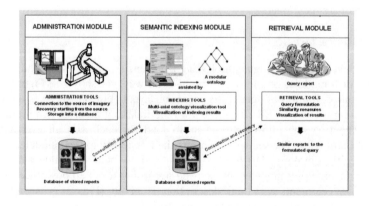

Fig. 3. An overview of our future medical CBIR system (IR-DICOM SR)

modular ontology the content of the database. As a result, each DICOM SR file will be described with a multi-dimensional vector called descriptor or index. In the retrieval mode, specialists will submit a query (an existing report extracted from the database) in search of similar reports. The system, with the help of a tool, will compute similarities between the descriptor of the query and those in the database. Finally, reports that are most similar to specialists query will be showed.

4 Related Work

A review of CBIR systems can be found in [9] [10] [11]. In the medical domain, many systems have been developed and most of them are dedicated to a very specific medical context, dealing with one given modality and interested in visual content of images (color, texture, shape, spatial layout, etc.) [9] [18]. But questions with respect to semantic indexing or querying are still unanswered and more desirable for medical applications. Moreover, CBIR systems that take into account DICOM and more particularly DICOM SR are relatively inexistent. Currently, existing applications (Osiris, IconoTech, DICOMEye, DICOMscope...) are centered on viewing and conventional databases [5] [6] [7] [8]. Moreover, in literature existing ontology-based semantic indexing tools [16] [17] don't take into account DICOM or DICOM SR files. This is partly due to imaging devices from which such format is relatively difficult to obtain. Specific equipments are often required.

5 Conclusion

The work presented here is a contribution towards a future medical CBIR system for clinical reports according to DICOM SR standard and related to patient imagery examinations. Currently, to improve diagnostic quality, the medical imaging community is increasingly aware of the potential benefit of these kinds of systems. We are currently planning to confront our approach with initial concrete experiences in bones and joints radiology. Clearly, our visualization prototype requires consolidating and several extensions are possible: visualization of other elements of the ontology (properties, relations ...), the structure of the index vector, a semi-automatic indexing process. Moreover, ontology reuse and more particularly ontology modularization is still a relatively new research domain. That's why several questions arise around our modular ontology. In the medical field, a classification for the considerable number of existing ontologies is imperative. Keeping a check on the reusing possibilities and then on existing ontologies heterogeneities are also initial research tracks.

References

1. Hussein, R., Engelmann, U., Schroeter, A., Meinzer, H.P.: DICOM Structured Reporting: Part 1. Overview and characteristics. Radiographics 24(3), 891–896 (2004)
2. Hussein, R., Engelmann, U., Schroeter, A., Meinzer, H.P.: DICOM Structured Reporting: Part 2. Problems and Challenges in Implementation for PACS Workstations. Radiographics 24(3), 897–909 (2004)

3. Clunie, D.: DICOM structured reporting. PixelMed, Bangor, Pa (2000)
4. Csipo, D., Dayhoff, R., Kuzmak, M.: Integrating Digital Imaging and Communications in Medicine (DICOM)-structured reporting into the hospital environment. Journal of Digital Imaging 14, 12–16 (2001)
5. Eichelberg, M., Riesmeier, J., Wilkens, T., Jensch, P.: One Decade of Medical Imaging Standardisation and Implementation: a Short Review of DICOM and the OFFIS DICOM Toolkit. In: Proceedings of EuroPACS-MIR in the Enlarged Europe, pp. 253–256 (2004)
6. Clunie, D.: DICOM structured reporting: implementation experience. ftp://medical.nema.org/SRWorkshopJune0801/5-Comview-SR-2001-06-08dac.pdf
7. Jonathan, L.: ACC 2001 demonstration of DICOM structured reporting. ftp://medical.nema.org/SRWorkshopJune0801
8. Software, http://www.adninformatique.net/data/doc_dicom.html and http://www.radio.univ-rennes1.fr/Sources/FR/Inform.html
9. Muller, H., Michous, N., Bandon, D., Geissbuhler, A.: A review of content- based image retrieval systems in medical applications: clinical benefits and future directions. International Journal of Medical Informatics 73(1), 1–23 (2004)
10. Smeulders, A.W.M., Worring, M., Santini, S., Gupta, A., Jain, R.: Content-based image retrieval at the end of the early years. IEEE Transactions On Pattern Analysis and Machine Intelligence 22(12), 1349–1380 (2000)
11. Veltkamp, R.C., Tanase, M.: Content-based image retrieval systems: a survey. Technical Report UU-CS-2000-34 (2002)
12. Gruber, T.R.: Towards Principles for the Design of Ontologies Used for Knowledge Sharing. In: Guarino, N., Poli, R. (eds.) Formal Ontology in Conceptual Analysis and Knowledge Representation, Kluwer Academic Publishers, Dordrecht (1993)
13. Menken, M., Stuckenschmidt, H., Wache, H., Serafini, L., Tamilin, A., Jarrar, M., Porto, F., Parent, C., Rector, A., Pan, J., d'Aquin, M., Lieber, J., Napoli, A., Stoilos, G., Tzouvaras, V., Stamou, G.: Report on modularization of ontologies. The Knowledge Web Network of Excellence (NoE) IST-2004-507482. Deliverable D2.1.3.1 (WP2.1) (2005)
14. Doran, P.: Ontology reuse via ontology modularization. Technical report. Knowledge Web Network of Excellence (NoE). Deliverable D2.1.3.1 (2005)
15. Hyvonën, E.R., Styrman, A., Saarela, S.: Ontology-based image annotation and retrieval, pp. 15–27. HIIT Publications (2002)
16. Tsinaraki, C., Polydoros, P., Kazasis, F., Christodoulakis, S.: Ontology-based Semantic Indexing for MPEG-7 and TV-Anytime Audiovisual Content. Multimedia Tools and Applications, Video Segmentation for Semantic Annotation and Transcoding (2004)
17. Indexing tools, http://www.w3.org/2005/Incubator/mmsem/resources/tools.html
18. Holzinger, A., Geierhofer, R., Errath, M.: Semantic Information in Medical Information Systems - from Data and Information to Knowledge: Facing Information Overload. In: Proc. of I-MEDIA 2007 and I-SEMANTICS 2007, pp. 323–330 (2007)

Fostering Creativity Thinking in Agile Software Development

Claudio León de la Barra[1] and Broderick Crawford[1,2]

[1] Pontificia Universidad Católica de Valparaíso, Chile
[2] Universidad Técnica Federico Santa María, Chile
{cleond, broderick.crawford}@ucv.cl

Abstract. Psychology and Computer Science are growing in a interdisciplinary relationship mainly because human and social factors are very important in developing software and hardware. The development of new software/hardware products requires the generation of novel and useful ideas. In this paper, the Agile method called eXtreme Programming (XP) is analyzed and evaluated from the perspective of the creativity, in particular the creative performance and structure required at the teamwork level. The conclusion is that XP has characteristics that ensure the creative performance of the team members, but we believe that it can be fostered from a creativity perspective.

Keywords: Creativity, New Products Development, Software Development, Agile Methodologies, eXtreme Programming.

1 Introduction

"Software is developed for people and by people" [1].

But surprisingly, most of software engineering research is technical and deemphasizes the human and social aspects. By other hand, the traditional development process of new products that is a fundamental part in the marketing, has been recently criticized by Kotler and Trías de Bes [2]. They point out that fundamental creative aspects are not considered at all and as a consequence this development is not useful, viable or innovative. In this context, it is interesting to consider the new proposals of agile methodologies for software development in order to analyse and evaluate them at the light of the existing creative expositions, mainly considering the teamwork practices.

The agile principles and values have emphasized the importance of collaboration and interaction in the software development and, by other hand, creative work commonly involves collaboration in some form and it can be understood as an interaction between an individual and a sociocultural context. In relation with the joint work between users and software developers, there are very interesting cases in healthcare enterprises and medicine. The relationship between agile approaches and the health sector has a notable case with the work of Jeff Sutherland, the inventor of the popular Scrum Agile Development Process [3-7].

A. Holzinger (Ed.): USAB 2007, LNCS 4799, pp. 415–426, 2007.
© Springer-Verlag Berlin Heidelberg 2007

Scrum, the most notorious competitor of eXtreme Programming XP [8], has attained worldwide fame for its ability to increase the productivity of software teams by several magnitudes through empowering individuals, fostering a team-oriented environment, and focusing on project transparency and results. Furthermore, there are important experiences with XP in the development of tele-health applications [9] and different works introducing innovations in health information systems [10-13].

We believe that the innovation and development of new products is an interdisciplinary issue [14], we are interested in the study of the potential of new concepts and techniques to foster creativity in software engineering [15]. This paper is organised as follows: in section 2 we explain the motivation of this work fixing the relevance of Creativity in Software Development. Section 3 is about central aspects in Creativity. Section 4 gives a brief overview of XP and its phases and roles. Section 5 presents a comparison between roles in creative teams and roles in XP teams. Finally, in Section 6 we conclude the paper and give some perspectives for future research.

2   Creativity in Software Development

Software engineering is a knowledge intensive process that includes human and social factors in all phases: eliciting requirements, design, construction, testing, implementation, maintenance, and project management [1]. No worker of a development project has all the knowledge required to fulfill all activities. This underlies the need for communication, collaboration and knowledge sharing support to share domain expertise between the customer and the development team [16].

Since human creativity is thought as the source to resolve complex problem or create innovative products, one possibility to improve the software development process is to design a process which can stimulate the creativity of the developers. There are few studies reported on the importance of creativity in software development. In management and business, researchers have done much work about creativity and obtained evidence that the employees who had appropriate creativity characteristics, worked on complex, challenging jobs, and were supervised in a supportive, noncontrolling fashion, produced more creative work. Then, according to the previous ideas the use of creativity in software development is undeniable, but requirements engineering is not recognized as a creative process in all the cases [17].

In a few publications the importance of creativity has been investigated in all the phases of software development process [18, 15, 19] and mostly focused in the requirements engineering [20-22]. Nevertheless, the use of techniques to foster creativity in requirements engineering is still shortly investigated. It is not surprising that the role of communication and interaction is central in many of the creativity techniques. The most popular creativity technique used for requirements identification is the classical brainstorming and more recently, role-playing-based scenarios, storyboard-illustrated scenarios, simulating and visualizing have been applied as an attempt to bring more creativity to requirements elicitation. These techniques try to address the problem of identifying the viewpoints of all the stakeholders [22].

However, in requirements engineering the answers do not arrive by themselves, it is necessary to ask, observe, discover, and increasingly create requirements. If the goal is to build competitive and imaginative products, we must make creativity part of the requirements process. Indeed, the importance of creative thinking is expected to increase over the next decade [23].

In [20, 24] very interesting open questions are proposed: Is inventing part of the requirements activity? It is if we want to advance. So who does the inventing? We cannot rely on the customer to know what to invent. The designer sees his task as deriving the optimal solution to the stated requirements. We can not rely on programmers because they are far away from the work of client to understand what needs to be invented. Requirements analysts are ideally placed to innovate. They understand the business problem, have updated knowledge of the technology, will be blamed if the new product does not please the customer, and know if inventions are appropriate to the work being studied. In short, requirements analysts are the people whose skills and position allows, indeed encourages, creativity.

In [25] the author, a leading authority on cognitive creativity, identifies basic types of creative processes: exploratory creativity explores a possible solution space and discovers new ideas, combinatorial creativity combines two or more ideas that already exist to create new ideas, and transformational creativity changes the solution space to make impossible things possible. Then, most requirements engineering activities are exploratory, acquiring and discovering requirements and knowledge about the problem domain. Requirements engineering practitioners have explicitly focused on combinatorial and transformational creativity.

In relation with the active participation of the end users in medical software development, Holzinger has very valuable work making usability practitioners first-class citizens in the process [26-29].

3   Creativity: Purposes, Performance and Structure

The creativity definitions are numerous [30-32], therefore, considering the object of analysis in the present paper: a software development teamwork, that must respond to the requirements of a specific client for a particular problem, a suitable definition is the one raised by Welsch [33]:

Creativity is the process of generating unique products by transformation of existing products. These products, tangible and intangible, must be unique only to the creator, and must meet the criteria of purpose and value established by the creator.

More specifically, and from an eminently creative perspective, it is possible to distinguish three aspects at the interior of a group developing new products:

a) The *purposes* that the team tries to reach, which demand two scopes of results [34-38]:
  − Those related to the creative result that must be original, elaborated, productive and flexible.
  − Those related to the creative team, so that it reaches its goals, developing cognitive abilities and presenting an improved disposition to the change. All this in order to obtain a better creative team performance in the future.

b) The *performance* shown by the team in connection with the main aspects of the complex dynamics that the persons build inside a team. We describe three aspects:

- The personal conditions of the members of the team, in terms of the styles and cognitives abilities, the personality, their intrinsic motivation and knowledge [32, 39, 30, 34].
- The *organizational conditions* in which the creative team is inserted, and that determines, at least partly, its functioning. These conditions, in the extent that present/display certain necessary particular characteristics -although non sufficient- for the creative performance. They emphasize in special the culture (communication, collaboration, trust, conflict handle, pressure and learning) [32, 40, 41]; the internal structure (formalization, autonomy and evaluation of the performance) [32, 40, 41, 39]; the team available resources (time disposition) [32, 40, 30] and the physical atmosphere of work [31].
- The *conditions of performance* of the creative team, mainly the creative process realized, which supposes the set of specific phases that allow to assure the obtaining of a concrete result (creative product) [31, 42].
- c) The *structure of the creative team,* particularly the group characteristics, such as norms, cohesiveness, size, diversity, roles, task and problem-solving approaches [32].

Of the mentioned aspects, we deepen in those referred to the structure and performance of the team for the development of new products, specially considering: the creative process and the roles surrounding this process.

3.1 Creative Process

The creative process constitutes the central aspect of team performance, because it supposes a serie of clearly distinguishable phases that had to be realized by one or more of the team members in order to obtain a concrete creative result.

The phases - on the basis of Wallas [42] and Leonard and Swap [31] - are the following ones:

- *Initial preparation:* the creativity will bloom when the mental ground is deep, fertile and it has a suitable preparation. Thus, the deep and relevant knowledge, and the experience precedes the creative expression.
- *Encounter:* the findings corresponding to the perception of a problematic situation. For this situation a solution does not exist. It is a new problem.
- *Final preparation:* it corresponds to the understanding and foundation of the problem. It's the immersion in the problem and the use of knowledge and analytical abilities. It includes search for data and the detailed analysis of factors and variables.
- *Generation of options:* referred to produce a menu of possible alternatives. It supposes the divergent thinking. It includes, on one hand, finding principles, lines or addresses, when making associations and uniting different marks of references and, on the other hand, to generate possible solutions, combinations and interpretations.

- *Incubation:* it corresponds to the required time to reflect about the elaborated alternatives, and "to test them mentally".
- *Options Choice:* it corresponds to the final evaluation and selection of the options. It supposes the convergent thinking.
- *Persuasion:* closing of the creative process and communication to other persons.

Considering the creativity as a "nonlinear" process some adjustments are necessary, redefinitions or discardings that force to return to previous phases, in a complex creative dynamic.

Therefore, for each one of the defined phases it is possible to associate feedbacks whose "destiny" can be anyone of the previous phases in the mentioned sequence.

3.2 Roles in a Creative Team

Lumsdaine and Lumsdaine [43] raise the subject of the required cognitives abilities (mindsets) for creative problem resolution. Their tipology is excellent for the creative team, and the different roles to consider. These roles are the following ones:

- *Detective.* In charge of collecting the greatest quantity of information related to the problem. It has to collect data without making judgements, even when it thinks that it has already understood the problem exactly.
- *Explorer.* Detects what can happen in the area of the problem and its context. It thinks on its long term effects and it anticipates certain developments that can affect the context (in this case, the team). The explorer perceives the problem in a broad sense.
- *Artist.* Creates new things, transforming the information. It must be able to break his own schemes to generate eccentric ideas, with imagination and feeling.
- *Engineer.* Is the one in charge of evaluating new ideas. It must make converge the ideas, in order to clarify the concepts and to obtain practical ideas that can be implemented for the resolution of problems.
- *Judge.* Must do a hierarchy of ideas and decide which of them will be implemented (and as well, which ones must be discarded). Additionally, it must detect possible faults or inconsistences, as well as raise the corresponding solutions. Its role must be critical and impartial, having to look for the best idea, evaluating the associated risks.
- *Producer.* In charge of implementing the chosen ideas.

Leonard and Swap [31] have mentioned additional roles, possible to be integrated with the previous ones, because they try to make more fruitful the divergence and the convergence in the creative process:

- The *provoker* who takes the members of the team "to break" habitual mental and procedural schemes to allow the mentioned divergence (in the case of the "artist") or even a better convergence (in the case of the "engineer").
- *Think tank* that it is invited to the team sessions to give a renewed vision of the problem-situation based on his/her experticia and experience.

- The *facilitator* whose function consists in helping and supporting the team work in its creative task in different stages.
- The *manager* who cares for the performance and especially for the results of the creative team trying to adjust them to the criteria and rules of the organization (use of resources, due dates).

Kelley and Littman [44], on the other hand, have raised a role tipology similar to Lumsdaine and Lumsdaine [43], being interesting that they group the roles in three categories:

1) those directed to the learning of the creative team (susceptible of corresponding with the detective, explorer, artist, provoker and think tank roles);

2) others directed to the internal organization and success of the team (similar to the judge, facilitator and manager roles) and,

3) finally, roles whose purpose is to construct the innovation (possibly related to the role of the engineer and judge).

4 eXtreme Programming XP

Extreme Programming is an iterative approach to software development [8]. The methodology is designed to deliver the software that customer needs when it's needed. This methodology emphasizes team work. Managers, customers, and developers are all part of a team dedicated to deliver quality software. XP implements a simple, yet effective way to enable groupware style development. XP improves a software project in four essential ways; communication, simplicity, feedback, and courage.

4.1 Roles in a Creative Team

XP defines the following roles for a software development process [8]:

- *Programmer.* The programmer writes source code for the software system under development. This role is at the technical heart of every XP project because it is responsible for the main outcome of the project: the application system.
- *Customer.* The customer writes user stories, which tell the programmer what to program. "The programmer knows how to program. The customer knows what to program".
- *Tester.* The tester is responsible for helping customers select and write functional tests. On the other side, the tester runs all the tests again and again to create an updated picture of the project state.
- *Tracker.* The tracker keeps track of all the numbers in a project. This role is familiar with the estimation reliability of the team. Whoever plays this role knows the facts and records of the project and should be able to tell the team whether they will finish the next iteration as planned.
- *Coach.* The coach is responsible for the development process as a whole. The coach notices when the team is getting "off track" and puts it "back on track." To do this, the coach must have experience with XP.

– *Consultant.* Whenever the XP team needs additional special knowledge, they "hire" a consultant who possesses this knowledge. The consultant transfers this knowledge to the team members, enabling them to solve the problem on their own.
– *Big boss.* The big boss or Manager provides the resources for the process. The big boss needs to have the general picture of the project, be familiar with the current project state, and know whether any interventions are needed to ensure the project's success.

5   Creativity in eXtreme Programming

Regarding to the structure dimension of a new product development team (in particular software), it is possible to relate the roles in creativity to the roles defined in the XP methodology distinguishing:

1) base roles, that is, those directly related to the creative processes and software development, and
2) support roles, whose function is to support or lead the other roles for a better performance.

In relation with the structure dimension it's important to considerate how the team can operate. In order to implement the functionality of each role, we must considerate two aspects: basic organizational conditions and the pertinent creative process.

5.1   Team Performance (Organizational Conditions)

The creative team performance is determined by the organizational conditions in which it's inserted [39, 31, 32, 40, 41]. Some conditions are necessary - although not sufficient - for the creative performance.

We are interested in explore the influence of autonomy, communication, cooperation and learning, the handling of possible conflicts, pressure, formalization, performance evaluation, available resources (time) and the physical atmosphere of work.

The *autonomy* refers to the capacity of the people and the team as a whole to act and make decisions. This aspect is related to the following XP practices: the actual client, since it is part of the team and, in addition, has decisional capacity delegated by its own organization; the use of metaphors, of codification standards and the existence of "right" rules really represent codes of shared thought and action, that make possible the autonomy of the team members; the small deliveries and the fact of the collective property allow that all the involved ones share official and explicit knowledge, that results in a greater independence of the members and the possibility of a minor coordination among them.

The team member's *communication*, cooperation and learning are fortified since the client is present and there exist opened spaces to work together and in a pair programming mode. The work dynamics is based on a game of planning and metaphors involving all the participants from the beginning (client and equipment developer). Also, the use of codification standards, the small deliveries, the collective property of the code and the simple design, allow that the person has clear performance codes and rules about what is expected and acceptable (internal culture) in order to establish the required communication and cooperation.

The *handling of possible conflicts* between the client and the development team, and internally at team level is favored by XP practices handling it (presence of the client, pairs programming, planning game, continuous integration, tests, collective property), or to reduce it and to avoid it (small deliveries, simple design, 40 hour a week and codification standard). Cooperation and trust are associated to this issue.

The *pressure* (that in creativity is appraised as favorable until certain degree, favoring the performance, and detrimental if it exceeds this degree), is susceptible then to favor in XP through the client in situ, the programming by pairs, the planning game, the tests and continuous integration. It's possible to avoid, or at least to reduce, the pressure through the refactorization, the small deliveries, the collective property, and the fact that surpassing the 40 weekly working hours is seen like an error.

The *formalization*, that gives account of all those formal aspects (norms, procedures) defined explicitly and that are known, and even shared, by the members of the team. It's assured in XP through planning game, metaphors, continuous integration, the collective property, the 40 hours per week and the codification standards guiding the desirable conduct and performance of the team.

The *evaluation of the performance* is made in XP through pair programming (self evaluation and pair evaluation), frequent tests and even through the 40 weekly hours (as a nonexceedable metric indicating limit of effectiveness), all at the light of the planning (including the standards). Finally, the presence of client constitutes the permanent and fundamental performance evaluation of the team and the products. The evaluation characteristics empower the learning processs.

The *time* dedicated has fundamental importance in XP team respecting the available resources. This aspect is strongly stressed in creativity.

The pair programming and the developer multifunctional role allow to optimize the partial working-times, as well as the whole project time, ensuring a positive pressure.

The *physical atmosphere of work*, referred in creativity to the surroundings that favor or make difficult the creative performance (including aspects like available spaces, noise, colours, ventilation, relaxation places) are assured only partially in XP through the open spaces, as a way to assure the interaction between members of the team.

5.2 Team Performance (Process)

The team performance is directly determined by the creative process [31, 42]. It is important to correlate the phases defined in XP with the phases considered in a creative process.

- The *initial preparation* and "finding" defined in the creative process correspond to the exploration phase in XP, where the functionality of the prototype and familiarization with the methodology are established.
- The *final stage* of preparation is equivalent with the phases of exploration and planning in XP, defining more in detail the scope and limit of the development.
- The *option generation phases,* incubation and election of options defined in the creative process correspond to the iterations made in XP and also with the liberations of the production phase (small releases). In XP there is not a clear distinction of the mentioned creative phases, assuming that they occur to the interior of the team.

- The *feedback phase* (understanding this one as a final stage of the process, and not excluding that can have existed previous micro - feedbacks since the creative process is nonlinear) it could correspond in XP with the maintenance phase.
- The *persuasion phase* is related to the phase of death established in XP, constituting the close of the development project with the final liberation.

5.3 Team Structure (Base and Supporting Roles)

As previously mentioned in the creative process there are base and supporting roles. The base roles are directly related to the creative and software development process and the supporting roles support or lead the base roles to a better performance. The following is the correlation between creative and XP roles:

- The detective function consisting in collecting information related to a problem is made by the client himself in XP, because this one generates the first contact with the software development team.
- The function of explorer consisting in defining completely the problem is made in XP as much by the client as the manager of the team, all together they appreciate the reach of the identified problem, as well as of the possible solutions. The function of the artist consisting in transforming the information, creating new relations, and therefore generating interesting solutions is made by the developer, that in XP methodology is in charge of the analysis, design and programming of software.
- The function of the engineer referred to clarify and to evaluate the new ideas, in terms of its feasibility is made in XP by the tester and the tracker.
- The function of the judge, understood as the definitive selection of the solutions to implant, is made in XP by the tracker and the client.
- The function of the producer, referred to the implementation of the selected ideas (strictly speaking it is working software) is made in XP by the client in his organization, including the processes and procedures that this function implies.

The supporting roles considered are:

- The provoker; creativity demands that the divergence as well as convergence in the solutions be maximum and complete. There is not explicit reference in XP methodology about divergent thinking.
- The think tank who helps the team work "from outside" is equivalent completely to the role of the consultant.
- The facilitator, whose function is helping the team, corresponds in XP to the coach role.
- The manager whose function is to lead to the team in terms of its general efficiency and its effectiveness corresponds with XP's big boss or manager.

6   Conclusions and Future Research

The Extreme Programming methodology includes implicitly central aspects of a creative teamwork. These aspects can be organized according to the structure that the team adopts and the performance that characterizes to the team.

The structure that the team adopts and specially the different roles that the methodology advises to define, nearly correspond with the roles at the interior of a creative team.

The performance that characterizes the team through certain advisable practices, from the perspective of creativity, constitutes the necessary basic conditions, although nonsufficient, in order to favor the group creative performance.

These conditions - called practices in XP methodology - are accompanied by concrete phases of constituent activities of an agile software development process, which is possible to correspond with the creative process, which is fundamental to the creative performance.

In spite of the previous comments, we think that XP methodology should have a more explicit reference to:

- The provoker role that is thoroughly described in creativity as a fundamental factor to generate innovation. This can be explained because, in general, agile methodologies do not aim, as a central element, to generate an original software, but an effective one.
- The distinction and formalization of the creative phases to generate options incubation and option choices (that are fundamental in creativity). It is assumed that they take place in the iterative and production process. Again, XP is not focused in "originality", resulting that the divergence is not so fundamental in XP.
- A more direct mention to the physical atmosphere of work, that in creativity are considered as highly relevant to enhance the performance. These aspects should have a greater consideration since software development is a special case of product development.

As future analysis we have in mind:

- The study of basic personal characteristics improving team performance.
- To study in depth the possible improvements when an agile team explicitly is looking for creative goals. Goals in relation with the creative product attributes and goals concerning to the team members.
- Synergetic blending of cognitive psychology, organizational behavior and agile development processes.
- What is the penetration of agile methods and our ideas in the chilean software market?

References

1. John, M., Maurer, F., Tessem, B.: Human and social factors of software engineering: workshop summary. SIGSOFT Software Engineering Notes 30, 1–6 (2005)
2. Kotler, P., TríasdeBes, F.: Marketing Lateral. Editorial Pearson/Prentice Hall, Spain (2004)
3. Sutherland, J.: Agile can scale: Inventing and reinventing scrum in five companies. Cutter IT Journal 14, 5–11 (2001)
4. Sutherland, J.: Agile development: Lessons learned from the first scrum. Cutter Agile Project Management Advisory Service: Executive Update 5, 1–4 (2004)

5. Sutherland, J.: Recipe for real time process improvement in healthcare. In: 13th Annual Physician-Computer Connection Symposium, Rancho Bernardo, CA, American Society for Medical Directors of Information Systems (AMDIS) (2004)

6. Sutherland, J.: Future of scrum: Parallel pipelining of sprints in complex projects. In: AGILE, pp. 90–102. IEEE Computer Society, Los Alamitos (2005)

7. Sutherland, J., van den Heuvel, W.J.: Towards an intelligent hospital environment: Adaptive workflow in the future. In: HICSS, IEEE Computer Society, Los Alamitos (2006)

8. Beck, K.: Extreme programming explained: embrace change. Addison-Wesley Longman Publishing Co., Inc, Boston, MA, USA (2000)

9. Fruhling, A.L., Tyser, K., de Vreede, G.J.: Experiences with extreme programming in telehealth: Developing and implementing a biosecurity health care application. In: HICSS, IEEE Computer Society, Los Alamitos (2005)

10. Christensen, C., Bohmer, R., Kenagy, J.: Will disruptive innovations cure health care. Harvard Business Review, 102–111 (2000)

11. Dadam, P., Reichert, M., Kuhn, K.: Clinical workflows - the killer application for process oriented information systems. In: BIS 2000. Proceedings of the 4th International Conference on Business Information Systems, pp. 36–59 (2000)

12. Fruhling, A.: Examining the critical requirements, design appoaches and evaluation methods for a public health emergency response system. Communications of the Association for Information Systems 18 (2006)

13. Fruhling, A.L., Steinhauser, L., Hoff, G., Dunbar, C.: Designing and evaluating collaborative processes for requirements elicitation and validation. In: HICSS, p. 15. IEEE Computer Society, Los Alamitos (2007)

14. Takeuchi, H., Nonaka, I.: The new product development game. Harvard Business Review (1986)

15. Gu, M., Tong, X.: Towards hypotheses on creativity in software development. In: Bomarius, F., Iida, H. (eds.) PROFES 2004. LNCS, vol. 3009, pp. 47–61. Springer, Heidelberg (2004)

16. Chau, T., Maurer, F., Melnik, G.: Knowledge sharing: Agile methods vs tayloristic methods. In: WETICE. Twelfth International Workshop on Enabling Technologies: Infrastructure for Collaborative Enterprises, pp. 302–307. IEEE Computer Society, Los Alamitos, CA, USA (2003)

17. Maiden, N., Gizikis, A., Robertson, S.: Provoking creativity: Imagine what your requirements could be like. IEEE Software 21, 68–75 (2004)

18. Glass, R.L.: Software creativity. Prentice-Hall, Inc., Upper Saddle River, NJ, USA (1995)

19. Crawford, B., de la Barra, C.L.: Enhancing creativity in agile software teams. In: Concas, G., Damiani, E., Scotto, M., Succi, G. (eds.) XP 2007. LNCS, vol. 4536, pp. 161–162. Springer, Heidelberg (2007)

20. Robertson, J.: Requirements analysts must also be inventors. IEEE Software 22, 48–50 (2005)

21. Maiden, N., Robertson, S.: Integrating creativity into requirements processes: Experiences with an air traffic management system. In: 13th IEEE International Conference on Requirements Engineering, Paris, France, pp. 105–116. IEEE Computer Society Press, Los Alamitos (2005)

22. Mich, L., Anesi, C., Berry, D.M.: Applying a pragmatics-based creativity-fostering technique to requirements elicitation. Requirements Engineering 10, 262–275 (2005)

23. Maiden, N., Gizikis, A.: Where do requirements come from? IEEE Software 18, 10–12 (2001)

24. Robertson, J.: Eureka! Why analysts should invent requirements. IEEE Software 19, 20–22 (2002)
25. Boden, M.: The Creative Mind. Abacus (1990)
26. Memmel, T., Reiterer, H., Holzinger, A.: Agile methods and visual specification in software development: a chance to ensure universal access. In: Coping with Diversity in Universal Access, Research and Development Methods in Universal Access. LNCS, vol. 4554, pp. 453–462. Springer, Heidelberg (2007)
27. Holzinger, A.: Rapid prototyping for a virtual medical campus interface. IEEE Software 21, 92–99 (2004)
28. Holzinger, A., Errath, M.: Designing web-applications for mobile computers: Experiences with applications to medicine. In: Stary, C., Stephanidis, C. (eds.) User-Centered Interaction Paradigms for Universal Access in the Information Society. LNCS, vol. 3196, pp. 262–267. Springer, Heidelberg (2004)
29. Holzinger, A., Errath, M.: Mobile computer Web-application design in medicine: some research based guidelines. Universal Access in the Information Society International Journal 6(1), 31–41 (2007)
30. Amabile, T., Conti, R., Coon, H., Lazenby, J., Herron, M.: Assessing the work environment for creativity. Academy of Management Journal (39), 1154–1184
31. Leonard, D.A., Swap, W.C.: When Sparks Fly: Igniting Creativity in Groups. Harvard Business School Press, Boston (1999)
32. Woodman, R.W., Sawyer, J.E., Griffin, R.W.: Toward a theory of organizational creativity. The Academy of Management Review 18, 293–321 (1993)
33. Welsh, G.: Personality and Creativity: A Study of Talented High School Students. Unpublished doctoral dissertation, Chapel Hill, University of North Carolina (1967)
34. Csikszentmihalyi, M.: Creativity: Flow and the Psychology of Discovery and Invention. Harper Perennial, New York (1998)
35. Guilford, J.P.: Intelligence, Creativity and Their Educational Implications. Edits Pub. (1968)
36. Hallman, R.: The necessary and sufficient conditions of creativity. Journal of Humanistic Psychology 3 (1963) Also reprinted in Gowan, J.C. et al., Creativity: Its Educational Implications. New York: John Wiley and Co. (1967)
37. Hallman, R.: Aesthetic pleasure and the creative process. Journal of Humanistic Psychology 6, 141–148 (1966)
38. Hallman, R.: Techniques of creative teaching. Journal of Creative Behavior I (1966)
39. Amabile, T.: How to kill creativity. Harvard Business Review, pp. 77–87 (September-October 1998)
40. Kotler, P., Armstrong, G.: Principles of Marketing, 10th edn. Prentice-Hall, Englewood Cliffs (2003)
41. Isaksen, S.G., Lauer, K.J., Ekvall, G.: Situational outlook questionnaire: A measure of the climate for creativity and change. Psychological Reports, 665–674
42. Wallas, G.: The art of thought. Harcourt Brace, New York (1926)
43. Lumsdaine, E., Lumsdaine, M.: Creative Problem Solving: Thinking Skills for a Changing World. McGraw-Hill, Inc., New York (1995)
44. Kelley, T., Littman, J.: The Ten Faces of Innovation: IDEO's Strategies for Defeating the Devil's Advocate and Driving Creativity Throughout Your Organization. Currency (2005)

An Analytical Approach for Predicting and Identifying Use Error and Usability Problem

Lars-Ola Bligård and Anna-Lisa Osvalder

Division of Design, Chalmers University of Technology
SE-412 96 Göteborg, Sweden
lars-ola.bligard@chalmers.se

Abstract. In health care, the use of technical equipment plays a central part. To achieve high patient safety and efficient use, it is important to avoid use errors and usability problems when handling the medical equipment. This can be achieved by performing different types of usability evaluations on prototypes during the product development process of medical equipment. This paper describes an analytical approach for predicting and identifying use error and usability problems. The approach consists of four phases; (1) Definition of Evaluation, (2) System Description, (3) Interaction Analysis, and (4) Result Compilation and Reflection. The approach is based on the methods Hierarchical Task Analysis (HTA), Enhanced Cognitive Walkthrough (ECW) and Predictive Use Error Analysis (PUEA).

Keywords: Usability Engineering, Usability Evaluation, Analytical Methods, Medical Equipment.

1 Introduction

In safety-critical technical systems, it is important that the systems are simple and safe to handle for the users when they perform intended tasks in the intended context. This is especially true for medical equipment, where a possibility of harm to patients can arise from erroneous use of the devices [19, 38]. Several studies have shown that there is a clear connection between problems of usability and human mistakes (use errors) [29, 24].

A step in creating safe and efficient medical equipment is therefore to try in advance to identify and counteract usability problems and use errors before they cause serious consequences. This is conducted by evaluating occasions in the interaction between user and product, when there is a risk of errors arising [37, 1]. To find the problems that can give rise to errors in handling a product, evaluations are normally made of the product's user interface with realistic tasks.

Evaluation of user interfaces can proceed according to two different approaches: empirical (test methods) and analytical (inspection methods) [15, 16]. Empirical evaluation involves studies of users who interact with the user interface by carrying out different tasks, which can be performed in usability tests [27]. Usability tests have

A. Holzinger (Ed.): USAB 2007, LNCS 4799, pp. 427–440, 2007.

been employed to study the usability of medical equipment, such as infusion pumps [10] and clinical information systems [20].

In an analytical evaluation, no users are present as test subjects, and the evaluation of the interface is made by one or more analysts with the help of theoretical methods such as Heuristic Evaluation [28, 37], Cognitive Walkthrough [22, 35], and Systematic Human Error Reduction and Prediction Approach [8, 14].

Heuristic evaluation of medical equipment has been made on infusion pumps [12]. Cognitive Walkthrough has been employed to evaluate medical equipment, e.g. dialysis machines and patient surveillance systems [23, 25]. A Systematic Human Error Reduction and Prediction Approach has been used to study medication errors [21].

Analytical usability evaluation methods are suitable to, in advance, predict and identify use errors and usability problems. Since they are not based on empirical studies, they actively seek for problems, and they can easily be applied on prototypes. Furthermore, the analytical usability evaluation methods are proactive, which make it possible to detect and counteract usability problems and use errors before they may be realised. A reactive method only studies use errors which already have occurred, and thereby such a method is inappropriate to use for finding measures to be taken to prevent use errors. Besides, a reactive method can thereby not be used early in the product development.

The scope of this paper is to theoretically present an analytical usability evaluation approach for predicting and identifying use error and usability problems during the development process for medical equipment. Even though it is developed for medical equipment, this approach is generic and can be applied to other cases.

The novelty of the approach is the combination of the integrated analysis of use error and usability problems, the grading and categorisation in the analysis, and the result presentations in the form of matrixes. This paper first presents a theoretical frame covering use error and usability problems, followed by a theoretical description of the approach. The paper ends with a short discussion about pros and cons of the approach and summarizing conclusions.

2 Theoretical Frame

Central to the analytical approach is to define use error and usability problems. Use error is thus not the individual user's mistake, but an error that arises within the system. The use error may be the result of a mismatch between the different parts of the system consisting of the user, equipment, task, and environment [9]. Use error is defined according to IEC [17] as *"an act or omission of an act that has a different result than intended by the manufacturer or expected by the operator"*. The incorrect act or the omission of an act can arise at different levels of human performance, e.g. slip or mistake [30, 31].

A usability problem is a factor or property in the human-machine system that decreases the system's usability. Nielsen [27] describes a usability problem as any aspect of the design that is expected, or observed, to cause user problems with respect to some relevant usability measure that can be attributed to the design of the device. Thus, a usability problem in a system can get the result that; the user does not attain a goal; the use is ineffective; the user becomes dissatisfied with the use, and/or the user commits and error.

The relationship between usability problems and use errors is the same as the relation between active failure and latent conditions [31, 32]. A usability problem is a latent weakness in the system of human, machine, environment and tasks that triggers, under certain circumstances, a use error in the system. A use error need, however, not always be caused by a usability problem, just as not all usability problems need to cause use errors.

A further description of the relationship between usability problems and use errors can be made by connecting these with the terms 'sharp end' and 'blunt end', which have been employed in regard to complex systems [36]. The sharp end of a system is the part that directly interacts with the hazardous process, while the blunt end is the part that controls and regulates the system without direct interaction with the hazardous process. In the medical care system, nurses, physicians, technicians and pharmacists are located at the sharp end, whereas administrators, economic policy makers, and technology suppliers are located at the blunt end [36]. A use error is thus something which arises at the sharp end in the medical care system, and usability problems originate at the blunt end – more precisely from the developers of medical equipment.

To summarise the theory about use error and usability problems, deficient usability in medical equipment can cause injury to the patient in varying ways [6]. Three main ways can be distinguished:

1. The user makes a mistake that results in injury to the patient – a use error.
2. The user becomes stressed and anxious, diminishing the user's capacity for giving the patients care.
3. The user can not use the technology and therefore the treatment does not benefit the patients.

Hence, to improve safety for the patient and efficiency in use, both the direct cause, use error, and the indirect cause, usability problems, must be counteracted. It is both easier and less costly to change and improve the equipment with regard to usability during the development process, than to modify the developed device when it is in use at hospitals. Usability had to be attended to during the development process.

3 Analytical Approach

The analytical approach consists of four phases, (1) Definition of evaluation, (2) System description, (3) Interaction analysis, and (4) Result compilation and reflection. The goal of the approach is to predict and identify use errors and usability problems. With prediction means investigating when, where and how use errors may arise and where usability problems exist. With identification means determining the type and properties of the predicted use errors and usability problems.

The approach can be employed on three different purposes in the development process. The first purpose is to investigate existing equipment on the market in order to find existing usability problems and potential use errors. This information is then used as input data in the design of new equipment. The second purpose is to analyse prototypes during the development process, so that problems and errors can be detected and mitigated (this is the main use for the approach). The last purpose is to confirm that the equipment released does not contain potential use errors or usability problems that can cause unacceptable risks for the patients, i.e. to use the approach as a validation tool.

The analytical approach is conducted by a human factors expert or a group that may consist of designers, software developers, marketing staff, and people with knowledge in human factors. Most important, however, is that knowledge about the users and the usage of the equipment is present among those who conduct the assessment.

3.1 Definition of Evaluation

The first phase in the analytical approach is to set the frame for the analysis, which serves as a base for further analysis. The definition of the evaluation phase shall answer the following five questions: (1) what is the purpose of the evaluation? (2) What artefact shall be analysed? (3) Which tasks shall be analysed? (4) Who is the indented user? and (5) What is the context for the use?

3.2 System Description

The use takes place within a human-machine system, which consists of the human(s), the machine(s), the task(s) and the environment [33]. This phase describes the system. It is a preparation for the interaction analysis and very important for the further analysis. If the information specified in the system description is deficient, incomplete or wrong, the results of the analysis will suffer. The phase consists of *User Profiling*, *Context Description*, *Task Analysis* and *Interface Specification*. These parts are described below in a sequence, but they should be taken in parallel and jointly during the system description phase.

User Profiling. An analytical usability evaluation is based on a simulated user. It must therefore be careful described to get the evaluation accurate. The user profiling should define which knowledge and experience of the artefact the intended user has. Examples of user profiles are described by Janhager [18] and IEC [17].

Context Description. The context of use also needs de be defined and described. The context concerns both the physical, organisational, and psychosocial environment during use.

Task Analysis
Selection and grading of tasks. The first step in the preparation is to choose which tasks are to be evaluated, and then to grade them. The tasks chosen for evaluation naturally depend on the aim and goal of the study. Above all, it is important that the tasks are realistic. The aim of the study may be to evaluate tasks that are carried out often, or tasks that are carried out more seldom but which are safety-critical.

Each task to be evaluated is given a unique number, or *task number*. The tasks are to be graded from 1 to 5, based on how important they are in the intended use of the artefact. The most important tasks are graded 1 and the least important 5. The grading is called *task importance*. To allow a comparison between different user interfaces with this analytical approach, it is important that the tasks which are selected for comparison should have the same task importance for all user interfaces. The selection of tasks must be based on the intended use, not on the design or function of the equipment.

Specification of the tasks. An essential part of the approach is the task analysis which is performed with Hierarchical Task Analysis (HTA) [34]. HTA breaks down a task into elements or sub-tasks. These become ever more detailed as the hierarchy is divided into smaller sub-tasks. The division continues until a stop criterion is reached, frequently when the sub-task consists of only one single operation (progressive re-description). HTA thus describes how the overall goal of the working task can be achieved through sub-tasks and plans. The result is usually presented in a hierarchical tree diagram.

In this approach the bottom level in the HTA, i.e. the individual steps in the interaction between user and interface, are termed *operations*. The tasks and sub-tasks that lie above the bottom level in the HTA are termed *nodes*. A node, together with underlying nodes and operations, is termed a *function* (Figure 1).

Often a task can be performed in several different ways to reach the goal described in the HTA with associated plans. When the interaction analysis is performed, only one of the correct ways in which a user can perform a task, is chosen. The chosen correct sequence should match the common or critical real use.

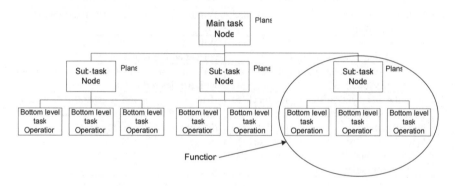

Fig. 1. HTA hierarchy with nodes, operations and functions

Description of user interface and interaction. Given the correct way in which the tasks are to be performed, as described in a HTA diagram, it should then be specified how the interface looks for the different operations. In this way it becomes possible to evaluate the user interface against each task. The interface specification can be made by combining screen dumps with the HTA diagram. A more advanced way is to use the User-Technical process suggested by Janhager [18].

3.3 Interaction Analysis

The interaction analysis consists of the methods Enhanced Cognitive Walkthrough [2] and Predictive Use Error Analysis [2]. They are performed in parallel when analysing the selected task. In the interaction analysis use errors and usability problems are predicted and identified.

Enhanced Cognitive Walkthrough. Enhanced Cognitive Walkthrough (ECW) is an inspection method based on the third version of Cognitive Walkthrough [35, 22]. ECW employs a clearly detailed procedure for simulating the user's problem-solving process in each step of the interaction. Throughout the interaction, it is checked whether the supposed user's established goal and previous experience will lead to the next correct action.

Prediction of usability problems. To predict usability problems, the analyst works through the question process in ECW for all the selected tasks. The interaction analysis is based on the described correct handling sequences in the HTA. The question process generates conceivable problems with use, which are then graded and categorised. This question process is divided into two levels. The first level of questions is applied to tasks/functions, while the second is applied to operations (Table 1).

Table 1. Analysis questions for ECW

	Level 1: Analysis of tasks/functions	Level 2: Analysis of operations
1	Will the user know that the evaluated function is available?	Will the user try to achieve the right effect?
2	Will the user interface give clues that show that the function is available?	Will the user be able to notice that the correct action is available?
3	Will the user associate the right clue with the desired function?	Will the user associate the correct action with the desired effect?
4	Will the user get sufficient feedback to understand that the desired function has been chosen	If the correct action is performed, will the user see that progress being made towards the solution of the task?
5	Will the user get sufficient feedback to understand that the desired function has been performed?	-

Table 2. Grading of the failure/success stories

Grade	Grade in words	Explanation
5	Yes	A very good chance of success
4	Yes, probably	Probably successful
3	Do not know	Impossible to decide if success or not
2	No, uncertain	Small chance of success
1	No	A very small chance of success

Each question is answered with a grade (a number between 1 and 5; Table 2) and a motivation for the grade. The motivations, called *failure/success stories,* are the assumptions underlying the choice of grades, such as that the user can not interpret a displayed symbol. The grading, called *problem seriousness,* from 1 to 5 represents different levels of success (Table 2). The grade makes it easier to determine what is most important to rectify in the subsequent reworking of the interface.

During the prediction of usability problem, each question is answered – assuming that the preceding questions are answered YES (grade 5) – independently of what the real answer was for the last question. In certain cases, however, the questions may be impossible to answer, and these must be marked with a dash in the protocol.

Identification of usability problems. The next step is to identify the predicted problems. If the problem seriousness is between 1 and 4, i.e. not with "A very good chance of success", it points at the existence of a potential usability problem. Based on the *failure story,* the *usability problem* is then described. The problem is the cause which restrains the user from performing the correct action.

Each problem is further categorised by a *problem type.* The categorisation is made from the *failure stories* and the description of the problem. Depending on the user interface and the task that the user should solve with the product, different problem types can be used. Suggestions of problem types are given in Table 3.

Table 3. Problem types employed in ECW

Problem type	Explanation
User	The problem is due to the user's experience and knowledge, perhaps because the user is accustomed to a different apparatus.
Hidden	The interface gives no indications that the function is available or how it should be used.
Text/icon	Placement, appearance and content can easily be misinterpreted or not understood
Sequence	Functions and operations have to be executed in an unnatural sequence
Feedback	The interface gives insufficient signals of what the user is doing or has done

The prediction and identification of usability problems are conducted in parallel, i.e. a problem is investigated deeply directly after being identified. The results of the identification are then reported in a tabular form.

Predictive Use Error Analysis. Predictive Use Error Analysis (PUEA) is a theoretical analysis method for Human Error Identification. It is based on the methods Action Error Analysis (AEA) [13], Systematic Human Error Reduction and Prediction Approach (SHERPA) [7], and Predictive Human Error Analysis (PHEA). PUEA employs a detailed process for breaking down the user's tasks in steps and, for each step, then predicting and identifying potential errors of use.

Prediction of Use Errors. To predict use errors, the analyst works through all the selected tasks. The interaction analysis is based on the described correct handling sequences describe with an HTA. To predict the potential incorrect actions, a question process is employed. This generates conceivable use errors and is divided into two levels of questions, the first being applied to tasks/functions and the second to operations (Table 4).

Table 4. Analysis questions for PUEA

Level 1: Analysis of tasks/functions	Level 2: Analysis of operations
What happens if the user performs an incomplete operation or omits an operation?	What can the user do wrongly in this operation?
What happens if the user performs an error in the sequence of operations?	What happens if the user performs the operation at the wrong time?
What happens if the user performs functions/tasks correctly at the wrong time?	

The analysts, guided by the questions, try to predict as many use errors as possible. Each predicted use error is noted. During this process, errors are eliminated that are considered too implausible to occur. This elimination is done in relation to how the simulated user is expected to work and think, in view of the artefact, the social, the organisational and the physical contexts. However, one must be careful about dismissing improbable errors that would have serious consequences without further investigation, as these can also constitute a risk. If there are no use errors corresponding to the answers to the questions, this too can be noted.

Table 5. Items of investigation for PUEA

#	Items of investigation	Explanation
1	Type	What is the type of use error? (categorisation)
2	Cause	Why does the use error occur? (description and categorisation)
3	Primary consequence	What is the direct effect of the use error? (description)
4	Secondary consequences	What effects can the use error have that lead to a hazardous situation for the user or other people, or to risk of machine damage or economic loss? (description and judgment of severity by a grade)
5	Detection	Can the user detect a use error before it has any secondary consequences? (description and judgment of probability by grade)
6	Recovery	Can the user recover from the error before any severe consequences arise? (description)
7	Protection from consequences	Which measures does the technical system employ to protect the user and the environment from the secondary consequences? (description)
8	Prevention of error	Which measures does the technical system employ to prevent occurrence of use errors? (description)

Use Error Identification. For each one predicted use error, an investigation is made according to eight items (Table 5). The first two of these items concern the error itself, the next two its potential consequences, and the last four items concern mitigations of the error and consequences. Four of the items also contain a categorisation, a judgment of probability, or a judgment of severity. This is done to facilitate a compilation and assessment of the investigation.

The prediction and identification of use errors are conducted in parallel, i.e. the error is investigated directly after being identified. The results of the identification are then reported in a tabular form.

3.4 Result Compilation and Reflection

Compilation in Matrixes. Matrixes are employed to present the semi-quantitative results from the interaction analysis. The collected answers from the prediction and identification are ordered in different ways within the matrixes so as to emphasise different aspects of the analysis.

The matrixes can be combined in various ways, and the numbers in the matrix cells show the number of detected problems distributed according to the two types of data that are compared.

Matrixes from ECW. The information employed from the ECW consists of: *task number*, *task importance*, *problem severity* and *problem type*. The matrixes can be combined in several ways (Table 6).

Table 6. Matrixes for presenting the results from the ECW analysis

Name	Content	Explanation
Matrix A:	Problem seriousness versus task importance	*This shows the interface's general condition.*
Matrix B	Problem seriousness versus problem type	*This shows the overall problems with the interface.*
Matrix C	Problem type versus task importance	*This shows which problems are most important to rectify.*
Matrix D	Problem seriousness versus task number	*This shows which tasks have the most problems.*
Matrix E	Problem type versus task number	*This shows which types of problems are most common in the tasks.*

			Prob. Severity	
Task importance	1	2	3	4
1	0	0	0	1
2	0	1	1	8
3	2	2	8	1
4	1	2	3	5
5	1	0	0	0

Fig. 2. Example of matrix for problem severity versus task importance

The example of a matrix in Figure 2, showing problem severity versus task importance, reveals whether a large problem exists in the user interface. If many problems are entered in the upper left corner, this means that there are severe problems in important tasks. If the problems are found in the lower part of the matrix, they originate from less important tasks, and if they are found in the right part there are not very severe problems.

Compilation in Matrixes

Matrixes from PUEA. The information employed from each investigation of use error comprises: secondary consequences, error type, error cause, detection, and task number. The matrixes can be combined in various ways below are listed ten variants of useful matrixes (Table 7).

Table 7. Matrixes for presenting the results from the PUEA analysis

Name	Content	Explanation
Matrix A:	Consequence versus task number	*Shows in which tasks the most serious consequences of errors exist.*
Matrix B:	Error type versus task number	*Shows which type of use error exist in the various tasks.*
Matrix C:	Error cause versus task number	*Shows the causes of the use errors in the different tasks.*
Matrix D:	Error type versus secondary consequence	*Shows which error type gives rise to the highest risks.*
Matrix E:	Error cause versus secondary consequence	*Shows which error cause gives rise to the highest risks.*
Matrix F:	Error cause versus error type	*Shows what coupling exists between error cause and error type.*
Matrix G:	Detection versus task number	*Shows in which tasks there are errors difficult to detect.*
Matrix H:	Detection versus error type	*Shows which type of error is difficult to detect.*
Matrix I:	Detection versus error cause	*Shows the causes of the errors that are difficult to detect.*
Matrix J:	Detection versus secondary consequence	*Shows how serious the consequences are for errors that are difficult to detect.*

	Secondary Consequences				
Detection	1	2	3	4	5
1	0	0	6	0	3
2	0	0	0	0	0
3	0	0	0	0	0
4	0	0	0	0	0
5	0	0	10	0	0

Fig. 3. Example of matrix for detection versus secondary consequences

The matrix in Figure 3, detection versus consequences, shows error detection probability versus error consequence severity. If the majority of the numbers are found on the left side of the upper corner in the matrix, the error consequences are severe and the errors are very hard to detect. On the contrary, if the numbers are found in the right lower corner of the matrix, the error consequences are not severe and detection of the errors is probable. The matrix uses grey shading to make the results easier to read and show which errors are the most serious.

Result reflection. The last step in the proposed analytical approach is the result reflection. Since both ECW and PUEA are methods that analyse potential use errors and usability problems, the predicted problems and errors may not occur during a real use situation. The grading systems are also a subjective judgment. The elicited potential problems and errors need to be reviewed and confirmed in interaction with real users, before changes are made in the analysed user interface.

One way of reviewing the results is to triangulate the analytical approach with usability evaluation methods such as heuristic evaluation [28] and usability testing [27]. By using these methods, the potential usability problems and use errors can be confirmed or dismissed. Another possibility is to let actual users in a focus group discussion [5] decide whether the potential problems and errors are relevant or not.

4 Discussion

The scope of this paper was to present an analytical approach which predicts and identifies use errors and usability problems. The prediction is achieved by a question-based process, and the identification is achieved by categorising and grading the detected use errors and usability problems. The results are presented with the help of matrixes.

Through the use of matrixes for result presentations it is possible to evaluate several aspects of use errors and usability problems. For example, it is feasible to read the types of errors and problems that have arisen and the categorisation of their causes. Then a summary can be made of which tendencies and patterns exist among use errors. By showing the probability of detecting errors versus the seriousness of the consequences of the error, a risk assessment can be made. Further, the matrixes' summary simplifies the redesign of the interface by providing an overview of the dominant characteristic of the errors and problems that may arise in specific tasks. Moreover, the matrix presentation of results enables a comparison of interfaces to be made in a simple way, both for completely different interfaces and in redesign.

The scope for this paper was to present an analytical approach particularly suitable for use on medical equipment. The usability of safety-critical technical systems such as medical equipment is essential since use errors in handling can have serious consequences.

When evaluating medical equipment, the most important aim is thus not to perform an evaluation rapidly, but to make it as good as possible. For an analysis of presumptive use errors and usability problems, it is more important to find as many errors and problems as possible, than to avoid finding the errors and problems which probably do not occur in a real situation [2].

Moreover, it is only after a use error or a usability problem has been identified that it is possible to decide whether the error is plausible or not. Exposing even improbable errors and problems for further evaluation is also beneficial, as these may have serious consequences that can otherwise be overlooked if only the plausible use errors and usability problems are investigated.

The drawbacks of the approach are that it might take much time to carry out , is tedious to apply, and it can detect usability problems and use errors which may not be plausible, and thus less prominent when the approach is applied to medical equipment [2].

However, the analytical approach is never meant to replace validation with real users or to be the only method used in the development process. The approach can and should be combined with other usability evaluation methods. Examples of methods that can be employed for usability engineering in medical equipment and include the users are described by Garmer et al. [11] and Moric et al. [26].

The approach have been used in two evaluations of user interfaces on dialysis machines [3, 4]. In the first study, the approach was employed for a comparison between three different design solutions. In the second study, the purpose was to elicit design improvements. Both evaluations gave the company in question more specific information about use errors and usability problems of their products and prototypes than the company previously possessed.

5 Conclusions

To conclude, the suggested analytical approach is suitable for predicting and identifying use errors and usability problems in medical equipment. The approach can therefore be seen as a necessary, but not comprehensive, segment of usability evaluation for medical equipment. The strength of the approach is that usability problems and use errors can be discovered before any empirical trials with real users are conducted. This simplifies the work in the development process.

References

1. Basnyat, A., Palanque, P.: Softaware hazard and barriers for informing the design of safety-critical interactive systems. In: Zio, G.S. (ed.) Safety and Reliability for Managing Risk, pp. 257–265. Taylor & Francis Group, London (2006)
2. Bligård, L.-O.: Prediction of Medical Device Usability Problems and Use Errors – An Improved Analytical Methodical Approach, Chalmers University of Technology, Göteborg (2007)
3. Bligård, L.-O., Eriksson, M., Osvalder, A.-L.: Internal Report Gambro Lundia AB, Classified (2006)
4. Bligård, L.-O., Osvalder, A.-L.: Internal Report Gambro Lundia AB, Classified (2006)
5. Cooper, L., Baber, C.: Focus Groups. In: Stanton, N.A., Hedge, A., Brookhuis, K., Salas, E., Hendrick, H. (eds.) Handbook of Human Factors and Ergonomics Methods, CRC Press, London (2005)
6. Crowley, J.J., Kaye, R.D.: Identifying and understanding medical device use errors. Journal of Clinical Engineering 27, 188–193 (2002)
7. Embrey, D.E.: SHERPA: a Systematic Human Error Reduction and Prediction Approach, International Topical Meeting on Advances in human factors in nuclear power system, American Nuclear Society, Knoxville, pp. 184–193 (1986)

8. Embrey, D.E., Reason, J.T.: The Application of Cognitive Models to the Evaluation and Prediction of Human Reability, International Topical Meeting on Advances in human factors in nuclear power system, American Nuclear Society, Knoxville (1986)

9. FDA, Proposal for Reporting of Use Errors with Medical Devices (1999)

10. Garmer, K., Liljegren, E., Osvalder, A.-L., Dahlman, S.: Application of usability testing to the development of medical equipment. Usability testing of a frequently used infusion pump and a new user interface for an infusion pump developed with a human factors approach, International Journal of Industrial Ergonomics 29, 145–159 (2002)

11. Garmer, K., Ylvén, J., Karlsson, I.C.M.: User participation in requirements elicitation comparing focus group interviews and usability tests for eliciting usability requirements for medical equipment: A case study. International Journal of Industrial Ergonomics 33, 85–98 (2004)

12. Graham, M.J., Kubose, T.K., Jordan, D., Zhang, J., Johnson, T.R., Patel, V.L.: Heuristic evaluation of infusion pumps: Implications for patient safety in Intensive Care Units. International Journal of Medical Informatics 73, 771–779 (2004)

13. Harms-Ringdahl, L.: Safety Analysis - Principles and Practice in Occupational Safety. Taylor & Francis, London (2001)

14. Harris, D., Stanton, N.A., Marshall, A., Young, M.S., Demagalski, J., Salmon, P.: Using SHERPA to predict design-induced error on the flight deck. Aerospace Science and Technology 9, 525–532 (2005)

15. Hartson, H.R., Andre, T.S., Williges, R.C.: Criteria for evaluating usability evaluation methods. International Journal of Human-Computer Interaction 13, 373–410 (2001)

16. Holzinger, A.: Usability engineering methods for software developers. Communications of the ACM 48, 71–74 (2005)

17. IEC, IEC 60601-1-6:2004 Medical electrical equipment - Part 1-6: General requirements for safety - Collateral standard: Usability IEC, Geneva (2004)

18. Janhager, J.: User Consideration in Early Stages of Product Development – Theories and Methods, The Royal Institute of Technology, Stockholm (2005)

19. Kaufman, D.R., Patel, V.L., Hilliman, C., Morin, P.C., Pevzner, J., Weinstock, R.S., Goland, R., Shea, S., Starren, J.: Usability in the real world: assessing medical information technologies in patients' homes. Journal of Biomedical Informatics 36, 45–60 (2003)

20. Kushniruk, A.W., Patel, V.L.: Cognitive and usability engineering methods for the evaluation of clinical information systems. Journal of Biomedical Informatics 37, 56–76 (2004)

21. Lane, R., Stanton, N.A., Harrison, D.: Applying hierarchical task analysis to medication administration errors. Applied Ergonomics 37, 669–679 (2006)

22. Lewis, C., Wharton, C.: Cognitive Walkthrough. In: Helander, M., Landauer, T.K., Prabhu, P. (eds.) Handbook of Human-computer Interaction, Elsevier Science BV, New York (1997)

23. Liljegren, E., Osvalder, A.-L.: Cognitive engineering methods as usability evaluation tools for medical equipment. International Journal of Industrial Ergonomics 34, 49–62 (2004)

24. Lin, L., Isla, R., Doniz, K., Harkness, H., Vicente, K.J., Doyle, D.J.: Applying human factors to the design of medical equipment: Patient-controlled analgesia. Journal of Clinical Monitoring and Computing 14, 253–263 (1998)

25. Liu, Y., Osvalder, A.-L., Dahlman, S.: Exploring user background settings in cognitive walkthrough evaluation of medical prototype interfaces: A case study. International Journal of Industrial Ergonomics 35, 379–390 (2005)

26. Moric, A., Bligård, L.-O., Osvalder, A.-L.: Usability of Reusable SpO2 Sensors: A Comparison between two Sensor Types. In: NES. 36th Annual Congress of the Nordic Ergonomics Society Conference, Kolding, Denmark (2004)
27. Nielsen, J.: Usability engineering. Academic Press, Boston (1993)
28. Nielsen, J., Mack, R.L. (eds.): Usability inspection methods. Wiley, New York (1994)
29. Obradovich, J.H., Woods, D.D.: Users as designers: How people cope with poor HCI design in computer-based medical devices. Human Factors 38, 574–592 (1996)
30. Rasmussen, J.: Skills, rules and knowledge; signals, signs and symbols, and other distinctions in human performance models. IEEE Transactions on Systems, Man and Cybernetics SMC-13, 257–266 (1983)
31. Reason, J.: Human error. Cambridge Univ. Press, cop., Cambridge (1990)
32. Reason, J.: Managing the Risks of Organizational Accidents, Ashgate, Aldershot (1997)
33. Sanders, M.S., McCormick, E.J.: Human Factors in Engineering and Design. McGraw-Hill, New York (1993)
34. Stanton, N.A.: Hierarchical task analysis: Developments, applications, and extensions. Applied Ergonomics 37, 55–79 (2006)
35. Wharton, C., Rieman, J., Lewis, C., Polson, P.G.: The Cognitive Walkthrough Method: A Practitioner's Guide. In: Nielsen, J., Mack, R.L. (eds.) Usability Inspection Methods, John Wiley And Sons Ltd, New York, UK (1994)
36. Woods, D., Cook, R.I.: The New Look at Error, Safety, and Failure: A Primer for Health Care (1999)
37. Zhang, J., Johnson, T.R., Patel, V.L., Paige, D.L., Kubose, T.: Using usability heuristics to evaluate patient safety of medical devices. Journal of Biomedical Informatics 36, 23–30 (2003)
38. Zhang, J., Patel, V.L., Johnson, T.R., Shortliffe, E.H.: A cognitive taxonomy of medical errors. Journal of Biomedical Informatics 37, 193–204 (2004)

User's Expertise Differences When Interacting with Simple Medical User Interfaces

Yuanhua Liu, Anna-Lisa Osvalder, and MariAnne Karlsson

Chalmers University of Technology, 412 96 Gothenburg, Sweden
Department of Product and Production Development
Division Design
yuanhua@hfe.chalmers.se

Abstract. In order to provide helpful proposals for future redesign of insulin pump interfaces, a study was carried out to investigate the expertise difference between novice users and expert users when interacting with a simple user interface of insulin pumps. In this study, two user groups with 13 participants in each, evaluated an insulin pump interface on a computer demo in usability tests. The results showed there was no significant difference between the novice users and expert users regarding the task completion time and the number of failures in performance. As for the cause of failures, the novice users showed weakness in domain knowledge, while the expert users showed weakness in task knowledge. No significant difference was shown on users' satisfaction between the two user groups. The results also implied that the novice users elaborated their redesign suggestions in a deductive and summaric way, while the expert users proposed suggestions in an inductive and thorough way.

Keywords: Simple Medical User Interface, Expertise differences, Usability Test, Novice User, Expert User, Insulin Pump.

1 Introduction

Diabetes is a rapidly growing disease in the world. The number of diabetics among the total population has increased dramatically in recent years. Although there are several different ways to treat diabetes, the most popular treatment and recently available advance in insulin delivery is delivery by means of an insulin pump. By using an insulin pump, diabetics can keep their blood glucose levels within their target ranges in a very convenient way.

Despite the advantages and better treatment offered by insulin pumps, there are still many diabetics who choose to use insulin pens or traditional syringe injections instead [1]. Studies made by Bligård et al. [2,3] exposed that the design of insulin pumps and the user interfaces might be one possible reason for non-usage.

Insulin pumps are typically shaped like a deck of poker cards, and about the size as some compact mobile phones. The interface of the pump is generally made up of two parts: a small crystal display panel, and 3-5 functional buttons that are used to program the computer in the pump, in order to control the amount of insulin to be

A. Holzinger (Ed.): USAB 2007, LNCS 4799, pp. 441–446, 2007.

delivered. Insulin is stored in a reservoir inside the pump, and it is delivered and pumped into the patient's body through an infusion set, 24 hours everyday. Although insulin pumps are relatively simple devices compared to other complex medical devices, the improper use or wrong insulin delivery can be dangerous to diabetics' health or even lethal. Therefore, it is necessary to improve the usability of the interface design of the pump, in order to make the interface easier to understand and usable for all groups of diabetics (both younger and elderly users, novices as well as experts).

The objective of this study was to investigate whether there were the possible differences between novice and expert users when interacting with a simple user interface of a medical device, in this case, an insulin pump. The purpose was to provide proposals for future redesign of pump interface in order to reach a more inclusive design. More specifically, the focus of this study was to investigate (1) if there were any differences between novice and expert users when interacting with a simple user interface; (2) if there were any differences, in which way they differed; and (3) how the user groups presented their proposals for future redesign.

2   Methods and Materials

In this study, the prototype interface of an insulin pump was created and set up as an interactive demo on a touch-screen computer. The demo was a two-dimensional and vertical prototype interface with full functionality for a few features [4].

The size of the prototype interface was 86×53mm, which is similar to normal pump interfaces. The prototype interface included two parts: a small rectangular display panel that showed relevant information during the treatment procedure; and five functional buttons that helped activate, operate and control the treatment. The five functional buttons were grouped into two areas: one pair of buttons with up and down arrow icons and used for parameter settings, were placed on the right side of the display panel; the other three functional buttons (the shortcut button for 'Bolus' function with the icon '→B', the backward-to-upper-menu button with the icon 'Esc', and the activating/confirming button with the icon 'ACT') were placed below the display panel.

2.1   Literature Review

In the first phase of the study, a literature review was made. Relevant materials and internet websites were browsed, in order to get the general information about diabetes and its physiology treatments for diabetes, insulin pump treatment, as well as information about several popular insulin pump products.

This part of the work provided the evaluator with general and theoretical domain knowledge about pump treatment. In addition, it also provided helpful information for the design of the computer demo.

2.2   Interview Study

In the second phase of the study, interviews were carried out with nurses at the Diabetes Center at Sahlgrenska University Hospital, Mölndal Hospital and Östra

Hospital, in Gothenburg, Sweden. The interviews were made in order to understand nurses' views on insulin pump treatment, and also information regarding how to discriminate between novice and expert pump users. Additionally, relevant information regarding treatment training, existing pump products in use, and pump users at hospitals were also discussed with the nurses.

Besides interviews with nurses at hospitals, interviews were also made with several diabetics who had used insulin pump treatment. The purpose was to understand patients' views on insulin pump treatment and their experience as pump users.

From the interviews, it was found that users' attitude towards pump treatment is the main factor that influences users' performance during treatment. This information was used to exclude confounding factors when searching for test subjects to include in the usability tests.

2.3 Usability Test

The usability tests were conducted according to a general test procedure proposed by Nielsen [4] and McClelland [5]. The tests were made in the usability laboratory at Division Design, Chalmers University of Technology, Sweden. Prior to the usability tests, three pilot tests were carried out to check the test procedure.

Seven typical task scenarios were selected to represent the actual situations of different functions during insulin pump treatment in the usability tests. 26 diabetics who are using pump treatment participated as test subjects. Half of the test subjects were novice users who had used pumps less than one year, while the other half were expert users who had used at least two different types of pumps in the past five years. The average age of the novice user group was 39.8, and that of the expert user group was 51.1. The average length of pump application history was 5.3 months for novice user group and 15.2 years for expert user group.

A pre-session interview was made before the test session, with the purpose of collecting a user profile of the respective test subjects. At the beginning of the test session, each test subject was informed about the purpose of the test and given instructions about the test procedure. Then the test subject was asked to repeat the test instructions. Each test session was video recorded; however, the test subject was notified that only their hands would be shown on the video recordings, thus ensuring anonymity. During the evaluation of the task scenarios, the test subjects were asked to operate the demo on the touch screen and explain reasons for their decision in problem solving. For each task scenario, data on task completion time and the number of failures defined as catastrophes in task completions were collected. The task completion time was measured by a digital stopwatch.

At the end of each test session, each test subject was requested to give comments freely on the design of the demo, as well as their proposals for future modification and redesign. Afterwards, each test subject was asked to rank his or her satisfaction on a SUS questionnaire proposed by Brooke [6].

3 Results

The task completion for each test subject was calculated. The hypothesis was that the difference in expertise had an effect on the task completion time in performance. A t-test was conducted on the data collected from each scenario. The p value was larger than 0.05, that is, $p = 0.610$, $p > 0.05$, and the hypothesis was consequently rejected. There was no statistically significant difference between the novice and expert users in terms of task completion time.

The number of failures for each test subject was calculated. The hypothesis was that the difference in expertise had an effect on the number of failures in performance. A t-test was conducted on the data collected from each scenario. The p value was larger than 0.05, that is, $p = 0.234$, $p > 0.05$, and the hypothesis was consequently rejected. There was no statistically significant difference between the novice and expert users in terms of number of failures in performance.

The failures made could be attributed to three types of knowledge: (1) Interaction knowledge, which indicates the general knowledge about interacting style or ways with the interface and the relevant software packages, e.g. menu system, functional buttons, icons, symbols; (2) Task knowledge, which indicates the knowledge of the task domain addressed by the interface/system, e.g. terms; (3) Domain knowledge, which indicates the theoretical knowledge or background knowledge relating to a task which independent of the product/system being used to complete that task, e.g. settings and treatment in some situations [4]. Based on the analysis of the causes of failures, an interesting finding was that lack of domain knowledge was the main and important reason for the failures in the novice user group; while weakness in task knowledge was the main and important reason for failures in the expert user group.

As for comparing the different users' satisfaction of the pump interface design, a Mann-Whitney test was carried out. The hypothesis was that there is a difference between the novice group and expert group regarding their respective satisfaction of the interface. Since the p value was larger than 0.05, that is, $p = 0.479$, the hypothesis was rejected. There was no significant difference between the novice and expert users regarding their satisfaction of the interface design.

All the test subjects gave comments and provided proposals for the redesign of the pump interface. The proposals from both groups touched upon similar aspects, for instance, functionality, task structure, aesthetic aspect, assembly, accessory, and explicitness etc. However, there existed some differences between the two groups. The expert users addressed, e.g. the clarity of the menu design, while the novice users did not touch this aspect. The expert users preferred less multifunctionality, while novice users expected more new functions. When the expert users proposed design modifications, they always gave a very detailed description on how the changes should be made. However, the novice users proposed design modifications in general terms, without any further concrete examples or descriptions.

4 Discussion

The focus of this study was to investigate (1) if there were any differences between novice and expert users when interacting with a simple user interface; (2) if there were any differences, in which way they differed; and (3) how the user groups presented their proposals for future redesign. The statistical analysis implied that the expert users' previous experience did not help much on performance accuracy and task completion time when interacting with an unknown new simple interface. One possible explanation could be that the interface was very simple to operate, that is, no special or long time practice was required for skill acquisition. Thus the difference in use expertise (between approximately 5 months and 15 years) did not seem to matter. However, a trend in the data implied that age and educational level affected the users' behavior in performance. These aspects need to be further investigated.

According to the analysis of the cause of failures, the results implied that novice users made more failures due to poorer domain knowledge; while expert users made more failures due to poorer task knowledge. Since novice users lack experience in their treatment, they need a long time to become proficient in practice. Although the expert users in this study had used at least 2-3 different pumps, the accumulated experience with different types of pumps did not help much when dealing with a new interface. On the contrary, when expert users get used to one system after a long period of time, they usually set up a stable mental model on task performance, a habit. Neisser [7] stated that human experience depends on stored mental models, which guide explorative behavior and the perception of an external context. The results of this study implied that the expert users relied on their latest mental model during problem solving, although they might have several mental models for different types of pump interface. This inference was proved by the after-session interviews. Many of the expert users mentioned that their latest mental model on the task performance always affected their performance on a new system. This happened every time when they started to use a new pump. In other words, it takes time for expert users to get rid of their old mental models and then develop a new mental model for the new system. This is an explanation to why expert users had more errors/ failures due to poorer task knowledge.

Another interesting cause of failures was that expert users were more careless than novice users. The explanation for this could be that the difference in experience level influenced the users' attitude and concentration during the interaction. During the evaluation test, the expert users showed more concentration on investigating the new system and were inclined to compare or evaluate the new interface system with their own in-use interface system. This lack of focus made the expert users overlook small and trivial problems, which led to relatively more failure due to carelessness.

Although the proposals for a future redesign of the pump interface overlapped between the two user groups, there were still some differences between the novices and experts' proposals. The results implied that the proposals given by the expert users were more concrete and thorough than those given by the novice users. The expert users' proposals appeared to be inductive, i.e. the expert users summarized their redesign suggestions based on their long-term practical experience, while the novice users' proposals appeared to be deductive, i.e. summarized their redesign suggestion either based on a couple of incidents experienced in the short-term period or on their subjective reasoning.

5   Conclusions

The following conclusions can be drawn based on the usability study results:

   (1) There was no significant difference regarding the task completion time and the number of failures between novice and expert users when interacting with a simple medical user interface;

(2) In terms of the cause of failures, the novice users showed their weakness in domain knowledge, while the expert users showed their weakness in task knowledge;

(3) The expert users were more careless and less focused than the novice users during task completion;

(4) There was no significant difference regarding satisfaction of the usability of the interface between the novice and expert users;

(5) The novice users proposed their redesign suggestions in a deductive and summaric way, while the expert users propose their redesign suggestions in a inductive and thorough way.

References

1. Eriksson, M.: Insulin Pump with Integrated Continuous Glucose Monitoring - A Design Proposal with the User in Focus. Master Thesis. Department of Product and Production Development, Division of Design, Chalmers University of Technology, Gothenburg, Sweden (2006)

2. Bligård, L.-O., Jönsson, A., Osvalder, A.-L.: Utvärdering och konceptdesign av användargränssnitt för insulinpumpar. (in Swedish) Internal Report 19, Department of Product and Production Development, Division of Human Factors Engineering, Chalmers University of Technology, Gothenburg, Sweden (2003)

3. Bligård, L.-O., Jönsson, A., Osvalder, A.-L.: Eliciting User Requirements and evaluation of Usability for Insulin Pumps. In: NES. Proceedings of the Nordic Ergonomic Society's 36th Annual Conference, Kolding, Denmark (2004)

4. Nielsen, J.: Usability Engineering. Academic Press, San Diego (1993)

5. McClelland, I.: Product assessment and user trials. In: Wilson, J.R., Corlett, E.N. (eds.) Evaluation of Human Work, Taylor & Francis Ltd., London (1995)

6. Brooke, J.: SUS - A Quick and Dirty Usability Scale. In: Jordan, P.W., et al. (eds.) Usability Evaluation in Industry, Taylor & Francis Ltd., London (1996)

7. Neisser, U.: Cognition and Reality. Freeman, San Francisco (1976)

Usability-Testing Healthcare Software with Nursing Informatics Students in Distance Education: A Case Study

Beth Meyer[1] and Diane Skiba[2]

[1] McKesson Corporation
285 Century Place
Louisville, CO 80027
Beth.Meyer@mckesson.com
[2] School of Nursing
University of Colorado at Denver & Health Sciences Center
Denver, CO 80262
Diane.Skiba@UCHSC.edu

Abstract. For two years, the University of Colorado Health Sciences Center nursing informatics program has joined with McKesson, a leading vendor of health care provider software, to simultaneously teach distance education students about user-centered design and to improve the usability of McKesson's products. This paper describes lessons learned in this industry-education partnership. We have found that usability testing with nursing informatics students who are also experienced nurses compares well to testing with nonstudent nurses in terms of data collected, although there can be differences in how the data are interpreted organizationally and in the constraints on the data collection process. The students find participation in a remote usability test of health care software to be an engaging and helpful part of their coursework.

Keywords: Remote usability testing, health care provider software, distance education, human-computer interaction.

1 Introduction

Participation in real-world user-centered design processes, such as usability tests, has long been a component of courses in Human–Computer interaction (HCI) - Wise and Bellaver [7] provide an example. Trying out the methods of the user interface design profession, either as a designer or as a test participant, is an extremely useful way to learn the skills needed to resolve complex design issues in a real development environment. There is also a long history of drawing general conclusions based on research performed exclusively on students; unfortunately, when this practice has been applied to health care, the results are not always as one would expect. (For example, when Effken et al [2] tested both student nurses and critical care nurses, the student nurses benefited far more than the critical care nurses from a graphical display.)

A. Holzinger (Ed.): USAB 2007, LNCS 4799, pp. 447–452, 2007.

However, different contexts in which design is being taught could affect both the educational value of usability testing and the applicability of the results to the target user population. The introduction of distance education to HCI education means that course design methods and exercises must support remote students. Also, adapting HCI education to the health care industry means that students have to represent and design for a very specialized user population with very specific needs.

In this paper, we address the question of how usability testing in HCI education can be adapted to the needs of remote students and of vendors. In this case, the students were part of the Master of Science in nursing program at the University of Colorado School of Nursing, which collaborates with McKesson to both teach the user-centered design process and improve the usability of health care provider software.

2   The Educational Experience of the Collaboration

The University of Colorado School of Nursing offers a Master of Science program in nursing designed to prepare nurses for advanced practice roles. One specialty offered in the program is focused on health care informatics. The informatics Masters degree program has been offered as a totally online program since 1997.

The partnership with McKesson is a unique aspect of this nursing informatics program. McKesson is a Fortune 18 healthcare services and information technology company dedicated to helping its customers deliver high-quality healthcare. The goal of the partnership is to provide leadership in the field of health care informatics and the further development of nursing informatics as a discipline.

This paper focuses on facilitating the development of a learning community between our program and personnel within McKesson and facilitating joint research and evaluation endeavors. This partnership is involved in a graduate-level course on Human Computer Interaction Design. In this course, nursing informatics students do the following:

1. Compare and contrast models of human and computer processing for their applicability to health care informatics solutions
2. Apply models of human-computer interaction to create an optimal HCI design for health care informatics solutions
3. Design a usability test of an information system based upon a selected strategy
4. Participate in a usability test of a health care information system based upon a selected strategy
5. Design an interactive clinical information system that incorporates design principles for use by health care providers or patients

Participation in a usability test is part of each student's coursework. Despite their geographic distance, learners work in design teams, and each team has a mentor from McKesson or the vendor community. Mentors provide a real-world view of the process, help in the translation of theory into practice and serve to socialize learners into the informatics specialist's role.

We have used a variety of standard usability testing methods in the tests with students.

Holzinger [3, 4] identifies several industry-standard usability techniques, such as think-aloud protocols, field observation, data logging, and subjective evaluation through questionnaires and interviews. Our tests have primarily featured performance measurement combined with think-aloud protocols and questionnaires for subjective evaluation; however, for projects that are in an earlier stage of design at the time of the class, we cannot always collect performance measures and must rely more on think-aloud data and questionnaire results. In Rubin's [6] classification of usability tests, most of these studies have been assessment tests designed to identify and rank design problems, though at least one was more of an exploratory test designed to ensure that the overall design matched the user's model of the task.

In the two times this course has been offered along with the McKesson partnership, the learners positively rated this course and appreciate McKesson's involvement. It is an exciting opportunity for our learners to have hands-on experience with real-world usability testing, and to receive expert advice from mentors as they learn to apply design principles to the creation of a prototype health care project. The following are examples of comments about the usability testing process from the 2006 course evaluations:

• Usability testing experience provided was very helpful to learning process for the class. Good move to include in class learning experience.
• The McKesson usability tests were GREAT!

3   The Vendor Experience of the Collaboration

The user interface design group at McKesson has performed a variety of usability tests on our hospital software, both with nursing informatics students and with non-student clinicians. The primary goal of the McKesson group, of course, is to improve the usability of the company's clinical software as much as possible. Toward this end, there were both advantages and disadvantages of using students in a distance education course on HCI as our participants.

3.1   Quality and Representativeness of Results

One of the most encouraging results we've found is that, so far, we have not seen any reliable differences in usability test results between the student participants and the non-student clinicians. When testing different versions of the same product with both a student group and a non-student group, we saw many of the same qualitative themes emerge about how they preferred to approach their tasks, and features of the product that they would be more or less likely to use. One might have expected that the students, who were partway through an HCI course, would be consistently more likely to comment on the interface in terms of major usability principles or industry-standard design conventions. In fact, participants in our non-student studies have spoken just as eloquently about the design principles and conventions as those in the studies with students. And we found that many participants in the student group were still very concerned about patient care in their regular work, and focused more on the implications of the product on patient care workflow than they did on issues of layout or conformity to HCI conventions.

What was sometimes very different about the two types of participants was how their results were perceived by our client organizations. In some cases, test results were eagerly taken by the business organizations as simply another source of useful design feedback. This acceptance was based on the fact that all participants were experienced nurses, not unlike the very analysts who were writing design requirements. However, other organizations expressed concern about how representative these students were after having worked in the informatics arena and taken courses such as this one. Certain client organizations were opposed to recruiting usability participants from a nursing informatics course, insisting that only nurses recruited directly from a hospital would truly represent our target user population. When testing with informatics students, we always gave a pretest questionnaire assessing demographic and job information for each participant. The information from this questionnaire proved to be critical in making client organizations comfortable that our participants would share the concerns of ordinary nurses.

The McKesson User Interface Design group has conducted remote usability testing with nurses who were informatics students, and with nurses who were not. Having done both, we have found that despite some expectations, the non-student nurses were often quite sophisticated in what they expected from information technology. This observation is consistent with an observation by Kujala and Kauppinen [5], who found that development organizations often underestimated the true diversity of their user population. While it is important not to assume extensive computer experience when designing nursing applications, we have found that stereotyping nurses as computer novices is not appropriate either.

To improve the applicability of the usability test results even further, we may in the future devote more effort to assigning particular subgroups in the class to particular tests. For example, we may recruit the students with the most experience in their facility's IT department for tests of the configuration features of a new product, while those with the most recent staff nursing experience could be assigned to tests of product features intended for more general use on the nursing floor.

Usability tests with distance education students can be assumed to be remote tests, and ours were no exception. Our group has also conducted in-person tests of similar software. Remote tests of software that is still being designed are most feasible with a synchronous test, so that the facilitator can give the participant access to software that is not yet publicly available (or even help run the sequence of displays in very early low-fidelity testing). Consistent with Andreasen et al [1], we found that this sort of remote testing provided us with excellent feedback comparable to what we get with in-person tests, with qualitative results that were extremely useful for design.

3.2 Test Procedures

The process of testing remote nursing informatics students went smoothly, but there are some logistical factors to bear in mind. There are definite advantages to conducting usability tests with nurses as part of their graduate nursing informatics course. Recruiting is greatly simplified, there is no concern about participant compensation, and the participants were highly motivated. The technical infrastructure of the distance learning course provides an easy way to allow participants to sign up for time slots and to distribute advance materials such as pretest questionnaires or session instructions.

As with all synchronous remote usability testing, we must pay careful attention to scheduling details. If a participant forgets, is confused about the session time (a distinct risk when in different time zones), or if a last-minute conflict arises, that time window is usually lost. This is not so much the case with testing that can be done at the health care facility – if one planned participant is suddenly unable to attend, the facilitator may be able to recruit another on short notice.

One potential disadvantage of working through a course for usability testing is that the timing of the course can constrain what we can accomplish in our test. We sometimes would have wished for more time to plan a test, but all tests obviously had to be completed before the end of the course. However, we find that external factors are just as likely to constrain usability tests conducted in other contexts – for example, the testing must be complete by a certain date to be able to make changes to the software, or so that a particular customer or stakeholder can be involved.

Similarly, we found that some tests were larger in scale than they really needed to be to identify the major usability issues, simply to ensure that all students in the course had the opportunity to participate. If the number of students enrolling in the course expanded significantly, one of the potential problems in scaling up the attendance would be that a single vendor might not have enough new products ready to test for all students to participate in a traditional one-on-one test. Another possible approach to this issue is to use the additional participants to evaluate different features and tasks supported by one product being tested – but this requires additional planning time, meaning that the vendor needs more advance notice to collaborate successfully with a larger class.

Synchronous usability testing is also more of a technical challenge for international students than most other online course activities. While we succeeded in using the same technical tools to test with international students as with those based in the U.S., we found that the response time was a real issue in allowing the student to control a remote desktop. The frequent pauses to allow the system to catch up to the participants' actions made for a very different usability test experience from that of other students. On the positive side, these pauses allowed the participant to spend more time reflecting on and discussing their actions and their expectations for the user interface. However, it was a far more frustrating and tiring experience for all involved, and we could not complete the usual test procedure in the time allowed.

A distinct advantage of doing usability testing with nurses who were already learning about the user-centered design process was that we had to spend less time explaining our procedures. These participants already knew why we needed them to think aloud, why we wanted them to focus on what was unexpected, and that we were not critiquing their own performance (though naturally, we reminded them of this).

A potential issue in the use of think-aloud techniques in remote usability testing is the challenge of collecting quality verbal data from the participant in the absence of any nonverbal cues. However, we have gotten very detailed think-aloud results from our student participants, perhaps due partly to their familiarity with user-centered design methods (although of course a few required extra prompting). Poor audio quality from certain participants' phones was at least as much of a challenge as participants forgetting to voice their thoughts.

Because these participants were studying the very same methods that those of us in the McKesson User Interface Design team were using, we sometimes found that we wanted to spend additional time in helping to teach the participant about the finer points of what we did in the test. We also were particularly interested in post-test feedback from the students. Was it a useful educational experience? Did the test raise any questions for them that we could answer? There was not always ongoing communication from individual students, but it appeared from comments in the overall course evaluations that the testing has had the desired effect.

4 Conclusions and Recommendations

We have found that collaboration between real-world user interface designers in the health care industry and nursing informatics educators teaching the user-centered design process was highly informative for both organizations. For software vendors considering such a relationship, we would particularly recommend:

- Be sure to include a pretest questionnaire with all student tests to address concerns about how well the student population represents the target users, and focus on programs that include students with sufficient experience in the health care workplace.
- Supplement the data gathered from students with tests in other populations.
- Develop the relationship with the educational institution well in advance of the course and keep in close communication, to be sure that you can plan tests that will meet the needs of both organizations.

References

1. Andreasen, M.S., Nielsen, H.V., Schrøder, S.O., Stage, J.: What happened to remote usability testing? An empirical study of three methods. In: CHI 2007 Proceedings, pp. 1405–1414. ACM Press, New York (2007)
2. Effken, J.A., Kim, M.-G., Shaw, R.E.: Making relationships visible: Testing alternative display design strategies for teaching principles of hemodynamic monitoring and treatment. Symposium on Computer Applications 11, 949–953 (1994)
3. Holzinger, A.: Usability Engineering for Software Developers. Communications of the ACM 48(1), 71–74 (2005)
4. Holzinger, A.: Application of Rapid Prototyping to the User Interface Development for a Virtual Medical Campus. IEEE Software 21(1), 92–99 (2004)
5. Kujala, S., Kauppinen, M.: Identifying and selecting users for user-centered design. In: Proceedings of the Third Nordic Conference on Human-Computer interaction, pp. 297–303. ACM Press, New York (2004)
6. Rubin, J.: Handbook of usability testing: How to plan, design, and conduct effective tests. John Wiley & Sons, New York, NY (1994)
7. Wise, M., Bellaver, R.: Usability testing as a teaching tool. Ergonomics in Design 5(2), 11–17 (1997)

Tutorial: Introduction to Visual Analytics

Wolfgang Aigner, Alessio Bertone, and Silvia Miksch

Danube University Krems, Austria
Department of Information and Knowledge Engineering
Dr.-Karl-Dorrek-Str. 30, 3500 Krems, Austria
{wolfgang.aigner,alessio.bertone,silvia.miksch}@donau-uni.ac.at
http://www.donau-uni.ac.at/ike

Abstract. Visual Analytics is an emerging area of research and practice that aims for supporting analytical reasoning by interactive visual interfaces. The basic idea is the integration of the outstanding capabilities of humans in terms of visual information exploration and the enormous processing power of computers to form a powerful knowledge discovery environment. In the course of our half-day tutorial we will introduce this multi-disciplinary field by discussing its key issues of analytical reasoning, perception & cognition, visualization interaction, computation mining, the visual analysis process, and show potential application areas.

Keywords: Visual Analytics, Information Visualization, Human-Computer Interaction, Analytical Reasoning, User-Centered Design.

1 Motivation

London 1854. A deadly cholera epidemic broke out and killed 93 people only within the first week of September. Physicians and municipality are helpless and do not know how the disease is transmitted at that time. Nobody knows how to get grip on this situation. Many people believe in so-called "miasms" transmitted through air. But Dr. John Snow has his own theory and supposes that cholera might be transmitted via contaminated water. Because of this, he walks from door to door in Soho, gathers data, and plots the locations of all deaths on a map of central London. Additionally, he marks all places where water pumps are located. He analyzes the graphed data and spots a clear pattern: most deaths occurred near the water pump in Broad Street. He interprets his analysis and reasons that contaminated water from this pump must be the cause for the epidemic. Based on this graphic evidence, he convinces the municipality to remove the handle of this pump and within days the cholera epidemic has ended.

Observe – analyze – interpret. The knowledge crystallization and problem solving process of Dr. Snow was facilitated and driven by visual methods. Visual Analytics is a continuation of this concept in the information age. In business as well as everyday life we are faced with ever growing amounts of complex data and information that need to be interpreted and analyzed.

A. Holzinger (Ed.): USAB 2007, LNCS 4799, pp. 453–456, 2007.

To support humans in coping with these huge amounts of complex information structures lies at the core of Visual Analytics science and technology [1–3].

The basic idea of Visual Analytics is the integration of the outstanding capabilities of humans in terms of visual information exploration and the enormous processing power of computers to form a powerful knowledge discovery environment. Main goals are to make complex information structures more comprehensible, facilitate new insights, and enable knowledge discovery. At this, human abilities as well as users' needs and tasks are central issues to assist in situations where complex decisions need to be made.

Visual Analytics is an inherently multi-disciplinary field ranging from cognitive psychology to database research. Potential application areas for Visual Analytics Technology are medicine & biotechnology, business, security & risk management, and environment & climate research.

Duration: Half day.

Level: Introductory.

Intended Audience:

- Researchers, students, and practitioners from various disciplines who want to get an introduction into an emerging, multi-disciplinary research area.
- Information professionals, who must analyze, interpret, present, and explore large amounts of complex data.
- Designers of advanced tools for data analysis and business intelligence.

Objectives: Attendees will

- learn what Visual Analytics is,
- see potential application areas of Visual Analytics Technologies,
- understand the advantages of the integrated approach of Visual Analytics,
- learn about basic principles of
    - analytical reasoning,
    - human perception & cognition,
    - visualization & interaction, and
    - computation & mining,
- get to know what the Visual Analytics Mantra is, and
- see demos of Visual Analytics applications.

2 Content

The human perceptual system is highly sophisticated and specifically suited to spot visual patterns. For this reason, visualization is successfully applied in aiding the task of transforming data into information and finally, synthesize knowledge.

But facing the huge volumes of data to be analyzed today, applying purely visual techniques is often not sufficient. Visual Analytics systems aim to bridge this gap by combining both, interactive visualization and computational analysis. The basic idea of Visual Analytics is the integration of the outstanding capabilities of humans in terms of visual information exploration and the enormous processing power of computers to form a powerful knowledge discovery environment. Both visual as well as analytical methods are combined intertwinedly to fully support this process. Most importantly, the user is not merely a passive element who interprets the outcome of visual and analytical methods but she is the core entity who drives the whole process.

Visual Analytics denotes the science of analytical reasoning facilitated by interactive visual interfaces and appropriate visualization techniques [1]. To achieve the main goal of Visual Analytics – the facilitation of deeper insights into huge datasets – it is crucial to consider both, the characteristics of the data and the needs of the analyst [4, 5].

Visual Analytics is an inherently multi-disciplinary field that aims to combine the findings of various research areas as Human-Computer Interaction (HCI), Usability Engineering, Cognitive and Perceptual Science, Information Visualization, Scientific Visualization, Databases, Data Mining, Statistics, Knowledge Discovery, Data Management & Knowledge Representation, Presentation, Production, and Dissemination.

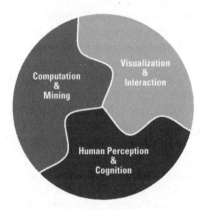

Fig. 1. *"Visual analytics strives to facilitate the analytical reasoning process by creating software that maximizes human capacity to perceive, understand, and reason about complex and dynamic data and situations."* [1]

2.1 Agenda

- Motivation & Introduction
    - What is Visual Analytics?, Application domains, Examples
- Analytical Reasoning, Perception & Cognition
    - Sense-Making, Analytical Reasoning, Perception vs. Cognition, Visual System, Color, Gestalt Laws, Preattentive processing
- Visualization & Interaction
    - Visualization Design Principles, Visual Variables, Interaction & Exploration, Usability & Evaluation
- Computation & Mining

-   Statistics (PCA, MDS), Data Mining (Clustering, Classification, Association-Rule Discovery, Generalization, Prediction)
- Visual Analysis Process
    -   Visual Analytics Mantra: "Analyze First - Show the Important - Zoom and Filter, and Analyze Further - Details on Demand." [6]
- Examples & Applications

References

1. Thomas, J.J., Cook, K.A.: Illuminating the Path: The Research and Development Agenda for Visual Analytics. IEEE Computer Science, Los Alamitos (2005)
2. Thomas, J.J., Cook, K.A.: A Visual Analytics Agenda. IEEE Computer Graphics and Applications 26(1), 10–13 (2006)
3. Thomas, J.J., Wong, P.C.: Visual Analytics. IEEE Computer Graphics and Applications 24(6), 10–13 (2004)
4. Aigner, W., Bertone, A., Miksch, S., Schumann, H., Tominski, C.: Towards a Conceptual Framework for Visual Analytics of Time and Time-Oriented Data. In: Henderson, S.G., Biller, B., Hsieh, M.H., Shortle, J., Tew, J.D., Barton, R.R. (eds.) Proceedings of the 2007 Winter Simulation Conference (invited paper)(in print)
5. Aigner, W., Miksch, S., Müller, W., Schumann, H., Tominski, C.: Visual Methods for Analyzing Time-Oriented Data. Transactions on Visualization and Computer Graphics (in print, 2008)
6. Keim, D.: Scaling Visual Analytics to Very Large Data Sets. In: Workshop on Visual Analytics, Darmstadt (2005)

Brief Presenter Biographies

WOLFGANG AIGNER is scientific researcher at the Department of Information and Knowledge Engineering, Danube University Krems, Austria and lecturer at Vienna University of Technology. He received his PhD from Vienna University of Technology and his main research interests include Visual Analytics, Information Visualization, Human-Computer Interaction, and User Centered Design.

ALESSIO BERTONE is scientific researcher at the Department of Information and Knowledge Engineering, Danube University Krems, Austria. He is PhD candidate at Vienna University of Technology and his main research interests include Visual Analytics, Information Visualization, Temporal Data Mining, Multi-relational Data Mining, and Semantic Web.

SILVIA MIKSCH is full professor and head of the Department of Information and Knowledge Engineering at Danube University Krems, Austria. Since 1998 she is head of the Information and Knowledge Engineering research group, Institute of Software Technology and Interactive Systems, Vienna University of Technology. Her main research interests include Information Visualization, Visual Analytics, Plan Management, and evaluation of Knowledge-Based Systems in real-world environments (health care).

Author Index

Ahamed, Sheikh I.   227
Ahlström, David   241
Aigner, Wolfgang   453
Albert, Dietrich   165
Alesanco Iglesias, Álvaro   69
Alfano, Bruno   389
Álvarez, J.A.   155

Basnyat, Sandra   21
Baumann, Thomas   97
Beck, Lydia   335
Becker, Annette   41
Behringer, Reinhold   255
Bertone, Alessio   453
Bligård, Lars-Ola   427
Bogner, Marilyn Sue   323
Borgert, Anfried   83
Breitenecker, Felix   213
Brezinka, Veronika   357
Brunetti, Arturo   389
Büchel, Dirk   97

Carrigan, Neil   273
Christian, Johannes   255
Conner, Mark   273
Crawford, Broderick   415

Daraiseh, Nancy   373
de la Barra, Claudio León   415
De Pietro, Giuseppe   389
Despres, Sylvie   409

Esposito, Amalia   389

Fernández-Montes, A.   155
Furler, Lukas   55

García Moros, José   69
Gardner, Peter H.   273
Geierhofer, Regina   403
Gil-Rodríguez, Eva Patrícia   69
Gonzalez, Cleotilde   289
Gruchmann, Torsten   83

Herzberg, Dominikus   41
Hessinger, Michael   213

Hitz, Martin   241
Hockemeyer, Cord   165
Holzinger, Andreas   119, 199, 213, 255, 403
Hovestadt, Ludger   357
Hunziker, Stefan   55
Huth, Myra M.   373

Islam, Rezwan   227
Ito, Kyoko   365

Janß, Armin   349
Javahery, Homa   185
Jung, Hartmut   41

Kalteis, Karin   133
Karlsson, MariAnne   441
Keller, Rochus   55, 113
Klein, Ulrike   303
Kryspin-Exner, Ilse   133
Kübler, Peter   41
Kuhlen, Torsten   335
Kulcsar, Zsuzsanna   165
Kurose, Hiroyuki   365

Lauer, Wolfgang   349
Leitner, Daniel   213
Leitner, Gerhard   241
Leonhardt, Corinna   41
Liu, Yuanhua   441
Loss, Christina   119

Madrid, Natividad Martínez   155
Marsden, Nicola   41
Martínez Ruiz, Ignacio   69
Matern, Ulrich   97
Maule, John   273
McGee, Susan   373
Meixner, Gerrit   303
Meyer, Beth   447
Mhiri, Sonia   409
Miksch, Silvia   453
Moore Jackson, Melody   267
Mungard, Nan   335
Murray, Michael D.   315

458   Author Index

Naef, Rahel   55, 113
Navarre, David   21
Nischelwitzer, Alexander   119
Nishida, Shogo   365
Norrie, Moira   113

Obermiller, Ian   227
Oppenauer, Claudia   133
Ortega, J.A.   155
Osvalder, Anna-Lisa   427, 441

Palanque, Philippe   21
Pintoffl, Klaus   119
Pohl, Margit   171
Preschl, Barbara   133
Proctor, Robert W.   315

Radermacher, Klaus   349
Rester, Markus   171
Reuss, Elke   55, 113
Roswarski, Todd Eric   315

Saigí Rubió, Francesc   69
Sambasivan, Nithya   267
Schubö, Anna   143
Seepold, Ralf   155

Seffah, Ahmed   185
Shorten, George   165
Silva, Paula Alexandra   105
Skiba, Diane   447
Stork, Sonja   143
Sturm, Walter   335

Takami, Ai   365
Talukder, Nilothpal   227
Thiels, Nancy   303
Thimbleby, Harold   1
Thomanek, Sabine   41
Tilz, Gernot P.   199

Van Laerhoven, Kristof   105
Vesper, Cordula   143
Vrbin, Colleen   289

Wassertheurer, Siegfried   213
Wiesbeck, Mathey   143
Wilkinson, Steve   255
Wiltgen, Marco   199
Wiltner, Sylvia   171
Wolter, Marc   335
Workman, Michael   375

Printing: Mercedes-Druck, Berlin
Binding: Stein+Lehmann, Berlin

Lecture Notes in Computer Science

Sublibrary 2: Programming and Software Engineering

For information about Vols. 1– 4192
please contact your bookseller or Springer

Vol. 4834: R. Cerqueira, R.H. Campbell (Eds.), Middleware 2007. XIII, 451 pages. 2007.

Vol. 4824: A. Paschke, Y. Biletskiy (Eds.), Advances in Rule Interchange and Applications. XIII, 243 pages. 2007.

Vol. 4807: Z. Shao (Ed.), Programming Languages and Systems. XI, 431 pages. 2007.

Vol. 4799: A. Holzinger (Ed.), HCI and Usability for Medicine and Health Care. XVI, 458 pages. 2007.

Vol. 4789: M. Butler, M.G. Hinchey, M.M. Larrondo-Petrie (Eds.), Formal Methods and Software Engineering. VIII, 387 pages. 2007.

Vol. 4767: F. Arbab, M. Sirjani (Eds.), International Symposium on Fundamentals of Software Engineering. XIII, 450 pages. 2007.

Vol. 4764: P. Abrahamsson, N. Baddoo, T. Margaria, R. Messnarz (Eds.), Software Process Improvement. XI, 225 pages. 2007.

Vol. 4762: K.S. Namjoshi, T. Yoneda, T. Higashino, Y. Okamura (Eds.), Automated Technology for Verification and Analysis. XIV, 566 pages. 2007.

Vol. 4758: F. Oquendo (Ed.), Software Architecture. XVI, 340 pages. 2007.

Vol. 4757: F. Cappello, T. Herault, J. Dongarra (Eds.), Recent Advances in Parallel Virtual Machine and Message Passing Interface. XVI, 396 pages. 2007.

Vol. 4753: E. Duval, R. Klamma, M. Wolpers (Eds.), Creating New Learning Experiences on a Global Scale. XII, 518 pages. 2007.

Vol. 4749: B.J. Krämer, K.-J. Lin, P. Narasimhan (Eds.), Service-Oriented Computing – ICSOC 2007. XIX, 629 pages. 2007.

Vol. 4748: K. Wolter (Ed.), Formal Methods and Stochastic Models for Performance Evaluation. X, 301 pages. 2007.

Vol. 4741: C. Bessière (Ed.), Principles and Practice of Constraint Programming – CP 2007. XV, 890 pages. 2007.

Vol. 4735: G. Engels, B. Opdyke, D.C. Schmidt, F. Weil (Eds.), Model Driven Engineering Languages and Systems. XV, 698 pages. 2007.

Vol. 4716: B. Meyer, M. Joseph (Eds.), Software Engineering Approaches for Offshore and Outsourced Development. X, 201 pages. 2007.

Vol. 4680: F. Saglietti, N. Oster (Eds.), Computer Safety, Reliability, and Security. XV, 548 pages. 2007.

Vol. 4670: V. Dahl, I. Niemelä (Eds.), Logic Programming. XII, 470 pages. 2007.

Vol. 4652: D. Georgakopoulos, N. Ritter, B. Benatallah, C. Zirpins, G. Feuerlicht, M. Schoenherr, H.R. Motahari-Nezhad (Eds.), Service-Oriented Computing ICSOC 2006. XVI, 201 pages. 2007.

Vol. 4634: H. Riis Nielson, G. Filé (Eds.), Static Analysis. XI, 469 pages. 2007.

Vol. 4615: R. de Lemos, C. Gacek, A. Romanovsky (Eds.), Architecting Dependable Systems IV. XIV, 435 pages. 2007.

Vol. 4610: B. Xiao, L.T. Yang, J. Ma, C. Muller-Schloer, Y. Hua (Eds.), Autonomic and Trusted Computing. XVIII, 571 pages. 2007.

Vol. 4609: E. Ernst (Ed.), ECOOP 2007 – Object-Oriented Programming. XIII, 625 pages. 2007.

Vol. 4608: H.W. Schmidt, I. Crnković, G.T. Heineman, J.A. Stafford (Eds.), Component-Based Software Engineering. XII, 283 pages. 2007.

Vol. 4591: J. Davies, J. Gibbons (Eds.), Integrated Formal Methods. IX, 660 pages. 2007.

Vol. 4589: J. Münch, P. Abrahamsson (Eds.), Product-Focused Software Process Improvement. XII, 414 pages. 2007.

Vol. 4574: J. Derrick, J. Vain (Eds.), Formal Techniques for Networked and Distributed Systems – FORTE 2007. XI, 375 pages. 2007.

Vol. 4556: C. Stephanidis (Ed.), Universal Access in Human-Computer Interaction, Part III. XXII, 1020 pages. 2007.

Vol. 4555: C. Stephanidis (Ed.), Universal Access in Human-Computer Interaction, Part II. XXII, 1066 pages. 2007.

Vol. 4554: C. Stephanidis (Ed.), Universal Acess in Human Computer Interaction, Part I. XXII, 1054 pages. 2007.

Vol. 4553: J.A. Jacko (Ed.), Human-Computer Interaction, Part IV. XXIV, 1225 pages. 2007.

Vol. 4552: J.A. Jacko (Ed.), Human-Computer Interaction, Part III. XXI, 1038 pages. 2007.

Vol. 4551: J.A. Jacko (Ed.), Human-Computer Interaction, Part II. XXIII, 1253 pages. 2007.

Vol. 4550: J.A. Jacko (Ed.), Human-Computer Interaction, Part I. XXIII, 1240 pages. 2007.

Vol. 4542: P. Sawyer, B. Paech, P. Heymans (Eds.), Requirements Engineering: Foundation for Software Quality. IX, 384 pages. 2007.

Vol. 4536: G. Concas, E. Damiani, M. Scotto, G. Succi (Eds.), Agile Processes in Software Engineering and Extreme Programming. XV, 276 pages. 2007.

Vol. 4530: D.H. Akehurst, R. Vogel, R.F. Paige (Eds.), Model Driven Architecture - Foundations and Applications. X, 219 pages. 2007.

Vol. 4523: Y.-H. Lee, H.-N. Kim, J. Kim, Y.W. Park, L.T. Yang, S.W. Kim (Eds.), Embedded Software and Systems. XIX, 829 pages. 2007.

Vol. 4498: N. Abdennahher, F. Kordon (Eds.), Reliable Software Technologies - Ada-Europe 2007. XII, 247 pages. 2007.

Vol. 4486: M. Bernardo, J. Hillston (Eds.), Formal Methods for Performance Evaluation. VII, 469 pages. 2007.

Vol. 4470: Q. Wang, D. Pfahl, D.M. Raffo (Eds.), Software Process Dynamics and Agility. XI, 346 pages. 2007.

Vol. 4468: M.M. Bonsangue, E.B. Johnsen (Eds.), Formal Methods for Open Object-Based Distributed Systems. X, 317 pages. 2007.

Vol. 4467: A.L. Murphy, J. Vitek (Eds.), Coordination Models and Languages. X, 325 pages. 2007.

Vol. 4454: Y. Gurevich, B. Meyer (Eds.), Tests and Proofs. IX, 217 pages. 2007.

Vol. 4444: T. Reps, M. Sagiv, J. Bauer (Eds.), Program Analysis and Compilation, Theory and Practice. X, 361 pages. 2007.

Vol. 4440: B. Liblit, Cooperative Bug Isolation. XV, 101 pages. 2007.

Vol. 4408: R. Choren, A. Garcia, H. Giese, H.-f. Leung, C. Lucena, A. Romanovsky (Eds.), Software Engineering for Multi-Agent Systems V. XII, 233 pages. 2007.

Vol. 4406: W. De Meuter (Ed.), Advances in Smalltalk. VII, 157 pages. 2007.

Vol. 4405: L. Padgham, F. Zambonelli (Eds.), Agent-Oriented Software Engineering VII. XII, 225 pages. 2007.

Vol. 4401: N. Guelfi, D. Buchs (Eds.), Rapid Integration of Software Engineering Techniques. IX, 177 pages. 2007.

Vol. 4385: K. Coninx, K. Luyten, K.A. Schneider (Eds.), Task Models and Diagrams for Users Interface Design. XI, 355 pages. 2007.

Vol. 4383: E. Bin, A. Ziv, S. Ur (Eds.), Hardware and Software, Verification and Testing. XII, 235 pages. 2007.

Vol. 4379: M. Südholt, C. Consel (Eds.), Object-Oriented Technology. VIII, 157 pages. 2007.

Vol. 4364: T. Kühne (Ed.), Models in Software Engineering. XI, 332 pages. 2007.

Vol. 4355: J. Julliand, O. Kouchnarenko (Eds.), B 2007: Formal Specification and Development in B. XIII, 293 pages. 2006.

Vol. 4354: M. Hanus (Ed.), Practical Aspects of Declarative Languages. X, 335 pages. 2006.

Vol. 4350: M. Clavel, F. Durán, S. Eker, P. Lincoln, N. Martí-Oliet, J. Meseguer, C. Talcott, All About Maude - A High-Performance Logical Framework. XXII, 797 pages. 2007.

Vol. 4348: S. Tucker Taft, R.A. Duff, R.L. Brukardt, E. Plödereder, P. Leroy, Ada 2005 Reference Manual. XXII, 765 pages. 2006.

Vol. 4346: L. Brim, B.R. Haverkort, M. Leucker, J. van de Pol (Eds.), Formal Methods: Applications and Technology. X, 363 pages. 2007.

Vol. 4344: V. Gruhn, F. Oquendo (Eds.), Software Architecture. X, 245 pages. 2006.

Vol. 4340: R. Prodan, T. Fahringer, Grid Computing. XXIII, 317 pages. 2007.

Vol. 4336: V.R. Basili, H.D. Rombach, K. Schneider, B. Kitchenham, D. Pfahl, R.W. Selby (Eds.), Empirical Software Engineering Issues. XVII, 193 pages. 2007.

Vol. 4326: S. Göbel, R. Malkewitz, I. Iurgel (Eds.), Technologies for Interactive Digital Storytelling and Entertainment. X, 384 pages. 2006.

Vol. 4323: G. Doherty, A. Blandford (Eds.), Interactive Systems. XI, 269 pages. 2007.

Vol. 4322: F. Kordon, J. Sztipanovits (Eds.), Reliable Systems on Unreliable Networked Platforms. XIV, 317 pages. 2007.

Vol. 4309: P. Inverardi, M. Jazayeri (Eds.), Software Engineering Education in the Modern Age. VIII, 207 pages. 2006.

Vol. 4294: A. Dan, W. Lamersdorf (Eds.), Service-Oriented Computing – ICSOC 2006. XIX, 653 pages. 2006.

Vol. 4290: M. van Steen, M. Henning (Eds.), Middleware 2006. XIII, 425 pages. 2006.

Vol. 4279: N. Kobayashi (Ed.), Programming Languages and Systems. XI, 423 pages. 2006.

Vol. 4262: K. Havelund, M. Núñez, G. Roşu, B. Wolff (Eds.), Formal Approaches to Software Testing and Runtime Verification. VIII, 255 pages. 2006.

Vol. 4260: Z. Liu, J. He (Eds.), Formal Methods and Software Engineering. XII, 778 pages. 2006.

Vol. 4257: I. Richardson, P. Runeson, R. Messnarz (Eds.), Software Process Improvement. XI, 219 pages. 2006.

Vol. 4242: A. Rashid, M. Aksit (Eds.), Transactions on Aspect-Oriented Software Development II. IX, 289 pages. 2006.

Vol. 4229: E. Najm, J.-F. Pradat-Peyre, V.V. Donzeau-Gouge (Eds.), Formal Techniques for Networked and Distributed Systems - FORTE 2006. X, 486 pages. 2006.

Vol. 4227: W. Nejdl, K. Tochtermann (Eds.), Innovative Approaches for Learning and Knowledge Sharing. XVII, 721 pages. 2006.

Vol. 4218: S. Graf, W. Zhang (Eds.), Automated Technology for Verification and Analysis. XIV, 540 pages. 2006.

Vol. 4214: C. Hofmeister, I. Crnković, R. Reussner (Eds.), Quality of Software Architectures. X, 215 pages. 2006.

Vol. 4204: F. Benhamou (Ed.), Principles and Practice of Constraint Programming - CP 2006. XVIII, 774 pages. 2006.

Vol. 4199: O. Nierstrasz, J. Whittle, D. Harel, G. Reggio (Eds.), Model Driven Engineering Languages and Systems. XVI, 798 pages. 2006.